Microwave Devices and Circuits

Microwave Devices and Circuits

SAMUEL Y. LIAO

Professor, School of Electrical Engineering
California State University-Fresno

PRENTICE-HALL, INC., *Englewood Cliffs,* New Jersey, *07632*

Library of Congress Cataloging in Publication Data

Liao, Samuel Y
    Microwave devices and circuits.

    Includes bibliographical references and index.
    1. Microwave devices.   2. Microwave circuits.
I. Title.
TK7876.L43 1980   621.281'3   79-19643
ISBN 0-13-581207-0

The author sincerely dedicates this book
to the memory of his
mother, Hu Chi, who loved him
and
father, Cheng Kung, who inspired him

Editorial/production supervision and
interior design by *Virginia Huebner*
Cover design by *Suzanne Behnke*
Manufacturing buyer: *Gordon Osbourne*

© 1980 by Prentice-Hall, Inc., Englewood Cliffs, N.J.   07632

Printed in the United States of America

10   9   8   7   6   5   4   3   2   1

PRENTICE-HALL INTERNATIONAL, INC., *London*
PRENTICE-HALL OF AUSTRALIA PTY. LIMITED, *Sydney*
PRENTICE-HALL OF CANADA, LTD., *Toronto*
PRENTICE-HALL OF INDIA PRIVATE LIMITED, *New Delhi*
PRENTICE-HALL OF JAPAN, INC., *Tokyo*
PRENTICE-HALL OF SOUTHEAST ASIA PTE. LTD., *Singapore*
WHITEHALL BOOKS LIMITED, *Wellington, New Zealand*

Contents

Crossed-Field Amplifiers (*FWCFA*): Physical
Description, Principles of Operation, Microwave
Characteristics ● *M*-Carcinotron Oscillators

Microwave Transistors: Physical Structures, Principles of
Operation, Microwave Characteristics, Power-Frequency
Limitations ● Microwave Tunnel Diodes: Principles of
Operation, Microwave Characteristics, State of the Art,
Power-Frequency Limitations ● Microwave Field-Effect
Transistors (*FETs*): Physical Structures, Principles of
Operation, Microwave Characteristics (Drain Current,
Cutoff-Frequency, Maximum Frequency of Oscillation,
Power Gain, Noise Figure), Power-Frequency
Limitations

Gunn-Effect Diodes—GaAs Diode: Background, Gunn
Effect, Principles of Operation (Two-Valley Model
Theory, High-Field Domain), Modes of Operation
(Gunn Oscillation Modes, Transit-Time Domain Mode,
Delayed Domain Mode, Quenched Domain Mode, *LSA*
Mode, Stable Amplification Mode), Microwave
Generation and Amplification ● *LSA* Diodes ● *InP*
Diodes ● *CdTe* Diodes

Read Diode: Physical Description, Avalanche
Multiplication, Carrier Current and External Current,
Output Power and Quality Factor Q ● IMPATT
Diodes: Characteristics, Negative Resistance, Power
Output and Efficiency ● TRAPATT Diodes:
Introduction, Principles of Operation, Power Output
and Efficiency ● BARITT Diodes: Introduction,
Principles of Operation, Microwave Performance

Introduction: Evolution of Masers and Lasers,
Transition Processes, Population Inversion, Resonant

Preface

This book is intended as a text in a first course in microwave devices and circuits at the senior or beginning graduate level in electrical engineering. It's primary objectives are to provide the student with a clear understanding of the commonly used microwave devices, and with some techniques for analyzing microwave circuits and measurements. The content of the book is an outgrowth of lecture notes used for several years in a one-semester course taught by the author, and from his long industrial experience in microwave electronics. It is assumed that the student has had previous courses in electromagnetics and solid-state electronics. Because the book is, to a large extent, self-contained, it can also be used as a reference book by electronics engineers working the microwave area.

It is the author's belief that the basic philosophy of teaching and writing is to explore clarity of exposition. Based on this philosophy, mathematical analysis is used throughout the text to clarify, if possible, the fundamental principles of the subjects.

The book is organized into ten chapters. The first two chapters are introductory, and deal with the interactions between electrons and fields and with plane waves. Chapter 3 treats transmission lines and waveguides, which are still very useful for high-power transmission. Chapter 4 describes microwave components such as cavity resonators, slow-wave structures, waveguide Tees, directional couplers, and circulators, which are commonly used in the microwave circuits. Chapter 5 deals with microwave tubes, which are classified into linear-beam tubes (O-type) and crossed-field tubes (M-type) according to their operations and characteristics. Chapter 6 analyzes the solid-state microwave devices which have been developed in the past decade. In Section 6-1, microwave transistors, tunnel diodes and Schottky barrier-gate field-effect transistors (MESFETs) are discussed. In Sections 6-2 through 6-5 transferred electron devices (TEDs), avalanche transit-time devices, quantum-electronic solid-state devices, and parametric amplifiers are well studied. The newly developed infrared devices and systems, which operate in the wavelength range of 5 to 14 micrometers, are described in Section 6-6. In recent years, with the advent of microwave integrated circuits, microstrip lines have been extensively used in the microwave circuits, since they provide one free and accessible surface upon which solid-state devices may be placed. Chapters 7 and 8 discuss microstrip lines and microwave integrated circuits, respectively. It is the experience of the author that microwave enclosures and microwave power measurements are also important parts of a microwave course. Chapters 9 and 10 examine the subjects of electromagnetic fields in shielded rooms and anechoic chambers, and of microwave electric field and power-density measurements.

The arrangement of topics is by no means inflexible, and the instructor has a choice in the selection or order of the topics to suit either a one-semester or possibly a one-quarter course. Problems for each chapter will aid the reader to understand further the subjects discussed in the text. A solutions manual may be obtained from the publisher by instructors who have adopted the book for their courses.

The author is indebted to his senior colleagues, Professor Owen F. Foin, Jr., and Professor Mac Jarrett, who read the manuscript and offered many valuable comments. The major portion of the manuscript was originally typed by Mrs. Virginia Sappington, to whom the author expresses his thanks. Special thanks are extended to Mrs. Elsie Taylor who skillfully typed the final version of the manuscript. The author would also like to acknowledge his appreciation for the publication of this book by Prentice-Hall, Inc., and especially thank Mrs. Virginia Huebner for her supervision in the production of this book.

Grateful acknowledgment is expressed to Dr. James D. Matheny, Dean of the School of Engineering, for his constant encouragement and helpful counsel. The author dedicates this book to the memory of his mother, Hu Chi, who loved him, and his father, Cheng Kung, who inspired him. Finally, the author wishes to express his deep appreciation to his wife, Lucia Hsiao Chuang, for her constant encouragement in the preparation of this book.

SAMUEL Y. LIAO

Fresno, California

Introduction

The central theme of this book concerns the basic principles and applications of microwave devices and circuits. Microwave techniques have been increasingly adopted in such diverse applications as radio astronomy, long-distance communications, space navigation, radar systems, and missile electronic systems. As a result of the accelerating rate of growth of microwave technology in research and industry, students who are preparing themselves for, and electronics engineers who are working in, the microwave area are faced with the need to understand the theoretical and experimental design and analysis of microwave devices and circuits.

c h a p t e r

0

1

0-1 Microwave Frequencies

The term *microwave frequencies* is generally used for those wavelengths measured in centimeters, roughly from 30 centimeters to 1 millimeter (1 GHz to 300 GHz). The microwave band designation that derived from World War II radar security considerations has never been officially sanctioned by any industrial, professional, or government organization. In August 1969 the United States Department of Defense, Office of Joint Chiefs of Staff, by message to all Services, directed the use of a new frequency band breakdown as shown in Table 0-1. In May 24, 1970, the Department of Defense adopted another band designation for microwave frequencies as listed in Table 0-2. At the present state of the art microwave amplifiers and oscillator tubes can operate in a frequency range up to 40 GHz and solid-state microwave devices up to 100 GHz.

TABLE 0-1. U.S. Military Microwave Bands.

Designation	Frequency range in Gigahertz
P band	0.225– 0.390
L band	0.390– 1.550
S band	1.550– 3.900
C band	3.900– 6.200
X band	6.200– 10.900
K band	10.900– 36.000
Q band	36.000– 46.000
V band	46.000– 56.000
W band	56.000–100.000

TABLE 0-2. U.S. New Military Microwave Bands.

Designation	Frequency range in Gigahertz	Designation	Frequency range in Gigahertz
A Band	0.100–0.250	H Band	6.000– 8.000
B Band	0.250–0.500	I Band	8.000– 10.000
C Band	0.500–1.000	J Band	10.000– 20.000
D Band	1.000–2.000	K Band	20.000– 40.000
E Band	2.000–3.000	L Band	40.000– 60.000
F Band	3.000–4.000	M Band	60.000–100.000
G Band	4.000–6.000		

0-2 Microwave Devices

In the late 1930s it became evident that as the wavelength approached the physical dimensions of the vacuum tubes, the electron transit angle, inter-electrode capacitance, and lead inductance appeared to limit the operation of vacuum tubes in microwave frequencies. In 1935 A. A. Heil and O. Heil suggested that microwave voltages be generated by using transit-time effects together with lumped tuned circuits. In 1939 W. C. Hahn and G. F. Metcalf proposed a theory of velocity modulation for microwave tubes. Four months later R. H. Varian and S. F. Varian described a two-cavity klystron amplifier and oscillator by using velocity modulation. In 1944 R. Kompfner invented the helix-type traveling-wave tube (TWT). Ever since then the concept of microwave tubes has deviated from that of conventional vacuum tubes as a result of the application of new principles in the amplification and generation of microwave energy.

Historically microwave generation and amplification were accomplished by means of velocity-modulation theory. In recent years, however, microwave solid-state devices, such as tunnel diodes, Gunn diodes, transferred electron devices (TEDs), avalanche transit-time devices, and quantum-electronic devices, such as masers and lasers, have been developed to perform these functions. The conception and subsequent development of TEDs and avalanche transit-time devices were among the outstanding technical achievements of the past decade. B. K. Ridley and T. B. Watkins in 1961 and C. Hilsum in 1962 independently predicted that the transferred electron effect would occur in GaAs (Gallium Arsenide). In 1963 J. B. Gunn reported his "Gunn Effect." The common characteristic of all microwave solid-state devices is the negative resistance that can be used for microwave oscillation and amplification. The progress of TEDs and avalanche transit-time devices has been so swift that today they are firmly established as one of the most important classes of microwave solid-state devices.

0-3 Microwave Systems

A microwave system normally consists of a transmitter subsystem, including a microwave oscillator, waveguides, and a transmitting antenna, and a receiver subsystem that includes a receiving antenna, transmission line or waveguide, a microwave amplifier, and a receiver. Figure 0-1 shows a typical microwave system.

In order to design a microwave system and conduct a proper test of it, an adequate knowledge of the components involved is desirable. Besides microwave devices, microwave components, such as resonators, cavities, microstrip lines, hybrids, and microwave integrated circuits are therefore described in the text.

0-4 Microwave Measurements

The accuracy of microwave measurements is a major concern of electronics engineers. Theoretically a transmitter should only transmit electric energy and a receiver should only receive energy. Yet this hardly ever happens in

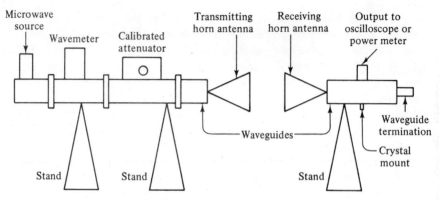

FIG. 0-1. Microwave system.

the ordinary laboratory because a piece of electronic equipment can either transmit or receive electric energy. Consequently, electronic equipment is constantly subjected to unwanted sources of energy and is constantly producing energy that adjacent equipment is not designed to accept. This is the basic problem of electromagnetic compatibility. The electronic equipment must operate in conjunction with other equipment without causing malfunction or degradation of operation of any of the associated equipment. An ideally designed piece of equipment should not radiate any unwanted energy; nor should it be susceptible to any unwanted energy. To accomplish this, a medium would have to enclose the equipment so that unwanted energy either leaving or attempting to enter the equipment is effectively attenuated. Shielded rooms and anechoic chambers are the ideal enclosures for microwave measurements and are described in Chapter 9. Microwave measurements are discussed in Chapter 10.

MKS Units, Prefixes, and Physical Constants

The rationalized meter-kilogram-second (RMKS) units (The International System of Units) are used throughout unless otherwise indicated. Table 0-3 lists the most commonly used MKS units.

TABLE 0-3. MKS Units.

Quantity	Unit	Symbol
Angstrom	10^{-10} m	Å
Capacitance	farad $= \dfrac{C}{V}$	F
Charge	coulomb: $A - s$	C

TABLE 0-3. (cont.)

Quantity	Unit	Symbol
Conductance	mhos	℧
Current	ampere = coulomb per second	A
Cycles	hertz per second	Hz
Energy	joule	J
Field	volt per meter	E
Flux linkage	weber = volt-second	ψ
Inductance	henry = $\dfrac{V - s}{A}$	H
Length	meter	m
Micron	10^{-6} m	μm
Power	watt = joule per second	W
Resistance	ohm	Ω
Time	second	s
Velocity	meter per second	v
Voltage	volt	V

The prefixes as tabulated in Table 0-4 are those recommended by the International Committee on Weights and Measures. They have been adopted by the National Bureau of Standards and are used by the Institute of Electrical and Electronics Engineers, Inc.

TABLE 0-4. Prefixes.

Prefix	Factor	Symbol
tera	10^{12}	T
giga	10^{9}	G
mega	10^{6}	M
kilo	10^{3}	K
hecto	10^{2}	h
deka	10	da
deci	10^{-1}	d
centi	10^{-2}	c
milli	10^{-3}	m
micro	10^{-6}	μ
nano	10^{-9}	n
pico	10^{-12}	p
femto	10^{-15}	f
atto	10^{-18}	a

The physical constants commonly used in the text are listed in Table 0-5.

TABLE 0-5. Physical Constants

Constant	Symbol	Value
Boltzmann constant	k	1.381×10^{-23} J/°K
Electron volt	eV	1.602×10^{-19} J
Electron charge	q	1.602×10^{-19} C
Electron mass	m	9.109×10^{-31} kg
Ratio of charge to mass of an electron	e/m	1.759×10^{11} C/kg
Permeability of free space	μ_0	1.257×10^{-6} H/m, or $4\pi \times 10^{-7}$ H/m
Permittivity of free space	ϵ_0	8.854×10^{-12} F/m
Planck's constant	h	6.626×10^{-34} J-s
Velocity of light in vacuum	c	2.998×10^{8} m/s

Interactions Between Electrons

and Fields

1-0 Introduction

In this chapter we are concerned with electron-field interactions. The motion of the electron beam is assumed to be in a uniform electric field, or a uniform magnetic field, or a uniform electromagnetic field because the inhomogeneous differential equations governing the motion of an electron beam in a field involve three dimensions and their solutions in a nonuniform field are, in most cases, extremely difficult to obtain and usually cannot be determined exactly. On the other hand, fortunately, all current microwave devices employ a uniform field for the electron-field interactions.

Our primary purpose here is to provide the reader with a background for understanding the electron-field interactions in microwave devices that will be discussed in later chapters.

chapter

1

1-1 Electron Motion in an Electric Field

In describing fields and electron-field interactions, certain experimental laws of electricity and magnetism are covered first. The fundamental force law of charges is Coulomb's law, which states that between two charges there exists either an attractive or a repulsive force, depending on whether the charges are of opposite or like sign. That is,

$$\mathbf{F} = \frac{Q_1 Q_2}{4\pi\epsilon_0 R^2} \mathbf{u}_{R_{12}} \quad \text{newtons} \tag{1-1-1}$$

where $Q =$ the charge in coulombs

$\epsilon_0 = 8.854 \times 10^{-12} \simeq \dfrac{1}{36\pi} \times 10^{-9}$ F/m is the permittivity of
free space

$R =$ the separation between the charges in meters

$\mathbf{u} =$ the unit vector

It should be noted that since the MKS system is used throughout this text, a factor of 4π appears in the preceding equation.

The electric field intensity produced by the charges is defined as the force per unit charge—that is,

$$\mathbf{E} \equiv \frac{\mathbf{F}}{Q} = \frac{Q}{4\pi\epsilon_0 R^2} \mathbf{u}_R \quad \text{volt/m} \tag{1-1-2}$$

If there are n charges, the electric field becomes

$$\mathbf{E} = \sum_{m=1}^{n} \frac{Q_m}{4\pi\epsilon_0 R_m^2} \mathbf{u}_{R_m} \tag{1-1-3}$$

In order to determine the path of an electron in an electric field, the force must be related to the mass and acceleration of the electron by Newton's second law of motion. So

$$\mathbf{F} = -e\mathbf{E} = ma = m\frac{d\mathbf{v}}{dt} \tag{1-1-4}$$

where $m = 9.109 \times 10^{-31}$ kg, mass of electron

$a =$ acceleration in m/s^2

$\mathbf{v} =$ velocity of electron in m/s

$e = 1.602 \times 10^{-19}$ coulomb, charge of electron that is
negative

It can be seen that the force is in the opposite direction of the field because the electron has a negative charge. Thus when an electron moves in an electric field $\mathbf{E}$, it experiences a force $-e\mathbf{E}$ newtons. The differential equations of motion for an electron in an electric field in rectangular coordinates are given by

$$\frac{d^2x}{dt^2} = -\frac{e}{m}E_x \tag{1-1-5a}$$

$$\frac{d^2y}{dt^2} = -\frac{e}{m}E_y \tag{1-1-5b}$$

$$\frac{d^2z}{dt^2} = -\frac{e}{m}E_z \tag{1-1-5c}$$

where $\frac{e}{m} = 1.759 \times 10^{11}$ coul/kg is the ratio of charge to mass of electron

E_x, E_y, E_z = the components of **E** in rectangular coordinates

In many cases, the equations of motion for electrons in an electric field in cylindrical coordinates are useful. The cylindrical coordinates (r, ϕ, z) are defined as in Fig. 1-1-1.

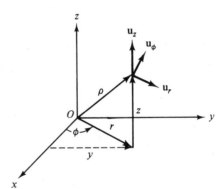

FIG. 1-1-1. Cylindrical coordinates.

It can be seen that

$$x = r \cos \phi \tag{1-1-6a}$$
$$y = r \sin \phi \tag{1-1-6b}$$
$$z = z \tag{1-1-6c}$$

and, conversely,

$$r = (x^2 + y^2)^{1/2} \tag{1-1-7a}$$

$$\phi = \tan^{-1}\left(\frac{y}{z}\right) = \sin^{-1}\frac{y}{(x^2 + y^2)^{1/2}} = \cos^{-1}\frac{x}{(x^2 + y^2)^{1/2}} \tag{1-1-7b}$$

$$z = z \tag{1-1-7c}$$

A system of unit vectors, $\mathbf{u}_r, \mathbf{u}_\phi, \mathbf{u}_z$, in the directions of increasing r, ϕ, z, respectively, is also shown in the same diagram. While $\mathbf{u}_z$ is constant, $\mathbf{u}_r$ and $\mathbf{u}_\phi$ are functions of ϕ; that is,

$$\mathbf{u}_r = \cos \phi\, \mathbf{u}_x + \sin \phi\, \mathbf{u}_y \tag{1-1-8a}$$
$$\mathbf{u}_\phi = -\sin \phi\, \mathbf{u}_x + \cos \phi\, \mathbf{u}_y \tag{1-1-8b}$$

Differentiation of Eqs. (1-1-8) with respect to ϕ yields

$$\frac{d\mathbf{u}_r}{d\phi} = \mathbf{u}_\phi \tag{1-1-9a}$$

$$\frac{d\mathbf{u}_\phi}{d\phi} = -\mathbf{u}_r \tag{1-1-9b}$$

The position vector $\boldsymbol{\rho}$ can be expressed in cylindrical coordinates in the form

$$\boldsymbol{\rho} = r\mathbf{u}_r + z\mathbf{u}_z \tag{1-1-9c}$$

Differentiation of Eq. (1-1-9c) with respect to t once for velocity and twice for acceleration yields

$$\mathbf{v} = \frac{d\boldsymbol{\rho}}{dt} = \frac{dr}{dt}\mathbf{u}_r + r\frac{d\mathbf{u}_r}{dt} + \frac{dz}{dt}\mathbf{u}_z = \frac{dr}{dt}\mathbf{u}_r$$

$$+ r\frac{d\phi}{dt}\cdot\frac{d\mathbf{u}_r}{d\phi} + \frac{dz}{dt}\mathbf{u}_z = \frac{dr}{dt}\mathbf{u}_r + r\frac{d\phi}{dt}\mathbf{u}_\phi + \frac{dz}{dt}\mathbf{u}_z \tag{1-1-10}$$

$$\mathbf{a} = \frac{d\mathbf{v}}{dt} = \left[\frac{d^2r}{dt^2} - r\left(\frac{d\phi}{dt}\right)^2\right]\mathbf{u}_r + \left(r\frac{d^2\phi}{dt^2} + 2\frac{dr}{dt}\frac{d\phi}{dt}\right)\mathbf{u}_\phi + \frac{d^2z}{dt^2}\mathbf{u}_z$$

$$= \left[\frac{d^2r}{dt^2} - r\left(\frac{d\phi}{dt}\right)^2\right]\mathbf{u}_r + \frac{1}{r}\frac{d}{dt}\left(r^2\frac{d\phi}{dt}\right)\mathbf{u}_\phi + \frac{d^2z}{dt^2}\mathbf{u}_z \tag{1-1-11}$$

Therefore the equations of motion for electrons in an electric field in cylindrical coordinates are given by

$$\frac{d^2r}{dt^2} - r\left(\frac{d\phi}{dt}\right)^2 = -\frac{e}{m}E_r \tag{1-1-12a}$$

$$\frac{1}{r}\frac{d}{dt}\left(r^2\frac{d\phi}{dt}\right) = -\frac{e}{m}E_\phi \tag{1-1-12b}$$

$$\frac{d^2z}{dt^2} = -\frac{e}{m}E_z \tag{1-1-12c}$$

where E_r, E_ϕ, and E_z are the components of $\mathbf{E}$ in cylindrical coordinates.

From Eq. (1-1-4) the work done by the field in carrying a unit positive charge from point A to point B is

$$-\int_A^B \mathbf{E}\cdot d\boldsymbol{\ell} = \frac{m}{e}\int_{v_A}^{v_B} v\,dv \tag{1-1-13}$$

However, by definition, the potential V of point B with respect to point A is the work done against the field in carrying a unit positive charge from A to B. That is,

$$V \equiv -\int_A^B \mathbf{E}\cdot d\boldsymbol{\ell} \tag{1-1-14}$$

Substitution of Eq. (1-1-14) in Eq. (1-1-13) and integration of the resultant yield

$$eV = \frac{1}{2}m(v_B^2 - v_A^2) \tag{1-1-15}$$

The left side of Eq. (1-1-15) is the potential energy, and the right side represents the change in kinetic energy. The unit of work or energy is called the *electron volt (eV)*, which means that if an electron falls through a potential of one volt, its kinetic energy will increase 1 *eV*. That is,

$$1 \ eV = (1.60 \times 10^{-19} \text{ coul})(1 \text{ volt}) = 1.60 \times 10^{-19} \text{ joule} \quad (1\text{-}1\text{-}16)$$

If an electron starts from rest and is accelerated through a potential rise of V volts, its final velocity is

$$v = \left(\frac{2eV}{m}\right)^{1/2} = 0.593 \times 10^6 \sqrt{V} \quad \text{m/s} \quad (1\text{-}1\text{-}17)$$

Since $d\ell$ is the increment of distance in the direction of an electric field E, the change in potential dV over the distance $d\ell$ can be expressed as

$$|dV| = E \, d\ell \quad (1\text{-}1\text{-}18)$$

In vector notation it is

$$\mathbf{E} = -\nabla V \quad (1\text{-}1\text{-}19)$$

The minus sign implies that the field is directed from regions of higher potential to those of lower potential. Equation (1-1-19) is valid in regions in which there is space charge as well as in regions that are free of charge.

1-2   Electron Motion in a Magnetic Field

A charged particle in motion in a magnetic field of flux density **B** is experimentally found to experience a force that is directly proportional to the charge Q, its velocity v, the flux density **B**, and the sine of the angle between the vectors **v** and **B**. The direction of the force is perpendicular to the plane of both **v** and **B**. Therefore the force exerted on the charged particle by the magnetic field can be expressed in vector form as

$$\mathbf{F} = Q\mathbf{v} \times \mathbf{B} \quad (1\text{-}2\text{-}1)$$

Since the electron has negative charge, then

$$\mathbf{F} = -e\mathbf{v} \times \mathbf{B} \quad (1\text{-}2\text{-}2)$$

The motion equations of an electron in a magnetic field in rectangular coordinates can be written

$$\frac{d^2x}{dt^2} = -\frac{e}{m}\left(B_z \frac{dy}{dt} - B_y \frac{dz}{dt}\right) \quad (1\text{-}2\text{-}3a)$$

$$\frac{d^2y}{dt^2} = -\frac{e}{m}\left(B_x \frac{dz}{dt} - B_z \frac{dx}{dt}\right) \quad (1\text{-}2\text{-}3b)$$

$$\frac{d^2z}{dt^2} = -\frac{e}{m}\left(B_y \frac{dx}{dt} - B_x \frac{dy}{dt}\right) \quad (1\text{-}2\text{-}3c)$$

Since

$$\mathbf{v} \times \mathbf{B} = (B_z r v_\phi - B_\phi v_z)\mathbf{u}_r + (B_r v_z - B_z v_r)\mathbf{u}_\phi + (B_\phi v_r - B_r r v_\phi)\mathbf{u}_z$$

$$(1\text{-}2\text{-}4)$$

the equations of motion for electrons in magnetic field for cylindrical coordinates can be given by

$$\frac{d^2r}{dt^2} - r\left(\frac{d\phi}{dt}\right)^2 = -\frac{e}{m}\left(B_z r \frac{d\phi}{dt} - B_\phi \frac{dz}{dt}\right) \qquad (1\text{-}2\text{-}5a)$$

$$\frac{1}{r}\frac{d}{dt}\left(r^2 \frac{d\phi}{dt}\right) = -\frac{e}{m}\left(B_r \frac{dz}{dt} - B_z \frac{dr}{dt}\right) \qquad (1\text{-}2\text{-}5b)$$

$$\frac{d^2z}{dt^2} = -\frac{e}{m}\left(B_\phi \frac{dr}{dt} - B_r r \frac{d\phi}{dt}\right) \qquad (1\text{-}2\text{-}5c)$$

Consider next an electron moving with a velocity of v_x to enter a constant uniform magnetic field that is perpendicular to v_x as shown in Fig. 1-2-1. The velocity of the electron is assumed to be

$$\mathbf{v} = v_x \mathbf{u}_x \qquad (1\text{-}2\text{-}6)$$

where $\mathbf{u}_x$ is a unit vector in the x direction.

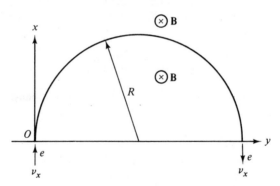

FIG. 1-2-1. Circular motion of an electron in a transverse magnetic field.

Since the force exerted on the electron by the magnetic field is normal to the motion at every instant, no work is done on the electron and its velocity remains constant. The magnetic field is assumed to be

$$\mathbf{B} = B_z \mathbf{u}_z \qquad (1\text{-}2\text{-}7)$$

Then the magnetic force at the instant when the electron just enters the magnetic field is given by

$$\mathbf{F} = -e\mathbf{v} \times \mathbf{B} = evB\mathbf{u}_y \qquad (1\text{-}2\text{-}8)$$

This means that the force remains constant in magnitude but changes the direction of motion because the electron is pulled by the magnetic force in a circular path. This type of magnetic force is analogous to the problem of a mass tied to a rope and twirled around with constant velocity. The force in the rope remains constant in magnitude and is always directed toward the center of the circle and so is perpendicular to the motion. At any point on the circle the outward centrifugal force is equal to the pulling force. That is,

$$\frac{mv^2}{R} = evB \qquad (1\text{-}2\text{-}9)$$

where R is the radius of the circle.

From Eq. (1-2-8) the radius of the path is given by

$$R = \frac{mv}{eB} \quad \text{meters} \qquad (1\text{-}2\text{-}10)$$

The cyclotron angular frequency of the circular motion of the electron is

$$\omega = \frac{v}{R} = \frac{eB}{m} \quad \text{radians/s} \qquad (1\text{-}2\text{-}11)$$

The period for one complete revolution is expressed by

$$T = \frac{2\pi}{\omega} = \frac{2\pi m}{eB} \quad \text{seconds} \qquad (1\text{-}2\text{-}12)$$

It should be noted that the radius of the path is directly proportional to the velocity of the electron but that the angular frequency and the period are independent of velocity or radius. This means that faster-moving electrons or particles will traverse larger circles in the same time that a slower-moving particle moves in a smaller circle. This very important result is the operating basis of such microwave devices as magnetic-focusing apparatus.

1-3 Electron Motion in an Electromagnetic Field

If both electric and magnetic fields exist simultaneously, the motion of the electrons will depend on the orientation of the two fields. If the two fields are in the same or in opposite directions, the magnetic field exerts no force on the electron, and the electron motion depends only on the electric field, which has been described in Section 1-1. Linear-beam tubes (*O*-type devices) use a magnetic field whose axis coincides with that of the electron beam to hold the beam together as it travels the length of the tube. In these tubes the electrons receive the full potential energy of the electric field but are not influenced by the magnetic field.

When the electric field **E** and the magnetic field **B** are at right angle to each other, a magnetic force is exerted on the electron beam. This type of field is called a *crossed field*. In a crossed-field tube (*M*-type device), electrons emitted by the cathode are accelerated by the electric field and gain velocity;

but the greater their velocity, the more their path is bent by the magnetic field. The Lorentz force acting on an electron due to the presence of both the electric field **E** and the magnetic flux **B** is given by

$$\mathbf{F} = -e(\mathbf{E} + \mathbf{v} \times \mathbf{B}) = m\frac{d\mathbf{v}}{dt} \qquad (1\text{-}3\text{-}1)$$

The equations of motion for electrons in a crossed field are expressed in rectangular coordinates and cylindrical coordinates, respectively, as

$$\frac{d^2x}{dt^2} = -\frac{e}{m}\left(E_x + B_z\frac{dy}{dt} - B_y\frac{dz}{dt}\right) \qquad (1\text{-}3\text{-}2a)$$

$$\frac{d^2y}{dt^2} = -\frac{e}{m}\left(E_y + B_x\frac{dz}{dt} - B_z\frac{dx}{dt}\right) \qquad (1\text{-}3\text{-}2b)$$

$$\frac{d^2z}{dt^2} = -\frac{e}{m}\left(E_z + B_y\frac{dx}{dt} - B_x\frac{dy}{dt}\right) \qquad (1\text{-}3\text{-}2c)$$

$$\frac{d^2r}{dt^2} - r\left(\frac{d\phi}{dt}\right)^2 = -\frac{e}{m}\left(E_r + B_z r\frac{d\phi}{dt} - B_\phi\frac{dz}{dt}\right) \qquad (1\text{-}3\text{-}3a)$$

$$\frac{1}{r}\frac{d}{dt}\left(r^2\frac{d\phi}{dt}\right) = -\frac{e}{m}\left(E_\phi + B_r\frac{dz}{dt} - B_z\frac{dr}{dt}\right) \qquad (1\text{-}3\text{-}3b)$$

$$\frac{d^2z}{dt^2} = -\frac{e}{m}\left(E_z + B_\phi\frac{dr}{dt} - B_r r\frac{d\phi}{dt}\right) \qquad (1\text{-}3\text{-}3c)$$

where

$$\frac{d\phi}{dt} = \omega_c = \frac{e}{m}B \text{ is the cyclotron frequency.}$$

It is, of course, difficult to solve these equations for solutions in three dimensions. In microwave devices and circuits, however, only one dimension is involved in most cases. So the equations of motion become simple and can easily be solved.

An example may help to verify some of the preceding equations.

EXAMPLE 1-3-1: Referring to the diagram of Fig. 1-3-1:

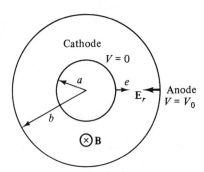

FIG. 1-3-1. Cylindrical magnetron.

Solution.

1. Write the equations of motion for electrons in cylindrical coordinates.

(a) $$\frac{d^2r}{dt^2} - r\left(\frac{d\phi}{dt}\right)^2 = +\frac{e}{m}E_r - \frac{e}{m}r\frac{d\phi}{dt}B_z$$

(b) $$\frac{1}{r}\frac{d}{dt}\left(r^2\frac{d\phi}{dt}\right) = \frac{e}{m}B_z\frac{dr}{dt}$$

2. From (b)

$$\frac{d}{dt}\left(r^2\frac{d\phi}{dt}\right) = \frac{1}{2}\omega_c\frac{d}{dt}(r^2) \qquad \left(\text{where } \omega_c = \frac{e}{m}B_0\right)$$

$$r^2\frac{d\phi}{dt} = \frac{1}{2}\omega_c r^2 + \text{constant}$$

3. Applications of the boundary conditions: At $r = a$,

$$a^2\frac{d\phi}{dt} = \frac{1}{2}\omega_c a^2 + \text{constant}$$

$$\frac{d\phi}{dt} = 0, \qquad \text{constant} = -\frac{1}{2}\omega_c a^2$$

Hence

$$r^2\frac{d\phi}{dt} = \frac{1}{2}\omega_c(r^2 - a^2)$$

4. The magnetic field does no work on the electrons.

$$\frac{1}{2}mv^2 = eV$$

$$v^2 = \frac{2e}{m}V = v_r^2 + v_\phi^2 = \left(\frac{dr}{dt}\right)^2 + \left(r\frac{d\phi}{dt}\right)^2$$

5. For grazing the anode,

$$r = b, \quad V = V_0, \quad \frac{dr}{dt} = 0$$

$$b^2\left(\frac{d\phi}{dt}\right)^2 = \frac{2e}{m}V_0 \quad \text{and} \quad b^2\frac{d\phi}{dt} = \frac{1}{2}\omega_c(b^2 - a^2)$$

$$b^2\left[\frac{1}{2}\omega_c\left(1 - \frac{a^2}{b^2}\right)\right]^2 = \frac{2e}{m}V_0$$

6. The Hull cutoff voltage is

$$V_{0c} = \frac{e}{8m}B_0^2 b^2\left(1 - \frac{a^2}{b^2}\right)^2 \tag{1-3-3d}$$

This means that if $V_0 < V_{0c}$ for a given B_0, the electrons will not reach the anode.

Conversely, the cutoff magnetic field can be expressed in terms of V_0.

$$B_{0c} = \frac{(8V_0 m/e)^{1/2}}{b(1 - a^2/b^2)} \tag{1-3-4}$$

This implies that if $B_0 > B_{0c}$ for a given V_0, the electrons will not reach the anode.

1-4 Summary

Three sets of equations of motion for electrons in three different coordinates were derived.

1. The equations of motion for an electron in electric field are given by

$$\frac{d^2x}{dt^2} = -\frac{e}{m}E_x$$

$$\frac{d^2y}{dt^2} = -\frac{e}{m}E_y \qquad \text{(rectangular coordinates)}$$

$$\frac{d^2z}{dt^2} = -\frac{e}{m}E_z$$

and

$$\frac{d^2r}{dt^2} - r\left(\frac{d\phi}{dt}\right)^2 = -\frac{e}{m}E_r$$

$$\frac{1}{r}\frac{d}{dt}\left(r^2\frac{d\phi}{dt}\right) = -\frac{e}{m}E_\phi \qquad \text{(cylindrical coordinates)}$$

$$\frac{d^2z}{dt^2} = -\frac{e}{m}E_z$$

2. The equations of motion for an electron in magnetic field are given by

$$\frac{d^2x}{dt^2} = -\frac{e}{m}\left(B_z\frac{dy}{dt} - B_y\frac{dz}{dt}\right)$$

$$\frac{d^2y}{dt^2} = -\frac{e}{m}\left(B_x\frac{dz}{dt} - B_z\frac{dx}{dt}\right) \qquad \text{(rectangular coordinates)}$$

$$\frac{d^2z}{dt^2} = -\frac{e}{m}\left(B_y\frac{dx}{dt} - B_x\frac{dy}{dt}\right)$$

and

$$\frac{d^2r}{dt^2} - r\left(\frac{d\phi}{dt}\right)^2 = -\frac{e}{m}\left(B_z r\frac{d\phi}{dt} - B_\phi\frac{dz}{dt}\right)$$

$$\frac{1}{r}\frac{d}{dt}\left(r^2\frac{d\phi}{dt}\right) = -\frac{e}{m}\left(B_r\frac{dz}{dt} - B_z\frac{dr}{dt}\right) \qquad \text{(cylindrical coordinate)}$$

$$\frac{d^2z}{dt^2} = -\frac{e}{m}\left(B_\phi\frac{dr}{dt} - B_r r\frac{d\phi}{dt}\right)$$

3. The equations of motion for an electron in electromagnetic field (crossed field) are given by

$$\frac{d^2x}{dt^2} = -\frac{e}{m}\left(E_x + B_z\frac{dy}{dt} - B_y\frac{dz}{dt}\right)$$

$$\frac{d^2y}{dt^2} = -\frac{e}{m}\left(E_y + B_x\frac{dz}{dt} - B_z\frac{dx}{dt}\right) \qquad \text{(rectangular coordinates)}$$

$$\frac{d^2z}{dt^2} = -\frac{e}{m}\left(E_z + B_y\frac{dx}{dt} - B_x\frac{dy}{dt}\right)$$

and

$$\frac{d^2r}{dt^2} - r\left(\frac{d\phi}{dt}\right)^2 = -\frac{e}{m}\left(E_r + B_z r\frac{d\phi}{dt} - B_\phi\frac{dz}{dt}\right)$$

$$\frac{1}{r}\frac{d}{dt}\left(r^2\frac{d\phi}{dt}\right) = -\frac{e}{m}\left(E_\phi + B_r\frac{dz}{dt} - B_z\frac{dr}{dt}\right) \qquad \begin{array}{l}\text{(cylindrical}\\ \text{coordinate)}\end{array}$$

$$\frac{d^2z}{dt^2} = -\frac{e}{m}\left(E_z + B_\phi\frac{dr}{dt} - B_r r\frac{d\phi}{dt}\right)$$

SUGGESTED READINGS

1. BRONWELL, A. B., and R. E. BEAM, *Theory and Application of Microwaves*, Chapter 2. McGraw-Hill Book Company, New York, 1947.

2. GEWARTOWSKI, J. W., and H. A. WATSON, *Principles of Electron Tubes*, Chapters 1 and 2. D. Van Nostrand Company, Princeton, N.J., 1965.

3. HARMAN, W. W., *Fundamentals of Electronic Motion*, Chapters 1, 2, and 4. McGraw-Hill Book Company, New York, 1953.

4. MILLMAN, JACOB, and C. C. HALKIAS, *Electronic Devices and Circuits*, Chapters 1 and 2. McGraw-Hill Book Company, New York, 1967.

Electromagnetic Plane Waves

2-0 Introduction

Since Maxwell's fundamental concepts of electromagnetic wave theory have been established, the electric and magnetic wave equations can readily be derived from Faraday's electromotive force law, Ampère's circuital law, and Gauss' law for the electric and magnetic fields. In Chapter 1 interactions between electron and field were discussed. Here electromagnetic plane waves are described in detail. Many topics associated with electromagnetic waves, such as Poynting theory, reflection theory, attenuation concepts, and plane-wave propagation in metallic-film coating on plastic substrates, are also analyzed, for these basic principles are used frequently in later chapters.

The principles of electromagnetic plane waves

are based on the relationships between electricity and magnetism. A changing magnetic field will induce an electric field, and a changing electric field will induce a magnetic field. Also, the induced fields are not confined but ordinarily extend outward into space. The sinusoidal form of the wave causes energy to be interchanged between the magnetic and electric fields in the direction of the wave propagation.

A plane wave is a wave that has a plane front, a cylindrical wave one that has a cylindrical front, and a spherical wave one that has a spherical front. The front of a wave is sometimes referred to as an *equiphase surface*. In the far field of free space, electric and magnetic waves are always perpendicular to each other, and both are normal to the direction of propagation of the wave. This type of wave is known as the *transverse electromagnetic* (TEM) wave. If only the transverse electric wave exists, the wave is called *TE-mode* wave. That means there is no component of the electric wave in the direction of propagation. In *TM-modes* only the transverse magnetic wave exists.

2-1 Electric and Magnetic Wave Equations

The electric and magnetic wave equations can be basically derived from Maxwell's equations, which in time domain are expressed as

$$\nabla \times \mathbf{E} = -\frac{\partial \mathbf{B}}{\partial t} \tag{2-1-1}$$

$$\nabla \times \mathbf{H} = \mathbf{J} + \frac{\partial \mathbf{D}}{\partial t} \tag{2-1-2}$$

$$\nabla \cdot \mathbf{D} = \rho_v \tag{2-1-3}$$

$$\nabla \cdot \mathbf{B} = 0 \tag{2-1-4}$$

It should be noted that boldface roman letters indicate vector quantities or complex quantities. The units of these field variables are

E is called the *electric field intensity* in volts per meter,
H is called the *magnetic field intensity* in amperes per meter,
D is called the *electric flux density* in coulombs per square meter,
B is called the *magnetic flux density* in webers per square meter,
J is called the *electric current density* in amperes per square meter, and
ρ_v is called the *electric charge density* in coulombs per cubic meter.

The electric current density includes two components—that is,

$$\mathbf{J} = \mathbf{J}_c + \mathbf{J}_0 \tag{2-1-5}$$

where $J_c = \sigma E$ is called the *conduction current density*

J_0 = the *impressed current density*, which is independent of the field

In addition to Maxwell's four equations, the characteristics of the medium in which the fields exist are needed to specify the flux in terms of the fields in a specific medium. These constitutive relationships are

$$D = \epsilon E \qquad (2\text{-}1\text{-}6)$$

$$B = \mu H \qquad (2\text{-}1\text{-}7)$$

$$J_c = \sigma E \qquad (2\text{-}1\text{-}8)$$

$$\epsilon = \epsilon_r \epsilon_0 \qquad (2\text{-}1\text{-}9)$$

and

$$\mu = \mu_r \mu_0 \qquad (2\text{-}1\text{-}10)$$

where ϵ = the dielectric permittivity or capacitivity of the medium in farad per meter

ϵ_r = the relative dielectric constant (dimensionless)

$\epsilon_0 = 8.854 \times 10^{-12} \cong \dfrac{1}{36\pi} \times 10^{-9}$ farad per meter is the dielectric permittivity of vacuum or free space

μ = the magnetic permeability or inductivity of the medium in henry per meter

μ_r = the relative permeability or relative inductivity (dimensionless)

$\mu_0 = 4\pi \times 10^{-7}$ henry per meter is the permeability of vacuum or free space

σ = the conductivity of the medium in mhos per meter

If a sinusoidal time function in the form of $e^{j\omega t}$ is assumed, $\dfrac{\partial}{\partial t}$ can be replaced by $j\omega$. Then Maxwell's equations in frequency domain are given by

$$\nabla \times E = -j\omega \mu H \qquad (2\text{-}1\text{-}11)$$

$$\nabla \times H = (\sigma + j\omega \epsilon)E \qquad (2\text{-}1\text{-}12)$$

$$\nabla \cdot D = \rho_v \qquad (2\text{-}1\text{-}13)$$

$$\nabla \cdot B = 0 \qquad (2\text{-}1\text{-}14)$$

Taking the curl of Eq. (2-1-11) on both sides yields

$$\nabla \times \nabla \times E = j\omega \mu \nabla \times H \qquad (2\text{-}1\text{-}15)$$

Substitution of Eq. (2-1-12) for the right-hand side of Eq. (2-1-15) gives

$$\nabla \times \nabla \times E = j\omega \mu (\sigma + j\omega \epsilon)E \qquad (2\text{-}1\text{-}16)$$

The vector identity for the curl of the curl of a vector quantity **A** is expressed as

$$\nabla \times \nabla \times \mathbf{A} = -\nabla^2 \mathbf{A} + \nabla(\nabla \cdot \mathbf{A}) \qquad (2\text{-}1\text{-}17)$$

In free space the space-charge density is zero, and in a perfect conductor time-varying or static fields do not exist. So

$$\nabla \cdot \mathbf{D} = \rho_v = 0 \qquad (2\text{-}1\text{-}18)$$

$$\nabla \cdot \mathbf{E} = 0 \qquad (2\text{-}1\text{-}19)$$

Substitution of Eq. (2-1-17) for the left-hand side of Eq. (2-1-16) and replacement of Eq. (2-1-19) yield the electric wave equation as

$$\nabla^2 \mathbf{E} = \gamma^2 \mathbf{E} \qquad (2\text{-}1\text{-}20)$$

where $\gamma = \sqrt{j\omega\mu(\sigma + j\omega\epsilon)} = \alpha + j\beta$ is called the *intrinsic propagation constant* of a medium

α = the *attenuation constant* in nepers per meter

β = the *phase constant* in radians per meter

Similarly, the magnetic wave equation is given by

$$\nabla^2 \mathbf{H} = \gamma^2 \mathbf{H} \qquad (2\text{-}1\text{-}21)$$

It should be noted that the "double del" or "del squared" is a scalar operator—that is,

$$\nabla \cdot \nabla = \nabla^2 \qquad (2\text{-}1\text{-}22)$$

which is a second-order operator in three different coordinate systems.

In rectangular (cartesian) coordinates,

$$\nabla^2 = \frac{\partial^2}{\partial x^2} + \frac{\partial^2}{\partial y^2} + \frac{\partial^2}{\partial z^2} \qquad (2\text{-}1\text{-}23)$$

In cylindrical (circular) coordinates,

$$\nabla^2 = \frac{1}{r}\frac{\partial}{\partial r}\left(r\frac{\partial}{\partial r}\right) + \frac{1}{r^2}\frac{\partial^2}{\partial \phi^2} + \frac{\partial^2}{\partial z^2} \qquad (2\text{-}1\text{-}24)$$

Also, the solutions of Eqs. (2-1-1) and (2-1-2) solved simultaneously yield the electric and magnetic wave equations in the time domain as

$$\nabla^2 \mathbf{E} = \mu\sigma\frac{\partial \mathbf{E}}{\partial t} + \mu\epsilon\frac{\partial^2 \mathbf{E}}{\partial t^2} \qquad (2\text{-}1\text{-}25)$$

$$\nabla^2 \mathbf{H} = \mu\sigma\frac{\partial \mathbf{H}}{\partial t} + \mu\epsilon\frac{\partial^2 \mathbf{H}}{\partial t^2} \qquad (2\text{-}1\text{-}26)$$

2-2 Poynting Theorem

At what rate will electromagnetic energy be transmitted through free space or any medium, be stored in the electric and magnetic fields, and be dissipated as heat? From the standpoint of complex power in terms of the complex field vectors, the time average of any two complex vectors is equal to the real part of the product of one complex vector multiplied by the complex conjugate of the other vector. Hence the time average of the instantaneous Poynting vector in steady state is given by

$$\langle P \rangle = \langle \mathbf{E} \times \mathbf{H} \rangle = \tfrac{1}{2} \operatorname{Re} (\mathbf{E} \times \mathbf{H^*}) \qquad (2\text{-}2\text{-}1)$$

where the notation $\langle \; \rangle$ stands for the average and the factor of $\tfrac{1}{2}$ appears in the equation for complex power when peak values are used for the complex quantities $\mathbf{E}$ and $\mathbf{H}$. **Re** represents the real part of the complex power, and the asterisk * indicates the complex conjugate.

It is necessary to define a complex Poynting vector as

$$\mathbf{P} = \tfrac{1}{2}(\mathbf{E} \times \mathbf{H^*}) \qquad (2\text{-}2\text{-}2)$$

Maxwell's equations in frequency domain for the electric and magnetic fields are

$$\nabla \times \mathbf{E} = -j\omega\mu\mathbf{H} \qquad (2\text{-}2\text{-}3)$$

$$\nabla \times \mathbf{H} = \mathbf{J} + j\omega\epsilon\mathbf{E} \qquad (2\text{-}2\text{-}4)$$

Dot multiplication of Eq. (2-2-3) by $\mathbf{H^*}$ and of the conjugate of Eq. (2-2-4) by $\mathbf{E}$ yields

$$(\nabla \times \mathbf{E}) \cdot \mathbf{H^*} = -j\omega\mu\mathbf{H} \cdot \mathbf{H^*} \qquad (2\text{-}2\text{-}5)$$

$$(\nabla \times \mathbf{H^*}) \cdot \mathbf{E} = (\mathbf{J^*} - j\omega\epsilon\mathbf{E^*}) \cdot \mathbf{E} \qquad (2\text{-}2\text{-}6)$$

Then subtraction of Eq. (2-2-5) from Eq. (2-2-6) results in

$$\mathbf{E} \cdot (\nabla \times \mathbf{H^*}) - \mathbf{H^*} \cdot (\nabla \times \mathbf{E}) = \mathbf{E} \cdot \mathbf{J^*} - j\omega(\epsilon E^2 - \mu H^2) \qquad (2\text{-}2\text{-}7)$$

where $\mathbf{E} \cdot \mathbf{E^*}$ is replaced by E^2 and $\mathbf{H} \cdot \mathbf{H^*}$ by H^2.

The left-hand side of Eq. (2-2-7) is equal to $-\nabla \cdot (\mathbf{E} \times \mathbf{H^*})$ by the vector identity. So

$$\nabla \cdot (\mathbf{E} \times \mathbf{H^*}) = -\mathbf{E} \cdot \mathbf{J^*} + j\omega(\epsilon E^2 - \mu H^2) \qquad (2\text{-}2\text{-}8)$$

Substituting Eqs. (2-1-5) and (2-2-2) into Eq. (2-2-8), we have

$$-\tfrac{1}{2}\mathbf{E} \cdot \mathbf{J_0^*} = \tfrac{1}{2}\sigma\mathbf{E} \cdot \mathbf{E^*} + j\omega(\tfrac{1}{2}\mu\mathbf{H} \cdot \mathbf{H^*} - \tfrac{1}{2}\epsilon\mathbf{E} \cdot \mathbf{E^*}) + \nabla \cdot \mathbf{P} \qquad (2\text{-}2\text{-}9)$$

Integration of Eq. (2-2-9) over a volume and application of Gauss' theorem to the last term on the right-hand side give

$$-\int_v \tfrac{1}{2}(\mathbf{E} \cdot \mathbf{J}_0^*) \, dv = \int_v \tfrac{1}{2}\sigma E^2 \, dv + j2\omega \int_v (w_m - w_e) \, dv + \oint_s \mathbf{P} \cdot d\mathbf{s}$$

$$(2\text{-}2\text{-}10)$$

where $\tfrac{1}{2}\sigma \, E^2 = \sigma \langle |E|^2 \rangle$ is the time-average dissipated power

$\quad\quad\quad \tfrac{1}{4}\mu \mathbf{H} \cdot \mathbf{H}^* = \tfrac{1}{2}\mu \langle |H|^2 \rangle = w_m$ is the time-average magnetic stored energy

$\quad\quad\quad \tfrac{1}{4}\epsilon \mathbf{E} \cdot \mathbf{E}^* = \tfrac{1}{2}\epsilon \langle |E|^2 \rangle = w_e$ is the time-average electric stored energy,

$\quad\quad\quad -\tfrac{1}{2}\mathbf{E} \cdot \mathbf{J}_0^* =$ the complex power impressed by the source J_0 into the field

Equation (2-2-10) is well known as the *complex Poynting theorem* or the *Poynting theorem* in frequency domain.

Furthermore, let

$$P_{\text{in}} = -\int_v \tfrac{1}{2}(\mathbf{E} \cdot \mathbf{J}_0^*) \, dv \text{ be the total complex power supplied by a source within a region}$$

$$\langle P_d \rangle = \int_v \tfrac{1}{2}\sigma E^2 \, dv \text{ be the time-average power dissipated as heat inside the region}$$

$$\langle W_m - W_e \rangle = \int_v (w_m - w_e) \, dv \text{ be the difference between time-average magnetic and electric energies stored within the region}$$

$$P_{\text{tr}} = \oint \mathbf{P} \cdot d\mathbf{s} \text{ be the complex power transmitted from the region}$$

The complex Poynting theorem shown in Eq. (2-2-10) can be simplified to

$$P_{\text{in}} = \langle P_d \rangle + j2\omega \, [\langle W_m - W_e \rangle] + P_{\text{tr}} \quad\quad (2\text{-}2\text{-}11)$$

This theorem states that the total complex power fed into a volume is equal to the algebraic sum of the active power dissipated as heat, plus the reactive power proportional to the difference between time-average magnetic and electric energies stored in the volume, and plus the complex power transmitted across the surface enclosed by the volume.

2-3 Uniform Plane Waves and Reflection

2-3-1. The Uniform Plane Waves

A plane wave is a wave whose phase is constant over a set of planes. A uniform plane wave is a wave whose magnitude and phase are both constant. A spherical wave in free space is a uniform plane wave as observed at a far distance. Its equiphase surfaces are concentric spheres, expanding as the wave travels outward from the source, and its magnitude is constant.

Electromagnetic waves in free space are typical uniform plane waves. The electric and magnetic fields are mutually perpendicular to each other and to the direction of propagation of the waves. The phases of the two fields are always in time phase and their magnitudes are always constant. The stored energies are equally divided between the two fields, and the energy flow is transmitted by the two waves in the direction of propagation. Thus a uniform plane wave is a transverse electromagnetic wave or a TEM wave.

A nonuniform plane wave is a wave whose amplitude (not phase) may vary within a plane normal to the direction of propagation. Consequently, the electric and magnetic fields are no longer in time phase.

Since a uniform plane wave of electric or magnetic field has no variation of intensity in a plane normal to the direction of propagation of the wave, then

$$\frac{\partial \mathbf{E}}{\partial x} = \frac{\partial \mathbf{E}}{\partial y} = 0 \quad \text{or} \quad \frac{\partial \mathbf{H}}{\partial x} = \frac{\partial \mathbf{H}}{\partial y} = 0$$

if the direction of propagation is assumed in the positive z direction.

With the preceding assumptions and lossless dielectric—that is, $\sigma = 0$ —the wave equations (2-1-25) and (2-1-26) in time domain for the electric and magnetic intensities in rectangular coordinates reduce to

$$\frac{\partial^2 E_x}{\partial z^2} = \mu_0 \epsilon_0 \frac{\partial^2 E_x}{\partial t^2} \tag{2-3-1}$$

$$\frac{\partial^2 H_y}{\partial z^2} = \mu_0 \epsilon_0 \frac{\partial^2 H_y}{\partial t^2} \tag{2-3-2}$$

in which the electric intensity is arbitrarily chosen in the x direction and the magnetic intensity in the y direction. With no loss in generality, it can be assumed that the electric intensity is given by

$$E_x = E_0 e^{j(\omega t - \beta z)} = E_0 e^{j\omega[t - (\beta/\omega)z]} \tag{2-3-3}$$

The magnetic intensity can be obtained by inserting Eq. (2-3-3) into the curl equation

$$\nabla \times \mathbf{E} = -j\omega\mu_0 \mathbf{H} \tag{2-3-4}$$

For the assumed conditions, the curl equation reduces to

$$\frac{\partial E_x}{\partial z} = -j\omega\mu_0 H_y \tag{2-3-5}$$

Differentiation of Eq. (2-3-3) with respect to z and substitution of the result in Eq. (2-3-5) yield

$$H_y = \sqrt{\frac{\epsilon_0}{\mu_0}} E_x \tag{2-3-6}$$

where $\dfrac{\beta}{\omega} = \sqrt{\mu_0 \epsilon_0} = \dfrac{1}{v_p}$ is accounted for in the derivation

$$v_p = \frac{1}{\sqrt{\mu_0 \epsilon_0}} = 3 \times 10^8 \text{ m/s} = \text{phase velocity}$$

$$= c, \text{ which is the velocity of light in vacuum}$$

$$\alpha = 0$$

$$\beta = \omega \sqrt{\mu_0 \epsilon_0} = \text{phase constant}$$

The ratio of electric to magnetic intensities is given by

$$\frac{E_x}{H_y} = \eta_0 = \sqrt{\frac{\mu_0}{\epsilon_0}} = 377 \quad \text{ohms} \qquad (2\text{-}3\text{-}7)$$

It is called the *intrinsic impedance* of free space. It must be noted that this impedance applies only for the far field. The relationship of the electric field **E** over the magnetic field **H** in the near field is not known and is unmeasurable.

Figure 2-3-1 shows uniform electric and magnetic plane waves in rectangular coordinates.

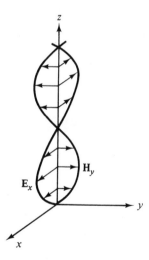

FIG. 2-3-1. Uniform plane waves traveling in the *z* direction.

In general, for a uniform plane wave propagating in a lossless dielectric medium ($\sigma = 0$), the characteristics of wave propagation would become

$$\alpha = 0 \qquad (2\text{-}3\text{-}8a)$$

$$\beta = \omega \sqrt{\mu_0 \epsilon} \qquad (2\text{-}3\text{-}8b)$$

$$\eta = \sqrt{\frac{\mu_0}{\epsilon}} = \frac{\eta_0}{\sqrt{\epsilon_r}} \qquad (2\text{-}3\text{-}8c)$$

$$v_p = \frac{1}{\sqrt{\mu_0 \epsilon}} = \frac{c}{\sqrt{\epsilon_r}} \qquad (2\text{-}3\text{-}8d)$$

It should be noted that all dielectrics have approximately the same permeability as free space.

2-3-2. Boundary Conditions

Since Maxwell's equations are in the form of differential rather than algebraic equations, boundary conditions must be applied to a given problem if a specific solution is required.

There are four basic rules for boundary conditions at the surface between two different materials.

1. The tangential components of electric field intensity are continuous across the boundary.
2. The normal components of electric flux density are discontinuous at the boundary by an amount equal to the surface-charge density on the boundary.
3. The tangential components of magnetic field intensity are discontinuous at the boundary by an amount equal to the surface-current density on the boundary.
4. The normal components of magnetic flux intensity are continuous across the boundary.

The four statements can be proved by applying Faraday's law, Gauss' law, Ampère's law, and $\nabla \cdot \mathbf{B} = 0$ to the boundaries of Fig. 2-3-2(a) and (b). It can be seen from the diagrams that

$$\oint_\ell \mathbf{E} \cdot d\ell = E_{t1} \, \Delta\ell - E_{t2} \, \Delta\ell = 0$$

$$\oint_s \mathbf{D} \cdot d\mathbf{s} = D_{n2} \, \Delta s - D_{n1} \, \Delta s = \rho_s \, \Delta s$$

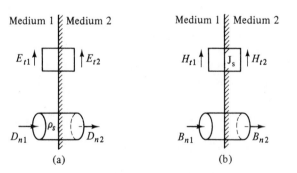

FIG. 2-3-2. Boundary conditions. (a) Electric intensity. (b) Magnetic intensity.

$$\oint_\ell \mathbf{H} \cdot d\boldsymbol{\ell} = H_{t1}\,\Delta\ell - H_{t2}\,\Delta\ell = J_s\,\Delta\ell$$

$$\oint_s \mathbf{B} \cdot d\mathbf{s} = B_{n1}\,\Delta s - B_{n2}\,\Delta s = 0$$

So the boundary equations are

$$E_{t1} = E_{t2} \tag{2-3-9a}$$

$$D_{n2} = D_{n1} + \rho_s \tag{2-3-9b}$$

$$H_{t1} = H_{t2} + J_s \tag{2-3-9c}$$

$$B_{n1} = B_{n2} \tag{2-3-9d}$$

If medium 1 is a perfect conductor ($\sigma = \infty$, $\epsilon_r = 1$, $\mu_r = 1$) and medium 2 is a perfect dielectric (vacuum or free space, $\sigma = 0$, ϵ_0, μ_0), then

$$E_{t1} = D_{t1}/\epsilon_0 = 0 \tag{2-3-10a}$$

$$D_n = \epsilon_0 E_n = \rho_s \tag{2-3-10b}$$

$$H_{t1} = H_{t2} \quad \text{if } J_s = 0 \tag{2-3-10c}$$

$$B_{n1} = B_{n2} \tag{2-3-10d}$$

2-3-3. Uniform Plane Wave Reflection

(*a*) *Normal-Incidence Reflection.* The simplest reflection problem is that of a uniform plane wave normally incident on a plane boundary between two media with no surface-charge density and surface-current density. This situation is shown in Fig. 2-3-3.

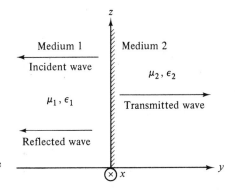

FIG. 2-3-3. Uniform plane -wave reflection.

In medium 1 the fields will be the sum of an incident wave plus a reflected wave, respectively.

$$E_x^{(1)} = E_0(e^{-j\beta_1 z} + \Gamma e^{j\beta_1 z}) \tag{2-3-11}$$

$$H_y^{(1)} = \frac{E_0}{\eta_1}(e^{-j\beta_1 z} - \Gamma e^{j\beta_1 z}) \tag{2-3-12}$$

where $\beta_1 = \omega\sqrt{\mu_1\epsilon_1}$

$$\eta_1 = \sqrt{\frac{\mu_1}{\epsilon_1}} = \frac{\eta_0}{\sqrt{\epsilon_{r1}}} = \text{intrinsic wave impedance of medium 1,}$$

$\Gamma = $ the reflection coefficient

In medium 2 there will be transmitted waves.

$$E_x^{(2)} = E_0 T e^{-j\beta_2 z} \qquad (2\text{-}3\text{-}13)$$

$$H_y^{(2)} = \frac{E_0}{\eta_2} T e^{-j\beta_2 z} \qquad (2\text{-}3\text{-}14)$$

where $\beta_2 = \omega\sqrt{\mu_2\epsilon_2}$

$$\eta_2 = \sqrt{\frac{\mu_2}{\epsilon_2}} = \frac{\eta_0}{\sqrt{\epsilon_{r2}}} = \text{intrinsic wave impedance of medium 2}$$

$T = $ the transmission coefficient

For continuity of wave impedance at the boundary, the wave impedance is

$$Z_z = \frac{E_x^{(1)}}{H_y^{(1)}}\bigg|_{z=0} = \eta_1 \frac{1+\Gamma}{1-\Gamma} = \frac{E_x^{(2)}}{H_y^{(2)}}\bigg|_{z=0} = \eta_2 \qquad (2\text{-}3\text{-}15)$$

So the reflection coefficient is given by

$$\Gamma = \frac{\eta_2 - \eta_1}{\eta_2 + \eta_1} \qquad (2\text{-}3\text{-}16)$$

From the boundary condition the tangential components of electric field intensity are continuous across the interface. Then

$$E_x^{(1)}|_{z=0} = E_0(1+\Gamma) = E_x^{(2)}|_{z=0} = E_0 T \qquad (2\text{-}3\text{-}17)$$

Hence the transmission coefficient is expressed as

$$T = 1 + \Gamma = \frac{2\eta_2}{\eta_2 + \eta_1} \qquad (2\text{-}3\text{-}18)$$

If medium 1 is lossless dielectric (i.e., $\sigma = 0$), the standing-wave ratio is defined as

$$\textbf{SWR} = \rho = \frac{|E_{\max}^{(1)}|}{|E_{\min}^{(2)}|} = \frac{1 + |\Gamma|}{1 - |\Gamma|} \qquad (2\text{-}3\text{-}19)$$

The power density transmitted across the boundary is

$$p_{\text{tr}} = \frac{1}{2}(\mathbf{E} \times \mathbf{H}^*)\bigg|_{z=0} \cdot \mathbf{u}_z = \frac{E_0^2}{2\eta_1}(1 - |\Gamma|^2) \qquad (2\text{-}3\text{-}20)$$

Then

$$p_{\text{tr}} = p_{\text{inc}}(1 - |\Gamma|^2) \qquad (2\text{-}3\text{-}21)$$

where $p_{\text{inc}} = $ incident power density.

The incident power density minus the transmitted power density would yield the reflected power density as

$$p_{\text{ref}} = p_{\text{inc}}|\Gamma|^2 \qquad (2\text{-}3\text{-}22)$$

(b) Oblique-Incidence Reflection.

(1) **E** IS IN THE PLANE OF INCIDENCE.   The plane of incidence is defined by the direction of propagation and the line normal to the boundary. The linearly polarized uniform plane waves **E** lying in and **H** normal to the plane of incidence are impinging obliquely on a boundary between two lossless dielectric materials as shown in Fig. 2-3-4.

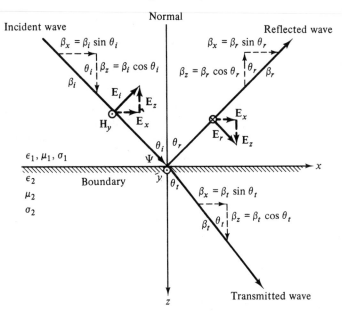

FIG. 2-3-4. Reflection and transmission of oblique incidence.

Whenever a wave is incident obliquely on the boundary surface between two media, the polarization of the wave is vertical or horizontal if the electric field is normal to or parallel to the boundary surface. The terms horizontal and vertical polarizations refer to the phenomenon of waves from horizontal and vertical antennas, respectively, producing the corresponding orientations of wave polarization when the waves strike the surface of the earth. For guided waves in waveguides, the terms *transverse electric* (TE) and *transverse magnetic* (TM) are used to designate the fact that either the electric or the magnetic field is parallel or normal to the direction of propagation. The polarization of a wave is an extremely useful concept for computing electromagnetic power flow. For example, a Poynting vector indicates that the power-flow density is the cross product of an electric and a magnetic field with the specific direction determined by the polarizations of the two fields.

As Fig. 2-3-4 shows, for a lossless dielectric media, the phase constants (or propagation constants) of the two media in the x direction on the interface are equal as required by the continuity of tangential **E** and **H** on the boundary. Thus

$$\beta_i \sin \theta_i = \beta_r \sin \theta_r, \qquad (2\text{-}3\text{-}23a)$$

$$\beta_i \sin \theta_i = \beta_t \sin \theta_t \qquad (2\text{-}3\text{-}23b)$$

From Eq. (2-3-23a), since $\beta_i = \beta_r = \beta_1$, the angle of reflection is equal to the angle of incidence. This is

$$\theta_i = \theta_r \qquad (2\text{-}3\text{-}24)$$

From Eq. (2-3-23b)

$$\frac{\sin \theta_t}{\sin \theta_i} = \frac{\beta_i}{\beta_t} = \frac{v_2}{v_1} = \sqrt{\frac{\mu_1 \epsilon_1}{\mu_2 \epsilon_2}} \qquad (2\text{-}3\text{-}25)$$

where v represents the phase velocity. This is well known as Snell's law. In general, all low-loss dielectrics have equal permeability—that is, $\mu_1 = \mu_2 = \mu_0$. If medium 2 is free space and medium 1 is a nonmagnetic dielectric, the right-hand side of Eq. (2-3-25) becomes $\sqrt{\epsilon_r}$, which is called the *index of refraction* of the dielectric.

The components of electric intensity **E** are

$$E_x = E_0 \cos \theta_i e^{-j\beta_1(x \sin \theta_i + z \cos \theta_i)} \qquad (2\text{-}3\text{-}26)$$

$$E_y = 0 \qquad (2\text{-}3\text{-}27)$$

$$E_z = -E_0 \sin \theta_i e^{-j\beta_1(x \sin \theta_i + z \cos \theta_i)} \qquad (2\text{-}3\text{-}28)$$

The components of magnetic intensity **H** are

$$H_x = 0 \qquad (2\text{-}3\text{-}29)$$

$$H_y = \frac{E_0}{\eta_1} e^{-j\beta_1(x \sin \theta_i + z \cos \theta_i)} \qquad (2\text{-}3\text{-}30)$$

$$H_z = 0 \qquad (2\text{-}3\text{-}31)$$

The wave impedance in the z direction is given by

$$Z_z = \frac{E_x}{H_y} = \eta \cos \theta \qquad (3\text{-}2\text{-}32)$$

It should be noted that the subscripts of η and θ have been dropped because the wave impedances of the two regions in the z direction are the same.

The wave impedance can be expressed in terms of the reflection coefficient of the normal components. In medium 1

$$E_x^{(1)} = E_0 \cos \theta_i [e^{-j\beta_1 z \cos \theta_i} + \Gamma e^{j\beta_1 z \cos \theta_r}] \qquad (2\text{-}3\text{-}33)$$

$$H_y^{(1)} = \frac{E_0}{\eta_1} [e^{-j\beta_1 z \cos \theta_i} - \Gamma e^{j\beta_1 z \cos \theta_r}] \qquad (2\text{-}3\text{-}34)$$

$$Z_z = \frac{E_x^{(1)}}{H_y^{(1)}} \bigg|_{z=0} = \eta_1 \cos \theta_i \frac{1 + \Gamma}{1 - \Gamma} \qquad (2\text{-}3\text{-}35)$$

The impedance must be equal to the z-directed wave impedance in region 2 at the boundary. Substitution of $Z_z = \eta_2 \cos \theta_t$ in Eq. (2-3-35) yields

$$\Gamma = \frac{\eta_2 \cos \theta_t - \eta_1 \cos \theta_i}{\eta_2 \cos \theta_t + \eta_1 \cos \theta_i} \tag{2-3-36}$$

Then the transmission coefficient is given by

$$T = \frac{2\eta_2 \cos \theta_t}{\eta_2 \cos \theta_t + \eta_1 \cos \theta_i} \tag{2-3-37}$$

The preceding two equations are known as Fresnel's formulas for **E** in the plane of incidence.

(2) H IS IN THE PLANE OF INCIDENCE.   If **H** is in the plane of incidence, the components of **H** are

$$H_x = H_0 \cos \theta_i e^{-j\beta_1(x \sin \theta_i + z \cos \theta_i)} \tag{2-3-38}$$

$$H_y = 0 \tag{2-3-39}$$

$$H_z = -H_0 \sin \theta_i e^{-j\beta_1(x \sin \theta_i + z \cos \theta_i)} \tag{2-3-40}$$

The components of electric intensity **E** normal to the plane of incidence are

$$E_x = 0 \tag{2-3-41}$$

$$E_y = -\eta_1 H_0 e^{-j\beta_1(x \sin \theta_i + z \cos \theta_i)} \tag{2-3-42}$$

$$E_z = 0 \tag{2-3-43}$$

The wave impedance in the z direction is given by

$$Z_z = -\frac{E_y}{H_x} = \frac{\eta}{\cos \theta} = \eta \sec \theta \tag{2-3-44}$$

It should be noted that the subscripts of η and θ have been dropped for the same reason stated previously.

The Fresnel's formulas for **H** in the plane of incidence are

$$\Gamma = \frac{\eta_2 \sec \theta_t - \eta_1 \sec \theta_i}{\eta_2 \sec \theta_t + \eta_1 \sec \theta_i} \tag{2-3-45}$$

$$T = \frac{2\eta_2 \sec \theta_t}{\eta_2 \sec \theta_t + \eta_1 \sec \theta_i} \tag{2-3-46}$$

2-4   Plane Wave Propagation in Free Space and Lossless Dielectric

2-4-1. Plane Wave Propagation in Free Space

The electromagnetic wave being propagated in free space near the surface of the earth is divided into two parts: the ground wave and the sky wave or ionosphere wave. The ground wave is further divided into a direct wave, an earth-reflected wave, and a surface wave. Figure 2-4-1 shows the wave

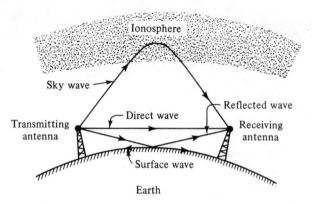

FIG. 2-4-1. Wave components near the surface of the earth.

components of electromagnetic wave from a nondirectional antenna to a receiving station.

The ionosphere is that region of the earth's atmosphere in which the constituent gases are ionized by solar radiation. This region extends from about 50 km above the earth to several earth radii and has different layers designated as C, D, E, and F layers in order of height. The electron-density distribution of each layer varies with the time of day, season, year, and the geographical location. During the day the electron density N is approximately 10^{12} electrons per cubic meter at an altitude between 90 and 1000 km. The E and F layers have a permanent existence, but the D layer is present only during the day. The electron density determines the reflection and refraction of microwaves. For vertical incidence, the critical frequency is given by

$$F_{cr} = \sqrt{81 N_{max}} \qquad \text{Hz} \qquad (2\text{-}4\text{-}0)$$

This means that a microwave of frequency F_{cr} will be reflected back to the earth if the electron density is equal to or higher than the required maximum electron density N_{max} (electrons per cubic meter).

The sky wave reaches the receiving station after reflection from the ionosphere. Although important in many communication systems, the sky wave need not be considered in most microwave applications because a wavelength shorter than about 4 m will not return to the earth from the ionosphere. The reflected wave is reflected from the earth in such a way as to reach the receiver. Energy radiated from the nondirectional antenna of Fig. 2-4-1 strikes the earth at all points between the base of the antenna and the horizon, but only that wave which leaves the antenna in the direction shown reaches the receiver. The surface wave is a wave diffracted around the surface of the earth or guided by the ground-air interface. This component is important at broadcast frequencies; at microwave frequencies, however, the surface wave is rapidly attenuated, and at a distance of two km from the antenna it has an amplitude of only a fraction of one percent of the direct wave. This component must be considered in blind-landing systems in which ranges of less than two km are important. The direct wave travels a nearly straight path from the antenna to the receiving station. It will be the only wave to be considered in this book. The term *free space* will be used to denote vacuum or any other media having essentially the same characteristics as vacuum, such as open air, anechoic chamber, and shielded enclosure. When power radiates from the antenna, the power density carried by the spherical wave will decrease with distance as the energy in the wave spreads out over an ever-increasing surface area as the wave progresses.

The power density is given by

$$p_d = \frac{p_t g_t}{4\pi R^2} \qquad \text{watt/m}^2 \qquad (2\text{-}4\text{-}1)$$

where p_t = transmitting power in watts

g_t = transmitting antenna gain (numerical)

R = distance between the antenna and the field point in meters

The power received by the receiving antenna will be given as

$$p_r = p_d A_e = \left(\frac{p_t g_t}{4\pi R^2}\right)\left(\frac{\lambda^2}{4\pi} g_r\right) \qquad \text{watt} \qquad (2\text{-}4\text{-}2)$$

where $A_e = \dfrac{\lambda^2}{4\pi} g_r$ = effective antenna aperture in square meters

$\dfrac{\lambda^2}{4\pi} = A_a$ = antenna aperture in square meters

g_r = receiving antenna gain (numerical)

Figure 2-4-2 shows the relationships of electromagnetic energy transmission in free space between two antennas.

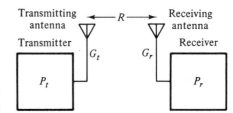

FIG. 2-4-2. Electromagnetic energy transmission between two antennas.

If the received power is expressed in terms of decibels, Eq. (2-4-2) will become

$$P_r = P_t + G_t + G_r - 20 \log \left(\frac{4\pi R}{\lambda}\right) \qquad \text{dB} \qquad (2\text{-}4\text{-}3)$$

where $P_t, G_t,$ and G_r are in dB(decibels). The term $20 \log (4\pi R/\lambda)$ is well known as the free-space attenuation in decibels. It can easily be found from the standard nomograph as shown in Fig. 2-4-3. For example, if the wave-

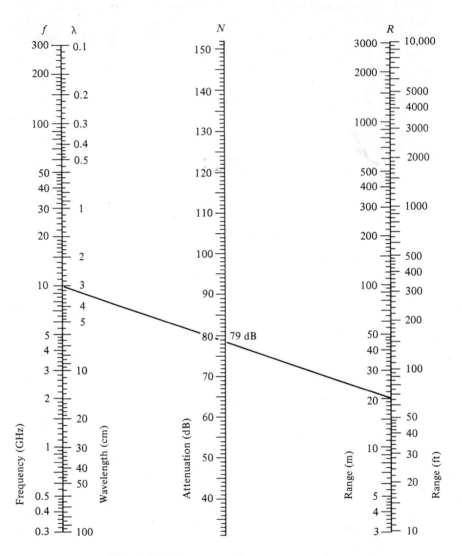

FIG. 2-4-3. Nomogram of free space attenuation.

length of a signal is 0.03 m and the range is 20 m, the free-space attenuation is about 79 dB.

It should be noted that the free-space attenuation is entirely different from the dissipative attenuation of a medium such as atmosphere that absorbs energy from the wave. The factor $(4\pi R^2)$ in Eq. (2-4-2) simply accounts for the fact that the power density is inversely proportionally decreasing with the squared distance when the energy spreads out over free space.

The factor $(\lambda^2/4\pi)$ is the aperture of a receiving antenna. It does not imply that a higher-frequency wave decreases in magnitude more rapidly than a lower-frequency wave. It is simply a consequence of the fact that, for a given antenna gain, the aperture of a higher-frequency antenna is smaller than that of a lower-frequency antenna so that it intercepts a smaller amount of power from the wave.

2-4-2. Plane Wave Propagation in Lossless Dielectric

The lossless dielectric, which is often called the good or perfect dielectric, is characterized by $\sigma = 0$. Hence the intrinsic impedance for a lossless dielectric can be expressed in terms of air. This is

$$\eta = \sqrt{\frac{\mu}{\epsilon}} = \sqrt{\frac{\mu_0}{\epsilon_r \epsilon_0}} = \frac{377}{\sqrt{\epsilon_r}} \quad \text{ohms} \qquad (2\text{-}4\text{-}4)$$

The attenuation constant α is zero, and the phase constant β is given by

$$\beta = \omega\sqrt{\mu\epsilon} \qquad (2\text{-}4\text{-}5)$$

The phase velocity is expressed by

$$v_p = \frac{1}{\sqrt{\mu\epsilon}} \qquad (2\text{-}4\text{-}6)$$

2-5 Plane Wave Propagation in Lossy Media

The lossy media are characterized by $\sigma \neq 0$. There are three types of lossy media: good conductor, poor conductor, and lossy dielectric, which are discussed in this section. The presence of a loss in the medium introduces wave dispersion by conductivity. Dispersion makes a general solution in the time domain impossible except by Fourier expansion methods. Thus only solutions for the frequency domain (or steady state) will be given.

The electric and magnetic wave equations in the frequency domain as shown in Eqs. (2-1-20) and (2-1-21) are repeated here:

$$\nabla^2 \mathbf{E} = j\omega\mu(\sigma + j\omega\epsilon)\mathbf{E} \qquad (2\text{-}5\text{-}1)$$

$$\nabla^2 \mathbf{H} = j\omega\mu(\sigma + j\omega\epsilon)\mathbf{H} \qquad (2\text{-}5\text{-}2)$$

For one dimension in the positive z direction, they become

$$\frac{\partial^2 E_x}{\partial z^2} = j\omega\mu(\sigma + j\omega\epsilon)E_x \qquad (2\text{-}5\text{-}3\text{a})$$

$$\frac{\partial^2 H_y}{\partial z^2} = j\omega\mu(\sigma + j\omega\epsilon)H_y \qquad (2\text{-}5\text{-}3\text{b})$$

The complex-frequency solutions would be given by

$$E_x = E_0 e^{-\alpha z} \cos(\omega t - \beta z) \qquad (2\text{-}5\text{-}4)$$

$$H_y = \frac{E_0}{\eta} e^{-\alpha z} \cos(\omega t - \beta z) \qquad (2\text{-}5\text{-}5)$$

where $\gamma = \sqrt{j\omega\mu(\sigma + j\omega\epsilon)} = \alpha + j\beta$

$$\eta = \sqrt{\frac{\mu}{\epsilon}}$$

2-5-1. Plane Wave in Good Conductor

A good conductor is defined as one having a very high conductivity; consequently, the conduction current is much larger than the displacement current. The energy transmitted by the wave traveling through the medium will decrease continuously as the wave propagates because ohmic losses are present. Expressed mathematically, a good conductor requires the criterion

$$\sigma \gg \omega\epsilon \qquad (2\text{-}5\text{-}6)$$

The propagation constant γ is expressed as

$$\gamma = \sqrt{j\omega\mu(\sigma + j\omega\epsilon)} = j\omega\sqrt{\mu\epsilon}\sqrt{1 - j\frac{\sigma}{\omega\epsilon}}$$

$$= j\omega\sqrt{\mu\epsilon}\sqrt{-j\frac{\sigma}{\omega\epsilon}} \qquad \text{for } \frac{\sigma}{\omega\epsilon} \gg 1$$

$$\doteq j\sqrt{\omega\mu\sigma}\sqrt{-j} = j\sqrt{\omega\mu\sigma}\left(\frac{1}{\sqrt{2}} - j\frac{1}{\sqrt{2}}\right) \qquad (2\text{-}5\text{-}7)$$

$$= (1 + j)\sqrt{\pi f\mu\sigma}$$

Hence

$$\alpha = \beta = \sqrt{\pi f\mu\sigma} \qquad (2\text{-}5\text{-}8)$$

The exponential factor $e^{-\alpha z}$ of the traveling wave will become $e^{-1} = 0.368$ when

$$z = \frac{1}{\sqrt{\pi f\mu\sigma}} \qquad (2\text{-}5\text{-}9)$$

This distance is called the *skin depth* and is denoted by

$$\delta = \frac{1}{\sqrt{\pi f\mu\sigma}} = \frac{1}{\alpha} = \frac{1}{\beta} \qquad (2\text{-}5\text{-}10)$$

Interestingly, at microwave frequencies the skin depth is extremely short and a piece of glass with an evaporated silver coat 5.40 microns thick is an excellent conductor at these frequencies.

Table 2-5-1 lists the conductivities of materials.

TABLE 2-5-1. Table of Conductivities.

Substance	Type	Conductivity, mhos/meter
Quartz, fused	insulator	10^{-17} approx.
Ceresin wax	insulator	10^{-17} approx.
Sulfur	insulator	10^{-15} approx.
Mica	insulator	10^{-15} approx.
Paraffin	insulator	10^{-15} approx.
Rubber, hard	insulator	10^{-15} approx.
Glass	insulator	10^{-12} approx.
Bakelite	insulator	10^{-9} approx.
Distilled water	insulator	10^{-4} approx.
Seawater	conductor	4 approx.
Tellurium	conductor	5×10^2 approx.
Carbon	conductor	3×10^4 approx.
Graphite	conductor	10^5 approx.
Cast iron	conductor	10^6 approx.
Mercury	conductor	10^6
Nichrome	conductor	10^6
Constantan	conductor	2×10^6
Silicon steel	conductor	2×10^6
German silver	conductor	3×10^6
Lead	conductor	5×10^6
Tin	conductor	9×10^6
Phosphor bronze	conductor	10^7
Brass	conductor	1.1×10^7
Zinc	conductor	1.7×10^7
Tungsten	conductor	1.8×10^7
Duralumin	conductor	3×10^7
Aluminum, hard-drawn	conductor	3.5×10^7
Gold	conductor	4.1×10^7
Copper	conductor	5.7×10^7
Silver	conductor	6.1×10^7

The intrinsic impedance of a good conductor is given as

$$\eta = \sqrt{\frac{j\omega\mu}{\sigma + j\omega\epsilon}} = \sqrt{\frac{j\omega\mu}{\sigma}} \quad \text{for } \sigma \gg \omega\epsilon$$

$$= \sqrt{\frac{\omega\mu}{\sigma}} \,\underline{/45^\circ} = (1+j)\sqrt{\frac{\omega\mu}{2\sigma}} \tag{2-5-11}$$

$$= (1+j)\frac{1}{\sigma\,\delta} = (1+j)R_s$$

in which $R_s = \sqrt{\omega\mu/(2\sigma)}$ is known as the *skin effect* or surface resistance per unit area of conductor. The average power density for a good conductor is given by

$$p = \tfrac{1}{2}|H|^2 R_s \tag{2-5-12}$$

and the phase velocity within a good conductor is

$$v = \omega \, \delta \qquad (2\text{-}5\text{-}13\text{a})$$

The reflectivity and transmittance of a good conductor in vertical and horizontal polarizations are usually measured in terms of the grazing angle. The grazing angle ψ is defined as the angle between the incident ray and the media boundary.

(a) *Vertical Polarization.* From Fig. 2-3-4 it can be seen that $\psi = 90° - \theta_i$; then $\sin \psi = \cos \theta_i$, $\sin \theta_t = \cos \psi$, $\sin^2 \theta_t + \cos^2 \theta_t = 1$, and $v_1 \sin \theta_t = v_2 \sin \theta_i$. The vertical reflectivity of a good conductor for the tangential components of electric intensity as shown in Eq. (2-3-36) is simplified to

$$\Gamma_v = \frac{\eta_2[1 - (v_2/v_1 \cos \psi)^2]^{1/2} - \eta_1 \sin \psi}{\eta_2[1 - (v_2/v_1 \cos \psi)^2]^{1/2} + \eta_1 \sin \psi} \qquad (2\text{-}5\text{-}13\text{b})$$

For vertical polarization, the normal components of the electric fields are generally used to determine the reflection coefficient. From Fig. 2-3-4 it can be seen that the vertical components of the incident and reflected electric fields are in opposite directions. Therefore the reflectivity of a good conductor in vertical polarization is

$$\Gamma_v = \frac{\eta_1 \sin \psi - \eta_2[1 - (v_2/v_1 \cos \psi)^2]^{1/2}}{\eta_1 \sin \psi + \eta_2[1 - (v_2/v_1 \cos \psi)^2]^{1/2}} \qquad (2\text{-}5\text{-}13\text{c})$$

Similarly, the vertical transmittance of a good conductor for electric fields as shown in Eq. (2-3-37) is given by

$$T_v = \frac{2\eta_2[1 - (v_2/v_1 \cos \psi)^2]^{1/2}}{\eta_2[1 - (v_2/v_1 \cos \psi)^2]^{1/2} + \eta_1 \sin \psi} \qquad (2\text{-}5\text{-}13\text{d})$$

(b) *Horizontal Polarization.* The reflectivity of a good conductor for electric fields in horizontal polarization as shown in Eq. (2-3-45) is simplified in terms of ψ as

$$\Gamma_h = \frac{\eta_2 \sin \psi[1 - (v_2/v_1 \cos \psi)^2]^{-1/2} - \eta_1}{\eta_2 \sin \psi[1 - (v_2/v_1 \cos \psi)^2]^{-1/2} + \eta_1} \qquad (2\text{-}5\text{-}13\text{e})$$

Similarly, the transmittance of a good conductor for electric fields in horizontal polarization as shown in Eq. (2-3-46) can be expressed as

$$T_h = \frac{2\eta_2[1 - (v_2/v_1 \cos \psi)^2]^{-1/2}}{\eta_2 \sin \psi[1 - (v_2/v_1 \cos \psi)^2]^{-1/2} + \eta_1} \qquad (2\text{-}5\text{-}13\text{f})$$

In Fig. 2-3-4 it is assumed that medium 1 is free space or air and that medium 2 is copper; then

$$\eta_1 = 377 \text{ ohms} \qquad \eta_2 = (1 + j)\sqrt{\frac{\omega \mu}{2\sigma}}$$

$$v_1 = 3 \times 10^8 \text{ m/s} \qquad v_2 = \omega \, \delta = \sqrt{\frac{2\omega}{\mu\sigma}}$$

The conductivity σ of copper is 5.8×10^7 mhos/m and its relative permeability is unity. The magnitudes of reflectivity of copper for vertical and horizontal polarizations are computed by Eqs. (2-5-13c) and (2-5-13e) against the grazing angle ψ of 0 to $90°$ at a frequency range of 0.1 to 40 GHz. This result indicates that copper is a perfect reflector for electromagnetic waves.

2-5-2. Plane Wave in Poor Conductor

Some conducting materials with low conductivity normally cannot be considered either good conductors or good dielectrics. Seawater is a good example. It has a conductivity of 4 mhos/m and a dielectric constant of 20. At some low frequencies the conduction current is greater than the displacement current, whereas at some high frequencies the reverse is true.

In general, the propagation constant and intrinsic impedance for a poor conductor are given by

$$\gamma = \sqrt{j\omega\mu(\sigma + j\omega\epsilon)} \tag{2-5-14}$$

$$\eta = \sqrt{\frac{j\omega\mu}{\sigma + j\omega\epsilon}} \tag{2-5-15}$$

2-5-3. Plane Wave in Lossy Dielectric

All dielectric materials have some conductivity, but the conductivity is very small ($\sigma \ll \omega\epsilon$). When the conductivity cannot be neglected, the electric and magnetic fields in the dielectric are no longer in time phase. This fact can be seen from the intrinsic impedance of the dielectric as

$$\eta = \sqrt{\frac{j\omega\mu}{\sigma + j\omega\epsilon}} = \sqrt{\frac{\mu}{\epsilon}}\left[1 - j\frac{\sigma}{\omega\epsilon}\right]^{-1/2} \tag{2-5-16}$$

The term $\sigma/(\omega\epsilon)$ is referred to as the *loss tangent* and is defined by

$$\tan\theta = \frac{\sigma}{\omega\epsilon} \tag{2-5-17}$$

This relationship is the result of displacement current density leading conduction current density by $90°$, just as the current flowing through a capacitor leads the current through a resistor in parallel with it by $90°$ in an ordinary electric circuit. This phase relationship is shown in Fig. 2-5-1.

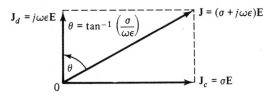

FIG. 2-5-1. Loss tangent for lossy dielectric.

If the loss tangent is very small $[\sigma/(\omega\epsilon) \ll 1]$, the propagation constant and intrinsic impedance can be calculated approximately by a binomial expansion.

Since

$$\gamma = j\omega\sqrt{\mu\epsilon}\sqrt{1 - j\frac{\sigma}{\omega\epsilon}} \tag{2-5-18}$$

then

$$\gamma = j\omega\sqrt{\mu\epsilon}\left[1 - j\frac{\sigma}{2\omega\epsilon} + \frac{1}{8}\left(\frac{\sigma}{\omega\epsilon}\right)^2 + \cdots\right] \tag{2-5-19}$$

Hence

$$\alpha \doteq j\omega\sqrt{\mu\epsilon}\left(-j\frac{\sigma}{\omega\epsilon}\right) = \frac{\sigma}{2}\sqrt{\frac{\mu}{\epsilon}} \tag{2-5-20}$$

$$\beta \doteq \omega\sqrt{\mu\epsilon}\left[1 + \frac{1}{8}\left(\frac{\sigma}{\omega\epsilon}\right)^2\right] = \omega\sqrt{\mu\epsilon} \tag{2-5-21}$$

Similarly,

$$\eta = \sqrt{\frac{\mu}{\epsilon}}\left[1 + j\frac{\sigma}{2\omega\epsilon} - \frac{3}{8}\left(\frac{\sigma}{\omega\epsilon}\right)^2 + \cdots\right] \tag{2-5-22}$$

or

$$\eta \doteq \sqrt{\frac{\mu}{\epsilon}}\left(1 + j\frac{\sigma}{2\omega\epsilon}\right) \tag{2-5-23}$$

It should be noted that the attenuation constant for a given dielectric is not independent of frequency as shown in Eq. (2-5-20). For high-frequency applications, the loss tangent is usually given in curves that vary slightly with frequencies.

The reflectivity and transmittance of a lossy dielectric in vertical and horizontal polarizations are commonly measured in terms of the grazing angle.

(a) *Vertical Polarization.* In Fig. 2-3-4 it is assumed that medium 1 has constants ϵ_1 and μ_1 and that medium 2 has ϵ_2 and μ_2. Then

$$\eta_1 = \sqrt{\frac{\mu_1}{\epsilon_1}}, \qquad \frac{\sin\theta_t}{\sin\theta_i} = \frac{\sqrt{\mu_1\epsilon_1}}{\sqrt{\mu_2\epsilon_2}}, \qquad \cos\theta_t = \sqrt{1 - \sin^2\theta_t}$$

$$\eta_2 = \sqrt{\frac{\mu_2}{\epsilon_2}}, \qquad \mu_1 = \mu_2, \qquad\qquad = \sqrt{1 - \frac{\epsilon_1}{\epsilon_2}\sin^2\theta_i}$$

The vertical reflectivity of a lossy dielectric for the tangential components of electric fields as shown in Eq. (2-3-36) is simplified to

$$\Gamma_v = \frac{\sqrt{\epsilon_2/\epsilon_1 - \sin^2\theta_i} - \epsilon_2/\epsilon_1\cos\theta_i}{\sqrt{\epsilon_2/\epsilon_1 - \sin^2\theta_i} + \epsilon_2/\epsilon_1\cos\theta_i} \tag{2-5-24a}$$

If $\Gamma_v = 0$, then

$$\theta_i = \arctan \sqrt{\frac{\epsilon_2}{\epsilon_1}} \tag{2-5-24b}$$

This angle is called the *Brewster angle*. Furthermore, if

$$\epsilon_2 = \epsilon \left(1 - j\frac{\sigma}{\omega\epsilon} \right) \qquad \text{for} \qquad \frac{\sigma}{\omega\epsilon} \ll 1$$

and $\epsilon_1 = \epsilon_0$ for air, then Eq. (2-5-24a) becomes

$$\Gamma_v = \frac{\sqrt{(\epsilon_r - jx) - \cos^2\psi} - (\epsilon_r - jx)\sin\psi}{\sqrt{(\epsilon_r - jx) - \cos^2\psi} + (\epsilon_r - jx)\sin\psi} \tag{2-5-25}$$

where $\epsilon_r = \dfrac{\epsilon}{\epsilon_0}$

$$x = \frac{\sigma}{\omega\epsilon_0} = \frac{18\sigma}{f_{\text{GHz}}}$$

$\psi = 90° - \theta_i$ is called the *pseudo-Brewster angle*.

Equation (2-5-25) is applicable only for the tangential components of incident and reflected fields that are in the same directions as shown in Fig. 2-3-4. For vertical polarization, the normal components of electric fields are usually used to determine the reflection coefficient. Therefore the reflectivity of a lossy dielectric in vertical polarization is given by

$$\Gamma_v = \frac{(\epsilon_r - jx)\sin\psi - \sqrt{(\epsilon_r - jx) - \cos^2\psi}}{(\epsilon_r - jx)\sin\psi + \sqrt{(\epsilon_r - jx) - \cos^2\psi}} \tag{2-5-26}$$

Similarly, the vertical transmittance of a lossy dielectric for electric fields as shown in Eq. (2-3-37) is

$$T_v = \frac{2\sqrt{(\epsilon_r - jx) - \cos^2\psi}}{\sqrt{(\epsilon_r - jx) - \cos^2\psi} + (\epsilon_r - jx)\sin\psi} \tag{2-5-27}$$

(b) Horizontal Polarization. The reflectivity of a lossy dielectric for electric fields in horizontal polarization as shown in Eq. (2-3-45) becomes

$$\Gamma_h = \frac{\sin\psi - \sqrt{(\epsilon_r - jx) - \cos^2\psi}}{\sin\psi + \sqrt{(\epsilon_r - jx) - \cos^2\psi}} \tag{2-5-28}$$

Similarly, the transmittance of a lossy dielectric for electric fields in horizontal polarization as shown in Eq. (2-3-46) is expressed as

$$T_h = \frac{2\sin\psi}{\sin\psi + \sqrt{(\epsilon_r - jx) - \cos^2\psi}} \tag{2-5-29}$$

The reflections of electromagnetic waves by such lossy dielectric materials as seawater, dry sands, and concrete cement are often of concern to many electronics engineers. The conductivities σ and relative dielectric constants ϵ_r of seawater, dry sands, and concrete cement are tabulated in Table 2-5-2.

TABLE 2-5-2. Conductivities and Relative Dielectric Constants of Seawater, Dry Sands, and Cement.

	Seawater	Dry sands	Concrete cement
σ (mhos/m)	4	2×10^{-4}	2×10^{-5}
ϵ_r	20	4	3

Figures 2-5-2 to 2-5-5 show, respectively, the magnitudes of reflectivity of seawaters, dry sands, and concrete cement for vertical and horizontal polarizations against the grazing angle ψ of 0 to 90° at a frequency range of 0.1 to 40 GHz [9].

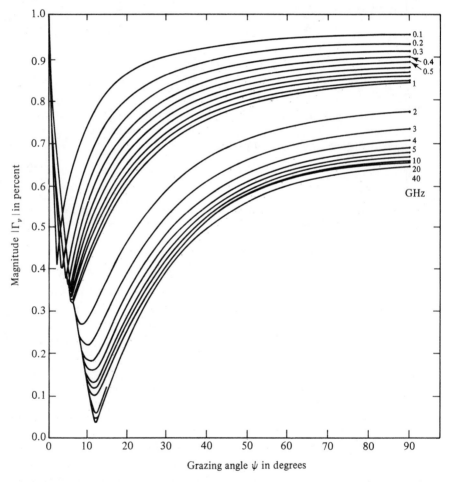

FIG. 2-5-2. Magnitudes of the reflectivity in vertical polarization vs. grazing angle for seawater.

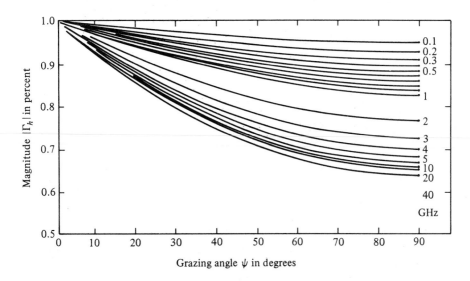

FIG. 2-5-3. Magnitude of the reflectivity in horizontal polarization vs. grazing angle for seawater.

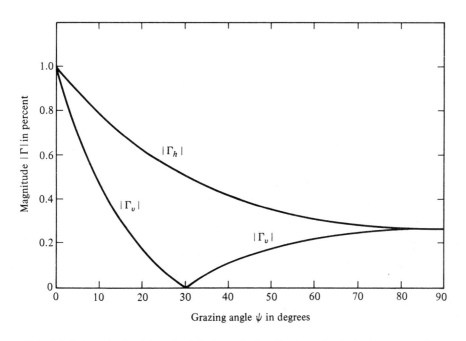

FIG. 2-5-4. Magnitude of the reflectivity in vertical and horizontal polarizations vs. grazing angle for dry sands.

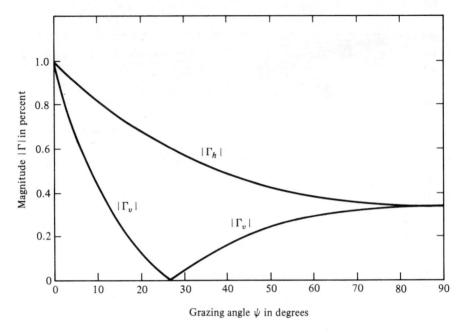

FIG. 2-5-5. Magnitude of the reflectivity in vertical and horizontal polarizations vs. grazing angle for concrete cement.

2-6 Plane Wave Propagation in Metallic-Film Coating on Plastic Substrate*

In certain engineering applications, it is often desirable to use a metallic-film-coated glass to attenuate optimum electromagnetic radiation at microwave frequencies and also to transmit as much light intensity as possible at visible-light frequencies. Generally the coated metallic film should have a high melting point, a high electrical conductivity, high adhesion to glass, high resistance to oxidation, and insensitivity to light and water, as well as the capability of dissipating some power for de-icing, de-fogging, or maintaining certain temperature levels. The metallic-film coatings on a plastic substrate are used in such applications as windshields on airplanes or automobiles, medical equipment in hospitals, and on dome windows of space vehicles or military devices.

2-6-1. Surface Resistances of Metallic Films

Very thin metallic films have a much higher resistivity than a bulk metal because of electron scattering from the film surface. If the film thickness is very large compared to the electron mean-free-path, the resistivity is

*Copyright © 1975 by the *IEEE*, Inc. Reprinted with permission from the *IEEE*, Inc. "Light transmittance and microwave attenuation of a gold-film coating on a plastic substrate" by S. Y. Liao. Vol. **MTT-23**, No. 10, pp. 846–849, October 1975.

expected to be nearly the same as that of a bulk metal. When the film thickness is on the order of the electron mean-free-path, then the role of electron scattering becomes dominant. Fuchs [1] and Sondheimer [2] considered the general form of the solution of the Boltzmann equation for the case of a conducting film and found the film conductivity σ_f in terms of the bulk conductivity σ, the film thickness t, and the electron mean-free-path p:

$$\sigma_f = \frac{3t\sigma}{4p}\left[\ell n\left(\frac{p}{t}\right) + 0.4228\right] \quad \text{for } t \ll p \quad (2\text{-}6\text{-}1)$$

The surface resistance of conducting films is generally quoted in units of ohms per square because in the equation for resistance

$$R = \frac{\text{specific resistivity} \times \text{length}}{\text{thickness} \times \text{width}} = \frac{\rho \ell}{tw} \quad (2\text{-}6\text{-}2)$$

when units of length ℓ and width w are chosen to have equal magnitude (i.e., resulting in a square), the resistance R in ohms per square is independent of the dimensions of the square and equals

$$R_s = \frac{\rho_f}{t} = \frac{1}{t\sigma_f} \quad \text{ohms/square} \quad (2\text{-}6\text{-}3)$$

According to the Fuchs–Sondheimer theory, the surface resistance of a metallic film is decreased as the thickness of the film is increased.

2-6-2. Optical Constants of Plastic Substrates and Metallic Films

The optical properties of materials are usually characterized by two constants, the refractive index n and the extinction index k. The refractive index is defined as the ratio of the phase velocities of light in a medium and in vacuum. The extinction index is related to the exponential decay of the wave as it passes through a medium. Most optical plastics are suitable as substrate materials for a dome window and for metallic-film applications. Table 2-6-1 lists the values of the refractive index n of several nonabsorbing plastic substrate materials in common use. [3].

TABLE 2-6-1. Substrate Materials.

Substrate material	Refractive index n
Corning vycor	1.458
Crystal quartz	1.540
Fused silica	1.458
Plexiglass	1.490
Polycyclohexyl metacrylate	1.504
Polyester glass	1.500
Polymethyl methacrylate	1.491
Zinc crown glass	1.508

The measured values of the refractive index n and the extinction index k of thin metallic-film coatings deposited in a vacuum [3] are tabulated in Table 2-6-2.

TABLE 2-6-2.* Refractive Index n and Extinction Index k of Thin Metallic Films.

Wavelength (Å)	Copper film		Gold film		Silver film	
	n	k	n	k	n	k
2000			1.427	1.215	1.13	1.23
2200					1.32	1.29
2300					1.38	1.31
2400					1.37	1.33
2500					1.39	1.34
2600					1.45	1.35
2700					1.51	1.33
2800					1.57	1.27
2900					1.60	1.17
3000					1.67	0.96
3100					1.54	0.54
3200					1.07	0.32
3300					0.30	0.55
3400					0.16	1.14
3500					0.12	1.35
3600					0.09	1.52
3700					0.06	1.70
3800						
4000						
4500	0.870	2.200	1.400	1.880		
4920						
5000	0.880	2.420	0.840	1.840		
5460						
5500	0.756	2.462	0.331	2.324		
6000	0.186	2.980	0.200	2.897		
6500	0.142	3.570	0.142	3.374		
7000	0.150	4.049	0.131	3.842		
7500	0.157	4.463	0.140	4.266		
8000	0.170	4.840	0.149	4.654		
8500	0.182	5.222	0.157	4.993		
9000	0.190	5.569	0.166	5.335		
9500	0.197	5.900	0.174	5.691		
10,000	0.197	6.272	0.179	6.044		

2-6-3. Microwave Radiation Attenuation
of Metallic-Film Coating on Plastic Substrate

A conductor of high conductivity and low permeability has low intrinsic impedance. When a radio wave propagates from a medium of high intrinsic impedance into a medium of low intrinsic impedance, the reflection coefficient is high. From electromagnetic plane wave theory in the far field, high attenuation occurs in a medium made of material having high conductivity and low permeability. Good conductors, such as gold, silver, and copper, have high conductivity and are often used as the material for attenuating electromagnetic energy. Microwave radiation attenuation by a metallic-film coating on substrate consists of three parts [4]:

$$\text{Attenuation} = A + R + C \qquad \text{dB} \qquad (2\text{-}6\text{-}4)$$

where A = absorption or attenuation loss in decibels inside the metallic-film coating while the substrate is assumed to be nonabsorbing plastic glass

R = reflection loss in decibels from the multiple boundaries of a metallic-film coating on substrate

C = correction term in decibels required to account for multiple internal reflections when the absorption loss A is much less than 10 dB for electrically thin film

Figure 2-6-1 shows the absorption and reflection of a metallic-film coating on a plastic substrate.

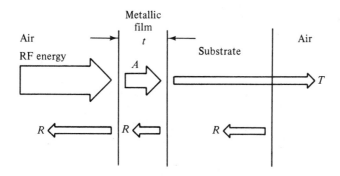

FIG. 2-6-1. Absorption and reflection of film coating on plastic substrate.

(*a*) *Absorption Loss A.* As described in Section 2-5-1, the propagation constant γ for a uniform plane wave in a good conducting material is given in Eq. (2-5-7) as

$$\gamma = \alpha + j\beta = (1 + j)\sqrt{\pi f \mu \sigma_f} \qquad \text{for } \sigma_f \gg \omega\epsilon \qquad (2\text{-}6\text{-}5)$$

If the plastic substrate is assumed to be a nonabsorbing material, the absorption loss A of the metallic-film coating on a substrate is related only to the thickness t of the coated film and the attenuation α as shown:

$$A = 20 \log_{10} e^{\alpha t} = 20(\alpha t) \log_{10} e = 20(0.4343)(\alpha t)$$

$$= 8.686t\sqrt{\pi f \mu \sigma_f} \qquad \text{dB} \qquad (2\text{-}6\text{-}6)$$

where t = thickness of the film coating in meters
μ = permeability of the film in henry per meter
f = frequency in hertz
σ_f = conductivity of the coated film in mhos per meter

Since the thickness of the coated film is very thin—for example, 100 Å(Angstrom) ($\text{Å} = 10^{-10}$ m) at most—the absorption loss A is very small and can be ignored.

(b) **Reflection Loss R.** The reflection loss R due to the multiple boundaries of the substrate glass coated with a metallic film can be analyzed by means of the energy-transmission theory [see Eq. (2-3-18) in Section 2-3-3], and it is expressed as

$$R = -20 \log \frac{2|\eta_f|}{|\eta_a + \eta_f|} - 20 \log \frac{2|\eta_g|}{|\eta_f + \eta_g|} - 20 \log \frac{2|\eta_a|}{|\eta_g + \eta_a|}$$

$$= 20 \log \frac{|\eta_a + \eta_f||\eta_f + \eta_g||\eta_g + \eta_a|}{8|\eta_f||\eta_g||\eta_a|} \qquad \text{dB} \qquad (2\text{-}6\text{-}7)$$

where η_f = intrinsic impedance of the coated metallic film
η_g = intrinsic impedance of the glass substrate
η_a = intrinsic impedance of air or free space = 377 ohms

The intrinsic impedance of a metallic film is given by Eq. (2-5-11) as

$$|\eta_f| = \left|(1 + j)\sqrt{\frac{\mu\omega}{2\sigma_f}}\right| = \sqrt{\frac{\mu\omega}{\sigma_f}} \qquad (2\text{-}6\text{-}8)$$

and the intrinsic impedance of a glass substrate is expressed in Eq. (2-5-16) as

$$\eta_g \simeq \frac{\eta_a}{\sqrt{\epsilon_r}} = \frac{377}{\sqrt{3.78}} = 194 \qquad \text{ohms} \quad \text{for } \sigma_g \ll \omega\epsilon_g \qquad (2\text{-}6\text{-}9)$$

where σ_g = about 10^{-12} mho/m is the conductivity of the glass substrate
$\epsilon_g = 4.77 \times 10^{-11}$ farad/m is the permittivity of the glass substrate
$\epsilon_r = 3.78$ is the relative permittivity of the glass substrate

Substituting the values of the intrinsic impedances $\eta_f, \eta_g,$ and η_a in

Eq. (2-6-7) yields the reflection loss as

$$R \simeq 20 \log \left[28.33 \sqrt{\frac{\sigma_f}{\mu f}} \right] = 88 + 10 \log \left(\frac{\sigma_f}{f} \right) \qquad \text{dB} \qquad (2\text{-}6\text{-}10)$$

(c) *Correction Term C.* For very electrically thin film, the value of the absorption loss A is much less than 10 dB and the correction term is given by [5]

$$C = 20 \log |1 - p 10^{-A/10} (\cos \theta - j \sin \theta)| \qquad (2\text{-}6\text{-}11)$$

$$\text{where } p = \left(\frac{\eta_f - \eta_a}{\eta_f + \eta_a} \right)^2 \simeq 1 \qquad \text{for } \eta_a \gg \eta_f$$

$$\theta = 3.54 t \sqrt{f \mu \sigma_f}$$

Over the frequency range of 100 MHz to 40 GHz, the angle θ is much smaller than one degree so that $\cos \theta \simeq 1$ and $\sin \theta \simeq \theta$. Thus the correction term of Eq. (2-6-11) can be simplified to

$$C \simeq 20 \log [3.54 t \sqrt{f \mu \sigma_f}] = -48 + 20 \log [t \sqrt{f \sigma_f}] \qquad \text{dB} \qquad (2\text{-}6\text{-}12)$$

Finally, the total microwave radiation attenuation by a metallic-film coating on a glass substrate, defined in Eq. (2-6-4) in the far field, becomes

$$\text{Attenuation} = 40 - 20 \log (R_s) \qquad \text{dB} \qquad (2\text{-}6\text{-}13)$$

It is interesting to note that the microwave radiation attenuation due to the coated metallic film on a glass substrate in the far field is independent of frequency and is related only to the surface resistance of the coated metallic film.

2-6-4. Light Transmittance of Metallic-Film Coating on Plastic Substrate

The complex refractive index of an optical material is given by [3] as

$$N = n - jK \qquad (2\text{-}6\text{-}14)$$

It is assumed that light in air is normally incident on a thin absorbing film N_1 of thickness t_1 and that it is transmitted through an absorbing substrate of complex refractive index N_2 and then emerges into air. The incidence and the emergence media are dielectrics of refractive index n_0. The reflection loss between the substrate and the air is small and, for convenience, is taken as zero. Figure 2-6-2 shows light transmittance, reflection, and absorption through a thin absorbing metallic film and a plastic substrate.

Using the multireflection and transmission theory, the reflection loss

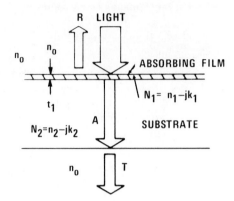

FIG. 2-6-2. Light transmittance, reflection, and absorption through a thin metallic film coated on plastic substrate.

is expressed by

$$R = \frac{a_1 e^{\alpha} + a_2 e^{-\alpha} + a_3 \cos v + a_4 \sin v}{b_1 e^{\alpha} + b_2 e^{-\alpha} + b_3 \cos v + b_4 \sin v} \qquad (2\text{-}6\text{-}15)$$

where $a_1 = [(n_0 - n_1)^2 + k_1^2][(n_1 + n_2)^2 + (k_1 + k_2)^2]$

$a_2 = [(n_0 + n_1)^2 + k_1^2][(n_1 - n_2)^2 + (k_1 - k_2)^2]$

$a_3 = 2\{[n_0^2 - (n_1^2 + k_1^2)][(n_1^2 + k_1^2) - (n_2^2 + k_2^2)]$
$\qquad + 4n_0 k_1 (n_1 k_2 - n_2 k_1)\}$

$a_4 = 4\{[(n_0^2 - (n_1^2 + k_1^2)](n_1 k_2 - n_2 k_1)$
$\qquad - n_0 k_1 [(n_1^2 + k_1^2) - (n_2^2 + k_2^2)]\}$

$\alpha = \dfrac{4\pi k_1 t_1}{\lambda_0}$

$\lambda_0 = \dfrac{c}{f}$ is the wavelength in a vacuum

$c = 3 \times 10^8$ m/s is the velocity of light in a vacuum: f is the frequency in hertz

$v = \dfrac{4\pi n_1 t_1}{\lambda_0}$

$b_1 = [(n_0 + n_1)^2 + k_1^2][(n_1 + n_2)^2 + (k_1 + k_2)^2]$

$b_2 = [(n_0 - n_1)^2 + k_1^2][(n_1 - n_2)^2 + (k_1 - k_2)^2]$

$b_3 = 2\{[n_0^2 - (n_1^2 + k_1^2)][(n_1^2 + k_1^2) - (n_2^2 + k_2^2)]$
$\qquad - 4n_0 k_1 (n_1 k_2 - n_2 k_1)\}$

$b_4 = 4\{[n_0^2 - (n_1^2 + k_1^2)](n_1 k_2 - n_2 k_1)$
$\qquad + n_0 k_1 [(n_1^2 + k_1^2) - (n_2^2 + k_2^2)]\}$

Transmittance T is given by [3] as

$$T = \frac{16 n_0 n_2 (n_1^2 + k_1^2)}{b_1 e^{\alpha} + b_2 e^{-\alpha} + b_3 \cos v + b_4 \sin v} \qquad (2\text{-}6\text{-}16)$$

Absorption loss A is given by

$$A = 1 - R - T \qquad (2\text{-}6\text{-}17)$$

and the total attenuation loss L is

$$L = A + R \qquad (2\text{-}6\text{-}18)$$

When the concave surface of a plastic dome is uniformly coated with an electromagnetic interference shield of metallic film, however, the light is normally incident on the plastic substrate N_2, transmits through the thin metallic film N_1, and then emerges into the air n_0. From the electromagnetic theory of luminous transmission in transparent media, the light transmittance is the same regardless of whether light is normally incident on the substrate medium N_2 or on the absorbing film N_1. Thus the total attenuation loss is the same in both cases.

2-6-5. Plane Wave in Gold-Film Coating on Plastic Glass

Metallic-film coatings on plastic glasses have many engineering applications [6]. A gold film, for example, is coated on the concave surface of a plastic-glass dome so that an optimum amount of microwave radiation is attenuated by the gold film while, at the same time, a sufficient light intensity is transmitted through the gold film.

(*a*) *Surface Resistance.* At room temperature the properties of bulk gold are

Conductivity: $\sigma = 4.10 \times 10^7$ mhos—m^{-1}

Resistivity: $\rho = 2.44 \times 10^{-8}$ ohm—m

Electron mean-free-path: $p = 570$ Å

It is assumed that the thickness t of the gold film varies from 10 to 100 Å. Its surface resistances are computed by using Eqs. (2-6-1) and (2-6-3) and are tabulated in Table 2-6-3.

TABLE 2-6-3. Surface Resistance of Gold Film.

Thickness t (Å)	*Conductivity σ_f* (mho-m^{-1} × 10⁷)	*Resistivity ρ_f* (ohm-m × 10⁻⁷)	*Surface resistance R_s* (ohms/square)
100	1.17	0.86	8.60
90	1.11	0.90	10.00
80	1.03	0.97	12.13
70	0.96	1.04	14.86
60	0.85	1.17	19.50
50	0.77	1.30	26.00
40	0.68	1.48	37.00
30	0.54	1.86	62.00
20	0.42	2.41	120.00
10	0.22	4.48	448.00

Figure 2-6-3 shows surface resistances of gold film in ohms per square against the thicknesses of gold film from 20 to 100 Å. According to the Fuchs–Sondheimer theory, gold films have a typical surface resistance at about 10 to 30 Ω/square for a thickness of about 90 to 45 Å. The surface resistance is decreased as the thickness of the gold film is increased.

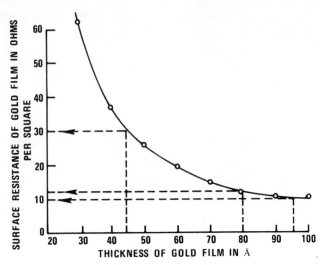

FIG. 2-6-3. Surface resistance of gold film vs. thickness of gold film.

(*b*) *Microwave Radiation Attenuation.* Substituting the values of the surface resistances for gold films in Eq. (2-6-13) yields the microwave radiation attenuation in decibels by the gold-film coating on a plastic glass. Figure 2-6-4 shows graphically the microwave radiation attenuation versus

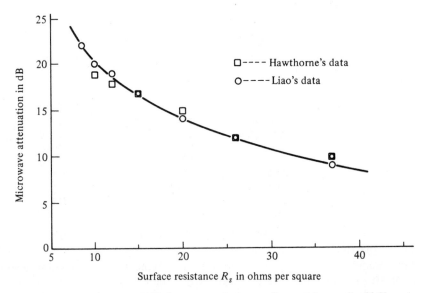

FIG. 2-6-4. Microwave radiation attenuation vs. surface resistance of gold film.

the surface resistance of the gold-film coating. For a coated gold film having a surface resistance of 12 Ω/square, the microwave radiation attenuation is about 19 dB. The data agree with Hawthorne's conclusion [7].

(c) *Light Transmittance.* For the visible-light region, the values of the refractive index n and the extinction index k of a gold-film coating on a plastic glass deposited in a vacuum are taken from Table 2-6-2. The refractive index n_0 of air or vacuum is unity. The refractive index n_2 of the non-absorbing plastic glass is taken as 1.50. Light transmittance T and light reflection loss R of a gold-film coating on a plastic glass are computed by using Eqs. (2-6-16) and (2-6-15), respectively. Then from the values of T and R absorption loss A and total attenuation L are calculated. The results are presented graphically in Fig. 2-6-5. It can be seen that for a light transmittance of 80% the thickness of the gold-film coating is about 80 Å. When the absorption loss in the substrate material is considered, however, the light transmittance may be a little less than 80%.

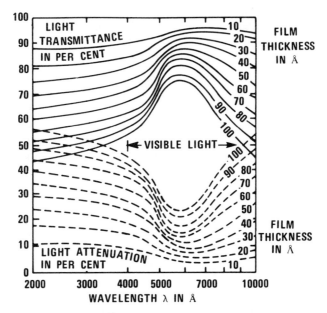

FIG. 2-6-5. Light transmittance T and light attenuation loss L vs. wavelength λ with film thickness t as parameter for gold film.

(d) *Optimum Condition.* The surface resistance of a metallic film decreases as the thickness of the film coating increases. However, the luminous transmittance is decreased as the surface resistance of the metallic film is decreased. This relationship for the visible-light region is shown in Fig. 2-6-6.

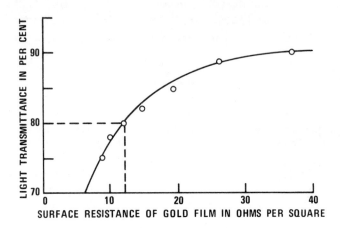

- LIGHT TRANSMITTANCE DECREASES, AS SURFACE
  RESISTANCE DECREASES
- FOR 80 PER CENT OF LIGHT TRANSMITTANCE,
  SURFACE RESISTANCE IS ABOUT 12 OHMS PER SQUARE

FIG. 2-6-6. Light transmittance vs. surface resistance of gold film.

Figure 2-6-7 illustrates the relationship of light transmittance versus wavelength for a given surface resistance of gold film. If a power dissipation of 5 W/square is allowed for de-icing and de-fogging or keeping warm by the gold-film coating on a plastic substrate and if the effective area of the coated film is 13 square inches in a missile, the surface resistance of the coated film must be 12 Ω/square. The power dissipation can be expressed as

$$P = \frac{V^2}{R_s} = \frac{(28)^2}{12 \times 13} = 5 \qquad \text{watts}$$

in which the voltage applied to the film-coating terminations is 28 V. The optimum condition occurs at 18 dB of microwave radiation attenuation and 90% of light transmittance.

2-6-6. Plane Wave in Silver- or Copper-Film Coating on Plastic Substrate*

Silver- or copper-film coating on a plastic substrate has many uses in engineering [8]. The surface resistance, microwave radiation attenuation, light transmittance, and optimum condition of both silver-film coating and copper-film coating can be described in the same way as for gold-film coating.

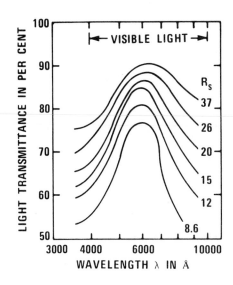

- LIGHT TRANSMITTANCE INCREASES AS SURFACE RESISTANCE INCREASES
- LIGHT TRANSMITTANCE OF 80 PER CENT OCCURS AT 12 OHMS PER SQUARE

FIG. 2-6-7. Light transmittance vs. wavelength with surface resistance R_s as parameter for gold film.

(*a*) *Surface Resistance.* At room temperature the properties of bulk silver and bulk copper are

Silver:
Conductivity: $\sigma = 0.617 \times 10^8$ mhos—m^{-1}
Resistivity: $\rho = 1.620 \times 10^{-8}$ ohm—m
Electron mean-free-path: $p = 570$ Å

Copper:
Conductivity: $\sigma = 0.580 \times 10^8$ mhos—m^{-1}
Resistivity: $\rho = 1.724 \times 10^{-8}$ ohm—m
Electron mean-free-path: $p = 420$ Å

It is assumed that the thickness t of the silver and copper films varies from 10 to 100 Å. The surface resistances of silver and copper films are computed by using Eqs. (2-6-1) and (2-6-3) and are tabulated in Tables 2-6-4 and 2-6-5, respectively.

TABLE 2-6-4. Surface Resistance R_s of Silver Film.

Thickness t ($\mathring{A}$)	Conductivity σ_f (mho-m^{-1} $\times$ 10^7)	Resistivity ρ_f (mho-m $\times$ 10^{-7})	Surface resistance R_s (ohms per square)
100	1.78	0.571	5.71
90	1.66	0.602	6.69
80	1.55	0.645	8.06
70	1.44	0.695	9.93
60	1.31	0.763	12.72
50	1.16	0.862	17.24
40	0.99	1.010	25.25
30	0.81	1.230	41.00
20	0.61	1.640	82.00
10	0.36	2.760	276.00

TABLE 2-6-5. Surface Resistance R_s of Copper Film.

Thickness t ($\mathring{A}$)	Conductivity σ_f (mho-m^{-1} $\times$ 10^7)	Resistivity ρ_f (mho-m $\times$ 10^{-7})	Surface resistance R_s (ohms per square)
100	1.93	0.52	5.20
90	1.85	0.54	6.00
80	1.73	0.58	7.25
70	1.62	0.62	8.86
60	1.47	0.68	11.33
50	1.33	0.75	15.00
40	1.17	0.86	21.50
30	0.95	1.05	35.00
20	0.73	1.37	68.50
10	0.43	2.31	231.00

Figure 2-6-8 graphically plots the surface resistances of silver and copper films, respectively, in ohms per square versus the thickness of the silver and copper films from 10 to 100 Å.

(*b*) *RF Radiation Attenuation.* Substitution of the values of the surface resistances of silver or copper films in Eq. (2-6-13) yields the microwave radiation attenuation in decibels by the silver- or copper-film coating on a plastic substrate. Figure 2-6-9 shows graphically the microwave radiation attenuation vs. the surface resistance of silver- or copper-film coating, respectively.

(*c*) *Light Transmittance.* Light transmittance T and light reflection loss R of silver- and copper-film coatings are computed by using Eqs. (2-6-16) and (2-6-15), respectively. The values of the refractive index n and the extinc-

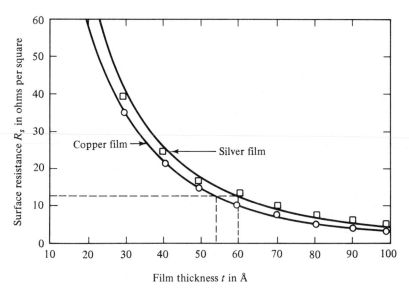

FIG. 2-6-8. Surface resistance of silver and copper film vs. thickness of the film.

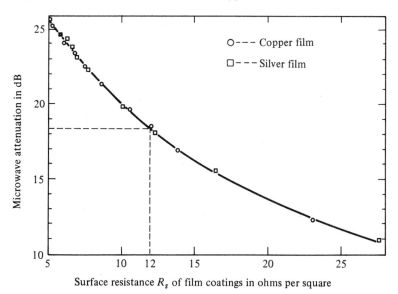

FIG. 2-6-9. Microwave radiation attenuation vs. surface resistance of silver and copper film.

tion index k of the silver- and copper-film coatings deposited in a vacuum for the light-frequency range are taken from Table 2-6-2. The refractive index n_0 of air or vacuum is unity. The refractive index n_2 of the nonabsorbing plastic glass is taken as 1.5. From the values of light transmittance T and

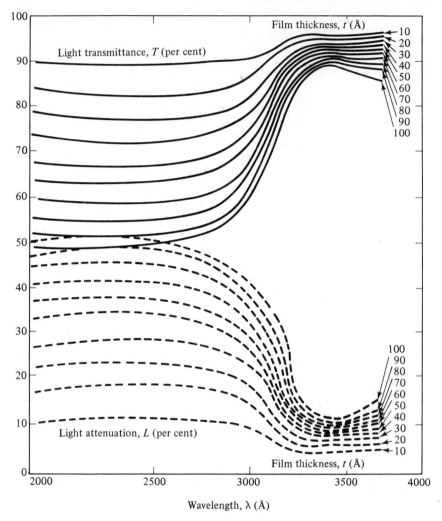

FIG. 2-6-10. Light transmittance T and light attenuation loss L of silver film vs. wavelength λ with film thickness t as parameter.

light reflection loss R absorption loss A and total attenuation L are calculated. The results are illustrated in Figs. 2-6-10 and 2-6-11 for silver- and copper-film coatings, respectively.

(**d**) *Optimum Condition.* The light transmittance is increased as the surface resistance is increased. This relationship is illustrated in Fig. 2-6-12 for silver film and copper film, respectively. The optimum condition occurs at 18 dB of microwave radiation attenuation and 94% of light transmittance with a surface resistance of about 12 Ω/square.

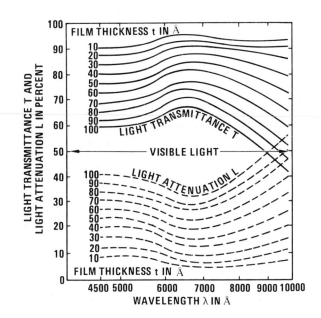

FIG. 2-6-11. Light transmittance T and light attenuation loss L of copper film vs. wavelength with film thickness t as parameter.

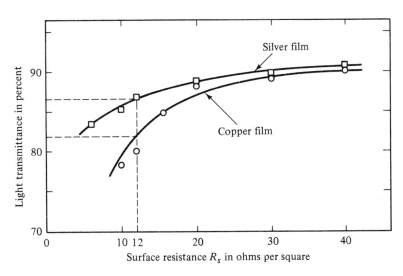

FIG. 2-6-12. Light transmittance vs. surface resistance of silver and copper films.

2-7 Summary

In this chapter the electric and magnetic wave equations, the Poynting theorem, and wave propagation in different media were analyzed. They are summarized as follows.

1. Wave equations

$$\nabla^2 \mathbf{E} = \mu\sigma\frac{\partial \mathbf{E}}{\partial t} + \mu\epsilon\frac{\partial^2 \mathbf{E}}{\partial t^2} \qquad \text{(electric wave equation)}$$

$$\nabla^2 \mathbf{H} = \mu\sigma\frac{\partial \mathbf{H}}{\partial t} + \mu\epsilon\frac{\partial^2 \mathbf{H}}{\partial t^2} \qquad \text{(magnetic wave equation)}$$

2. The Poynting theorem

$$P_{\text{in}} = \langle P_d \rangle + j2\omega[\langle W_m - W_e \rangle] + P_{\text{tr}}$$

3. Wave reflection and wave transmission between two media

(a) The **E** field is in the plane of incidence.
The reflection coefficient is given by

$$\Gamma = \frac{\eta_2 \cos\theta_t - \eta_1 \cos\theta_i}{\eta_2 \cos\theta_t + \eta_1 \cos\theta_i}$$

The transmission coefficient is given by

$$T = \frac{2\eta_2 \cos\theta_t}{\eta_2 \cos\theta_t + \eta_1 \cos\theta_i}$$

(b) The **H** field is in the plane of incidence.
The reflection coefficient is given by

$$\Gamma = \frac{\eta_2 \sec\theta_t - \eta_1 \sec\theta_i}{\eta_2 \sec\theta_t + \eta_1 \sec\theta_i}$$

The transmission coefficient is given by

$$T = \frac{2\eta_2 \sec\theta_t}{\eta_2 \sec\theta_t + \eta_1 \sec\theta_i}$$

4. Plane wave propagation in free space. The received power in decibels as expressed in terms of transmitter power, antenna gains, and antenna separation is given by

$$P_r = P_t + G_t + G_r - 20 \log\left(\frac{4\pi R}{\lambda}\right) \qquad \text{dB}$$

5. Wave propagation constant, wave impedance, and phase velocity in different media

(a) In a lossless dielectric

$$\alpha = 0$$

$$\beta = \omega\sqrt{\mu\epsilon}$$

$$\eta = \sqrt{\frac{\mu}{\epsilon}} = \sqrt{\frac{\mu_0}{\epsilon_r \epsilon_0}} = \frac{377}{\sqrt{\epsilon_r}}$$

$$v_p = \frac{1}{\sqrt{\mu\epsilon}}$$

(b) In a good conductor

$$\alpha = \beta = \sqrt{\mu f \mu \sigma}$$

$$\eta = \sqrt{\frac{j\omega\mu}{\sigma + j\omega\epsilon}} = \sqrt{\frac{j\omega\mu}{\sigma}} \qquad \text{for } \sigma \gg \omega\epsilon$$

$$v = \omega\delta$$

where

$$\delta = \frac{1}{\sqrt{\pi f \mu \sigma}} = \frac{1}{\alpha} = \frac{1}{\beta}$$

$$\Gamma_v = \frac{\eta_1 \sin\psi - \eta_2[1 - (v_2/v_1 \cos\psi)^2]^{1/2}}{\eta_1 \sin\psi + \eta_2[1 - (v_2/v_1 \cos\psi)^2]^{1/2}}$$

$$T_v = \frac{2\eta_2[1 - (v_2/v_1 \cos\psi)^2]^{1/2}}{\eta_2[1 - (v_2/v_1 \cos\psi)^2]^{1/2} + \eta_1 \sin\psi}$$

$$\Gamma_h = \frac{\eta_2 \sin\psi[1 - (v_2/v_1 \cos\psi)^2]^{-1/2} - \eta_1}{\eta_2 \sin\psi[1 - (v_2/v_1 \cos\psi)^2]^{-1/2} + \eta_1}$$

$$T_h = \frac{2\eta_2[1 - (v_2/v_1 \cos\psi)^2]^{-1/2}}{\eta_2 \sin\psi[1 - (v_2/v_1 \cos\psi)^2]^{-1/2} + \eta_1}$$

(c) In a poor conductor

$$\gamma = \sqrt{j\omega\mu(\sigma + j\omega\epsilon)}$$

$$\eta = \sqrt{\frac{j\omega\mu}{\sigma + j\omega\epsilon}}$$

(d) In a lossy dielectric

$$\alpha \doteq j\omega\sqrt{\omega\epsilon}\left(-j\frac{\sigma}{2\omega\epsilon}\right) = \frac{\sigma}{2}\sqrt{\frac{\mu}{\epsilon}} \qquad \text{for } \sigma \ll \omega\epsilon$$

$$\beta \doteq \omega\sqrt{\mu\epsilon}\left[1 + \frac{1}{8}\left(\frac{\sigma}{\omega\epsilon}\right)^2\right] = \omega\sqrt{\mu\epsilon}$$

$$\eta \doteq \sqrt{\frac{\mu}{\epsilon}}\left(1 + j\frac{\sigma}{2\omega\epsilon}\right)$$

$$\Gamma_v = \frac{(\epsilon_r - jx)\sin\psi - [(\epsilon_r - jx) - \cos^2\psi]^{1/2}}{(\epsilon_r - jx)\sin\psi + [(\epsilon_r - jx) - \cos^2\psi]^{1/2}}$$

$$T_v = \frac{2[(\epsilon_r - jx) - \cos^2\psi]^{1/2}}{[(\epsilon_r - jx) - \cos^2\psi]^{1/2} + (\epsilon_r - jx)\sin\psi}$$

$$\Gamma_h = \frac{\sin \psi - [(\epsilon_r - jx) - \cos^2 \psi]^{1/2}}{\sin \psi + [(\epsilon_r - jx) - \cos^2 \psi]^{1/2}}$$

$$T_h = \frac{2 \sin \psi}{\sin \psi + [(\epsilon_r - jx) - \cos^2 \psi]^{1/2}}$$

6. Wave propagation in metallic film

(a) Surface conductance and resistance.

$$\sigma_f = \frac{3t\sigma}{4p}\left[\ell n \left(\frac{p}{t}\right) + 0.4228\right]$$

$$R_s = \frac{\rho_f}{t} = \frac{1}{t\sigma_f} \qquad \text{ohms/square}$$

(b) Substrate materials

(c) Microwave attenuation of metallic film.

(i) Absorption loss

$$A = 8.686t\sqrt{\pi f \mu \sigma_f} \qquad \text{dB}$$

(ii) Reflection loss

$$R = 88 + 10 \log \left(\frac{\sigma_f}{f}\right) \qquad \text{dB}$$

(iii) Correction term

$$C = -48 + 20 \log [t\sqrt{f\sigma_f}] \qquad \text{dB}$$

Total microwave attenuation is

$$\text{Attenuation} = 40 - 20 \log (R_s) \qquad \text{dB}$$

(d) Light reflection and transmittance

$$R = \frac{a_1 e^\alpha + a_2 e^{-\alpha} + a_3 \cos v + a_4 \sin v}{b_1 e^\alpha + b_2 e^{-\alpha} + b_3 \cos v + b_4 \sin v}$$

$$T = \frac{16 n_0 n_2 (n_1^2 + k_1^2)}{b_1 e^\alpha + b_2 e^{-\alpha} + b_3 \cos v + b_4 \sin v}$$

REFERENCES

[1] FUCHS, K., The conductivity of thin metallic films according to the electron theory of metals. *Proc. Camb. Phil. Soc.*, **30**, 100 (1938).

[2] SONDHEIMER, E. H., The mean-free-path of electrons in metals. *Advances in Physics*, **1**, 1 (1952).

[3] *American Institute of Physics Handbook*, 1972. McGraw-Hill Book Co., New York, pp. 6–12, 6–119 to 6–121, and 6–138.

[4] SCHULZ, RICHARD B., et al., Shielding theory and practice. *Proc. 9th Tri-Service Conference on Electromagnetic Compatibility*, October 1963.

[5] VASAKA, C. S., Problems in shielding electrical and electronic equipments. U.S. Naval Air Development Center. Johnsville, Pa., *Rept No. NACD-EL-N5507*, June 1955.

[6] LIAO, SAMUEL Y.,
Design of a gold film on a glass substrate for maximum light transmittance and RF shielding effectiveness. *IEEE Electromagnetic Compatibility Symposium Records*, San Antonio, Texas, October 1975.
Light transmittance and microwave attenuation of a gold-film coating on a plastic substrate. *IEEE Trans. on Microwave Theory and Techniques*, **MTT-23**, No. 10, October 1975.
Light transmittance and RF shielding effectiveness of a gold film on a glass substrate. *IEEE Trans. on Electromagnetic Compatibility*, **EMC-17**, No. 4, November 1975.

[7] HAWTHORNE, E. I., Electromagnetic shielding with transparent coated glass. *Proc. IRE.*, **42**, 548–553, March 1954.

[8] LIAO, SAMUEL Y., RF shielding effectiveness and light transmitance of copper or silver coating on plastic substrate. *IEEE Trans. on Electromagnetic Compatibility*, **EMC-18**, No. 4, November 1976.

[9] LIAO, SAMUEL Y. Reflectivities of electromagnetic waves by seawater, dry sands, concrete cement, and dry ground. A report for the Naval Weapons Center, Department of The Navy, China Lake, Calif., August 1976.

SUGGESTED READINGS

1. ATWATER, H. A., *Introduction to Microwave Theory*, Chapter 3. McGraw-Hill Book Company, New York, 1962.

2. BRONWELL, A. B., and R. E. BEAM, *Theory and Application of Microwaves*, Chapters 13, 14, and 15. McGraw-Hill Book Company, New York, 1947.

3. COLLIN, ROBERT E., *Foundation for Microwave Engineering*, Chapter 2. McGraw-Hill Book Company, New York, 1966.

4. HAYT, WILLIAM H., *Engineering Electromagnetics*, 3rd ed., Chapters 10 and 11. McGraw-Hill Book Company, New York, 1974.

5. JORDAN, EDWARD, and K. G. BALMAIN, *Electromagnetic Waves and Radiating Systems*, 2nd ed., Chapters 4, 5, and 6. Prentice-Hall, Englewood Cliffs, N.J., 1968.

Transmission Lines

and

Microwave Waveguides

3-0 Introduction

Conventional two-conductor transmission lines are commonly used for transmitting microwave energy. If a line is properly matched to its characteristic impedance at each terminal, its efficiency can reach maximum at a very high frequency. A waveguide usually consists of a hollow metallic tube of rectangular or cylindrical shape that guides an electromagnetic wave. Waveguides are used principally at frequencies in the microwave range. The purpose of Chapter 3 is to describe microwave transmission lines and microwave waveguides for electromagnetic energy transmission.

3-1 Transmission Lines

In ordinary circuit theory it is assumed that all impedance elements are lumped constants. This is not true for a long transmission line over a wide range of frequencies. Frequencies of operation are so high that inductances of short lengths of conductors and capacitances between short conductors and their surroundings cannot be neglected. These inductances and capacitances are distributed along the length of a conductor, and their effects combine at each point of the conductor. Since the wavelength is short in comparison to the physical length of the line, distributed parameters cannot be represented accurately by means of a lumped-parameter equivalent circuit. Thus microwave transmission lines can be analyzed in terms of voltage, current, and impedance only by the distributed-circuit theory. If the spacing between the lines is smaller than the wavelength of the transmitted signal, the transmission line must be analyzed as a waveguide.

3-1-1. Transmission-Line Equations and Solutions

(1) *Transmission-Line Equations.* A transmission line can be analyzed either by the solution of Maxwell's field equations or by the methods of distributed-circuit theory. The solution of Maxwell's equations involves three space variables in addition to the time variable. The distributed-circuit method, however, involves only one space variable in addition to the time variable. In this section the latter method is used to analyze a transmission line in terms of the voltage, current, impedance, and power along the line.

Based upon uniformly distributed-circuit theory, the schematic circuit of a conventional two-conductor transmission line with constant parameters R, L, G, and C is shown in Fig. 3-1-1. The parameters are expressed in their respective names per unit length, and the wave propagation is assumed in the positive z direction.

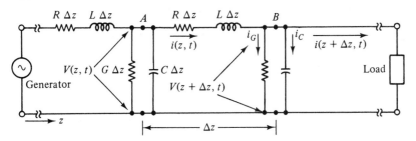

FIG. 3-1-1. Elementary section of a transmission line.

By Kirchhoff's voltage law, the summation of the voltage drops around the central loop is given by

$$v(z, t) = i(z, t)R\, \Delta z + L\, \Delta z\, \frac{\partial i(z, t)}{\partial t} + v(z, t) + \frac{\partial v(z, t)}{\partial z}\, \Delta z \quad (3\text{-}1\text{-}1)$$

Rearranging this equation, dividing it by Δz, and then omitting the argument (z, t), which is understood, we obtain

$$-\frac{\partial v}{\partial z} = Ri + L\frac{\partial i}{\partial t} \quad (3\text{-}1\text{-}2)$$

Using Kirchhoff's current law, the summation of the currents at point B in Fig. 3-1-1 can be expressed as

$$i(z, t) = v(z + \Delta z, t)G\, \Delta z + C\, \Delta z\, \frac{\partial v(z + \Delta z, t)}{\partial t} + i(z + \Delta z, t)$$

$$= \left[v(z, t) + \frac{\partial v(z, t)}{\partial z}\, \Delta z\right]G\, \Delta z + C\, \Delta z\, \frac{\partial}{\partial t}\left[v(z, t) + \frac{\partial v(z, t)}{\partial z}\, \Delta z\right]$$

$$+ i(z, t) + \frac{\partial i(z, t)}{\partial z}\, \Delta z \quad (3\text{-}1\text{-}3)$$

As Δz approaches zero, $\partial v/\partial t$ will also approach zero. By rearranging the preceding equation, dividing it by Δz, and omitting (z, t), we have

$$-\frac{\partial i}{\partial z} = Gv + C\frac{\partial v}{\partial t} \quad (3\text{-}1\text{-}4)$$

Then by differentiating Eq. (3-1-2) with respect to z and Eq. (3-1-4) with respect to t and combining the results, the final transmission-line equation in voltage form is found to be

$$\frac{\partial^2 v}{\partial z^2} = RGv + (RC + LG)\frac{\partial v}{\partial t} + LC\frac{\partial^2 v}{\partial t^2} \quad (3\text{-}1\text{-}5)$$

Also, by differentiating Eq. (3-1-2) with respect to t and Eq. (3-1-4) with respect to z and combining the results, the final transmission-line equation in current form is

$$\frac{\partial^2 i}{\partial z^2} = RGi + (RC + LG)\frac{\partial i}{\partial t} + LC\frac{\partial^2 i}{\partial t^2} \quad (3\text{-}1\text{-}6)$$

All these transmission-line equations are applicable to the general transient solution. The voltage and current on the line are the functions of both position z and time t. The instantaneous line voltage and current can be expressed as

$$v(z, t) = \text{Re } \mathbf{V}(z)e^{j\omega t} \quad (3\text{-}1\text{-}7)$$

$$i(z, t) = \text{Re } \mathbf{I}(z)e^{j\omega t} \quad (3\text{-}1\text{-}8)$$

where Re stands for "real part of." The factors $V(z)$ and $I(z)$ are complex quantities of the sinusoidal functions of position z on the line and are known as *phasors*. The phasors give the magnitudes and phases of the sinusoidal function at each position of z, and they can be expressed as

$$V(z) = V_+ e^{-\gamma z} + V_- e^{\gamma z} \tag{3-1-9}$$

$$I(z) = I_+ e^{-\gamma z} + I_- e^{\gamma z} \tag{3-1-10}$$

$$\gamma = \alpha + j\beta \text{ is the propagation constant.} \tag{3-1-11}$$

where V_+ and I_+ indicate complex quantities in the positive z direction
and V_- and I_- complex quantities in the negative z direction
α is the attenuation constant in nepers per unit length
β is the phase constant in radians per unit length

If we substitute $j\omega$ for $\partial/\partial t$ in Eqs. (3-1-2), (3-1-4), (3-1-5), and (3-1-6) and divide each equation by $e^{j\omega t}$, the transmission-line equations in phasor form of the frequency domain become

$$\frac{dV}{dz} = -ZI \tag{3-1-12}$$

$$\frac{dI}{dz} = -YV \tag{3-1-13}$$

$$\frac{d^2V}{dz^2} = \gamma^2 V \tag{3-1-14}$$

$$\frac{d^2I}{dz^2} = \gamma^2 I \tag{3-1-15}$$

in which the following substitutions have been made:

$$Z = R + j\omega L \quad \text{(ohms per unit length)} \tag{3-1-16}$$

$$Y = G + j\omega C \quad \text{(mhos per unit length)} \tag{3-1-17}$$

$$\gamma = \sqrt{ZY} = \alpha + j\beta \quad \text{(propagation constant)} \tag{3-1-18}$$

For a lossless line, $R = G = 0$, and the transmission-line equations are expressed as

$$\frac{dV}{dz} = -j\omega LI \tag{3-1-19}$$

$$\frac{dI}{dz} = -j\omega CV \tag{3-1-20}$$

$$\frac{d^2V}{dz^2} = -\omega^2 LCV \tag{3-1-21}$$

$$\frac{d^2I}{dz^2} = -\omega^2 LCI \tag{3-1-22}$$

It is interesting to note that Eqs. (3-1-14) and (3-1-15) for a transmission line are similar to equations of the electric and magnetic waves, respectively. The only difference is that the transmission-line equations are one dimensional.

(2) *Solutions of the Transmission-Line Equations.* The one possible solution for Eq. (3-1-14) is

$$V = V_+ e^{-\gamma z} + V_- e^{\gamma z} = V_+ e^{-\alpha z} e^{-j\beta z} + V_- e^{\alpha z} e^{j\beta z} \qquad (3\text{-}1\text{-}23)$$

The factors V_+ and V_- represent complex quantities. The term involving $e^{-j\beta z}$ shows a wave traveling in the positive z direction, and the term with the factor $e^{j\beta z}$ is a wave going in the negative z direction. The quantity βz is called the electrical length of the line and is measured in radians.

Similarly, the one possible solution for Eq. (3-1-15) is

$$I = Y_0(V_+ e^{-\gamma z} - V_- e^{\gamma z}) = Y_0(V_+ e^{-\alpha z} e^{-j\beta z} - V_- e^{\alpha z} e^{j\beta z}) \qquad (3\text{-}1\text{-}24)$$

In Eq. (3-1-24) the characteristic impedance of the line is defined as

$$Z_0 = \frac{1}{Y_0} \equiv \frac{Z}{Y} = \frac{R + j\omega L}{G + j\omega C} = R_0 \pm jX_0 \qquad (3\text{-}1\text{-}25)$$

The magnitude of both voltage and current waves on the line is shown in Fig. 3-1-2.

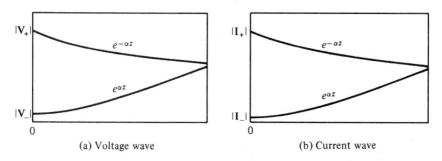

| (a) Voltage wave | (b) Current wave |

FIG. 3-1-2. Magnitude of voltage and current traveling waves.

At microwave frequencies it can be seen that

$$R \ll \omega L \quad \text{and} \quad G \ll \omega C \qquad (3\text{-}1\text{-}26)$$

By using the binomial expansion, the propagation constant can be expressed as

$$\gamma = \sqrt{(R + j\omega L)(G + j\omega C)}$$
$$= \sqrt{(j\omega)^2 LC} \sqrt{\left(1 + \frac{R}{j\omega L}\right)\left(1 + \frac{G}{j\omega C}\right)}$$

$$\simeq j\omega\sqrt{LC}\left[\left(1 + \frac{1}{2}\frac{R}{j\omega L}\right)\left(1 + \frac{1}{2}\frac{G}{j\omega C}\right)\right]$$

$$\simeq j\omega\sqrt{LC}\left[1 + \frac{1}{2}\left(\frac{R}{j\omega L} + \frac{G}{j\omega C}\right)\right]$$

$$= \frac{1}{2}\left(R\sqrt{\frac{C}{L}} + G\sqrt{\frac{L}{C}}\right) + j\omega\sqrt{LC} \tag{3-1-27}$$

Therefore the attenuation and phase constants are, respectively, given by

$$\alpha = \frac{1}{2}\left(R\sqrt{\frac{C}{L}} + G\sqrt{\frac{L}{C}}\right) \tag{3-1-28}$$

$$\beta = \omega\sqrt{LC} \tag{3-1-29}$$

Similarly, the characteristic impedance is found to be

$$\mathbf{Z}_0 = \sqrt{\frac{R + j\omega L}{G + j\omega C}}$$

$$= \sqrt{\frac{L}{C}}\left(1 + \frac{R}{j\omega L}\right)^{1/2}\left(1 + \frac{G}{j\omega C}\right)^{-1/2}$$

$$\simeq \sqrt{\frac{L}{C}}\left(1 + \frac{1}{2}\frac{R}{j\omega L}\right)\left(1 - \frac{1}{2}\frac{G}{j\omega C}\right)$$

$$\simeq \sqrt{\frac{L}{C}}\left[1 + \frac{1}{2}\left(\frac{R}{j\omega L} - \frac{G}{j\omega C}\right)\right]$$

$$\simeq \sqrt{\frac{L}{C}} \tag{3-1-30}$$

From Eq. (3-1-29) the phase velocity is

$$v_p = \frac{\omega}{\beta} = \frac{1}{\sqrt{LC}} \tag{3-1-31}$$

The product of LC is independent of the size and separation of the conductors and depends only on the permeability μ and permittivity ϵ of the insulating medium. If a lossless transmission line used for microwave frequencies has an air dielectric and contains no ferromagnetic materials, free-space parameters can be assumed. Thus the numerical value of $1/\sqrt{LC}$ for air-insulated conductors is approximately equal to the velocity of light. That is,

$$v_p = \frac{1}{\sqrt{LC}} = \frac{1}{\sqrt{\mu_0\epsilon_0}} = c = 3 \times 10^8 \text{ m/s} \tag{3-1-32}$$

When the dielectric of a lossy microwave transmission line is not air, the phase velocity is smaller than the velocity of light and is given by

$$v_\epsilon = \frac{1}{\sqrt{\mu\epsilon}} = \frac{c}{\sqrt{\mu_r\epsilon_r}} \tag{3-1-33}$$

In general, the relative phase velocity constant can be defined as

$$\text{Velocity constant} = \frac{\text{actual phase velocity}}{\text{velocity of light}}$$

$$v_r = \frac{v_\epsilon}{c} = \frac{1}{\sqrt{\mu_r \epsilon_r}} \tag{3-1-34}$$

A low-loss transmission line filled only with dielectric medium, such as a coaxial line with solid dielectric between conductors, has a velocity constant on the order of about 0.65.

3-1-2. Reflection Coefficient and Transmission Coefficient

(1) *Reflection Coefficient.* In the analysis of the solutions of transmission-line equations in Section 3-1-1, the traveling wave along the line contains two components: one traveling in the positive z direction and the other traveling in the negative z direction. If the load impedance is equal to the line characteristic impedance, however, the reflected traveling wave does not exist.

Figure 3-1-3 shows a transmission line terminated in an impedance Z_ℓ.

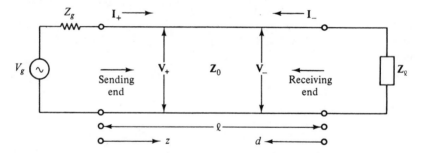

FIG. 3-1-3. Transmission line terminated in a load impedance.

It is usually more convenient to start solving the transmission-line problem from the receiving rather than the sending end, since the voltage-to-current relationship at the load point is fixed by the load impedance. The incident voltage and current waves traveling along the transmission line are given by

$$V = V_+ e^{-\gamma z} + V_- e^{+\gamma z} \tag{3-1-35}$$

$$I = I_+ e^{-\gamma z} + I_- e^{+\gamma z} \tag{3-1-36}$$

in which the current wave can be expressed in terms of the voltage by

$$I = \frac{V_+}{Z_0} e^{-\gamma z} - \frac{V_-}{Z_0} e^{\gamma z} \tag{3-1-37}$$

If the line has a length of ℓ, the voltage and current at the receiving end become

$$\mathbf{V}_\ell = \mathbf{V}_+ e^{-\gamma\ell} + \mathbf{V}_- e^{\gamma\ell} \qquad (3\text{-}1\text{-}38)$$

$$\mathbf{I}_\ell = \frac{1}{\mathbf{Z}_0}(\mathbf{V}_+ e^{-\gamma\ell} - \mathbf{V}_- e^{\gamma\ell}) \qquad (3\text{-}1\text{-}39)$$

The ratio of the voltage to the current at the receiving end is the load impedance. That is,

$$\mathbf{Z}_\ell = \frac{\mathbf{V}_\ell}{\mathbf{I}_\ell} = \mathbf{Z}_0 \frac{\mathbf{V}_+ e^{-\gamma\ell} + \mathbf{V}_- e^{\gamma\ell}}{\mathbf{V}_+ e^{-\gamma\ell} - \mathbf{V}_- e^{\gamma\ell}} \qquad (3\text{-}1\text{-}40)$$

The reflection coefficient, which is designated by Γ (gamma), is defined as

$$\text{Reflection coefficient} \equiv \frac{\text{reflected voltage or current}}{\text{incident voltage or current}}$$

$$\Gamma \equiv \frac{\mathbf{V}_{\text{ref}}}{\mathbf{V}_{\text{inc}}} = \frac{\mathbf{I}_{\text{ref}}}{\mathbf{I}_{\text{inc}}} \qquad (3\text{-}1\text{-}41)$$

If Eq. (3-1-40) is solved for the ratio of the reflected voltage at the receiving end, which is $\mathbf{V}_- e^{\gamma\ell}$, to the incident voltage at the receiving end, which is $\mathbf{V}_+ e^{\gamma\ell}$, the result is the reflection coefficient at the receiving end.

$$\Gamma_\ell = \frac{\mathbf{V}_- e^{\gamma\ell}}{\mathbf{V}_+ e^{-\gamma\ell}} = \frac{\mathbf{Z}_\ell - \mathbf{Z}_0}{\mathbf{Z}_\ell + \mathbf{Z}_0} \qquad (3\text{-}1\text{-}42)$$

If the load impedance and/or the characteristic impedance are complex quantities, as is usually the case, the reflection coefficient is generally a complex quantity that can be expressed as

$$\Gamma_\ell = |\Gamma_\ell| e^{j\theta_\ell} \qquad (3\text{-}1\text{-}43)$$

where $|\Gamma_\ell|$ is the magnitude and never greater than unity—that is, $|\Gamma_\ell| \leq 1$. θ_ℓ is the phase angle between the incident and reflected voltages at the receiving end. It is usually called the phase angle of the reflection coefficient.

The general solution of the reflection coefficient at any point on the line, then, corresponds to the incident and reflected waves at that point, each attenuated in the direction of its own progress along the line. The generalized reflection coefficient is defined as

$$\Gamma \equiv \frac{\mathbf{V}_- e^{\gamma z}}{\mathbf{V}_+ e^{-\gamma z}} \qquad (3\text{-}1\text{-}44)$$

From Figure 3-1-3 let $z = \ell - d$. Then the reflection coefficient at some point located a distance d from the receiving end is

$$\Gamma_d = \frac{\mathbf{V}_- e^{\gamma(\ell-d)}}{\mathbf{V}_+ e^{-\gamma(\ell-d)}} = \frac{\mathbf{V}_- e^{\gamma\ell}}{\mathbf{V}_+ e^{-\gamma\ell}} e^{-2\gamma d} = \Gamma_\ell e^{-2\gamma d} \qquad (3\text{-}1\text{-}45)$$

Next, the reflection coefficient at that point can be expressed in terms of the reflection coefficient at the receiving end as

$$\Gamma_d = \Gamma_\ell e^{-2\alpha d} e^{-j2\beta d} = |\Gamma_\ell| e^{-2\alpha d} e^{j(\theta_\ell - 2\beta d)} \qquad (3\text{-}1\text{-}46)$$

This is a very useful equation for determining the reflection coefficient at any point along the line. For a lossy line, both the magnitude and phase of the reflection coefficient are changing in an inward-spiral way as shown in Fig. 3-1-4. For a lossless line, $\alpha = 0$, the magnitude of the reflection coefficient remains constant, and only the phase of Γ is changing circularly toward the generator with an angle of $-2\beta d$ as shown in Fig. 3-1-5.

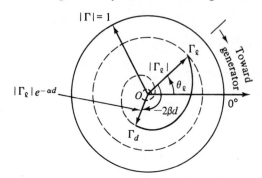

FIG. 3-1-4. Reflection coefficient for lossy line.

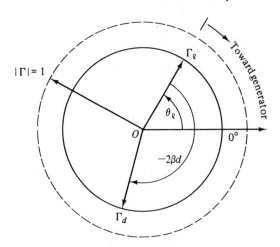

FIG. 3-1-5. Reflection coefficient for lossless line.

It is evident that Γ_ℓ will be zero and that there will be no reflection from the receiving end when the terminating impedance is equal to the characteristic impedance of the line. Thus a terminating impedance that differs from the characteristic impedance will create a reflected wave traveling toward the source from the termination. The reflection, upon reaching the sending end, will itself be reflected if the source impedance is different from the line impedance at the sending end.

(2) Transmission Coefficient. A transmission line terminated in its characteristic impedance Z_0 is called a *properly terminated* line or *nonresonant* line. Otherwise it is called an *improperly terminated* line or *resonant* line. As described earlier, there is a reflection coefficient Γ at any point along an improperly terminated line. According to the principle of conservation of energy, the incident power minus the reflected power must be equal to the power transmitted to the load. This can be expressed as

$$1 - |\Gamma_\ell|^2 = \frac{Z_0}{Z_\ell}|\mathbf{T}|^2 \qquad (3\text{-}1\text{-}47)$$

Equation (3-1-47) will be verified later. The letter $\mathbf{T}$ represents the transmission coefficient, which is defined as

$$\mathbf{T} \equiv \frac{\text{transmitted voltage or current}}{\text{incident voltage or current}} = \frac{\mathbf{V}_{tr}}{\mathbf{V}_{inc}} = \frac{\mathbf{I}_{tr}}{\mathbf{I}_{inc}} \qquad (3\text{-}1\text{-}48)$$

Figure 3-1-6 shows the transmission of power along a transmission line where P_{inc} is the incident power, P_{ref} the reflected power, and P_{tr} the transmitted power.

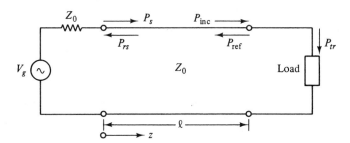

FIG. 3-1-6. Power transmission on a line.

Let the traveling waves at the receiving end be

$$\mathbf{V}_+ e^{-\gamma\ell} + \mathbf{V}_- e^{\gamma\ell} = \mathbf{V}_{tr} e^{-\gamma\ell} \qquad (3\text{-}1\text{-}49)$$

$$\frac{\mathbf{V}_+}{\mathbf{Z}_0} e^{-\gamma\ell} - \frac{\mathbf{V}_-}{\mathbf{Z}_0} e^{\gamma\ell} = \frac{\mathbf{V}_{tr}}{\mathbf{Z}_\ell} e^{-\gamma\ell} \qquad (3\text{-}1\text{-}50)$$

Multiplication of Eq. (3-1-50) by $\mathbf{Z}_\ell$ and substitution of the result in Eq. (3-1-49) yield

$$\Gamma_\ell = \frac{\mathbf{V}_- e^{\gamma\ell}}{\mathbf{V}_+ e^{-\gamma\ell}} = \frac{\mathbf{Z}_\ell - \mathbf{Z}_0}{\mathbf{Z}_\ell + \mathbf{Z}_0} \qquad (3\text{-}1\text{-}51)$$

which, in turn, on substitition back into Eq. (3-1-49), results in

$$\mathbf{T} = \frac{\mathbf{V}_{tr}}{\mathbf{V}_+} = \frac{2\mathbf{Z}_\ell}{\mathbf{Z}_\ell + \mathbf{Z}_0} \qquad (3\text{-}1\text{-}52)$$

The average power carried by the two waves in the side of the incident and reflected waves is

$$\langle P_{inr} \rangle = \langle P_{inc} \rangle - \langle P_{ref} \rangle = \frac{|V_+ e^{-\alpha \ell}|^2}{2|Z_0|} - \frac{|V_- e^{\alpha \ell}|^2}{2|Z_0|} \qquad (3\text{-}1\text{-}53)$$

The average power carried to the load by the transmitted waves is

$$\langle P_{tr} \rangle = \frac{|V_{tr} e^{-\alpha \ell}|^2}{2|Z_\ell|} \qquad (3\text{-}1\text{-}54)$$

By setting $\langle P_{inr} \rangle = \langle P_{tr} \rangle$ and using Eqs. (3-1-51) and (3-1-52), we have

$$|T|^2 = \frac{|Z_\ell|}{|Z_0|}(1 - |\Gamma_\ell|^2) \qquad (3\text{-}1\text{-}55)$$

This relation verifies the previous statement that the transmitted power is equal to the algebraic sum of the incident power and reflected power.

3-1-3. Standing Wave and Standing-Wave Ratio

(1) **Standing Wave.** The general solutions of the transmission-line equation consist of two waves traveling in opposite directions with unequal amplitude as shown in Eqs. (3-1-23) and (3-1-24). Equation (3-1-23) can be written

$$\begin{aligned} V &= V_+ e^{-\alpha z} e^{-j\beta z} + V_- e^{\alpha z} e^{j\beta z} \\ &= V_+ e^{-\alpha z}(\cos \beta z - j \sin \beta z) + V_- e^{\alpha z}(\cos \beta z + j \sin \beta z) \\ &= (V_+ e^{-\alpha z} + V_- e^{\alpha z}) \cos z - j(V_+ e^{-\alpha z} - V_- e^{\alpha z}) \sin \beta z \qquad (3\text{-}1\text{-}56) \end{aligned}$$

With no loss in generality it can be assumed that $V_+ e^{-\alpha z}$ and $V_- e^{\alpha z}$ are real. Then the voltage-wave equation can be expressed as

$$V_s = V_0 e^{-j\Phi} \qquad (3\text{-}1\text{-}57)$$

This is called the *equation of the voltage standing wave*, where

$$V_0 = [(V_+ e^{-\alpha z} + V_- e^{\alpha z})^2 \cos^2 \beta z + (V_+ e^{-\alpha z} - V_- e^{-\alpha z})^2 \sin^2 \beta z \qquad (3\text{-}1\text{-}58)$$

which is called the *standing-wave pattern* of the voltage wave or the amplitude of the standing wave, and

$$\Phi = \arctan \left(\frac{V_+ e^{-\alpha z} - V_- e^{\alpha z}}{V_+ e^{-\alpha z} + V_- e^{\alpha z}} \tan \beta z \right) \qquad (3\text{-}1\text{-}59)$$

which is called the *phase pattern of the standing wave*. The maximum and minimum values of Eq. (3-1-58) can be found as usual by differentiating the equation with respect to βz and equating the result to zero. By doing so and substituting the proper values of βz in the equation,

1. The maximum amplitude is

$$\mathbf{V}_{max} = \mathbf{V}_{-}e^{-\alpha z} + \mathbf{V}_{-}e^{\alpha z} = \mathbf{V}_{+}e^{-\alpha z}(1 + |\Gamma|) \qquad (3\text{-}1\text{-}60)$$

and this occurs at $\beta z = n\pi,\quad n = 0, \pm 1, \pm 2, \ldots.$

2. The minimum amplitude is

$$\mathbf{V}_{min} = \mathbf{V}_{+}e^{-\alpha z} - \mathbf{V}_{-}e^{\alpha z} = \mathbf{V}_{+}e^{-\alpha z}(1 - |\Gamma|) \qquad (3\text{-}1\text{-}61)$$

and this occurs at $\beta z = (2n - 1)\dfrac{\pi}{2}\quad n = 0, \pm 1, \pm 2, \ldots.$

3. The distance between any two successive maxima or minima is one-half wavelength, since

$$\beta z = n\pi, \qquad z = \frac{n\pi}{\beta} = \frac{n\pi}{2\pi/\lambda} = n\frac{\lambda}{2} \qquad n = 0, \pm 1, \pm 2, \ldots$$

Then

$$z_1 = \frac{\lambda}{2} \qquad (3\text{-}1\text{-}62)$$

It is evident that there are no zeros in the minimum. Similarly,

$$\mathbf{I}_{max} = \mathbf{I}_{+}e^{-\alpha z} - \mathbf{I}_{-}e^{\alpha z} = \mathbf{I}_{+}e^{-\alpha z}(1 + |\Gamma|) \qquad (3\text{-}1\text{-}63)$$

$$\mathbf{I}_{min} = \mathbf{I}_{+}e^{-\alpha z} - \mathbf{I}_{-}e^{\alpha z} = \mathbf{I}_{+}e^{-\alpha z}(1 - |\Gamma|) \qquad (3\text{-}1\text{-}64)$$

The standing-wave patterns of two oppositely traveling waves with unequal amplitude in lossy or lossless line are shown in Figs. 3-1-7 and 3-1-8.

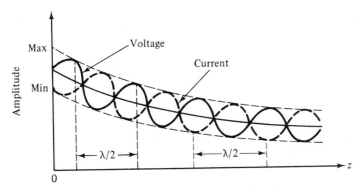

FIG. 3-1-7. Standing-wave pattern in a lossy line.

A further study of Eq. (3-1-58) reveals that

1. When $\mathbf{V}_{+} \neq 0$ and $\mathbf{V}_{-} = 0$, the standing-wave pattern becomes

$$\mathbf{V}_0 = \mathbf{V}_{+}e^{-\alpha z} \qquad (3\text{-}1\text{-}65)$$

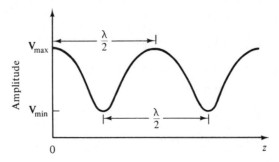

FIG. 3-1-8. Voltage standing-wave pattern in a lossless line.

and the standing wave is

$$\mathbf{V}_s = \mathbf{V}_+ e^{-\gamma z} \tag{3-1-66}$$

which is also called a *pure traveling wave*.

   2. When $\mathbf{V}_+ = 0$ and $\mathbf{V}_- \neq 0$, the standing-wave pattern becomes

$$\mathbf{V}_0 = \mathbf{V}_- e^{\alpha z} \tag{3-1-67}$$

and the standing wave is

$$\mathbf{V}_s = \mathbf{V}_- e^{\gamma z} \tag{3-1-68}$$

which is also called a pure traveling wave.

   3. When the positive wave and the negative wave have equal amplitudes, (i.e., $|\mathbf{V}_+ e^{-\alpha z}| = |\mathbf{V}_- e^{\alpha z}|$) or the magnitude of the reflection coefficient is unity, the standing-wave pattern with a zero phase is given by

$$\mathbf{V}_s = 2\mathbf{V}_+ e^{-\alpha z} \cos \beta z \tag{3-1-69}$$

which is called a *pure standing wave*.

   Similarly, the equation of a pure standing wave for the current is

$$\mathbf{I}_s = -j2\mathbf{Y}_0\mathbf{V}_+ e^{-\alpha z} \sin \beta z \tag{3-1-70}$$

Equation (3-1-69) and (3-1-70) show that the voltage and current standing waves are 90° out of phase along the line. The points of zero current are called the *current nodes*. The voltage nodes and current nodes are interlaced a quarter wavelength apart.

   The voltage and current may be expressed as real functions of time and space.

$$v_s(z, t) = \text{Re}\,[\mathbf{V}_s(z)e^{j\omega t}] = 2\mathbf{V}_+ e^{-\alpha z} \cos \beta z \cos \omega t \tag{3-1-71}$$

$$i_s(z, t) = \text{Re}\,[\mathbf{I}_s(z)e^{j\omega t}] = 2\mathbf{Y}_0\mathbf{V}_+ e^{-\alpha z} \sin \beta z \sin \omega t \tag{3-1-72}$$

The amplitudes of Eqs. (3-1-71) and (3-1-72) vary sinusoidally with time, the voltage being a maximum at time instant when the current is zero and vice

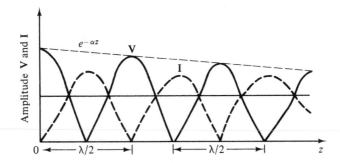

FIG. 3-1-9. Pure standing waves of voltage and current.

versa. Figure 3-1-9 shows the pure-standing-wave patterns of the phasor of Eqs. (3-1-49) and (3-1-70) for an open-terminal line.

(2) Standing-Wave Ratio. Standing waves result from the simultaneous presence of waves traveling in opposite directions on a transmission line. The ratio of the maximum of the standing-wave pattern to the minimum is defined as the standing-wave ratio, designated by ρ. That is,

$$\text{Standing-wave ratio} \equiv \frac{\text{maximum voltage or current}}{\text{minimum voltage or current}}$$

$$\rho \equiv \frac{|\mathbf{V}_{\max}|}{|\mathbf{V}_{\min}|} = \frac{|\mathbf{I}_{\max}|}{|\mathbf{I}_{\min}|} \qquad (3\text{-}1\text{-}73)$$

The standing-wave ratio results from the fact that the two traveling-wave components of Eq. (3-1-56) add in phase at some points and subtract at other points. The distance between two successive maxima or minima is $\lambda/2$. The standing-wave ratio of a pure traveling wave is unity and that of a pure standing wave is infinite. It should be noted that since the standing-wave ratios of voltage and current are identical, no distinctions are made between VSWR and ISWR.

When the standing-wave ratio is unity, there is no reflected wave and the line is called a *flat line*. The standing-wave ratio cannot be defined on a lossy line because the standing-wave pattern changes markedly from one position to another. On a low-loss line the ratio remains fairly constant, and it may be defined for some region. For a lossless line, the ratio stays the same throughout the line.

Since the reflected wave is defined as the product of an incident wave and its reflection coefficient, the standing-wave ratio ρ is related to the reflection coefficient Γ by

$$\rho = \frac{1 + |\Gamma|}{1 - |\Gamma|} \qquad (3\text{-}1\text{-}74)$$

and vice versa

$$|\Gamma| = \frac{\rho - 1}{\rho + 1} \qquad (3\text{-}1\text{-}75)$$

This relation is very useful for determining the reflection coefficient from the standing-wave ratio, which is usually found from the Smith chart. The curve in Fig. 3-1-10 shows the relationship between reflection coefficient $|\Gamma|$ and standing-wave ratio ρ.

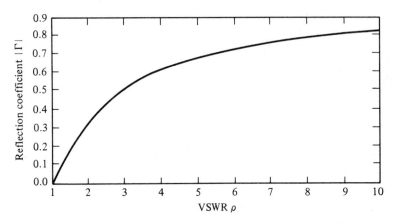

FIG. 3-1-10. SWR vs. reflection coefficient.

As a result of Eq. (3-1-74), since $|\Gamma| \leq 1$, the standing-wave ratio is a positive real number and never less than unity, $\rho \geq 1$. From Eq. (3-1-75) the magnitude of the reflected coefficient is never greater than unity.

3-1-4. Line Impedance and Impedance Matching

(1) Generalized Impedance. The generalized impedance of a transmission line is the complex ratio of the voltage phasor at any point to the current phasor at that point. It is defined as

$$Z \equiv \frac{V(z)}{I(z)} \qquad (3\text{-}1\text{-}76)$$

Figure 3-1-11 shows a diagram for a transmission line.

In general, the voltage or current along a line is the sum of the respective incident wave and reflected wave—that is,

$$V = V_{\text{inc}} + V_{\text{ref}} = V_{+}e^{-\gamma z} + V_{-}e^{\gamma z} \qquad (3\text{-}1\text{-}77)$$

$$I = I_{\text{inc}} + I_{\text{ref}} = Y_0(V_{+}e^{-\gamma z} - V_{-}e^{\gamma z}) \qquad (3\text{-}1\text{-}78)$$

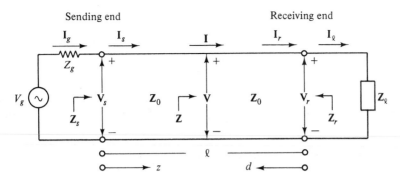

FIG. 3-1-11. Diagram of a transmission line showing notations.

At the sending end $z = 0$; then Eqs. (3-1-77) and (3-1-78) become

$$I_s Z_s = V_+ + V_- \tag{3-1-79}$$

$$I_s Z_0 = V_+ - V_- \tag{3-1-80}$$

By solving these two equations for V_+ and V_-, we obtain

$$V_+ = \frac{I_s}{2}(Z_s + Z_0) \tag{3-1-81}$$

$$V_- = \frac{I_s}{2}(Z_s - Z_0) \tag{3-1-82}$$

Substitution of V_+ and V_- in Eqs. (3-1-77) and (3-1-78) yields

$$V = \frac{I_s}{2}[(Z_s + Z_0)e^{-\gamma z} + (Z_s - Z_0)e^{\gamma z}] \tag{3-1-83}$$

$$I = \frac{I_s}{2Z_0}[(Z_s + Z_0)e^{-\gamma z} - (Z_s - Z_0)e^{\gamma z}] \tag{3-1-84}$$

Then the impedance of the line at any point z from the sending end in terms of Z_s and Z_0 is expressed as

$$Z = Z_0 \frac{(Z_s + Z_0)e^{-\gamma z} + (Z_s - Z_0)e^{\gamma z}}{(Z_s + Z_0)e^{-\gamma z} - (Z_s - Z_0)e^{\gamma z}} \tag{3-1-85}$$

At $z = \ell$ the impedance at the receiving end in terms of Z_s and Z_0 is given by

$$Z_r = Z_0 \frac{(Z_s + Z_0)e^{-\gamma \ell} + (Z_s - Z_0)e^{\gamma \ell}}{(Z_s + Z_0)e^{-\gamma \ell} - (Z_s - Z_0)e^{\gamma \ell}} \tag{3-1-86}$$

Alternatively, the generalized impedance can be expressed in terms of Z_ℓ and Z_0. At $z = \ell$, $V_r = I_\ell Z_\ell$; then

$$I_\ell Z_\ell = V_+ e^{-\gamma \ell} + V_- e^{\gamma \ell} \tag{3-1-87}$$

$$I_\ell Z_0 = V_+ e^{-\gamma \ell} - V_- e^{\gamma \ell} \tag{3-1-88}$$

Solving these two equations for $\mathbf{V}_+$ and $\mathbf{V}_-$, we have

$$\mathbf{V}_+ = \frac{\mathbf{I}_\ell}{2}(\mathbf{Z}_\ell + \mathbf{Z}_0)e^{\gamma\ell} \tag{3-1-89}$$

$$\mathbf{V}_- = \frac{\mathbf{I}_\ell}{2}(\mathbf{Z}_\ell - \mathbf{Z}_0)e^{-\gamma\ell} \tag{3-1-90}$$

Then substituting these results in Eqs. (3-1-77) and (3-1-78) and letting $z = \ell - d$ yield

$$\mathbf{V} = \frac{\mathbf{I}_\ell}{2}[(\mathbf{Z}_\ell + \mathbf{Z}_0)e^{\gamma d} + (\mathbf{Z}_\ell - \mathbf{Z}_0)e^{-\gamma d}] \tag{3-1-91}$$

$$\mathbf{I} = \frac{\mathbf{I}_\ell}{2\mathbf{Z}_0}[(\mathbf{Z}_\ell + \mathbf{Z}_0)e^{\gamma d} - (\mathbf{Z}_\ell - \mathbf{Z}_0)e^{-\gamma d}] \tag{3-1-92}$$

Next, the impedance of the line at any point from the receiving end in terms of $\mathbf{Z}_s$ and $\mathbf{Z}_0$ is expressed as

$$\mathbf{Z} = \mathbf{Z}_0 \frac{(\mathbf{Z}_\ell + \mathbf{Z}_0)e^{\gamma d} + (\mathbf{Z}_\ell - \mathbf{Z}_0)e^{-\gamma d}}{(\mathbf{Z}_\ell + \mathbf{Z}_0)e^{\gamma d} - (\mathbf{Z}_\ell - \mathbf{Z}_0)e^{-\gamma d}} \tag{3-1-93}$$

The line impedance at the sending end can also be found from Eq. (3-1-93) by setting $d = \ell$,

$$\mathbf{Z}_s = \mathbf{Z}_0 \frac{(\mathbf{Z}_\ell + \mathbf{Z}_0)e^{\gamma\ell} + (\mathbf{Z}_\ell - \mathbf{Z}_0)e^{-\gamma\ell}}{(\mathbf{Z}_\ell + \mathbf{Z}_0)e^{\gamma\ell} - (\mathbf{Z}_\ell - \mathbf{Z}_0)e^{-\gamma\ell}} \tag{3-1-94}$$

It is tedious work to solve Eqs. (3-1-85), (3-1-86), (3-1-93), or (3-1-94) for the line impedance. These equations can be simplified by replacing the exponential factors with either hyperbolic functions or cirular functions. The hyperbolic functions are obtained from

$$e^{\pm\gamma z} = \cosh \gamma z \pm \sinh \gamma z \tag{3-1-95}$$

Substitution of the hyperbolic functions in Eq. (3-1-85) yields the line impedance at any point from the sending end in terms of the hyperbolic functions.

$$\mathbf{Z} = \mathbf{Z}_0 \frac{\mathbf{Z}_s \cosh \gamma z - \mathbf{Z}_0 \sinh \gamma z}{\mathbf{Z}_0 \cosh \gamma z - \mathbf{Z}_s \sinh \gamma z} = \mathbf{Z}_0 \frac{\mathbf{Z}_s - \mathbf{Z}_0 \tanh \gamma z}{\mathbf{Z}_0 - \mathbf{Z}_s \tanh \gamma z} \tag{3-1-96}$$

Similarly, substitution of the hyperbolic functions in Eq. (3-1-93) yields the line impedance from the receiving end in terms of the hyperbolic function.

$$\mathbf{Z} = \mathbf{Z}_0 \frac{\mathbf{Z}_\ell \cosh \gamma d + \mathbf{Z}_0 \sinh \gamma d}{\mathbf{Z}_0 \cosh \gamma d + \mathbf{Z}_\ell \sinh \gamma d} = \mathbf{Z}_0 \frac{\mathbf{Z}_\ell + \mathbf{Z}_0 \tanh \gamma d}{\mathbf{Z}_0 + \mathbf{Z}_\ell \tanh \gamma d} \tag{3-1-97}$$

For a lossless line, $\gamma = j\beta$; and by using the following relationships between hyperbolic and circular functions

$$\sinh(j\beta z) = j \sin \beta z \tag{3-1-98}$$

$$\cosh(j\beta z) = \cos \beta z \tag{3-1-99}$$

the impedance of a lossless transmission line can be expressed in terms of the circular functions.

$$Z = Z_0 \frac{Z_s \cos \beta z - jZ_0 \sin \beta z}{Z_0 \cos \beta z - jZ_s \sin \beta z} = Z_0 \frac{Z_s - jZ_0 \tan \beta z}{Z_0 - jZ_s \tan \beta z} \quad (3\text{-}1\text{-}100)$$

and
$$Z = Z_0 \frac{Z_\ell \cos \beta d + jZ_0 \sin \beta d}{Z_0 \cos \beta d + jZ_\ell \sin \beta d} = Z_0 \frac{Z_\ell + jZ_0 \tan \beta d}{Z_0 + jZ_\ell \tan \beta d} \quad (3\text{-}1\text{-}101)$$

(2) Impedance in Terms of Reflection Coefficient or Standing-Wave Ratio. Rearrangement of Eq. (3-1-93) gives the line impedance—looking at it from the receiving end—as

$$Z = Z_0 \frac{1 + \Gamma_\ell e^{-2\gamma d}}{1 - \Gamma_\ell e^{-2\gamma d}} \quad (3\text{-}1\text{-}102)$$

in which the following substitution is made by

$$\Gamma_\ell = \frac{Z_\ell - Z_0}{Z_\ell + Z_0} \quad (3\text{-}1\text{-}103)$$

From Eq. (3-1-46) the reflection coefficient at a distance d from the receiving end is given by

$$\Gamma = \Gamma_\ell e^{-2\gamma d} = |\Gamma_\ell| e^{-2\alpha d} e^{j(\theta_\ell - 2\beta d)} \quad (3\text{-}1\text{-}104)$$

Then the simple equation for the impedance of a line at a distance d from the load is expressed by

$$Z = Z_0 \frac{1 + \Gamma}{1 - \Gamma} \quad (3\text{-}1\text{-}105)$$

The reflected coefficient is usually a complex quantity and can be written

$$\Gamma = |\Gamma| e^{j\phi} \quad (3\text{-}1\text{-}106)$$

where
$$|\Gamma| = |\Gamma_\ell| e^{-2\alpha d}$$
$$\phi = \theta_\ell - 2\beta d$$

The impedance variation along a lossless line can be found as follows:

$$Z(d) = Z_0 \frac{1 + |\Gamma| e^{j\phi}}{1 - |\Gamma| e^{j\phi}} = Z_0 \frac{1 + |\Gamma|(\cos \phi + j \sin \phi)}{1 - |\Gamma|(\cos \phi + j \sin \phi)}$$
$$= R(d) + jX(d) = |Z(d)| e^{j\theta_d} \quad (3\text{-}1\text{-}107a)$$

where
$$|Z(d)| = Z_0 \sqrt{\frac{1 + 2|\Gamma| \cos \phi + |\Gamma|^2}{1 - 2|\Gamma| \cos \phi + |\Gamma|^2}} \quad (3\text{-}1\text{-}107b)$$

$$R(d) = Z_0 \frac{1 - |\Gamma|^2}{1 - 2|\Gamma| \cos \phi + |\Gamma|^2} \quad (3\text{-}1\text{-}107c)$$

$$X(d) = Z_0 \frac{2|\Gamma| \sin \phi}{1 - 2|\Gamma| \cos \phi + |\Gamma|^2} \quad (3\text{-}1\text{-}107d)$$

$$\theta_d = \arctan\left(\frac{X}{R}\right) = \arctan\left(\frac{2|\Gamma| \sin \phi}{1 - |\Gamma|^2}\right) \quad (3\text{-}1\text{-}107e)$$

Since $\phi = \theta_\ell - 2\beta d$, then $\phi = \theta_\ell - 2\pi$ if $\beta d = \pi$. However, $\cos(\theta_\ell - 2\pi)$ $= \cos\theta_\ell$ and $\sin(\theta_\ell - 2\pi) = \sin\theta_\ell$; then

$$\mathbf{Z}(d) = \mathbf{Z}\left(d + \frac{\pi}{\beta}\right) = \mathbf{Z}\left(d + \frac{\lambda}{2}\right) \qquad (3\text{-}1\text{-}108)$$

It is concluded that the impedance along a lossless line will be repeated for every interval at a half-wavelength distance.

Furthermore, the magnitude of a reflection coefficient $|\Gamma|$ is related to the standing-wave ratio ρ by

$$|\Gamma| = \frac{\rho - 1}{\rho + 1} \quad \text{and} \quad \rho = \frac{1 + |\Gamma|}{1 - |\Gamma|} \qquad (3\text{-}1\text{-}109)$$

The impedance of a transmission line at any location from the receiving end can be written

$$\mathbf{Z} = \mathbf{Z}_0 \frac{(\rho + 1) + (\rho - 1)e^{j\phi}}{(\rho + 1) + (\rho - 1)e^{j\phi}} \qquad (3\text{-}1\text{-}110)$$

This is a very useful equation for determining the impedance of the line in terms of standing-wave ratio ρ, since ρ can easily be measured by a detector or a standing-wave meter.

    (3) Determination of Characteristic Impedance. A common procedure for determining the characteristic impedance and propagation constant of a given transmission line is to take two measurements.

First: Measure the sending-end impedance with the receiving end short-circuited and record the result.

$$\mathbf{Z}_{sc} = \mathbf{Z}_0 \tanh \gamma\ell \qquad (3\text{-}1\text{-}111)$$

Second: Measure the sending-end impedance with the receiving end open-circuited and record the result.

$$\mathbf{Z}_{oc} = \mathbf{Z}_0 \coth \gamma\ell \qquad (3\text{-}1\text{-}112)$$

Then the characteristic impedance of the measured transmission line is given by

$$\mathbf{Z}_0 = \sqrt{\mathbf{Z}_{sc}\mathbf{Z}_{oc}} \qquad (3\text{-}1\text{-}113)$$

and the propagation constant of the line can be computed from

$$\gamma = \alpha + j\beta = \frac{1}{\ell} \operatorname{arctanh} \sqrt{\frac{\mathbf{Z}_{sc}}{\mathbf{Z}_{oc}}} \qquad (3\text{-}1\text{-}114)$$

    (4) Normalized Impedance. The normalized impedance of a transmission line is defined as

$$\mathbf{z} \equiv \frac{\mathbf{Z}}{\mathbf{Z}_0} = \frac{1 + \Gamma}{1 - \Gamma} = r + jx \qquad (3\text{-}1\text{-}115)$$

It should be noted that the lowercase letters are commonly designated for normalized quantities in describing the distributed transmission-line circuits.

An examination of Eqs. (3-1-110), (3-1-111), and (3-1-115) shows that the normalized impedance for a lossless line has the following significant features.

1. The maximum normalized impedance is

$$z_{max} = \frac{Z_{max}}{Z_0} = \frac{|V_{max}|}{Z_0 |I_{min}|} = \frac{1 + |\Gamma|}{1 - |\Gamma|} = \rho \qquad (3\text{-}1\text{-}116)$$

z_{max} is a positive real value and it is equal to the standing-wave ratio ρ at the location of any maximum voltage on the line.

2. The minimum normalized impedance is

$$z_{min} = \frac{Z_{min}}{Z_0} = \frac{|V_{min}|}{Z_0 |I_{max}|} = \frac{1 - |\Gamma|}{1 + |\Gamma|} = \frac{1}{\rho} \qquad (3\text{-}1\text{-}117)$$

z_{min} is also a positive real number but equals the reciprocal of the standing-wave ratio at the location of any minimum voltage on the line.

3. z_{max} or z_{min} is repeated for every interval of a half-wavelength distance along the line.

$$z_{max}(z) = z_{max}\left(z \pm \frac{\lambda}{2}\right) \qquad (3\text{-}1\text{-}118)$$

and

$$z_{min}(z) = z_{min}\left(z \pm \frac{\lambda}{2}\right) \qquad (3\text{-}1\text{-}119)$$

4. Since V_{max} and V_{min} are separated by a quarter wavelength, z_{max} is equal to the reciprocal of z_{min} for every $\lambda/4$ separation.

$$z_{max}\left(z \pm \frac{\lambda}{4}\right) = \frac{1}{z_{min}(z)} \qquad (3\text{-}1\text{-}120)$$

(5) *Admittance.* When a transmission line is branched, it is better to solve the line equations for the line voltage, current, and transmitted power in terms of admittance rather than impedance. The characteristic admittance and the generalized admittance are defined as

$$\mathbf{Y}_0 = \frac{1}{\mathbf{Z}_0} = G_0 \pm jB_0 \qquad (3\text{-}1\text{-}121)$$

$$\mathbf{Y} = \frac{1}{\mathbf{Z}} = G \pm jB \qquad (3\text{-}1\text{-}122)$$

Then the normalized admittance can be written

$$y = \frac{\mathbf{Y}}{\mathbf{Y}_0} = \frac{\mathbf{Z}_0}{\mathbf{Z}} = \frac{1}{z} = g \pm jb \qquad (3\text{-}1\text{-}123)$$

(6) *Smith Chart.* Many of the computations required to solve transmission-line problems involve the use of rather complicated equations. The
solutions of such problems is tedious and difficult because the accurate
manipulation of numerous equations is necessary. To simplify their solutions, we need a graphic method of arriving at a quick answer.

A number of impedance charts have been designed to facilitate the
graphical solution of transmission-line problems. Basically all the charts are
derived from the fundamental relationships expressed in the transmission
equations. The most popular chart is that developed by Phillip H. Smith
[1]. The purpose of this section is to present the graphical solutions of
transmission-line problems by using the Smith chart.

The Smith chart consists of a plot of the normalized impedance or
admittance with the angle and magnitude of a generalized complex reflection
coefficient in a unity circle. The chart is applicable to the analysis of a lossless
line as well as a lossy line. By simple rotation on the chart, the effect of the
position on the line can be determined. To see how a Smith chart works,
consider the equation of reflection coefficient at the load for a transmission
line as shown in Eq. (3-1-42):

$$\Gamma_\ell = \frac{Z_\ell - Z_0}{Z_\ell + Z_0} = |\Gamma_\ell| e^{j\theta_\ell} = \Gamma_r + j\Gamma_i \qquad (3\text{-}1\text{-}124a)$$

Since $|\Gamma_\ell| \leq 1$, the value of Γ_ℓ must lie on or within the unity circle with a
radius of one. The reflection coefficient at any other location along a line as
shown in Eq. (3-1-46) is

$$\Gamma_d = \Gamma_\ell e^{-2\alpha d} e^{-j2\beta d} = |\Gamma_\ell| e^{-2\alpha d} e^{j(\theta_\ell - 2\beta d)} \qquad (3\text{-}1\text{-}124b)$$

Figure 3-1-12 shows a complete Smith chart. It is interesting to note that
the chart was set up with a reflection coefficient as the radial coordinate and
that the circles concentric with the center of the unity circle are circles of a
constant reflection coefficient. Since the standing-wave ratio is determined
only by the magnitude of a reflection coefficient, these circles are also the
contour of a constant standing-wave ratio. Since the standing-wave ratio is
never less than one, the scale for the standing-wave ratio varies from one to
infinity on the real axis. It should be noted that the distances are given in
wavelengths toward the generator and also toward the load and so it is easy
to determine which direction to advance as a position on the line is changed.

The characteristics of the Smith chart are summarized as follows.

1. The constant r and constant x loci form two families of orthogonal
   circles in the chart.
2. The constant r and constant x circles all pass through the point
   ($\Gamma_r = 1$, $\Gamma_i = 0$).
3. The upper half of the diagram represents $+jx$.
4. The lower half of the diagram represents $-jx$.

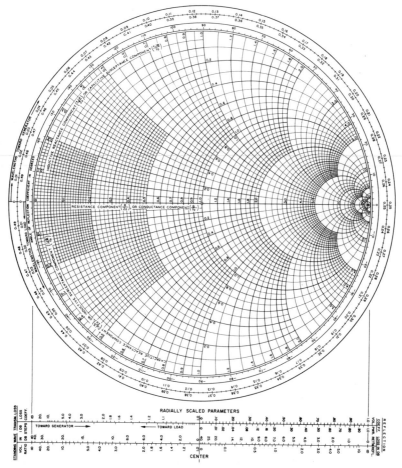

Fig. 3-1-12. Smith chart.

5. For admittance, the constant r circles become constant g circles and the constant x circles become constant susceptance b circles.

6. The distance around the Smith chart once is one-half wavelength $(\lambda/2)$.

7. At a point of $z_{min} = 1/\rho$ there is a V_{min} on the line.

8. At a point of $z_{max} = \rho$ there is a V_{max} on the line.

9. The horizontal radius to the right of the chart center corresponds to V_{max}, I_{min}, z_{max}, and ρ (SWR).

10. The horizontal radius to the left of the chart center corresponds to V_{min}, I_{max}, z_{min}, and $1/\rho$.

11. Since the normalized admittance y is a reciprocal of the normalized impedance z, the corresponding quantities in the admittance chart are $180°$ out of phase with those in the impedance chart.

12. The normalized impedance or admittance is repeated for every half-wavelength of distance.

85

The magnitude of the reflection coefficient is related to the standing-wave ratio by the following expression.

$$|\Gamma| = \frac{\rho - 1}{\rho + 1} \tag{3-1-124c}$$

A Smith chart or slotted line can be used to measure a standing-wave pattern directly and then the magnitudes of the reflection coefficient, reflected power, transmittted power, and the load impedance can be calculated from it. The use of the Smith chart is illustrated in the following examples.

EXAMPLE 3-1-1: Location Determination of Voltage Maximum and Minimum from Load. Given the normalized load impedance $z_\ell = 1 + j1$ ohms and the operating wavelength $\lambda = 5$ cm, determine the first V_{max}, first V_{min} from the load and the VSWR ρ as shown in Fig. 3-1-13a.

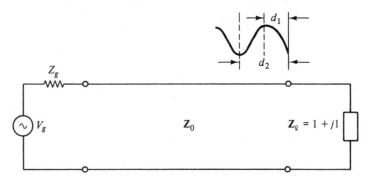

Fig. 3-1-13a. Diagram for Example 3-1-1.

Solution.

1. Enter $z_\ell = 1 + j1$ on the chart as shown in Fig. 3-1-13b.
2. Read 0.162λ on the distance scale by drawing a dashed-straight line from the center of the chart through the load point and intersecting the distance scale.
3. Move a distance from the point at 0.162λ toward the generator and first stop at the voltage maximum on the right-hand real axis at 0.25λ. Then

$$d_1(V_{\text{max}}) = (0.25 - 0.162)\lambda = (0.088)(5) = 0.44 \text{ cm}$$

4. Similarly, move a distance from the point of 0.162λ toward the generator and first stop at the voltage minimum on the left-hand real axis at 0.5λ. Then

$$d_2(V_{\text{min}}) = (0.5 - 0.162)\lambda = (0.338)(5) = 1.69 \text{ cm}$$

5. Make a standing-wave circle with the center at $(1, 0)$ and passing the

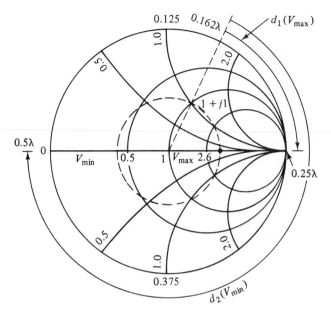

Fig. 3-1-13b. Graphical solution for Example 3-1-1.

point of $\sqrt{1+1}$, and the location intersecting by the circle at the right portion of the real axis indicate the SWR. This is $\rho = 2.6$.

EXAMPLE 3-1-2: Impedance Determination with Short-Circuit Minima Shift. The location of a minimum instead of a maximum is usually specified because it can be determined more accurately. Given that the characteristic impedance of the line $Z_0 = 50$ ohms, and the SWR $\rho = 2$ when the line is loaded. When the load is shorted, the minima shift 0.15λ toward the load. Determine the load impedance. Figure 3-1-14a shows the diagram for the example.

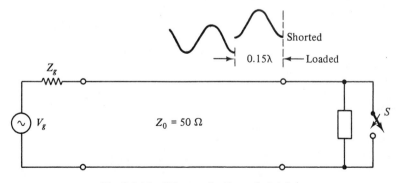

Fig. 3-1-14a. Diagram for Example 3-1-2.

Solution.

1. When the line is shorted, the first voltage minimum occurs at the place of the load as shown in Fig. 3-1-14b.

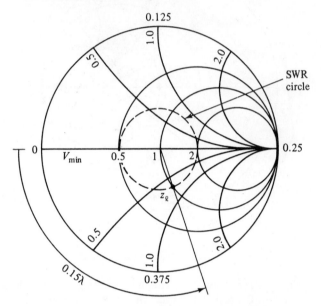

Fig. 3-1-14b. Graphical solution for Example 3-1-2.

2. When the line is loaded, the first voltage minimum shifts 0.15λ from the load. The distance between two successive minima is one-half wavelength.
3. Plot a SWR circle for $\rho = 2$.
4. Move a distance of 0.15λ from the minimum point along the distance scale toward the load and stop at 0.15λ.
5. Draw a line from this point to the center of the chart.
6. The intersection between the line and the SWR circle is

$$z_\ell = 1 - j0.65$$

7. The load impedance is

$$Z_\ell = (1 - j0.65)(50) = 50 - j32.5 \text{ ohms}$$

(7) Impedance Matching. Impedance matching is very desirable with radio frequency (RF) transmission lines. Standing waves lead to increased losses and frequently cause the transmitter to malfunction. A line terminated in its characteristic impedance has a standing-wave ratio of unity and transmits a given power without reflection. Also, transmission efficiency is optimum when there is no reflected power. A "flat" line is nonresonant; that is, its

input impedance always remains at the same value Z_0 when the frequency changes.

"Matching" a transmission line has a special meaning, one differing from that used in circuit theory to indicate equal impedance seen looking both directions from a given terminal pair for maximum power transfer. In circuit theory maximum power transfer requires the load impedance to be equal to the complex conjugate of the generator. This condition is sometimes referred to as a *conjugate match*. In transmission-line problems "matching" means simply terminating the line in its characteristic impedance.

A common application of RF transmission lines is the one in which there is a feeder connection between a transmitter and an antenna. Usually the input impedance to the antenna itself is not equal to the characteristic impedance of the line. Furthermore, the output impedance of the transmitter may not be equal to the Z_0 of the line. Matching devices are necessary to flatten the line. A complete matched transmission-line system is shown in Fig. 3-1-15.

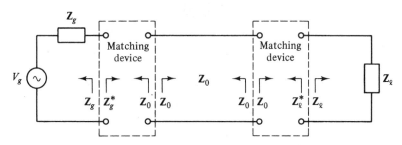

Fig. 3-1-15. Matched transmission line system.

For a low-loss or lossless transmission line at radio frequency, the characteristic impedance Z_0 of the line is resistive. At every point the impedances looking in opposite directions are conjugate. If Z_0 is real, it is its own conjugate. Matching can be tried first on the load side to flatten the line; then adjustment may be made on the transmitter side to provide maximum power transfer. At audio frequencies an iron-cored transformer is almost universally used as an impedance-matching device. Occasionally an iron-cored transformer is also used at radio frequencies. In a practical transmission-line system, the transmitter is ordinarily matched to the coaxial cable for a maximum power transfer. However, due to the variable loads, an impedance-matching technique is often required at the load side.

Since the matching problems involve parallel connections on the transmission line, it is necessary to work out the problems with admittances rather than impedances. The Smith chart itself can be used as a computer to convert the normalized impedance to admittance by a rotation of 180°, as described earlier.

I. SINGLE-STUB MATCHING. Although single-lumped inductors or capacitors can match the transmission line, it is more common to use the susceptive properties of short-circuited sections of transmission lines. Short-circuited sections are preferable to open-circuited ones because a good short circuit is easier to obtain than a good open circuit.

For a lossless line with $\mathbf{Y}_g = \mathbf{Y}_0$, maximum power transfer requires $\mathbf{Y}_{11} = \mathbf{Y}_0$, where $\mathbf{Y}_{11}$ is the total admittance of the line and stub looking to the right at the point 1-1 (see Fig. 3-1-16a). The stub must be located at that

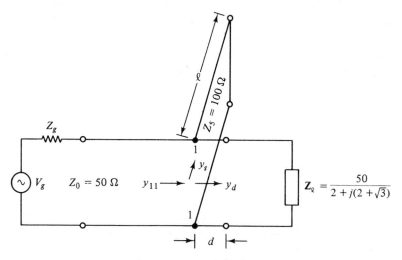

Fig. 3-1-16a. Single-stub matching for Example 3-1-3.

point on the line where the real part of the admittance, looking toward the load, is $\mathbf{Y}_0$. In a normalized unit y_{11} must be in the form

$$y_{11} = y_d \pm y_s = 1 \qquad (3\text{-}1\text{-}125)$$

if the stub has same characteristic impedance as that of the line. Otherwise

$$\mathbf{Y}_{11} = \mathbf{Y}_d \pm \mathbf{Y}_s = \mathbf{Y}_0 \qquad (3\text{-}1\text{-}126)$$

The stub length is then adjusted so that its susceptance just cancels out the susceptance of the line at the junction.

EXAMPLE 3-1-3: Single-Stub Matching. A lossless line of characteristic impedance $Z_0 = 50$ ohms is to be matched to a load $\mathbf{Z}_\ell = 50/[2 + j(2 + \sqrt{3})]$ ohms by means of a lossless short-circuited stub. The characteristic impedance of the stub is 100 ohms. Find the stub position (closest to the load) and length so that a match is obtained.

Solution.

1. Compute the normalized load admittance and enter it on the Smith chart (see Fig. 3-1-16b).

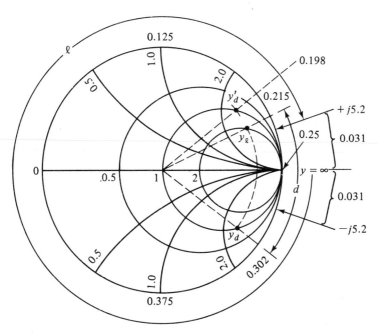

Fig. 3-1-16b. Graphical solution for Example 3-1-3.

$$y_\ell = \frac{1}{z_\ell} = \frac{Z_0}{\mathbf{Z}_\ell} = 2 + j(2 + \sqrt{3}) = 2 + j3.732$$

2. Draw a SWR circle through the point of y_ℓ and the circle intersects the unity circle at the point y_d.

$$y_d = 1 - j2.6$$

Note that there are an infinite number of y_d. Take the one that allows the stub to be attached as closely as possible to the load.

3. Since the characteristic impedance of the stub is different from that of the line, the condition for impedance matching at the junction requires

$$\mathbf{Y}_{11} = \mathbf{Y}_d + \mathbf{Y}_s$$

where $\mathbf{Y}_s$ is the susceptance that the stub will contribute.

It is clear that the stub and the portion of the line from the load to the junction are in parallel, as seen by the mainline extending to the generator. The admittances must be converted to normalized values for matching on the Smith chart. Then Eq. (3-1-126) becomes

$$y_{11}\mathbf{Y}_0 = y_d\mathbf{Y}_0 + y_s\mathbf{Y}_{0s}$$

$$y_s = (y_{11} - y_d)\left(\frac{\mathbf{Y}_0}{\mathbf{Y}_{0s}}\right) = [1 - (1 - j2.6)]\frac{100}{50} = +j5.20$$

4. The distance between the load and the stub position can be calculated from the distance scale as

$$d = (0.302 - 0.215)\lambda = 0.087\lambda$$

5. Since the stub contributes a susceptance of $+j5.20$, enter $+j5.20$ on the chart and determine the required distance ℓ from the short-circuited end ($z = 0$, $y = \infty$), which corresponds to the right side of the real axis on the chart, by transversing the chart toward the generator until the point of $+j5.20$ is reached. Then

$$\ell = (0.50 - 0.031)\lambda = 0.469\lambda$$

When a line is matched at the junction, there will be no standing wave in the line from the stub to the generator.

6. If an inductive stub is required,

$$y_d' = 1 + j2.6$$

and the susceptance of the stub will be

$$y_s' = -j5.2$$

7. The position of the stub from the load is

$$d' = [0.50 - (0.215 - 0.198)]\lambda = 0.483\lambda$$

and the length of the short-circuited stub is

$$\ell' = 0.031\lambda$$

II. DOUBLE-STUB MATCHING.   Since single-stub matching is sometimes impractical because the stub cannot be placed physically in the ideal location, double-stub matching is needed. Double-stub devices consist of two short-circuited stubs connected in parallel with a fixed length between them. The length of the fixed section is usually $\frac{1}{8}$, $\frac{3}{8}$, or $\frac{5}{8}$ of a wavelength. The stub that is nearest the load is used to adjust the susceptance and is located at a fixed wavelength from the constant conductance unity circle ($g = 1$) on an appropriate constant-standing-wave-ratio circle. Then the admittance of the line at the second stub as shown in Fig. 3-1-17 is

$$y_{22} = y_{d2} \pm y_{s2} = 1 \tag{3-1-127}$$

$$\mathbf{Y}_{22} = \mathbf{Y}_{d2} \pm \mathbf{Y}_{s2} = \mathbf{Y}_0 \tag{3-1-128}$$

If the positions and lengths of the stubs are chosen properly, there will be no standing wave on the line to the left of the second stub measured from the load.

EXAMPLE 3-1-4:   Double-Stub Matching. The terminating impedance $\mathbf{Z}_\ell$ is $100 + j100$ ohms, and the characteristic impedance $\mathbf{Z}_0$ of the line and stub is 50 ohms. The first stub is placed at 0.40λ away from the load. The spacing between the two stubs is $\frac{3}{8}\lambda$. Determine the length of the short-

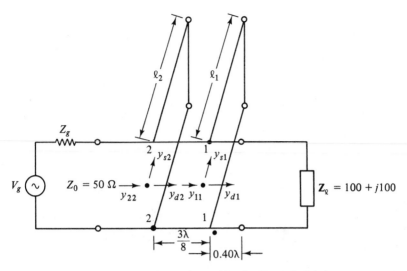

Fig. 3-1-17. Double-stub matching for Example 3-1-4.

circuited stubs when the match is achieved. What terminations are forbidden for matching the line by the double-stub device?

Solution.

1. Compute the normalized load impedance z_ℓ and enter it on the chart as shown in Fig. 3-1-18.

$$z_\ell = \frac{100 + j100}{50} = 2 + j2$$

2. Plot a SWR ρ circle and read the normalized load admittance $180°$ out of phase with z_ℓ on the SWR circle.

$$y_\ell = 0.25 - j0.25$$

3. Draw the spacing circle of $\frac{3}{8}\lambda$ by rotating the constant-conductance unity circle ($g = 1$) through a phase angle of $2\beta d = 2\beta\frac{3}{8}\lambda = \frac{3}{2}\pi$ toward the load. y_{11} must be on this spacing circle, since y_{d2} will be on the $g = 1$ circle (y_{11} and y_{d2} are $\frac{3}{8}\lambda$ apart).

4. Move y_ℓ for a distance of 0.40λ from 0.458 to 0.358 along the SWR ρ circle toward the generator and read y_{d1} on the chart.

$$y_{d1} = 0.55 - j1.08$$

5. There are two possible solutions for y_{11}. They can be found by carrying y_{d1} along the constant-conductance ($g = 0.55$) circle that intersects the spacing circle at two points.

$$y_{11} = 0.55 - j0.11$$
$$y'_{11} = 0.55 - j1.88$$

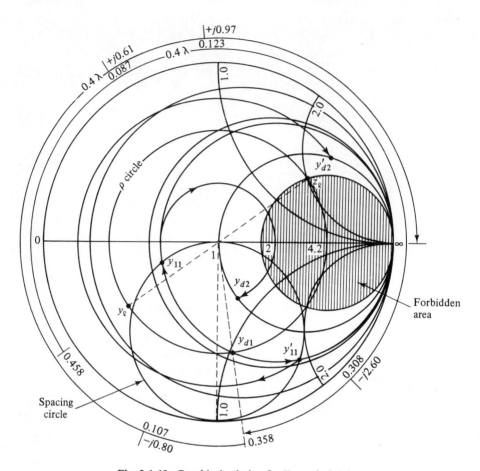

Fig. 3-1-18. Graphical solution for Example 3-1-4.

6. At the junction 1-1,

$$y_{11} = y_{d1} + y_{s1}$$

Then

$$y_{s1} = y_{11} - y_{d1} = (0.55 - j0.11) - (0.55 - j1.08) = +j0.97$$

Similarly,

$$y'_{s1} = -j0.80$$

7. The lengths of stub No. 1 are found as

$$\ell_1 = (0.25 + 0.123)\lambda = 0.373\lambda$$

$$\ell'_1 = (0.25 - 0.107)\lambda = 0.143\lambda$$

8. The $\frac{3}{8}\lambda$ section of line transforms y_{11} to y_{d2}, and y_{11} to y'_{d2} along their constant standing-wave circles, respectively. That is,

$$y_{d2} = 1 - j0.61$$

$$y'_{d2} = 1 + j2.60$$

94

9. Then stub No. 2 must contribute

$$y_{s2} = +j0.61$$
$$y'_{s2} = -j2.60$$

10. The lengths of stub No. 2 are found as

$$\ell_2 = (0.25 + 0.087)\lambda = 0.337\lambda$$
$$\ell'_2 = (0.308 - 0.25)\lambda = 0.058\lambda$$

11. It can be seen from Fig. 3-1-18 that a normalized admittance y_ℓ located inside the hatched area cannot be brought to lie on the locus of y_{11} or y'_{11} for a possible match by the parallel connection of any short-circuited stub because the spacing circle and $g = 2$ circle are mutually tangent. Thus the area of a $g = 2$ circle is called the *forbidden region* of the normalized load admittance for possible match.

Normally the solution of a double-stub-matching problem can be worked out backward from the load toward the generator, since the load is known and the distance of the first stub away from the load can be arbitrarily chosen. In quite a few practical matching problems, however, some stubs have a different Z_0 from that of the line, the length of a stub may be fixed, and so on. So it is hard to describe a definite procedure for solving the double-matching problems.

3-2 Microwave Waveguides

In general, a waveguide consists of a hollow metallic tube of a rectangular or cylindrical shape used to guide an electromagnetic wave. Waveguides are used principally at frequencies in the microwave range; inconveniently large guides would be required to transmit radio frequency power at longer wavelengths.

In waveguides the electric and magnetic fields are confined to the space within the guides. So no power is lost through radiation, and even the dielectric loss is negligible, since the guides are normally air filled. However, there is some power loss as heat in the walls of the guides, but the loss is very small.

It is possible to propagate several modes of electromagnetic waves within a waveguide. These modes correspond to solutions of Maxwell's equations for the particular waveguides. A given waveguide has a definite cutoff frequency for each allowed mode. If the frequency of the impressed signal is above the cutoff frequency for a given mode, the electromagnetic energy can be transmitted down the guide for that particular mode without attenuation. Otherwise the electromagnetic energy with a frequency below the cutoff frequency for that particular mode will be attenuated to a negligible value in a relatively short distance. *The dominant mode in a particular guide is the mode having the lowest cutoff frequency.* It is advisable to choose the dimensions of a guide in such a way that, for a given input signal, only the energy of the dominant mode can be transmitted through the guide.

The process of solving the waveguide problems may involve three steps.

Step 1. The desired wave equations are written in the form of either rectangular or cylindrical coordinate systems suitable to the problem on hand.

Step 2. The boundary conditions are then applied to the wave equations set up in step 1.

Step 3. The resultant equations usually are in the form of partial differential equations in either time or frequency domain. They can be solved by using the proper method as desired.

3-2-1. Rectangular Waveguides

A rectangular waveguide is a hollow metallic tube with a rectangular cross section. The conducting walls of the guide serve to confine the electromagnetic fields and thereby guide the electromagnetic wave. A number of distinct field configurations or modes can exist in waveguides. When the waves travel longitudinally down the guide, the plane waves are reflected from wall to wall. This process results in a component of either electric or magnet field in the direction of propagation of the resultant wave; therefore the wave is no longer a *transverse electromagnetic* (TEM) wave. Figure 3-2-1 shows that any uniform plane wave in a lossless guide may be resolved into TE and TM waves.

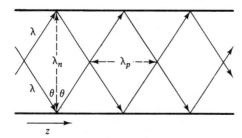

Fig. 3-2-1. Plane wave reflected in a waveguide.

It is clear that when the wavelength λ is in the direction of propagation of the incident wave, there will be one component λ_n in the direction normal to the reflecting plane and another λ_p parallel to the plane. These components are

$$\lambda_n = \frac{\lambda}{\cos \theta} \qquad (3\text{-}2\text{-}1)$$

$$\lambda_p = \frac{\lambda}{\sin \theta} \qquad (3\text{-}2\text{-}2)$$

where θ = angle of incidence
λ = wavelength of the impressed signal in unbounded medium

A plane wave in a waveguide resolves into two components: one standing wave in the direction normal to the reflecting walls of the guide and one traveling wave in the direction parallel to the reflecting walls. In lossless waveguides the modes may be classified as either *transverse electric* (TE) mode or *transverse magnetic* (TM) mode. In rectangular guides the modes are designated TE_{mn} or TM_{mn}. The integer m denotes the number of half waves of electric or magnetic intensity in the x direction, while n is the number of half waves in the y direction if the direction of propagation of the wave is assumed in the positive z direction.

(1) Solutions of Wave Equations in Rectangular Coordinates. As stated previously, there are time domain and frequency domain solutions for each wave equation. However, for the simplicity of the solution to the wave equation in three dimensions plus a time-varying variable, only the sinusoidal steady-state or the frequency domain solution will be given. A rectangular coordinate system is shown in Fig. 3-2-2.

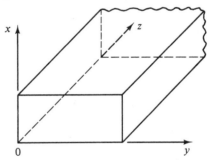

Fig. 3-2-2. Rectangular coordinates.

The electric and magnetic wave equations in frequency domain in Eqs. (2-1-20) and (2-1-21) are given by

$$\nabla^2 \mathbf{E} = \gamma^2 \mathbf{E} \tag{3-2-3}$$

$$\nabla^2 \mathbf{H} = \gamma^2 \mathbf{H} \tag{3-2-4}$$

where $\qquad \gamma = \sqrt{j\omega\mu(\sigma + j\omega\epsilon)} = \alpha + j\beta$

These are called the vector wave equations.

Rectangular coordinates are the usual right-hand system. The rectangular components of $\mathbf{E}$ or $\mathbf{H}$ satisfy the complex scalar wave equation or Helmholtz equation.

$$\nabla^2 \psi = \gamma^2 \psi \tag{3-2-5}$$

The Helmholtz equation in rectangular coordinates is

$$\frac{\partial^2 \psi}{\partial x^2} + \frac{\partial^2 \psi}{\partial y^2} + \frac{\partial^2 \psi}{\partial z^2} = \gamma^2 \psi \tag{3-2-6}$$

This is a linear and inhomogenous partial differential equation in three dimensions. By the method of separation of variables, the solution is assumed in the form of

$$\psi = X(x)\,Y(y)Z(z) \qquad (3\text{-}2\text{-}7)$$

where
$$X(x) = \text{a function of the } x \text{ coordinate only}$$
$$Y(y) = \text{a function of the } y \text{ coordinate only}$$
$$Z(z) = \text{a function of the } z \text{ coordinate only}$$

Substitution of Eq. (3-2-7) in Eq. (3-2-6) and division of the resultant by Eq. (3-2-7) yield

$$\frac{1}{X}\frac{d^2X}{dx^2} + \frac{1}{Y}\frac{d^2Y}{dy^2} + \frac{1}{Z}\frac{d^2Z}{dz^2} = \gamma^2 \qquad (3\text{-}2\text{-}8)$$

Since the sum of the three terms on the left-hand side is a constant and each term is independently variable, it follows that each term must be equal to a constant.

Let the three terms be k_x^2, k_y^2, and k_z^2 respectively; then the separation equation is given by

$$-k_x^2 - k_y^2 - k_z^2 = \gamma^2 \qquad (3\text{-}2\text{-}9)$$

The general solution of each differential equation in Eq. (3-2-8)

$$\frac{d^2X}{dx^2} = -k_x^2 X \qquad (3\text{-}2\text{-}10)$$

$$\frac{d^2Y}{dy^2} = -k_y^2 Y \qquad (3\text{-}2\text{-}11)$$

$$\frac{d^2Z}{dz^2} = -k_z^2 Z \qquad (3\text{-}2\text{-}12)$$

will be in the form of

$$X = A \sin k_x x + B \cos k_x x \qquad (3\text{-}2\text{-}13)$$
$$Y = C \sin k_y y + D \cos k_y y \qquad (3\text{-}2\text{-}14)$$
$$Z = E \sin k_z z + F \cos k_z z \qquad (3\text{-}2\text{-}15)$$

The total solution of the Helmholtz equation in rectangular coordinates is

$$\psi = (A \sin k_x x + B \cos k_x x)(C \sin k_y y + D \cos k_y y)$$
$$\times\ (E \sin k_z z + F \cos k_z z) \qquad (3\text{-}2\text{-}16)$$

The direction of propagation of the wave in the guide is conventionally assumed in the positive z direction. It should be noted that the propagation constant γ_g in the guide differs from the intrinsic propagation constant γ of the dielectric. Let

$$\gamma_g^2 = \gamma^2 + k_x^2 + k_y^2 = \gamma^2 + k_c^2 \qquad (3\text{-}2\text{-}17)$$

where $k_c = \sqrt{k_x^2 + k_y^2}$ is usually called the cutoff wave number. For a loss-less dielectric, $\gamma^2 = -\omega^2\mu\epsilon$. Then

$$\gamma_g = \pm j\sqrt{\omega^2\mu\epsilon - k_c^2} \tag{3-2-18}$$

There are three cases for the propagation constant γ_g in the waveguide.

CASE I: There will be no wave propagation (evanescence) in the guide if $\omega_c^2\mu\epsilon = k_c^2$ and $\gamma_g = 0$. This is the critical condition for cutoff propagation. The cutoff frequency is expressed as

$$f_c = \frac{1}{2\pi\sqrt{\mu\epsilon}}\sqrt{k_x^2 + k_y^2} \tag{3-2-19}$$

CASE II: The wave will be propagating in the guide if $\omega^2\mu\epsilon > k_c^2$ and

$$\gamma_g = \pm j\beta_g = \pm j\omega\sqrt{\mu\epsilon}\sqrt{1 - \left(\frac{f_c}{f}\right)^2} \tag{3-2-20}$$

This means that the operating frequency must be above the cutoff frequency in order for a wave to propagate in the guide.

CASE III: The wave will be attenuated if $\omega^2\mu\epsilon < k_c^2$ and

$$\gamma_g = \pm\alpha_g = \pm\omega\sqrt{\mu\epsilon}\sqrt{\left(\frac{f_c}{f}\right)^2 - 1} \tag{3-2-21}$$

This means that if the operating frequency is below the cutoff frequency, the wave will decay exponentially with respect to a factor of $-\alpha_g z$ and there will be no wave propagation because the propagation constant is a real quantity. Therefore the traveling-wave solution to the Helmholtz equation in rectangular coordinates is given by

$$\psi = (A \sin k_x x + B \cos k_x x)$$
$$\times (C \sin k_y y + D \cos k_y y)e^{j(\omega t - \beta_g z)} \tag{3-2-22}$$

(2) *TE Modes in Rectangular Waveguides.* It has been previously assumed that the waves are propagating in the positive z direction in the waveguide. Figure 3-2-3 shows the coordinates of a rectangular waveguide.

The TE_{mn} modes in a rectangular guide are characterized by $E_z = 0$. In other words, the z component of the magnetic field, H_z, must exist in order to have energy transmission in the guide. Consequently, from a given Helmholtz equation,

$$\nabla^2 H_z = \gamma^2 H_z \tag{3-2-23}$$

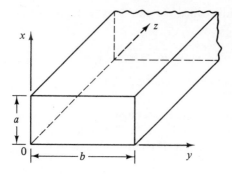

Fig. 3-2-3. Coordinates of rectangular guide.

a solution in the form of

$$H_z = \left(A_m \sin \frac{m\pi x}{a} + B_m \cos \frac{m\pi x}{a} \right)$$

$$\times \left(C_n \sin \frac{n\pi y}{b} + D_n \cos \frac{n\pi y}{b} \right) e^{j(\omega t - \beta_g z)} \qquad (3\text{-}2\text{-}24)$$

will be determined in accordance with the given boundary conditions, where $k_x = m\pi/a$ and $k_y = n\pi/b$ are replaced. For a lossless dielectric, Maxwell's curl equations in frequency domain are

$$\nabla \times \mathbf{E} = -j\omega\mu\mathbf{H} \qquad (3\text{-}2\text{-}25a)$$

$$\nabla \times \mathbf{H} = j\omega\epsilon\mathbf{E} \qquad (3\text{-}2\text{-}25b)$$

In rectangular coordinates, their components are

$$\frac{\partial E_z}{\partial y} - \frac{\partial E_y}{\partial z} = -j\omega\mu H_x \qquad (3\text{-}2\text{-}26)$$

$$\frac{\partial E_x}{\partial z} - \frac{\partial E_z}{\partial x} = -j\omega\mu H_y \qquad (3\text{-}2\text{-}27)$$

$$\frac{\partial E_y}{\partial x} - \frac{\partial E_x}{\partial y} = -j\omega\mu H_z \qquad (3\text{-}2\text{-}28)$$

$$\frac{\partial H_z}{\partial y} - \frac{\partial H_y}{\partial z} = j\omega\epsilon E_x \qquad (3\text{-}2\text{-}29)$$

$$\frac{\partial H_x}{\partial z} - \frac{\partial H_z}{\partial x} = j\omega\epsilon E_y \qquad (3\text{-}2\text{-}30)$$

$$\frac{\partial H_y}{\partial x} - \frac{\partial H_x}{\partial y} = j\omega\epsilon E_z \qquad (3\text{-}2\text{-}31)$$

With the substitution $\partial/\partial z = -j\beta_g$ and $E_z = 0$, the foregoing equations are simplified to

$$\beta_g E_y = -\omega\mu H_x \qquad (3\text{-}2\text{-}32)$$

$$\beta_g E_x = \omega\mu H_y \qquad (3\text{-}2\text{-}33)$$

$$\frac{\partial E_y}{\partial x} - \frac{\partial E_x}{\partial y} = -j\omega\mu H_z \qquad (3\text{-}2\text{-}34)$$

$$\frac{\partial H_z}{\partial y} + j\beta_g H_y = j\omega\epsilon E_x \qquad (3\text{-}2\text{-}35)$$

$$-j\beta_g H_x - \frac{\partial H_z}{\partial x} = j\omega\epsilon E_y \qquad (3\text{-}2\text{-}36)$$

$$\frac{\partial H_y}{\partial x} - \frac{\partial H_x}{\partial y} = 0 \qquad (3\text{-}2\text{-}37)$$

Solving these six equations for E_x, E_y, H_x, and H_y in terms of H_z will give the TE-mode field equations in rectangular waveguides as

$$E_x = \frac{j\omega\mu}{k_c^2} \frac{\partial H_z}{\partial y} \qquad (3\text{-}2\text{-}38)$$

$$E_y = \frac{j\omega\mu}{k_c^2} \frac{\partial H_z}{\partial x} \qquad (3\text{-}2\text{-}39)$$

$$E_z = 0 \qquad (3\text{-}2\text{-}40)$$

$$H_x = \frac{-\beta_g}{k_c^2} \frac{\partial H_z}{\partial x} \qquad (3\text{-}2\text{-}41)$$

$$H_y = \frac{-\beta_g}{k_c^2} \frac{\partial H_z}{\partial y} \qquad (3\text{-}2\text{-}42)$$

$$H_z = \text{Eq. } (3\text{-}2\text{-}24) \qquad (3\text{-}2\text{-}43)$$

where $k_c^2 = \omega^2\mu\epsilon - \beta_g^2$ has been replaced.

Differentiating Eq. (3-2-24) with respect to x and y and then substituting the results in Eqs. (3-2-38) through (3-2-42) will yield a set of field equations. The boundary conditions are applied to the newly found field equations in such a manner that either the tangent **E** field or the normal **H** field vanishes at the surface of the conductor. Since $E_x = 0$, then $\partial H_z/\partial y = 0$ at $y = 0, b$. Hence $C_n = 0$. Since $E_y = 0$, then $\partial H_z/\partial x = 0$ at $x = 0, a$. Hence $A_m = 0$.

It is generally concluded that the normal derivative of H_z must vanish at the conducting surfaces—that is,

$$\frac{\partial H_z}{\partial n} = 0 \qquad (3\text{-}2\text{-}44)$$

at the guide walls. Therefore the magnetic field in the positive z direction is given by

$$H_z = H_{0z} \cos\frac{m\pi x}{a} \cos\frac{n\pi y}{b} e^{j(\omega t - \beta_g z)} \qquad (3\text{-}2\text{-}45)$$

where H_{0z} is the amplitude constant.

Substitution of Eq. (3-2-45) in Eqs. (3-2-38) through (3-2-42) yields the TE$_{mn}$ field equations in rectangular waveguides as

$$E_x = E_{0x} \cos \frac{m\pi x}{a} \sin \frac{n\pi y}{b} e^{j(\omega t - \beta_g z)} \qquad (3\text{-}2\text{-}46)$$

$$E_y = E_{0y} \sin \frac{m\pi x}{a} \cos \frac{n\pi y}{b} e^{j(\omega t - \beta_g z)} \qquad (3\text{-}2\text{-}47)$$

$$E_z = 0 \qquad (3\text{-}2\text{-}48)$$

$$H_x = H_{0x} \sin \frac{m\pi x}{a} \cos \frac{n\pi y}{b} e^{j(\omega t - \beta_g z)} \qquad (3\text{-}2\text{-}49)$$

$$H_y = H_{0y} \cos \frac{m\pi x}{a} \sin \frac{n\pi y}{b} e^{j(\omega t - \beta_g z)} \qquad (3\text{-}2\text{-}50)$$

$$H_z = \text{Eq. (3-2-45)} \qquad (3\text{-}2\text{-}51)$$

where
$$m = 0, 1, 2, \ldots$$
$$n = 0, 1, 2, \ldots$$
$$m = n = 0 \text{ excepted}$$

The cutoff wave number k_c, as defined by Eq. (3-2-17) for the TE$_{mn}$ modes, is given by

$$k_c = \sqrt{\left(\frac{m\pi}{a}\right)^2 + \left(\frac{n\pi}{b}\right)^2} = \omega_c \sqrt{\mu\epsilon} \qquad (3\text{-}2\text{-}52)$$

where a and b are in meters. The cutoff frequency, as defined in Eq. (3-2-19) for the TE$_{mn}$ modes, is

$$f_c = \frac{1}{2\sqrt{\mu\epsilon}} \sqrt{\frac{m^2}{a^2} + \frac{n^2}{b^2}} \qquad (3\text{-}2\text{-}53)$$

The propagation constant (or the phase constant here) β_g, as defined in Eq. (3-2-18), is expressed by

$$\beta_g = \omega \sqrt{\mu\epsilon} \sqrt{1 - \left(\frac{f_c}{f}\right)^2} \qquad (3\text{-}2\text{-}54)$$

The phase velocity in the positive z direction for the TE$_{mn}$ modes is shown as

$$v_g = \frac{\omega}{\beta_g} = \frac{v_p}{\sqrt{1 - (f_c/f)^2}} \qquad (3\text{-}2\text{-}55)$$

where $v_p = 1/\sqrt{\mu\epsilon}$ is the phase velocity in an unbounded dielectric.

The characteristic wave impedance of TE$_{mn}$ modes in the guide can be derived from Eqs. (3-2-32) and (3-2-33):

$$Z_g = \frac{E_x}{H_y} = -\frac{E_y}{H_x} = \frac{\omega\mu}{\beta_g} = \frac{\eta}{\sqrt{1 - (f_c/f)^2}} \qquad (3\text{-}2\text{-}56)$$

where $\eta = \sqrt{\mu/\epsilon}$ is the intrinsic impedance in an unbounded dielectric. The wavelength λ_g in the guide for the TE_{mn} modes is given by

$$\lambda_g = \frac{\lambda}{\sqrt{1 - (f_c/f)^2}} \qquad (3\text{-}2\text{-}57)$$

where $\lambda = v_p/f$ is the wavelength in an unbounded dielectric.

Since the cutoff frequency shown in Eq. (3-2-53) is a function of the modes and guide dimensions, the physical size of the waveguide will determine the propogation of the modes. Table 3-2-1 tabulates the ratio of cutoff frequency of some modes with respect to that of the dominant mode in terms of the physical dimension.

TABLE 3-2-1.   Modes of $(f_c)_{mn}/f_c$ for $b \geq a$.

Modes f/f_{01} b/a	TE_{01}	TE_{10}	TE_{11} TM_{11}	TE_{02}	TE_{20}	TE_{12} TM_{12}	TE_{21} TM_{21}	TE_{22} TM_{22}	TE_{03}
1	1	1	1.414	2	2	2.236	2.236	2.828	3
1.5	1	1.5	1.803	2	3	2.500	3.162	3.606	3
2	1	2	2.236	2	4	2.828	4.123	4.472	3
3	1	3	3.162	2	6	3.606	6.083	6.325	3
∞	1	∞	∞	2	∞	∞	∞	∞	3

In a rectangular guide the corresponding TE_{mn} and TM_{mn} modes are always degenerate. In a square guide the TE_{mn}, TE_{nm}, TM_{mn}, and TM_{nm} modes form a foursome of degeneracy. Rectangular guides ordinarily have dimensions of $2a = b$ ratio. The mode with the lowest cutoff frequency in a particular guide is called the *dominant mode*. The dominant mode in a rectangular guide with $a < b$ is the TE_{01} mode. Each mode has a specific mode pattern (or field pattern).

It is possible for several modes to exist simultaneously in a given waveguide. The situation is not too serious, however. Actually, only the dominant mode will propagate, and the higher modes near the sources or discontinuities will decay very fast.

EXAMPLE 3-2-1: TE_{01} in Rectangular Waveguide. An air-filled rectangular waveguide of inside dimensions 3.5×7 cm operates in the dominant TE_{01} mode as shown in Fig. E3-2-1.
(a) Find the cutoff frequency.
(b) Determine the phase velocity of the wave in the guide at a frequency of 3.5 GHz.
(c) Determine the guided wavelength at the same frequency.

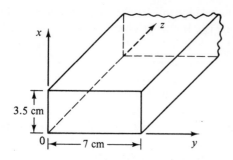

3.5 cm

$0 \vdash\!\!\!-\!\!\!-7\ cm\!-\!\!\!-\!\!\!\rightarrow\!\vdash$ y **Fig. E3-2-1.**

Solution.

(a) $f_c = \dfrac{c}{2b} = \dfrac{3 \times 10^8}{2 \times 7 \times 10^{-2}} = 2.14\ \text{GHz}$

(b) $v_g = \dfrac{c}{\sqrt{1 - (f_c/f)^2}} = \dfrac{3 \times 10^8}{\sqrt{1 - (2.14/3.5)^2}} = 3.78 \times 10^8\ \text{m/s}$

(c) $\lambda_g = \dfrac{\lambda_0}{\sqrt{1 - (f_c/f)^2}} = \dfrac{3 \times 10^8/(3.5 \times 10^9)}{\sqrt{1 - (2.14/3.5)^2}} = 10.8\ \text{cm}$

(3) TM Modes in Rectangular Waveguides. The TM_{mn} modes in a rectangular guide are characterized by $H_z = 0$. In other words, the z component of an electric field E must exist in order to have energy transmission in the guide. Consequently, the Helmholtz equation for E in the rectangular coordinates is given by

$$\nabla^2 E_z = \gamma^2 E_z \qquad (3\text{-}2\text{-}58)$$

A solution of the Helmholtz equation is in the form of

$$E_z = \left(A_m \sin \frac{m\pi x}{a} + B_m \cos \frac{m\pi x}{a}\right)\left(C_n \sin \frac{n\pi y}{b} + D_n \cos \frac{n\pi y}{b}\right)e^{j(\omega t - \beta_g z)}$$

$$(3\text{-}2\text{-}59)$$

which must be determined according to the given boundary conditions. The procedures for doing so are similar to those used in finding the TE-mode wave.

The boundary conditions on E_z require that the field vanish at the waveguide walls, since the tangent component of the electric field E_z is zero on the conducting surface. This requirement is that for $E_z = 0$ at $x = 0,\ a$, then $B_m = 0$; and for $E_z = 0$ at $y = 0,\ b$, then $D_n = 0$. Thus the solution as shown in Eq. (3-2-59) reduces to

$$E_z = E_{0z} \sin \frac{m\pi x}{a} \sin \frac{n\pi y}{b} e^{j(\omega t - \beta_g z)} \qquad (3\text{-}2\text{-}60)$$

where $m = 1, 2, 3, \ldots$
$\quad\quad\ \ n = 1, 2, 3, \ldots$

If either $m = 0$ or $n = 0$, the field intensities all vanish. So there is no TM_{01} or TM_{10} mode in a rectangular waveguide, which means that TE_{01} is the dominant mode in a rectangular wavequide for $a < b$. For $H_z = 0$, the field equations, after expanding $\nabla \times \mathbf{H} = j\omega\epsilon\mathbf{E}$, are given by

$$\frac{\partial E_z}{\partial y} + j\beta_g E_y = -j\omega\mu H_x \tag{3-2-61}$$

$$j\beta_g E_x + \frac{\partial E_z}{\partial x} = j\omega\mu H_y \tag{3-2-62}$$

$$\frac{\partial E_y}{\partial x} - \frac{\partial E_x}{\partial y} = 0 \tag{3-2-63}$$

$$\beta_g H_y = \omega\epsilon E_x \tag{3-2-64}$$

$$-\beta_g H_x = \omega\epsilon E_y \tag{3-2-65}$$

$$\frac{\partial H_y}{\partial x} - \frac{\partial H_x}{\partial y} = j\omega\epsilon E_z \tag{3-2-66}$$

These equations can be solved simultaneously for E_x, E_y, H_x, and H_y in terms of E_z. The resultant field equations are

$$E_x = \frac{-\beta_g}{k_c^2}\frac{\partial E_z}{\partial x} \tag{3-2-67}$$

$$E_y = \frac{-\beta_g}{k_c^2}\frac{\partial E_z}{\partial y} \tag{6-2-68}$$

$$E_z = \text{Eq. (3-2-60)} \tag{3-2-69}$$

$$H_x = \frac{j\omega\epsilon}{k_c^2}\frac{\partial E_z}{\partial y} \tag{3-2-70}$$

$$H_y = \frac{j\omega\epsilon}{k_c^2}\frac{\partial E_z}{\partial x} \tag{3-2-71}$$

$$H_z = 0 \tag{3-2-72}$$

where $\beta_g^2 - \omega^2\mu\epsilon = -k_c^2$ is replaced.

Differentiating Eq. (3-2-60) with respect to x or y and substituting the results in Eqs. (3-2-67) through (3-2-72) yield a new set of field equations. The TM_{mn}-mode field equations in rectangular waveguides are

$$E_x = E_{0x}\cos\frac{m\pi x}{a}\sin\frac{n\pi y}{b}e^{j(\omega t - \beta_g z)} \tag{3-2-73}$$

$$E_y = E_{0y}\sin\frac{m\pi x}{a}\cos\frac{n\pi y}{b}e^{j(\omega t - \beta_g z)} \tag{3-2-74}$$

$$E_z = \text{Eq. (3-2-60)} \tag{3-2-75}$$

$$H_x = H_{0x}\sin\frac{m\pi x}{a}\cos\frac{n\pi y}{b}e^{j(\omega t - \beta_g z)} \tag{3-2-76}$$

$$H_y = H_{0y} \cos \frac{m\pi x}{a} \sin \frac{n\pi y}{b} e^{j(\omega t - \beta_g z)} \tag{3-2-77}$$

$$H_z = 0 \tag{3-2-78}$$

Some of the TM-mode characteristic equations are identical to those of the TE modes, but some are different. For convenience, all are shown here.

$$f_c = \frac{1}{2\sqrt{\mu\epsilon}} \sqrt{\frac{m^2}{a^2} + \frac{n^2}{b^2}} \tag{3-2-79}$$

$$\beta_g = \omega\sqrt{\mu\epsilon} \sqrt{1 - \left(\frac{f_c}{f}\right)^2} \tag{3-2-80}$$

$$\lambda_g = \frac{\lambda}{\sqrt{1 - (f_c/f)^2}} \tag{3-2-81}$$

$$v_g = \frac{v_p}{\sqrt{1 - (f_c/f)^2}} \tag{3-2-82}$$

$$Z_g = \frac{\beta_g}{\omega\epsilon} = \eta\sqrt{1 - \left(\frac{f_c}{f}\right)^2} \tag{3-2-83}$$

(4) Power Transmission in Rectangular Guides. The power transmitted through a waveguide and the power loss in the guide walls can be calculated by means of the complex Poynting theorem as described in Chapter 2. It is assumed that the guide is terminated in such a way that there is no reflection from the receiving end or that the guide is infinitely long compared with the wavelength. From the Poynting theorem in Section 2-2, the power transmitted through a guide is given by

$$P_{tr} = \oint \mathbf{p} \cdot d\mathbf{s} = \oint \frac{1}{2} (\mathbf{E} \times \mathbf{H}^*) \cdot d\mathbf{s} \tag{3-2-84}$$

For a lossless dielectric, the time-average power flow through a rectangular guide is given by

$$P_{tr} = \frac{1}{2Z_g} \int_a |E|^2 \, da = \frac{Z_g}{2} \int_a |H|^2 \, da \tag{3-2-85}$$

where
$$Z_g = \frac{E_x}{H_y} = -\frac{E_y}{H_x}$$

$$|E|^2 = |E_x|^2 + |E_y|^2$$

$$|H|^2 = |H_x|^2 + |H_y|^2$$

For TE_{mn} modes, the average power transmitted through a rectangular waveguide is given by

$$P_{tr} = \frac{\sqrt{1 - (f_c/f)^2}}{2\eta} \int_0^b \int_0^a (|E_x|^2 + |E_y|^2) \, dx \, dy \tag{3-2-86}$$

For TM_{mn} modes, the average power transmitted through a rectangular waveguide is given by

$$P_{tr} = \frac{1}{2\eta\sqrt{1 - (f_c/f)^2}} \int_0^b \int_0^a (|E_x|^2 + |E_y|^2)\, dx\, dy \qquad (3\text{-}2\text{-}87)$$

where $\eta = \sqrt{\mu/\epsilon}$ is the intrinsic impedance in an unbounded dielectric.

(5) *Power Losses in Rectangular Guides.* There are three types of power losses in a rectangular waveguide:

1. Losses by the signal with frequency below the cutoff frequency
2. Losses in the dielectric
3. Losses in the guide walls

1. Power losses due to signal frequency. As described in Section 3-2-1, when the impressed frequency is below the cutoff frequency, the signal will be attenuated exponentially with respect to $(-\alpha_g z)$ and nonpropagation (or evanescence) occurs because the propagation constant is a real value. So waveguides with dimensions much smaller than the cutoff are often used as attenuators. The attenuation constant α_g for the TE_{mn} and TM_{mn} modes is given in Eq. (3-2-21) as

$$\alpha_g = \omega\sqrt{\mu\epsilon}\sqrt{\left(\frac{f_c}{f}\right)^2 - 1} \qquad (3\text{-}2\text{-}88)$$

2. Power losses due to dielectric attenuation. In a low-loss dielectric (i.e., $\sigma \ll \mu\epsilon$) the propagation constant for a plane wave traveling in an unbounded lossy dielectric is given in Eq. (2-5-20) by

$$\alpha = \frac{\sigma}{2}\sqrt{\frac{\mu}{\epsilon}} = \frac{\eta\sigma}{2} \qquad (3\text{-}2\text{-}89)$$

The attenuation due to the low-loss dielectric in the rectangular waveguide for the TE_{mn} and TM_{mn} modes is given by

$$\alpha_g = \frac{\sigma\eta}{2\sqrt{1 - (f_c/f)^2}} \qquad (3\text{-}2\text{-}90)$$

As $f \gg f_c$, the attenuation constant in the guide approaches that for the unbounded dielectric given by Eq. (3-2-89). However, if the operating frequency is way below the cutoff frequency, $f \ll f_c$, the attenuation constant becomes very large and nonpropagation occurs.

3. Power losses due to the guide walls. When the electric and magnetic intensities propagate through a lossy waveguide, their magnitudes may be written

$$|E| = |E_{0z}|\, e^{-\alpha_g z} \qquad (3\text{-}2\text{-}91)$$

$$|H| = |H_{0z}|\, e^{-\alpha_g z} \qquad (3\text{-}2\text{-}92)$$

where E_{0z} and H_{0z} are the field intensities at $z = 0$. It is interesting to note that, for a low-loss guide, the time-average power flow will decrease proportionally to $e^{-2\alpha_g z}$. Hence

$$P_{tr} = (P_{tr} + P_{loss})e^{-2\alpha_g z} \tag{3-2-93}$$

For $P_{loss} \ll P_{tr}$ and $2\alpha_g z \ll 1$,

$$\frac{P_{loss}}{P_{tr}} + 1 \doteq 1 + 2\alpha_g z \tag{3-2-94}$$

Finally,

$$\alpha_g = \frac{P_L}{2P_{tr}} \tag{3-2-95}$$

where P_L is the power loss per unit length. Consequently, the attenuation constant of the guide walls is equal to the ratio of the power loss per unit length to twice the power transmitted through the guide.

Since the electric and magnetic field intensities established at the surface of a low-loss guide wall will decay exponentially with respect to the skin depth while the waves progress into the walls, it is better to define a surface resistance of the guide walls as

$$R_s \equiv \frac{\rho}{\delta} = \frac{1}{\sigma \delta} = \frac{\alpha_g}{\sigma} = \sqrt{\frac{\pi f \mu}{\sigma}} \qquad \text{ohms/m}^2 \tag{3-2-96}$$

where ρ = resistivity of the conducting wall in ohms per cubic meter
$\quad\quad \sigma$ = conductitity in mhos per meter
$\quad\quad \delta$ = skin depth or depth of penetration in meters

The power loss per unit length of guide is obtained by integrating the power density over the surface of the conductor corresponding to the unit length of the guide. This is

$$P_L = \frac{R_s}{2} \int_s |H_t|^2 \, ds \qquad \text{watts/unit length} \tag{3-2-97}$$

where H_t is the tangential component of magnetic intensity at the guide walls. Substitution of Eqs. (3-2-85) and (3-2-97) in Eq. (3-2-95) yields

$$\alpha_g = \frac{R_s \int_s |H_t|^2 \, ds}{2Z_g \int_a |H|^2 \, da} \tag{3-2-98}$$

where

$$|H|^2 = |H_x|^2 + |H_y|^2 \tag{3-2-99}$$

$$|H_t|^2 = |H_{tx}|^2 + |H_{ty}|^2 \tag{3-2-100}$$

EXAMPLE 3-2-2: TE_{01} Mode in Rectangular Waveguide. An air-filled waveguide with a cross section 1×2 cm transports energy in the TE_{01} mode at the rate of 0.5 hp. The impressed frequency is 30 GHz. What is the peak value of the electric field occurring in the guide? (Refer to Fig. 3-2-4).

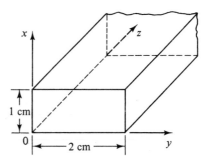

Fig. 3-2-4. Rectangular waveguide for Example 3-2-2.

Solution.

The field components of the dominant mode TE_{01} can be obtained by substituting $m = 0$ and $n = 1$ in Eqs. (3-2-46) through (3-2-51). Then

$$E_x = E_{0x} \sin \frac{\pi y}{b} e^{-j\beta_g z} \qquad H_x = 0$$

$$E_y = 0 \qquad\qquad H_y = \frac{E_{0x}}{Z_g} \sin \frac{\pi y}{b} e^{-j\beta_g z}$$

$$E_z = 0 \qquad\qquad H_z = H_{0z} \cos \frac{\pi y}{b} e^{-j\beta_g z}$$

where
$$Z_g = \frac{\omega \mu_0}{\beta_g}$$

The phase constant β_g can be found from

$$\beta_g = \sqrt{\omega^2 \mu_0 \epsilon_0 - \frac{\pi^2}{b^2}} = \pi \sqrt{\frac{(2f)^2}{c^2} - \frac{1}{b^2}} = \pi \sqrt{\frac{4 \times 9 \times 10^{20}}{9 \times 10^{16}} - \frac{1}{4 \times 10^{-4}}}$$

$$= 193.5\pi = 608.81 \text{ radians/m}$$

The power delivered in the z direction by the guide is

$$P = \text{Re} \left[\frac{1}{2} \int_0^b \int_0^a (\mathbf{E} \times \mathbf{H^*}) \right] \cdot dx \, dy \, \mathbf{u}_z$$

$$= \frac{1}{2} \int_0^b \int_0^a \left[\left(E_{0z} \sin \frac{\pi y}{b} e^{-j\beta_g z} \mathbf{u}_x \right) \times \left(\frac{\beta_g}{\omega \mu_0} E_{0x} \sin \frac{\pi y}{b} e^{+j\beta_g z} \mathbf{u}_y \right) \right] \cdot dx \, dy \, \mathbf{u}_z$$

$$= \frac{1}{2} E_{0x} \frac{\beta_g}{\omega \mu_0} \int_0^b \int_0^a \left(\sin \frac{\pi y}{a} \right)^2 dx \, dy$$

$$= \frac{1}{4} E_{0x} \frac{\beta_g}{\omega \mu_0} ab$$

$$373 = \frac{1}{4} E_{0x}^2 \frac{193.5\pi (10^{-2})(2 \times 10^{-2})}{2\pi (3 \times 10^{10})(4\pi \times 10^{-7})}$$

$$E_{0x} = 53.87 \quad \text{kV/m}$$

The peak value of the electric intensity is 53.87 kV/m.

(*6*) *Excitations of Modes in Rectangular Waveguides.* In general, the field intensities of the desired mode in a waveguide can be established by means of a probe or loop-coupling device. The probe may be called a monopole antenna; the coupling loop, the loop antenna. A probe should be located so as to excite the electric field intensity of the mode and a coupling loop in such a way as to generate the magnetic field intensity for the desired mode. If two or more probes or loops are to be used, care must be taken to ensure the proper phase relationship between the currents in the various antennas. This factor can be achieved by inserting additional lengths of transmission line in one or more of the antenna feeders. Impedance matching can be accomplished by varying the position and depth of the antenna in the quide or by using impedance matching stubs on the coaxial line feeding the waveguide. A device that serves to excite a given mode in the guide can also serve reciprocally as a receiver or collector of energy for that mode.

The methods of excitation for various modes in rectangular waveguides are shown in Fig. 3-2-5.

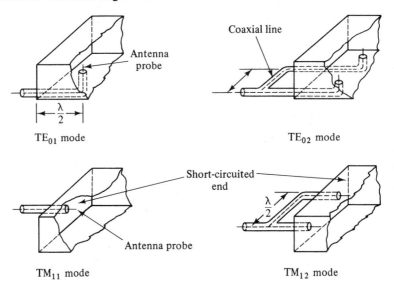

Fig. 3-2-5. Methods of exciting various modes in rectangular waveguides.

In order to excite a TE_{01} mode in one direction of the guide, the two exciting antennas should be arranged in such a way that the field intensities will cancel each other in one direction and reinforce in the other. Figure 3-2-6 shows an arrangement for launching a TE_{01} mode in one direction only. The two antennas are placed a quarter wavelength apart and their phases are in time quadrature. Phasing is compensated by use of an additional quarter-wavelength section of line connected to the antenna feeders. The field inten-

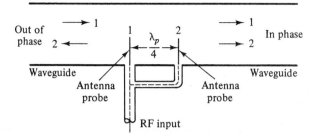

Fig. 3-2-6. A method of launching a TE_{01} mode in one direction only.

sities radiated by the two antennas are in phase opposition to the left of the antennas and cancel each other, whereas in the region to the right of the antennas the field intensities are in time phase and reinforce each other. The resulting wave thus propagates to the right in the guide.

Some higher modes are generated by discontinuities of the waveguide, such as obstacles, bends, and loads. However, the higher-order modes are, in general, more highly attenuated than the corresponding dominant mode. On the other hand, the dominant mode will tend to remain as a dominant wave even when the guide is large enough to support the higher modes.

3-2-2. Cylindrical Waveguides

A cylindrical waveguide is a tubular, circular conductor. A plane wave propagating through a circular waveguide results in a transverse electric (TE) or transverse magnetic (TM) mode. Several other types of waveguides, such as elliptical and reentrant guides, will also propagate electromagnetic waves.

(*1*) *Solutions of the Wave Equations in Cylindrical Coordinates.* As described in Section 3-2-1 for rectangular waveguides, only a sinusoidal steady-state or frequency domain solution will be attempted for cylindrical waveguides. A cylindrical coordinate system is shown in Fig. 3-2-7.

The scalar Helmholtz equation in cylindrical coordinates is given by

$$\frac{1}{r}\frac{\partial}{\partial r}\left(r\frac{\partial \psi}{\partial r}\right) + \frac{1}{r^2}\frac{\partial^2 \psi}{\partial \phi^2} + \frac{\partial^2 \psi}{\partial z^2} = \gamma^2 \psi \qquad (3\text{-}2\text{-}101)$$

Using the method of separation of variables, the solution is assumed in the form of

$$\Psi = R(r)\Phi(\phi)Z(z) \qquad (3\text{-}2\text{-}102)$$

where $R(r) =$ a function of the r coordinate only
$\Phi(\phi) =$ a function of the ϕ coordinate only
$Z(z) =$ a function of the z coordinate only

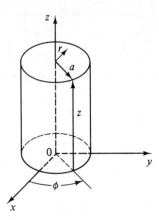

Fig. 3-2-7. Cylindrical coordinates.

Substitution of Eq. (3-2-102) in (3-2-101) and division of the resultant by (3-2-102) yield

$$\frac{1}{rR}\frac{d}{dr}\left(r\frac{dR}{dr}\right) + \frac{1}{r^2\Phi}\frac{d^2\Phi}{d\phi^2} + \frac{1}{Z}\frac{d^2Z}{dz^2} = \gamma^2 \qquad (3\text{-}2\text{-}103)$$

The sum of the three independent terms is a constant and so each of the three terms must be a constant. The third term may be set equal to a constant γ_g^2.

$$\frac{d^2Z}{dz^2} = \gamma_g^2 Z \qquad (3\text{-}2\text{-}104)$$

The solutions of this equation are given by

$$Z = Ae^{-\gamma_g z} + Be^{\gamma_g z} \qquad (3\text{-}2\text{-}105)$$

where γ_g = propagation constant of the wave in the guide.

Inserting γ_g^2 for the third term in the left-hand side in Eq. (3-2-103) and multiplying the resultant by r^2 yield

$$\frac{r}{R}\frac{d}{dr}\left(r\frac{dR}{dr}\right) + \frac{1}{\Phi}\frac{d^2\Phi}{d\phi^2} - (\gamma^2 - \gamma_g^2)r^2 = 0 \qquad (3\text{-}2\text{-}106)$$

The second term is a function of ϕ only; so equating the second term to a constant $(-n^2)$ yields

$$\frac{d^2\Phi}{d\phi^2} = -n^2\Phi \qquad (3\text{-}2\text{-}107)$$

The solution of this equation is also a harmonic function,

$$\Phi = A_n \sin n\phi + B_n \cos n\phi \qquad (3\text{-}2\text{-}108)$$

Replacing the Φ term by $(-n^2)$ in Eq. (3-2-106) and multiplying through by R, we have

$$r\frac{d}{dr}\left(r\frac{dR}{dr}\right) + [(K_c r)^2 - n^2]R = 0 \qquad (3\text{-}2\text{-}109)$$

This is Bessel's equation of order n in which

$$k_c^2 + \gamma^2 = \gamma_g^2 \qquad (3\text{-}2\text{-}110)$$

The above equation is called the *characteristic equation* of Bessel's equation. For a lossless guide, the characteristic equation reduces to

$$\beta_g = \pm\sqrt{\omega^2 \mu\epsilon - k_c^2} \qquad (3\text{-}2\text{-}111)$$

The solutions of Bessel's equation are

$$R = C_n J_n(k_c r) + D_n N_n(k_c r) \qquad (3\text{-}2\text{-}112)$$

where $J_n(k_c r)$ is the nth-order Bessel function of the first kind, representing a standard wave of $\cos(k_c r)$ for $r < a$ as shown in Fig. 3-2-8. $N_n(k_c r)$ is the nth-order Bessel function of the second kind, representing a standard wave of $\sin(k_c r)$ for $r > a$ as shown in Fig. 3-2-9.

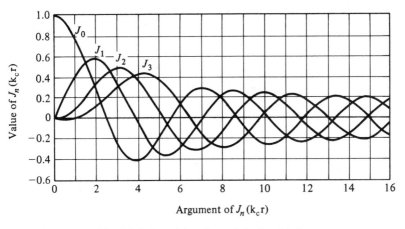

Fig. 3-2-8. Bessel functions of the first kind.

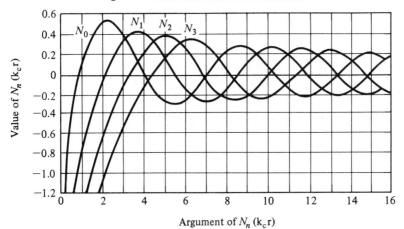

Fig. 3-2-9. Bessel functions of the second kind.

Therefore the total solution of the Helmholtz equation in cylindrical coordinates is given by

$$\Psi = [C_n J_n(k_c r) + D_n N_n(k_c r)](A_n \sin n\phi + B_n \cos n\phi)e^{\pm j\beta_g z} \quad (3\text{-}2\text{-}113)$$

However, at $r = 0$, $k_c r = 0$; then the function N_n approaches infinity, so $D_n = 0$. This means that at $r = 0$ on the z axis, the field must be finite. Also, by use of trigonometrical manipulations, the two sinusoidal terms will become

$$(A_n \sin n\phi + B_n \cos n\phi) = \sqrt{A_n^2 + B_n^2} \cos\left[n\phi + \tan^{-1}\left(\frac{A_n}{B_n}\right)\right]$$

$$= F_n \cos n\phi \quad (3\text{-}2\text{-}114)$$

Finally, the traveling-wave solution of the Helmholtz equation is reduced to

$$\Psi = \Psi_0 J_n(k_c r)\cos n\phi\, e^{j(\omega t - \beta_g z)} \quad (3\text{-}2\text{-}115)$$

(2) TE Modes in Cylindrical Waveguides. It is commonly assumed that the waves in a circular waveguide are propagating in the positive z direction. Figure 3-2-10 shows the coordinates of a circular guide.

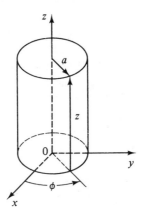

Fig. 3-2-10. Coordinates of a circular waveguide.

The TE_{np} modes in the circular guide are characterized by $E_z = 0$. This means that the z component of the magnetic field H_z must exist in the guide in order to have electromagnetic energy transmission. A Helmholtz equation for H_z in a circular guide is given by

$$\nabla^2 H_z = \gamma^2 H_z \quad (3\text{-}2\text{-}116)$$

Its solution is given in Eq. (3-2-115) by

$$H_z = H_{0z} J_n(k_c r) \cos n\phi\, e^{j(\omega t - \beta_g z)} \quad (3\text{-}2\text{-}117)$$

which is subject to the given boundary conditions.

For a lossless dielectric, Maxwell's curl equations in frequency domain are given by

$$\nabla \times \mathbf{E} = -j\omega\mu\mathbf{H} \quad (3\text{-}2\text{-}118)$$

$$\nabla \times \mathbf{H} = j\omega\epsilon\mathbf{E} \tag{3-2-119}$$

In cylindrical coordinates, their components are expressed as

$$\frac{1}{r}\frac{\partial E_z}{\partial \phi} - \frac{\partial E_\phi}{\partial z} = -j\omega\mu H_r \tag{3-2-120}$$

$$\frac{\partial E_r}{\partial z} - \frac{\partial E_z}{\partial r} = -j\omega\mu H_\phi \tag{3-2-121}$$

$$\frac{1}{r}\frac{\partial}{\partial r}(rE_\phi) - \frac{1}{r}\frac{\partial E_r}{\partial \phi} = -j\omega\mu H_z \tag{3-2-122}$$

$$\frac{1}{r}\frac{\partial H_z}{\partial \phi} - \frac{\partial H_\phi}{\partial z} = j\omega\epsilon E_r \tag{3-2-123}$$

$$-j\beta_g H_r - \frac{\partial H_z}{\partial r} = j\omega\epsilon E_\phi \tag{3-2-124}$$

$$\frac{1}{r}\frac{\partial}{\partial r}(rH_\phi) - \frac{1}{r}\frac{\partial H_r}{\partial \phi} = j\omega\epsilon E_z \tag{3-2-125}$$

When the differentiation $\partial/\partial z$ is replaced by $(-j\beta_g)$ and the z component of electric field E_z by zero, the TE-mode equations in terms of H_z in a circular waveguide are expressed as

$$E_r = -\frac{j\omega\mu}{k_c^2}\frac{1}{r}\frac{\partial H_z}{\partial \phi} \tag{3-2-126}$$

$$E_\phi = \frac{j\omega\mu}{k_c^2}\frac{\partial H_z}{\partial r} \tag{3-2-127}$$

$$E_z = 0 \tag{3-2-128}$$

$$H_r = \frac{-j\beta_g}{k_g^2}\frac{\partial H_z}{\partial r} \tag{3-2-129}$$

$$H_\phi = \frac{-j\beta_g}{k_c^2}\frac{1}{r}\frac{\partial H_z}{\partial \phi} \tag{3-2-130}$$

$$H_z = H_{0z}J_n(k_c r)\cos n\phi\, e^{j(\omega t - \beta_g z)} \tag{3-2-131}$$

where $k_c^2 = \omega^2\mu\epsilon - \beta_g^2$ has been replaced.

The boundary conditions require that the ϕ component of the electric field E_ϕ, which is tangential to the inner surface of the circular waveguide at $r = a$, must vanish or that the r component of the magnetic field H_r, which is normal to the inner surface of $r = a$, must vanish. Consequently,

$$E_\phi = 0 \text{ at } r = a, \quad \therefore \ \frac{\partial H_z}{\partial r}\bigg|_{r=a} = 0$$

or

$$H_r = 0 \text{ at } r = a, \quad \therefore \ \frac{\partial H_z}{\partial r}\bigg|_{r=a} = 0$$

This requirement is equivalent to that expressed in Eq. (3-2-117).

$$\frac{\partial H_z}{\partial r}\bigg|_{r=a} = H_{0z}J_n'(k_c a)\cos n\phi\, e^{j(\omega t - \beta_g z)} = 0 \tag{3-2-132}$$

Hence

$$J'_n(k_c a) = 0 \qquad (3\text{-}2\text{-}133)$$

where J'_n indicates the derivative of J_n.

Since the J_n are oscillatory functions, the $J'_n(k_c a)$ are also oscillatory functions. An infinite sequence of values of $(k_c a)$ satisfies Eq. (3-2-132). These points, the roots of Eq. (3-2-132), correspond to the maxima and minima of the curves $J'_n(k_c a)$, as shown in Fig. 3-2-8. Table 3-2-2 tabulates a few roots of $J'_n(k_c a)$ for some lower-order n.

TABLE 3-2-2. pth Zeros of $J'_n(k_c a)$ for TE$_{np}$ modes.

n p	0	1	2	3	4	5
1	3.832	1.841	3.054	4.201	5.317	6.416
2	7.016	5.331	6.706	8.015	9.282	10.520
3	10.173	8.536	9.969	11.346	12.682	13.987
4	13.324	11.706	13.170			

The permissible values of k_c can be written

$$k_c = \frac{X'_{np}}{a} \qquad (3\text{-}2\text{-}134)$$

Substitution of Eq. (3-2-117) in Eqs. (3-2-126) through (3-2-131) yields the complete field equations of the TE$_{np}$ modes in circular waveguides:

$$E_r = E_{0r} J_n\left(\frac{X'_{np} r}{a}\right) \sin n\phi\, e^{j(\omega t - \beta_g z)} \qquad (3\text{-}2\text{-}135)$$

$$E_\phi = E_{0\phi} J'_n\left(\frac{X'_{np} r}{a}\right) \cos n\phi\, e^{j(\omega t - \beta_g z)} \qquad (3\text{-}2\text{-}136)$$

$$E_z = 0 \qquad (3\text{-}2\text{-}137)$$

$$H_r = -\frac{E_{0\phi}}{Z_g} J'_n\left(\frac{X'_{np} r}{a}\right) \cos n\phi\, e^{j(\omega t - \beta_g z)} \qquad (3\text{-}2\text{-}138)$$

$$H_\phi = \frac{E_{0r}}{Z_g} J_n\left(\frac{X'_{np} r}{a}\right) \sin n\phi\, e^{j(\omega t - \beta_g z)} \qquad (3\text{-}2\text{-}139)$$

$$H_z = H_{0z} J_n\left(\frac{X'_{np} r}{a}\right) \cos n\phi\, e^{j(\omega t - \beta_g z)} \qquad (3\text{-}2\text{-}140)$$

where $Z_g = E_r/H_\phi = -E_\phi/H_r$ has been replaced for the wave impedance in the guide

$n = 0, 1, 2, 3, \ldots$

$p = 1, 2, 3, 4, \ldots$

The first subscript n represents the number of full cycles of field variation in one revolution through 2π radians of ϕ. The second subscript p indicates the number of zeros of E_ϕ—that is, $J'_n(X'_{np}r/a)$ along the radial of a guide, but the zero on the axis is excluded if it exists.

The mode propagation constant is determined by Eqs. (3-2-126) through (3-2-131) and (3-2-134).

$$\beta_g = \sqrt{\omega^2 \mu\epsilon - \left(\frac{X'_{np}}{a}\right)^2} \qquad (3\text{-}2\text{-}141)$$

The cutoff wave number of a mode is that for which the mode propagation constant vanishes. Hence

$$k_c = \frac{X'_{np}}{a} = \omega_c\sqrt{\mu\epsilon} \qquad (3\text{-}2\text{-}142)$$

The cutoff frequency for TE modes in a circular guide is then given by

$$f_c = \frac{X'_{np}}{2\pi a\sqrt{\mu\epsilon}} \qquad (3\text{-}2\text{-}143)$$

and the phase velocity for TE modes is

$$v_g = \frac{\omega}{\beta_g} = \frac{v_p}{\sqrt{1 - (f_c/f)^2}} \qquad (3\text{-}2\text{-}144)$$

where $v_p = \dfrac{1}{\sqrt{\mu\epsilon}} = \dfrac{c}{\sqrt{\mu_r\epsilon_r}}$ is the phase velocity in an unbounded dielectric.

The wavelength and wave impedance for TE modes in a circular guide are given, respectively, by

$$\lambda_g = \frac{\lambda}{\sqrt{1 - (f_c/f)^2}} \qquad (3\text{-}2\text{-}145)$$

and
$$Z_g = \frac{\omega\mu}{\beta_g} = \frac{\eta}{\sqrt{1 - (f_c/f)^2}} \qquad (3\text{-}2\text{-}146)$$

where $\lambda = \dfrac{v_p}{f} = $ the wavelength in an unbounded dielectric

$\eta = \sqrt{\dfrac{\mu}{\epsilon}} = $ the intrinsic impedance in an unbounded dielectric

EXAMPLE 3-2-3: TE Mode in Cylindrical Waveguide. A TE_{11} mode is propagating through a circular waveguide. The radius of the guide is 5 cm, and the guide contains an air dielectric (Refer to Fig. 3-2-11).
(a) Determine the cutoff frequency.
(b) Determine the wavelength λ_g in the guide for an operating frequency of 3 GHz.
(c) Determine the wave impedance Z_g in the guide.

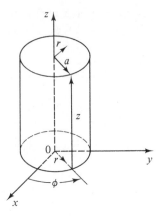

Fig. 3-2-11. Diagram for Example 3-2-3.

Solution.

(a) From Table 3-2-11 for TE$_{11}$ mode, $n = 1$, $p = 1$, $X'_{11} = 1.841 = k_c a$. The cutoff wave number is

$$k_c = \frac{1.841}{a} = \frac{1.841}{5 \times 10^{-2}} = 36.82$$

The cutoff frequency is

$$f_c = \frac{k_c}{2\pi\sqrt{\mu_0\epsilon_0}} = \frac{(36.82)(3 \times 10^8)}{2\pi} = 1.758 \times 10^9 \text{ Hz}$$

(b) The phase constant in the guide is

$$\beta_g = \sqrt{\omega^2\mu_0\epsilon_0 - k_c^2}$$
$$= \sqrt{(2\pi \times 3 \times 10^9)^2(4\pi \times 10^{-7} \times 8.85 \times 10^{-12}) - (36.82)^2}$$
$$= 50.9 \quad \text{radians/m}$$

The wavelength in the guide is

$$\lambda_g = \frac{2\pi}{\beta_g} = \frac{6.28}{50.9} = 12.3 \text{ cm}$$

(c) The wave impedance in the guide is

$$Z_g = \frac{\omega\mu_0}{\beta_g} = \frac{(2\pi \times 3 \times 10^9)(4\pi \times 10^{-7})}{50.9} = 465 \quad \text{ohms}$$

(3) TM Modes in Cylindrical Waveguides. The TM$_{np}$ modes in a circular guide are characterized by $H_z = 0$. However, the z component of the electric field E_z must exist in order to have energy transmission in the guide. Consequently, the Helmholtz equation for E_z in a cylindrical waveguide is given by

$$\nabla^2 E_z = \gamma^2 E_z \tag{3-2-147}$$

Its solution is given in Eq. (3-2-115) by

$$E_z = E_{0z}J_n(k_c r) \cos n\phi e^{j(\omega t - \beta_g z)} \tag{3-2-148}$$

which is subject to the given boundary conditions.

The boundary condition requires that the tangential component of electric field E_z at $r = a$ vanish. Consequently,

$$J_n(k_c a) = 0 \tag{3-2-149}$$

Since $J_n(k_c r)$ are oscillatory functions, as shown in Fig. 3-2-8, there are infinite numbers of roots of $J_n(k_c r)$. Table 3-2-3 tabulates a few of them for some lower-order n.

TABLE 3-2-3. pth Zeros of $J_n(k_c r)$ for TM$_{np}$ modes.

p \\ n	0	1	2	3	4	5
1	2.405	3.832	5.136	6.380	7.588	8.771
2	5.520	7.106	8.417	9.761	11.065	12.339
3	8.645	10.173	11.620	13.015	14.372	
4	11.792	13.324	14.796			

For $H_z = 0$ and $\partial/\partial z = -j\beta_g$, the field equations in the circular guide, after expanding $\nabla \times \mathbf{E} = -j\omega\mu\mathbf{H}$ and $\nabla \times \mathbf{H} = j\omega\epsilon\mathbf{E}$, are given by

$$E_r = \frac{-j\beta_g}{k_c^2} \frac{\partial E_z}{\partial r} \tag{3-2-150}$$

$$E_\phi = \frac{-j\beta_g}{k_c^2} \frac{1}{r} \frac{\partial E_z}{\partial \phi} \tag{3-2-151}$$

$$E_z = \text{Eq. (3-2-148)} \tag{3-2-152}$$

$$H_r = \frac{j\omega\epsilon}{k_c^2} \frac{1}{r} \frac{\partial E_z}{\partial \phi} \tag{3-2-153}$$

$$H_\phi = -\frac{j\omega\epsilon}{k_c^2} \frac{\partial E_z}{\partial r} \tag{3-2-154}$$

$$H_z = 0 \tag{3-2-155}$$

where $k_c^2 = \omega^2\mu\epsilon - \beta_g^2$ has been replaced.

Differentiation of Eq. (3-2-148) with respect to z and substitution of the result in Eqs. (3-2-150) through (3-2-155) yield the field equations of TM$_{np}$ modes in a circular waveguide:

$$E_r = E_{0r}J_n'\left(\frac{X_{np}r}{a}\right) \cos n\phi e^{j(\omega t - \beta_g z)} \tag{3-2-156}$$

$$E_\phi = E_{0\phi} J_n\left(\frac{X_{np} r}{a}\right) \sin n\phi e^{j(\omega t - \beta_g z)} \tag{3-2-157}$$

$$E_z = E_{0z} J_n\left(\frac{X_{np} r}{a}\right) \cos n\phi e^{j(\omega t - \beta_g z)} \tag{3-2-158}$$

$$H_r = \frac{E_{0\phi}}{Z_g} J_n\left(\frac{X_{np} r}{a}\right) \sin n\phi e^{j(\omega t - \beta_g z)} \tag{3-2-159}$$

$$H_\phi = \frac{E_{0r}}{Z_g} J_n'\left(\frac{X_{np} r}{a}\right) \cos n\phi e^{j(\omega t - \beta_g z)} \tag{3-2-160}$$

$$H_z = 0 \tag{3-2-161}$$

where $Z_g = \dfrac{E_r}{H_\phi} = -\dfrac{E_\phi}{H_r} = \dfrac{\beta_g}{\omega\epsilon}$ and $k_c = \dfrac{X_{np}}{a}$ have been replaced

$n = 0, 1, 2, 3, \ldots$
$p = 1, 2, 3, 4, \ldots$

Some of the TM-mode characteristic equations in the circular guide are identical to those of the TE mode, but some are different. For convenience, all are shown here.

$$\beta_g = \sqrt{\omega^2 \mu\epsilon - \left(\frac{X_{np}}{a}\right)^2} \tag{3-2-162}$$

$$k_c = \frac{X_{np}}{a} = \omega_c \sqrt{\mu\epsilon} \tag{3-2-163}$$

$$f_c = \frac{X_{np}}{2\pi a \sqrt{\mu\epsilon}} \tag{3-2-164}$$

$$v_g = \frac{\omega}{\beta_g} = \frac{v_p}{\sqrt{1 - (f_c/f)^2}} \tag{3-2-165}$$

$$\lambda_g = \frac{\lambda}{\sqrt{1 - (f_c/f)^2}} \tag{3-2-166}$$

$$Z_g = \frac{\beta_g}{\omega\epsilon} = \eta\sqrt{1 - \left(\frac{f_c}{f}\right)^2} \tag{3-2-167}$$

It should be noted that the dominant mode, or the mode of lowest cutoff frequency in a circular waveguide, is the mode of TE_{11} that has the smallest value of the product, $k_c a = 1.841$, as shown in Tables 3-2-2 and 3-2-3.

EXAMPLE 3-2-4: Wave Propagation in Cylindrical Waveguide. An air-filled circular waveguide has a radius of 2 cm and is to carry energy at a frequency of 10 GHz. Find all the TE_{np} and TM_{np} modes for which energy transmission is possible.

Solution. Since the physical dimension of the guide and the frequency of the wave remain constant, the product of $(k_c a)$ is also constant. Thus

$$k_c a = (\omega_0 \sqrt{\mu_0 \epsilon_0}) a = \frac{2\pi \times 10^{10}}{3 \times 10^8} (2 \times 10^{-2}) = 4.18$$

Any mode having a product of $(k_c a)$ less than or equal to 4.18 will propagate the wave with a frequency of 10 GHz. This is

$$k_c a \leqq 4.18$$

The possible modes are

$TE_{11}(1.841)$	$TM_{01}(2.405)$
$TE_{21}(3.054)$	$TM_{11}(3.832)$
$TE_{01}(3.832)$	

(4) TEM Modes in Cylindrical Waveguides. The transverse electric and transverse magnetic (TEM) modes or transmission-line modes are characterized by

$$E_z = H_z = 0$$

This means that the electric and magnetic fields are completely transverse to the direction of wave propagation. This mode cannot exist in hollow waveguides, since it requires two conductors, such as the coaxial transmission line and two-open wires.

Analysis of the TEM mode illustrates an excellent analogous relationship between the method of circuit theory and that of the field theory. Figure 3-2-12 shows a coaxial line.

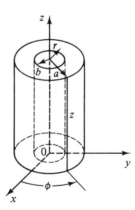

Fig. 3-2-12. Coordinates of a coaxial line.

Maxwell's curl equations in cylindrical coordinates

$$\nabla \times \mathbf{E} = -j\omega\mu\mathbf{H} \tag{3-2-168}$$

$$\nabla \times \mathbf{H} = j\omega\epsilon\mathbf{E} \tag{3-2-169}$$

will become

$$B_g E_r = \omega\mu H_\phi \tag{3-2-170}$$

$$B_g E_\phi = \omega\mu H_r \tag{3-2-171}$$

$$\frac{\partial}{\partial r}(rE_\phi) - \frac{\partial E_r}{\partial \phi} = 0 \tag{3-2-172}$$

$$\beta_g H_r = -\omega \epsilon E_\phi \qquad (3\text{-}2\text{-}173)$$

$$\beta_g H_\phi = \omega \epsilon E_r \qquad (3\text{-}2\text{-}174)$$

$$\frac{\partial}{\partial r}(rH_\phi) - \frac{\partial H_r}{\partial \phi} = 0 \qquad (3\text{-}2\text{-}175)$$

where $\partial/\partial r = -j\beta_g$ and $E_z = H_z = 0$ are replaced.

Substitution of Eq. (3-2-171) in (3-2-173) yields the propagation constant of the TEM mode in a coaxial line.

$$\beta_g = \omega\sqrt{\mu\epsilon} \qquad (3\text{-}2\text{-}176)$$

which is the phase constant of the wave in a lossless transmission line with a dielectric.

In comparing the preceding equation with the characteristic equation of the Helmholtz equation in cylindrical coordinates as given in Eq. (3-2-111) by

$$\beta_g^2 = \omega^2\mu\epsilon - k_c^2 \qquad (3\text{-}2\text{-}177)$$

it is evident that

$$k_c = 0 \qquad (3\text{-}2\text{-}178)$$

This means that the cutoff frequency of the TEM mode in a coaxial line is zero, which is the same as in the ordinary transmission lines.

The phase velocity of the TEM mode can be expressed from Eq. (3-2-176) as

$$v_p = \frac{\omega}{\beta_g} = \frac{1}{\sqrt{\mu\epsilon}} \qquad (3\text{-}2\text{-}179)$$

which is the velocity of light in an unbounded dielectric.

The wave impedance of the TEM mode is found from either Eqs. (3-2-170) and (3-2-173) or Eqs. (3-2-171) and (3-2-174) as

$$\eta(\text{TEM}) = \sqrt{\frac{\mu}{\epsilon}} \qquad (3\text{-}2\text{-}180)$$

which is the wave impedance of a lossless transmission line in a dielectric.

Ampère's law states that the line integral of **H** about any closed path is exactly equal to the current enclosed by that path. This is

$$\oint \mathbf{H} \cdot d\boldsymbol{\ell} = I = I_0 e^{j(\omega t - \beta_g z)} = 2\pi r H_\phi \qquad (3\text{-}2\text{-}181)$$

where I is the complex current that must be supported by the center conductor of a coaxial line. This clearly demonstrates that the TEM mode can only exist in the two-conductor system—not in the hollow waveguide because the center conductor does not exist.

In summary, the properties of TEM modes in a lossless medium are as follows.

1. Its cutoff frequency is zero.
2. Its transmission line is a two-conductor system.
3. Its wave impedance is the impedance in an unbounded dielectric.
4. Its propoagation constant is the constant in an unbounded dielectric.
5. Its phase velocity is the velocity of light in an unbounded dielectric.

(5) *Power Transmission in Circular Guides or Coaxial Lines.* In general, the power transmitted through circular waveguides and coaxial lines can be calculated by means of the complex Poynting theorem described in Section 2-2. For a lossless dielectric, the time-average power transmitted through a circular guide can be given by

$$P_{tr} = \frac{1}{2Z_g} \int_0^{2\pi} \int_0^a [|E_r|^2 + |E_\phi|^2] r \, dr \, d\phi \qquad (3\text{-}2\text{-}182)$$

$$P_{tr} = \frac{Z_g}{2} \int_0^{2\pi} \int_0^a [|H_r|^2 + |H_\phi|^2] r \, dr \, d\phi \qquad (3\text{-}2\text{-}183)$$

where $Z_g = \dfrac{E_r}{H_\phi} = -\dfrac{E_\phi}{H_r} =$ the wave impedance in the guide

$a =$ radius of the circular guide

Substitution of Z_g for a particular mode in Eq. (3-2-182) yields the power transmitted by that mode through the guide.

For TE_{np} modes, the average power transmitted through a circular guide is given by

$$P_{tr} = \frac{\sqrt{1 - (f_c/f)^2}}{2\eta} \int_0^{2\pi} \int_0^a [|E_r|^2 + |E_\phi|^2] r \, dr \, d\phi \qquad (3\text{-}2\text{-}184)$$

where $\eta = \sqrt{\mu/\epsilon}$ is the intrinsic impedance in an unbounded dielectric.

For TM_{np} modes, the average power transmitted through a circular guide is given by

$$P_{tr} = \frac{1}{2\eta\sqrt{1 - (f_c/f)^2}} \int_0^{2\pi} \int_0^a [|E_r|^2 + |E_\phi|^2] r \, dr \, d\phi \qquad (3\text{-}2\text{-}185)$$

For TEM modes in coaxial lines, the average power transmitted through a coaxial line or two open-wire lines is given by

$$P_{tr} = \frac{1}{2\eta} \int_0^{2\pi} \int_0^a [|E_r|^2 + |E_\phi|^2] r \, dr \, d\phi \qquad (3\text{-}2\text{-}186)$$

If the current carried by the center conductor of a coaxial line is assumed to be

$$I_z = I_0 e^{j(\omega t - \beta_g z)} \qquad (3\text{-}2\text{-}187)$$

the magnetic intensity induced by the current around the center conductor is given by Ampère's law as

$$H_\phi = \frac{I_0}{2\pi r} e^{j(\omega t - \beta_g z)} \tag{3-2-188}$$

The potential rise from the outer conductor to the center conductor is given by

$$V_r = -\int_b^a E_r \, dr = -\int_b^a \eta H_\phi \, dr = \frac{I_0 \eta}{2\pi} \ell n \left(\frac{b}{a}\right) e^{j(\omega t - \beta_g z)} \tag{3-2-189}$$

The characteristic impedance of a coaxial line is

$$Z_0 = \frac{V}{I} = \frac{\eta}{2\pi} \ell n \left(\frac{b}{a}\right) \tag{3-2-190}$$

where $\eta = \sqrt{\mu/\epsilon}$ is the intrinsic impedance in an unbounded dielectric.

The power transmitted by TEM modes in a coaxial line can be expressed from Eq. (3-2-186) as

$$P_{tr} = \frac{1}{2\eta} \int_0^{2\pi} \int_a^b |\eta H_\phi|^2 \, r \, dr \, d\phi = \frac{\eta I_0^2}{4\pi} \ell n \left(\frac{b}{a}\right) \tag{3-2-191}$$

Substitution of $|V_r|$ from Eq. (3-2-189) into Eq. (3-2-191) yields

$$P_{tr} = \tfrac{1}{2} V_0 I_0 \tag{3-2-192}$$

This shows that the power transmission derived from the Poynting theory is the same as from the circuit theory for an ordinary transmission line.

(6) *Power Losses in Cylindrical Waveguides or Coaxial Lines.* The theory and equations derived in Section 3-2-1 for TE and TE modes in rectangular waveguides are applicable to TE and TM modes in circular guides. The power losses for the TEM mode in coaxial lines can be computed from transmission-line theory by means of

$$P_L = 2\alpha P_{tr} \tag{3-2-193}$$

where P_L = power loss per unit length
P_{tr} = transmitted power
α = attenuation constant

For a low-loss conductor, the attenuation constant is given by

$$\alpha = \frac{1}{2}\left(R\sqrt{\frac{C}{L}} + G\sqrt{\frac{L}{C}}\right) \tag{3-2-194}$$

(7) *Excitations of Modes in Cylindrical Waveguides.* As described earlier, TE modes have no z component of an electric field and TM modes have no z component of magnetic intensity. If a device is inserted in a circular guide in such a way that it excites only a z component of electric intensity, the wave propagating through the guide will be the TM mode; on the other hand,

if a device is placed in a circular guide in such a manner that only the z component of magnetic intensity exists, the traveling wave will be the TE mode. The methods of excitation for various modes in circular waveguides are shown in Fig. 3-2-13.

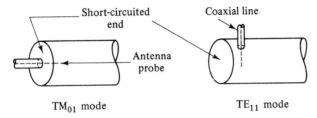

TM$_{01}$ mode   TE$_{11}$ mode

Fig. 3-2-13. Methods of exciting various modes in circular waveguides.

A common way to excite TM modes in a circular guide is by a coaxial line as shown in Fig. 3-2-14. At the end of the coaxial line a large magnetic intensity exists in the ϕ direction of wave propagation. The magnetic field from the coaxial line will excite the TM modes in the guide. However, when the guide is connected to the source by a coaxial line, a discontinuity problem at the junction will increase the standing-wave ratio on the line and eventually decrease the power transmission. It is often necessary to place a turning device around the junction in order to suppress the reflection.

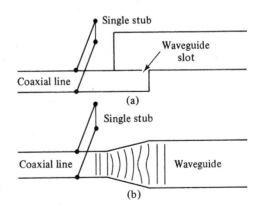

Fig. 3-2-14. Methods of exciting TM modes in a circular waveguide.

3-3 Summary

In this chapter transmission lines and microwave waveguides were described. These passive microwave components are very useful for high-power transmission.

Line impedance

$$\mathbf{Z} = \mathbf{Z}_0 \frac{\mathbf{Z}_\ell + j\mathbf{Z}_0 \tan \beta d}{\mathbf{Z}_0 + j\mathbf{Z}_\ell \tan \beta d} \qquad \text{for lossless line}$$

Reflection coefficient

$$\Gamma_\ell = \frac{\mathbf{Z}_\ell - \mathbf{Z}_0}{\mathbf{Z}_\ell + \mathbf{Z}_0}$$

Transmission coefficient

$$\mathbf{T} = \frac{2\mathbf{Z}_\ell}{\mathbf{Z}_\ell + \mathbf{Z}_0}$$

$$|\mathbf{T}|^2 = \frac{\mathbf{Z}_\ell}{\mathbf{Z}_0}(1 - |\Gamma_\ell|^2)$$

Standing-wave ratio

$$\rho = \frac{1 + |\Gamma|}{1 - |\Gamma|}$$

Characteristic equations of TE_{mn} and TM_{mn} modes in rectangular waveguides

$$f_c = \frac{1}{2\sqrt{\mu\epsilon}}\sqrt{\frac{m^2}{a^2} + \frac{n^2}{b^2}} \qquad \text{for } TE_{mn} \text{ and } TM_{mn} \text{ modes}$$

$$\beta_g = \omega\sqrt{\mu\epsilon}\sqrt{1 - \left(\frac{f_c}{f}\right)^2} \qquad \text{for } TE_{mn} \text{ and } TM_{mn} \text{ modes}$$

$$\lambda_g = \frac{\lambda}{\sqrt{1 - (f_c/f)^2}} \qquad \text{for } TE_{mn} \text{ and } TM_{mn} \text{ modes}$$

$$v_g = \frac{v_p}{\sqrt{1 - (f_c/f)^2}} \qquad \text{for } TE_{mn} \text{ and } TM_{mn} \text{ modes}$$

$$Z_g = \frac{\beta_g}{\omega\epsilon} = \eta\sqrt{1 - \left(\frac{f_c}{f}\right)^2} \qquad \text{for } TM_{mn} \text{ mode}$$

$$Z_g = \frac{\omega\mu}{\beta_g} = \frac{\eta}{\sqrt{1 - (f_c/f)^2}} \qquad \text{for } TE_{mn} \text{ mode}$$

Characteristic equations for TE_{np} and TM_{np} modes in cylindrical waveguides

$$\beta_g = \sqrt{\omega^2\mu\epsilon - \left(\frac{X'_{np}}{a}\right)^2} \qquad \text{for } TE_{np} \text{ mode}$$

$$\beta_g = \sqrt{\omega^2\mu\epsilon - \left(\frac{X_{np}}{a}\right)^2} \qquad \text{for } TM_{np} \text{ mode}$$

$$k_c = \frac{X'_{np}}{a} = \omega_c\sqrt{\mu\epsilon} \qquad \text{for } TE_{np} \text{ mode}$$

$$k_c = \frac{X_{np}}{a} = \omega_c\sqrt{\mu\epsilon} \qquad \text{for } TM_{np} \text{ mode}$$

$$f_c = \frac{X'_{np}}{2\pi a\sqrt{\mu\epsilon}} \qquad \text{for } TE_{np} \text{ mode}$$

$$f_c = \frac{X_{np}}{2\pi a \sqrt{\mu \epsilon}} \qquad \text{for TM}_{np} \text{ mode}$$

$$v_g = \frac{\omega}{\beta_g} = \frac{v_p}{\sqrt{1 - (f_c/f)^2}} \qquad \text{for TE}_{np} \text{ and TM}_{np} \text{ modes}$$

$$\lambda_g = \frac{\lambda}{\sqrt{1 - (f_c/f)^2}} \qquad \text{for TE}_{np} \text{ and TM}_{np} \text{ modes}$$

$$Z_g = \frac{\omega \mu}{\beta_g} = \frac{\eta}{\sqrt{1 - (f_c/f)^2}} \qquad \text{for TE}_{np} \text{ mode}$$

$$Z_g = \frac{\beta_g}{\omega \epsilon} = \eta \sqrt{1 - \left(\frac{f_c}{f}\right)^2} \qquad \text{for TM}_{np} \text{ mode}$$

REFERENCES

[1] SMITH, PHILIP H.,
Transmission Line Calculator, Electronics, Vol. 12, pp. 29–31, 1939.
An Improved Transmission Line Calculator, Electronics, Vol. 17, pp. 130–133 and 318–325, 1944.
Smith Charts—Their Development and Use, A series published at intervals by the Kay Electric Co.; No. 1 is dated March 1962, and No. 9 is dated December 1966.

SUGGESTED READINGS

1. ATWATER, H. A., *Introduction to Microwave Theory*. McGraw-Hill Book Company, New York, 1962.

2. BROWN, ROBERT G., et al., *Lines*, *Waves*, and *Antennas*, 2nd ed. Ronald Press, New York, 1973.

3. MAGNUSSON, PHILIP C., *Transmission Lines and Wave Propagation*, 2nd ed. Allyn and Bacon, Boston, Mass., 1970.

4. SOUTHWORTH, G. C., *Principles and Applications of Waveguide Transmission*. D. Van Nostrand, New York, 1950.

Microwave Components

4-0 Introduction

In Chapter 3 transmission lines and microwave waveguides were analyzed in some detail. These passive elements are often used in microwave systems. Here microwave components, such as cavity resonators, reentrant cavities, slow-wave structures, waveguide Tees, directional couplers, circulators, and isolators, are discussed. Such devices are part of the active microwave devices themselves. For instance, the cavity resonators are used in klystrons, the reentrant cavities in magnetrons, and the slow-wave structures in traveling-wave tubes (TWTs). Consequently, this discussion is an important prelude to the topic of microwave tubes in Chapter 5.

In general, a cavity resonator is a metallic enclosure that confines the electromagnetic energy.

The stored electric and magnetic energies inside the cavity determine its equivalent inductance and capacitance. The energy dissipated by the finite conductivity of the cavity walls determines its equivalent resistance. In practice, the rectangular-cavity resonator, cylindrical-cavity resonator, and reentrant-cavity resonator are commonly used as shown in Figs. 4-1-1 and 4-1-2.

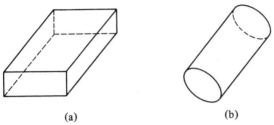

(a) (b)

Fig. 4-1-1. Resonators. (a) Rectangular cavity. (b) Cylindrical cavity.

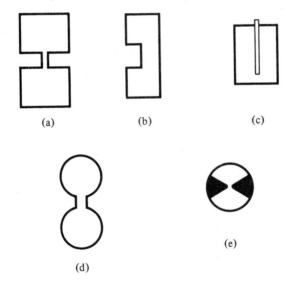

(a) (b) (c)

(d) (e)

Fig. 4-1-2. Reentrant cavities. (a) Coaxial cavity. (b) Radial cavity. (c) Tunable cavity. (d) Toroidal cavity. (e) Butterfly cavity.

Theoretically a given resonator has an infinite number of resonant modes, and each mode corresponds to a definite resonant frequency. When the frequency of an impressed signal is equal to a resonant frequency, a maximum amplitude of the standing wave occurs, and the peak energies stored in the electric and magnetic fields are equal. The mode having the lowest resonant frequency is known as the *dominant mode*.

As the operating frequency is increased, both the inductance and capacitance of the resonant circuit must be decreased in order to maintain resonance at the operating frequency. Because the gain-bandwidth product is limited by the resonant circuit, the ordinary resonator cannot generate a large output. Several nonresonant periodic circuits or slow-wave structures (see Fig. 4-1-3) are designed for a large gain over a wide bandwidth.

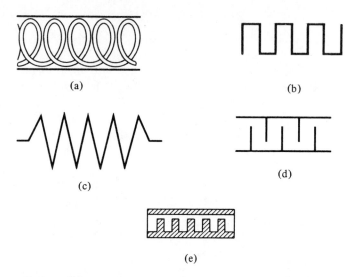

Fig. 4-1-3. Slow-wave structures. (a) Helical line. (b) Folded-back line. (c) Zigzag line. (d) Interdigital line. (e) Corrugated waveguide.

4-1   Rectangular-Cavity Resonator

The electromagnetic field inside the cavity should satisfy Maxwell's equations, subject to the boundary conditions that the electric field tangential to and the magnetic field normal to the metal walls must vanish. The geometry of a rectangular cavity is illustrated in Fig. 4-1-4.

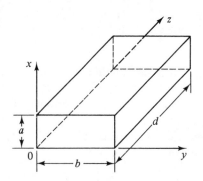

Fig. 4-1-4. Coordinates of a rectangular cavity.

The wave equations in the rectangular resonator should satisfy the boundary condition of the zero tangential **E** at four of the walls. It is merely necessary to choose the harmonic functions in z to satisfy this condition at the remaining two end-walls. These functions can be found if

$$H_z = H_{0z} \cos \frac{m\pi x}{a} \cos \frac{n\pi x}{b} \cos \frac{p\pi z}{d} \qquad (\text{TE}_{mnp}) \qquad (4\text{-}1\text{-}1)$$

where $m = 0, 1, 2, 3, \ldots$ represents the number of the half-wave periodicity in the x direction,

$\quad\quad n = 0, 1, 2, 3, \ldots$ represents the number of the half-wave periodicity in the y direction

$\quad\quad p = 1, 2, 3, 4, \ldots$ represents the number of the half-wave periodicity in the z direction

and

$$E_z = E_{0z} \sin \frac{m\pi x}{a} \sin \frac{n\pi x}{b} \cos \frac{p\pi z}{d} \qquad (\text{TM}_{mnp}) \qquad (4\text{-}1\text{-}2)$$

where $m = 1, 2, 3, 4, \ldots$

$\quad\quad n = 1, 2, 3, 4, \ldots$

$\quad\quad p = 0, 1, 2, 3, \ldots$

The separation equation for both TE and TM modes is given by

$$k^2 = \left(\frac{m\pi}{a}\right)^2 + \left(\frac{n\pi}{b}\right)^2 + \left(\frac{p\pi}{d}\right)^2 \qquad (4\text{-}1\text{-}3)$$

For a lossless dielectric, $k^2 = \omega^2 \mu \epsilon$; therefore the resonant frequency is expressed by

$$f_r = \frac{1}{2\sqrt{\mu\epsilon}} \sqrt{\left(\frac{m}{a}\right)^2 + \left(\frac{n}{b}\right)^2 + \left(\frac{p}{d}\right)^2} \qquad (\text{TE}_{mnp}, \text{TM}_{mnp}) \quad (4\text{-}1\text{-}4)$$

For $a < b < d$, the dominant mode is the TE_{011} mode.

In general, a straight-wire probe inserted at the position of maximum electric intensity is used to excite a desired mode, and the loop coupling placed at the position of maximum magnetic intensity is utilized to launch a specific mode. Figure 4-1-5 shows the methods of excitation for the rectangular resonator. The maximum amplitude of the standing wave occurs when the frequency of the impressed signal is equal to the resonant frequency.

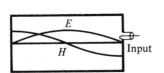 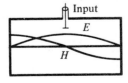

Fig. 4-1-5. Methods of exciting wave modes in a resonator.

4-2 Circular-Cavity Resonator and Semicircular-Cavity Resonator

4-2-1. Circular-Cavity Resonator

A circular-cavity resonator is a circular waveguide with two ends closed by a metal wall. (See Fig. 4-2-1.) The wave function in the circular resonator should satisfy Maxwell's equations, subject to the same boundary conditions

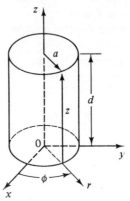

Fig. 4-2-1. Coordinates of a circular resonator.

described for a rectangular-cavity resonator. It is merely necessary to choose the harmonic functions in z to satisfy the boundary conditions at the remaining two end-walls. These can be achieved if

$$H_z = H_{0z} J_n\left(\frac{X'_{np}r}{a}\right) \cos n\phi \, \sin\left(\frac{q\pi z}{d}\right) \qquad (\text{TE}_{npq}) \qquad (4\text{-}2\text{-}1)$$

where $n = 0, 1, 2, 3, \ldots$ is the number of the periodicity in the ϕ direction

$p = 1, 2, 3, 4, \ldots$ is the number of zeros of the field in the radial direction

$q = 1, 2, 3, 4, \ldots$ is the number of the half wave in the axial direction

J_n = Bessell's function of the first kind,

H_{0z} = the amplitude of the magnetic field

and

$$E_z = E_{0z} J_n\left(\frac{X_{np}r}{a}\right) \cos n\phi \, \cos\left(\frac{q\pi z}{d}\right) \qquad (\text{TM}_{npq}) \qquad (4\text{-}2\text{-}2)$$

where $n = 0, 1, 2, 3, \ldots$

$p = 1, 2, 3, 4, \ldots$

$q = 0, 1, 2, 3, \ldots$

E_{0z} = the amplitude of the electric field

The separation equations for TE and TM modes are given by

$$k^2 = \left(\frac{X'_{np}}{a}\right)^2 + \left(\frac{q\pi}{d}\right)^2 \qquad \text{(Te mode)} \qquad \text{(4-2-3a)}$$

$$k^2 = \left(\frac{X_{np}}{a}\right)^2 + \left(\frac{q\pi}{d}\right)^2 \qquad \text{(TM mode)} \qquad \text{(4-2-3b)}$$

Substitution of $k^2 = \omega^2 \mu \epsilon$ in Eqs. (4-2-3) yields the resonant frequencies for TE and TM modes, respectively, as

$$f_r = \frac{1}{2\pi\sqrt{\mu\epsilon}} \sqrt{\left(\frac{X'_{np}}{a}\right)^2 + \left(\frac{q\pi}{d}\right)^2} \qquad \text{(TE)} \qquad \text{(4-2-4a)}$$

$$f_r = \frac{1}{2\pi\sqrt{\mu\epsilon}} \sqrt{\left(\frac{X_{np}}{a}\right)^2 + \left(\frac{q\pi}{d}\right)^2} \qquad \text{(TM)} \qquad \text{(4-2-4b)}$$

It is interesting to note that, for $2a > d$, the TM_{010} mode is dominant and that, for $d \geq 2a$, the TE_{111} mode is dominant.

4-2-2. Semicircular-Cavity Resonator

A semicircular-cavity resonator is shown in Fig. 4-2-2. The wave function of the TE_{npq} mode in the semicircular resonator can be written

$$H_z = H_{0z} J_n\left(\frac{X'_{np}r}{a}\right) \cos n\phi \sin\left(\frac{q\pi}{d}\right) \qquad \text{(TE mode)} \qquad \text{(4-2-5)}$$

where $n = 0, 1, 2, 3, \ldots$
$p = 1, 2, 3, 4, \ldots$
$q = 1, 2, 3, 4, \ldots$
$a = $ radius of the semicircular-cavity resonator
$d = $ length of the resonator

The wave function of the TM_{npq} mode in the semicircular-cavity resonator can be written

$$E_z = E_{0z} J_n\left(\frac{X_{np}r}{a}\right) \sin n\phi \cos\left(\frac{q\pi}{d}z\right) \qquad \text{(TM mode)} \qquad \text{(4-2-6)}$$

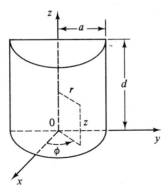

Fig. 4-2-2. Semicircular resonator.

where $n = 1, 2, 3, 4, \ldots$
$p = 1, 2, 3, 4, \ldots$
$q = 0, 1, 2, 3, \ldots$

With the separation equations as given in Eqs. (4-2-3), the equations of resonant frequency for TE and TM modes in a semicircular-cavity resonator are the same as in the circular-cavity resonator. They are repeated as follows:

$$f_r = \frac{1}{2\pi a\sqrt{\mu\epsilon}} \sqrt{(X'_{np})^2 + \left(\frac{q\pi a}{d}\right)^2} \qquad (\text{TE}_{npq} \text{ mode}) \qquad (4\text{-}2\text{-}7)$$

$$f_r = \frac{1}{2\pi a\sqrt{\mu\epsilon}} \sqrt{(X_{np})^2 + \left(\frac{q\pi a}{d}\right)^2} \qquad (\text{TM}_{npq} \text{ mode}) \qquad (4\text{-}2\text{-}8)$$

However, the values of the subscripts n, p, and q differ from those for the circular-cavity resonator. Also, it must be emphasized that the TE_{111} mode is dominant if $d > a$ and that the TM_{110} mode is dominant if $d < a$.

4-3   The Q of a Cavity Resonator

The quality factor Q is a measure of the frequency selectivity of a resonant or antiresonant circuit, and it is defined as

$$Q \equiv 2\pi \frac{\text{maximum energy stored}}{\text{energy dissipated per cycle}} = \frac{\omega W}{P} \qquad (4\text{-}3\text{-}1)$$

where $W =$ the maximum stored energy
$P =$ the average power loss

At resonant frequency, the electric and magnetic energies are equal and in time quadrature. When the electric energy is maximum, the magnetic energy is zero and vice versa. The total energy stored in the resonator is obtained by integrating the energy density over the volume of the resonator:

$$W_e = \int_v \frac{\epsilon}{2} |E|^2 \, dv = W_m = \int_v \frac{\mu}{2} |H|^2 \, dv = W \qquad (4\text{-}3\text{-}2)$$

where $|E|$ and $|H|$ are the peak values of the field intensities.

The average power loss in the resonator can be evaluated by integrating the power density as given in Eq. (2-5-12) over the inner surface of the resonator. Hence

$$P = \frac{R_s}{2} \int_s |H_t|^2 \, da \qquad (4\text{-}3\text{-}3)$$

where H_t is the peak value of the tangential magnetic intensity and R_s is the surface resistance of the resonator.

Substitution of Eqs. (4-3-2) and (4-3-3) in Eq. (4-3-1) yields

$$Q = \frac{\omega\mu \int_v |H|^2 \, dv}{R_s \int_s |H_t|^2 \, da} \tag{4-3-4}$$

Since the peak value of the magnetic intensity is related to its tangential and normal components by

$$|H|^2 = |H_t|^2 + |H_n|^2$$

where H_n is the peak value of the normal magnetic intensity, the value of $|H_t|^2$ at the resonator walls is approximately twice the value of $|H|^2$ averaged over the volume. So the Q of a cavity resonator as shown in Eq. (4-3-4) can be expressed approximately by

$$Q = \frac{\omega\mu \text{ (volume)}}{2R_s \text{ (surface areas)}} \tag{4-3-5}$$

An unloaded resonator can be represented by either a series or a parallel resonant circuit. The resonant frequency and the unloaded Q_0 of a cavity resonator are

$$f_0 = \frac{1}{2\pi\sqrt{LC}} \tag{4-3-6}$$

and

$$Q_0 = \frac{\omega_0 L}{R} \tag{4-3-7}$$

If the cavity is coupled by means of an ideal $N : 1$ transformer and a series inductance L_s to a generator having internal impedance Z_g, then the coupling circuit and its equivalent are as shown in Fig. 4-3-1. The loaded Q_ℓ of the system is given by

$$Q_\ell = \frac{\omega_0 L}{R + N^2 Z_g} \quad \text{for } |N^2 L_s| \ll |R + N^2 Z_g| \tag{4-3-8}$$

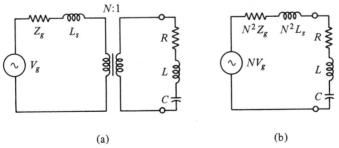

Fig. 4-3-1. Cavity coupled to a generator. (a) Coupling circuit. (b) Equivalent circuit.

The coupling coefficient of the system is defined as

$$K \equiv \frac{N^2 Z_g}{R} \tag{4-3-9}$$

and the loaded Q_ℓ would become

$$Q_\ell = \frac{\omega_0 L}{R(1 + K)} = \frac{Q_0}{1 + K} \tag{4-3-10}$$

Rearrangement of Eq. (4-3-10) yields

$$\frac{1}{Q_\ell} = \frac{1}{Q_0} + \frac{1}{Q_{ext}} \tag{4-3-11}$$

where $Q_{ext} = \frac{Q_0}{K} = \frac{\omega_0 L}{KR}$ is the external Q.

There are three types of coupling coefficients:

1. Critical coupling: If the resonator is matched to the generator, then

$$K = 1 \tag{4-3-12}$$

The loaded Q_ℓ is given by

$$Q_\ell = \tfrac{1}{2} Q_{ext} = \tfrac{1}{2} Q_0 \tag{4-3-13}$$

2. Undercoupling: If $K < 1$, the cavity terminals are at a voltage maximum in the input line at resonance. The normalized impedance at the voltage maximum is the standing-wave ratio ρ. That is,

$$K = \rho \tag{4-3-14}$$

The loaded Q_ℓ is given by

$$Q_\ell = \frac{Q_0}{1 + \rho} \tag{4-3-15}$$

3. Overcoupling: If $K > 1$, the cavity terminals are at a voltage minimum and the input terminal impedance is equal to the reciprocal of the standing-wave ratio. That is,

$$K = \frac{1}{\rho} \tag{4-3-16}$$

The loaded Q_ℓ is given by

$$Q_\ell = \frac{\rho}{\rho + 1} Q_0 \tag{4-3-17}$$

The relationship of the coupling coefficient K and the standing-wave ratio is shown in Fig. 4-3-2.

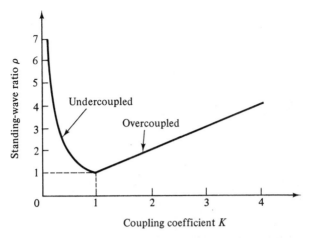

Fig. 4-3-2 Coupling coefficient versus standing-wave ratio.

4-4 Reentrant Cavities

At a frequency well below the microwave range, the cavity resonator can be represented by a lumped-constant resonant circuit. When the operating frequency is increased to several tens of megahertz, both the inductance and the capacitance must be reduced to a minimum in order to maintain resonance at the operating frequency. Ultimately the inductance is reduced to a minimum by short wire. Therefore the reentrant cavities are designed for use in klystrons and microwave triodes. A reentrant cavity is one in which the metallic boundaries extend into the interior of the cavity. Several types of reentrant cavities are shown in Fig. 4-1-2. One of the commonly used reentrant cavities is the coaxial cavity shown in Fig. 4-4-1.

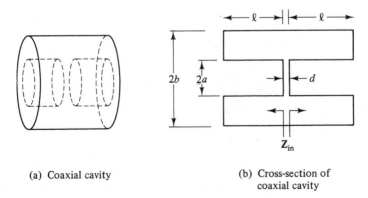

(a) Coaxial cavity

(b) Cross-section of coaxial cavity

Fig. 4-4-1. Coaxial cavity and its equivalent.

It is clear from the figure that not only has the inductance been considerably decreased but the resistance losses are markedly reduced as well, and the shelf-shielding enclosure prevents radiation losses. It is difficult to calculate the resonant frequency of a coaxial cavity. An approximation can be made, however, using transmission-line theory. The characteristic impedance of a coaxial line is given by

$$Z_0 = \frac{1}{2\pi} \sqrt{\frac{\mu}{\epsilon}} \ln \frac{b}{a} \qquad \text{ohms} \qquad (4\text{-}4\text{-}1)$$

The coaxial cavity is similar to a coaxial line shorted at two ends and joined at the center by a capacitor. The input impedance to each shorted coaxial line is given by

$$Z_{\text{in}} = jZ_0 \tan \beta\ell \qquad (4\text{-}4\text{-}2)$$

where ℓ is the length of the coaxial line.

Substitution of Eq. (4-4-1) in (4-4-2) results in

$$Z_{\text{in}} = j\frac{1}{2\pi} \sqrt{\frac{\mu}{\epsilon}} \ln \frac{b}{a} \tan \beta\ell \qquad (4\text{-}4\text{-}3)$$

The inductance of the cavity is given by

$$L = \frac{2X_{\text{in}}}{\omega} = \frac{1}{\pi\omega} \sqrt{\frac{\mu}{\epsilon}} \ln \frac{b}{a} \tan \beta\ell \qquad (4\text{-}4\text{-}4)$$

and the capacitance of the gap by

$$C_g = \frac{\epsilon\pi a^2}{d} \qquad (4\text{-}4\text{-}5)$$

At resonance the inductive reactance of the two shorted coaxial lines in series is equal in magnitude to the capacitive reactance of the gap. That is, $\omega L = 1/(\omega C_g)$. So

$$\tan \beta\ell = \frac{dv}{\omega a^2 \ln (b/a)} \qquad (4\text{-}4\text{-}6)$$

where $v = 1/\sqrt{\mu\epsilon}$ is the phase velocity in any medium.

The solution to this equation gives the resonant frequency of a coaxial cavity. Since Eq. (4-4-6) contains the tangent function, it has an infinite number of solutions with larger values of frequency. Therefore this type of reentrant cavity can support an infinite number of resonant frequencies or modes of oscillations. It can be shown that a shorted coaxial-line cavity stores more magnetic energy than electric energy. The balance of the electric stored energy appears in the gap, for at resonance the magnetic and electric stored energies are equal.

The radial reentrant cavity shown in Fig. 4-4-2 is also a commonly used reentrant resonator.

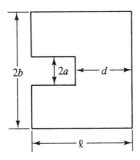

Fig. 4-4-2. Radial reentrant cavity.

The inductance and capacitance [1] of a radial reentrant cavity is expressed by

$$L = \frac{\mu \ell}{2\pi} \ell n \frac{b}{a} \qquad (4\text{-}4\text{-}7)$$

$$C = \epsilon_0 \left(\frac{\pi a^2}{d} - 4a \, \ell n \frac{0.765}{\sqrt{\ell^2 + (b-a)^2}} \right) \qquad (4\text{-}4\text{-}8)$$

The resonant frequency [2] is given by

$$f_r = \frac{c}{2\pi \sqrt{\epsilon_r}} \left[a\ell \left(\frac{a}{2d} - \frac{2}{\ell} \ell n \frac{0.765}{\sqrt{\ell^2 + (b-a)^2}} \right) \ell n \frac{b}{a} \right]^{-1/2} \qquad (4\text{-}4\text{-}9)$$

where $c = 3 \times 10^8$ m/s is the velocity of light in a vacuum.

4-5   Slow-Wave Structures

Slow-wave structures are special circuits that are used in microwave tubes to reduce the wave velocity in a certain direction so that the electron beam and the signal wave can interact. The phase velocity of a wave in ordinary waveguides is greater than the velocity of light in a vacuum. In the operation of traveling-wave and magnetron-type devices, the electron beam must keep in step with the microwave signal. Since the electron beam can be accelerated only to velocities that are about a fraction of the velocity of light, a slow-wave structure must be incorporated in the microwave devices in order that the phase velocity of the microwave signal can keep pace with that of the electron beam for effective interactions. Several types of slow-wave structures are shown in Fig. 4-1-3. The commonly used slow-wave structure is a helical coil with a concentric conducting cylinder (see Fig. 4-5-1).

It can be shown that the ratio of the phase velocity v_p along the pitch to the phase velocity along the coil is given by

$$\frac{v_p}{c} = \frac{p}{\sqrt{p^2 + (\pi d)^2}} = \sin \psi \qquad (4\text{-}5\text{-}1)$$

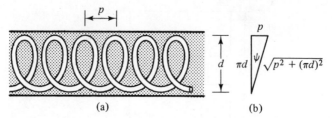

Fig. 4-5-1. Helical slow-wave structure. (a) Helical coil. (b) One turn of helix.

where $c = 3 \times 10^8$ m/s is the velocity of light in free space

p = the helix pitch

d = the diameter of the helix

ψ = the pitch angle

In general, the helical coil may be within a dielectric-filled cylinder. The phase velocity in the axial direction is expressed as

$$v_{p\epsilon} = \frac{p}{\sqrt{\mu\epsilon[p^2 + (\pi d)^2]}} \qquad (4\text{-}5\text{-}2)$$

If the dielectric constant is too large, however, the slow-wave structure may introduce sufficient loss to the microwave devices, thereby reducing their efficiency. For a very small pitch angle, the phase velocity along the coil in free space is approximately represented by

$$v_p \simeq \frac{pc}{\pi d} = \frac{\omega}{\beta} \qquad (4\text{-}5\text{-}3)$$

Figure 4-5-2 shows the ω-β (or Brillouin) diagram for a helical slow-wave structure. The helix ω-β diagram is very useful in designing a helix slow-wave structure. Once β is found, v_p can be computed from Eq. (4-5-3) for a given dimension of the helix. Furthermore, the group velocity of the wave is merely the slope of the curve as given by

$$v_{gr} = \frac{\partial \omega}{\partial \beta} \qquad (4\text{-}5\text{-}4)$$

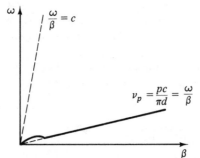

Fig. 4-5-2. ω-β diagram for a helix structure.

In order for a circuit to be a slow-wave structure, it must have the property of periodicity in the axial direction. The phase velocity of some of the spatial harmonics in the axial direction obtained by Fourier analysis of the waveguide field may be smaller than the velocity of light. In the helical slow-wave structure a translation back or forth through a distance of one pitch length results in identically the same structure again. Thus the period of a helical slow-wave structure is its pitch.

In general, the field of the slow-wave structure must be distributed according to Floquet's theorem for periodic boundaries. Floquet's periodicity theorem states that

The steady-state solutions for the electromagnetic fields of a single propagating mode in a periodic structure have the property that fields in adjacent cells are related by a complex constant.

Mathematically, the theorem can be stated

$$E(x, y, z - L) = E(x, y, z)e^{j\beta_0 L} \qquad (4\text{-}5\text{-}5)$$

where $E(x, y, z)$ is a periodic function of z with period L. β_0 is the phase constant in the axial direction. Therefore in a slow-wave structure β_0 is the phase constant of average electron velocity.

It is postulated that the solution to Maxwell's equations in a periodic structure can be written

$$E(x, y, z) = f(x, y, z)e^{-j\beta_0 z} \qquad (4\text{-}5\text{-}6)$$

where $f(x, y, z)$ is a periodic function of z with period L that is the period of the slow-wave structure.

For a periodic structure, Eq. (4-5-6) can be rewritten with z replaced by $z - L$.

$$E(x, y, z - L) = f(x, y, z - L)e^{-j\beta_0(z-L)} \qquad (4\text{-}5\text{-}7)$$

Since $f(x, y, z - L)$ is a periodic function with period L, then

$$f(x, y, z - L) = f(x, y, z) \qquad (4\text{-}5\text{-}8)$$

Substitution of Eq. (4-5-8) in (4-5-7) results in

$$E(x, y, z - L) = f(x, y, z)e^{-j\beta_0 z}e^{j\beta_0 L} \qquad (4\text{-}5\text{-}9)$$

and substitution of Eq. (4-5-6) in (4-5-9) gives

$$E(x, y, z - L) = E(x, y, z)e^{j\beta_0 L} \qquad (4\text{-}5\text{-}10)$$

This expression is the mathematical statement of Floquet's theorem, Eq. (4-5-5). Therefore Eq. (4-5-6) does indeed satisfy Floquet's theorem.

From the theory of Fourier series, any function that is periodic, single valued, finite, and continuous may be represented by a Fourier series. Hence the field distribution function $E(x, y, z)$ may be expanded into a Fourier series of fundamental period L as

$$E(x, y, z) = \sum_{n=-\infty}^{\infty} E_n(x, y)e^{-j(2\pi n/L)z}e^{-j\beta_0 z} = \sum_{n=-\infty}^{\infty} E_n(x, y)e^{-j\beta_n z} \quad (4\text{-}5\text{-}11)$$

where

$$E_n(x, y) = \frac{1}{L} \int_0^L E(x, y, z)e^{j(2\pi n/L)z} \, dz \quad (4\text{-}5\text{-}12)$$

are the amplitudes of n harmonics and

$$\beta_n = \beta_0 + \frac{2\pi n}{L} \quad (4\text{-}5\text{-}13)$$

is the phase constant of the nth modes, where $n = -\infty, \ldots, -2, -1, 0, 1, 2, 3, \ldots, \infty$.

The quantities $E_n(x, y)e^{-j\beta_n z}$ are known as spatial harmonics by analogy with time-domain Fourier series. The question is whether Eq. (4-5-11) can satisfy the wave equation, Eq. (2-1-20). Substitution of Eq. (4-5-11) into the wave equation results in

$$\nabla^2 \left[\sum_{n=-\infty}^{\infty} E_n(x, y)e^{-j\beta_n z} \right] - \gamma^2 \left[\sum_{n=-\infty}^{\infty} E_n(x, y)e^{-j\beta_n z} \right] = 0 \quad (4\text{-}5\text{-}14)$$

Since the wave equation is linear, Eq. (4-5-14) can be rewritten as

$$\sum_{n=-\infty}^{\infty} [\nabla^2 E_n(x, y)e^{-j\beta_n z} - \gamma^2 E_n(x, y)e^{-j\beta_n z}] = 0 \quad (4\text{-}5\text{-}15)$$

It is evident from the preceding equation that if each spatial harmonic is itself a solution of the wave equation for each value of n, the summation of space harmonics also satisfies the wave equation of Eq. (4-5-14). This means that only the complete solution of Eq. (4-5-14) can satisfy the boundary conditions of a periodic structure.

Furthermore, Eq. (4-5-11) shows that the field in a periodic structure can be expanded as an infinite series of waves, all at the same frequency but with different phase velocities v_{pn}. That is,

$$v_{pn} = \frac{\omega}{\beta_n} \equiv \frac{\omega}{\beta_0 + (2\pi n/L)} \quad (4\text{-}5\text{-}16)$$

The group velocity v_{gr}, defined by $v_{gr} = \dfrac{\partial \omega}{\partial \beta}$, is then given as

$$v_{gr} = \left\{ \frac{d[\beta_0 + (2\pi n/L)]}{d\omega} \right\}^{-1} = \frac{\partial \omega}{\partial \beta_0} \quad (4\text{-}5\text{-}17)$$

which is independent of n.

It is important to note that the phase velocity v_{pn} in the axial direction decreases for higher values of positive n and β_0. So it appears possible for a microwave of suitable n to have a phase velocity less than the velocity of light. It follows that interactions between the electron beam and microwave signal are possible and thus the amplification of active microwave devices can be achieved.

Figure 4-5-3 shows the ω-β (or Brillouin) diagram for a helix with several spatial harmonics. This ω-β diagram demonstrates some important properties needing more explanations. First, the second quadrant of the ω-β diagram indicates the negative phase velocity that corresponds to the negative n. This means that the electron beam moves in the positive z direction while the beam velocity coincides with the negative spatial harmonic's phase velocity. This type of tube is called a *backward-wave oscillator*. Secondly, the shaded areas are the forbidden regions for propagation. This situation occurs because if the axial phase velocity of any spatial harmonic exceeds the velocity of light, the structure radiates energy. This property has been verified by experiments [3].

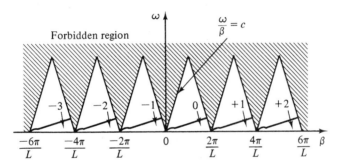

Fig. 4-5-3. ω-β diagram of spatial harmonics for helical structure.

4-6 Microwave Hybrid Circuits

A microwave circuit ordinarily consists of several microwave devices connected in some way to achieve the desired transmission of microwave signal. The interconnection of two or more microwave devices may be regarded as a microwave junction. Commonly used microwave junctions include such waveguide Tees as E-plane Tee, H-plane Tee, magic Tee, hybrid ring (rat-race circuit), directional coupler, and circulator. This section describes these microwave hybrids, which are shown in Fig. 4-6-1.

From network theory a two-port device (Fig. 4-6-2) can be described by a number of parameter sets, such as the H, Y, and Z parameters.

H parameters

$$V_1 = h_{11}I_1 + h_{12}V_2 \tag{4-6-1}$$

$$I_2 = h_{21}I_1 + h_{22}V_2 \tag{4-6-2}$$

Y parameters

$$I_1 = y_{11}V_1 + y_{12}V_2 \tag{4-6-3}$$

$$I_2 = y_{21}V_1 + y_{22}V_2 \tag{4-6-4}$$

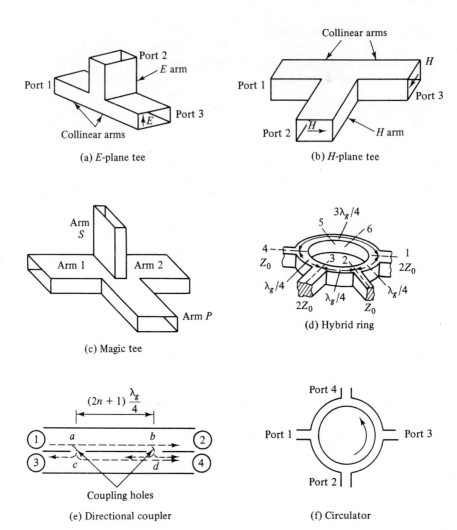

Port 2
E arm

Port 1

Port 3

Collinear arms

(a) *E*-plane tee

Collinear arms

H

Port 1

Port 3

Port 2

H arm

(b) *H*-plane tee

Arm
S

Arm 1

Arm 2

Arm *P*

(c) Magic tee

$3\lambda_g/4$

5

6

4 --

Z_0

3

2

1

$2Z_0$

$\lambda_g/4$

$2Z_0$

$\lambda_g/4$

Z_0

$\lambda_g/4$

(d) Hybrid ring

$(2n+1)\dfrac{\lambda_g}{4}$

1

a

b

2

3

c

d

4

Coupling holes

(e) Directional coupler

Port 4

Port 1

Port 3

Port 2

(f) Circulator

Fig. 4-6-1. Microwave hybrids. (a) *E*-plane tee. (b) *H*-plane tee. (c) Magic tee. (d) Hybrid ring. (e) Directional coupler. (f) Circulator.

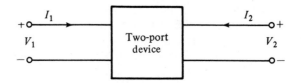

I_1

$+$

V_1

Two-port
device

I_2

$+$

V_2

$-$

$-$

Fig. 4-6-2. Two-port network.

Z parameters
$$V_1 = z_{11}I_1 + z_{12}I_2 \qquad (4\text{-}6\text{-}5)$$
$$V_2 = z_{21}I_1 + z_{22}I_2 \qquad (4\text{-}6\text{-}6)$$

All these network parameters relate total voltages and total currents at each of the two ports. For instance,

$$h_{11} = \frac{V_1}{I_1}\bigg|_{V_2=0} \qquad \text{(short circuit)} \qquad (4\text{-}6\text{-}7)$$

$$h_{12} = \frac{V_1}{V_2}\bigg|_{I_1=0} \qquad \text{(open circuit)} \qquad (4\text{-}6\text{-}8)$$

If the frequencies are in the microwave range, however, the H, Y and Z parameters cannot be measured because

1. equipment is not readily available to measure total voltage and total current at the ports of the network.
2. short and open circuits are difficult to achieve over a broad band of frequencies.
3. active devices, such as power transistors and tunnel diodes, frequently will not be short or open circuit stable.

Consequently, some new method of characterization is needed to overcome these problems. The logical variables to use at the microwave frequencies are traveling waves rather than total voltages and total currents. These are the S parameters, which are expressed as

$$b_1 = S_{11}a_1 + S_{12}a_2$$
$$b_2 = S_{21}a_1 + S_{22}a_2 \qquad (4\text{-}6\text{-}9)$$

Figure 4-6-3 shows the S parameters of a two-port network.

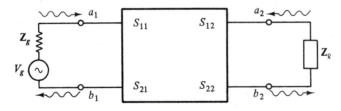

Fig. 4-6-3. Two-port network.

4-7 *S* Parameters

This section presents the S-parameter theory and its general properties that are applicable to any n-port devices in microwave-range frequencies. A microwave junction may have n ports, each of which is a lossless uniform transmission line (see Fig. 4-7-1).

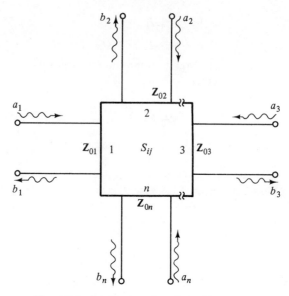

Fig. 4-7-1. A microwave junction with n-ports.

4-7-1. S-Parameter Theory

As shown in Fig. 4-7-1, a_j is the incident traveling wave coming toward the junction, and b_i is the reflected traveling wave coming outward from the junction. From transmission-line theory, the incident and reflected waves are related by

$$b_i = \sum_j^n S_{ij} a_j \qquad \text{for } i = 1, 2, 3, \ldots, n \qquad (4\text{-}7\text{-}1)$$

where $S_{ij} = \Gamma_{ij}$, is the reflection coefficient of the ith port if $i = j$ with all other ports matched

$S_{ij} = T_{ij}$, is the forward transmission coefficient of the jth port if $i > j$ with all other ports matched

$S_{ij} = T_{ij}$, is the reverse transmission coefficient of the jth port if $i < j$ with all other ports matched

In general, Eq. (4-7-1) can be written

$$b_2 = S_{11}a_1 + S_{12}a_2 + S_{13}a_3 + \cdots + S_{1n}a_n \qquad (4\text{-}7\text{-}2)$$
$$b_2 = S_{21}a_1 + S_{22}a_2 + S_{23}a_3 + \cdots + S_{2n}a_n$$
$$\cdots\cdots\cdots\cdots\cdots\cdots\cdots\cdots\cdots\cdots\cdots\cdots$$
$$b_n = S_{n1}a_1 + S_{n2}a_2 + S_{n3}a_3 + \cdots + S_{nn}a_n$$

In matrix notation, boldface roman letters are used to represent matrix quantities; Eq. (4-7-2) is then written

$$\mathbf{b} = \mathbf{Sa} \qquad (4\text{-}7\text{-}3)$$

where both b and a are column matrices that are usually written

$$\mathbf{b} = \begin{bmatrix} b_1 \\ b_2 \\ \cdot \\ \cdot \\ \cdot \\ b_n \end{bmatrix} \quad \text{and} \quad \mathbf{a} = \begin{bmatrix} a_1 \\ a_2 \\ \cdot \\ \cdot \\ \cdot \\ a_n \end{bmatrix} \tag{4-7-4}$$

The $n \times n$ matrix **S** is called the *scattering matrix*:

$$\mathbf{S} = \begin{bmatrix} S_{11} & S_{12} & S_{13} & \cdots & S_{1n} \\ S_{21} & S_{22} & S_{23} & \cdots & S_{2n} \\ \cdot & & & & \cdot \\ \cdot & & & & \cdot \\ \cdot & & & & \\ S_{n1} & S_{n2} & S_{n3} & \cdots & S_{nn} \end{bmatrix} \tag{4-7-5}$$

The coefficients $S_{11}, S_{12}, \ldots S_{nn}$ are called the *scattering parameters* (S parameters) or scattering coefficients. As a corollary to the S parameters, n-port voltages are linearly related to n-port currents by the impedance matrix of the junction. That is,

$$V_i = \sum_j^n Z_{ij} I_j \quad \text{for } i = 1, 2, 3, \ldots, n \tag{4-7-6}$$

In matrix form, Eq. (4-7-6) can be written

$$\begin{bmatrix} V_1 \\ V_2 \\ \cdot \\ \cdot \\ \cdot \\ V_n \end{bmatrix} = \begin{bmatrix} Z_{11} & Z_{12} & \cdots & Z_{1n} \\ Z_{21} & Z_{22} & \cdots & Z_{2n} \\ \multicolumn{4}{c}{\cdots\cdots\cdots\cdots\cdots} \\ Z_{n1} & Z_{n2} & \cdots & Z_{nn} \end{bmatrix} \begin{bmatrix} I_1 \\ I_2 \\ \cdot \\ \cdot \\ \cdot \\ I_n \end{bmatrix} \tag{4-7-7}$$

Symbolically,

$$\mathbf{V} = \mathbf{ZI} \tag{4-7-8}$$

4-7-2. Properties of S Parameters

Several properties of S parameters are described below.

PROPERTY 1: Symmetry of S parameters. If a microwave junction satisfies a reciprocity condition or if there are no vacuum tubes, transistors, and so on at the junction, the junction is a linear passive circuit and the S parameters are equal to their corresponding transposes. That is,

$$\mathbf{S} = \tilde{\mathbf{S}} \tag{4-7-9}$$

where $\tilde{\mathbf{S}} = S_{ji} = \mathbf{S} = S_{ij}$. $\tilde{S}$ is the transpose of matrix **S**.

The steady-state total voltage and current at the kth port are

$$V_k = V_k^+ + V_k^- \tag{4-7-10}$$

$$I_k = \frac{V_k^+}{Z_{0k}} - \frac{V_k^-}{Z_{0k}} \tag{4-7-11}$$

Therefore the incident and reflected voltages at the kth port are

$$V_k^+ = \tfrac{1}{2}(V_k + Z_{0k}I_k) \tag{4-7-12}$$

$$V_k^- = \tfrac{1}{2}(V_k - Z_{0k}I_k) \tag{4-7-13}$$

The average incident power (complex) of the kth port is

$$\frac{1}{2}V_k I_k^* = \frac{|V_k^+|^2}{2Z_{0k}^*} \tag{4-7-14}$$

The normalized incident and reflected voltages at the kth port can be defined as

$$a_k = \frac{V_k^+}{\sqrt{Z_{0k}}} = \frac{1}{2}\left[\frac{V_k}{\sqrt{Z_{0k}}} + \sqrt{Z_{0k}}I_k\right] \tag{4-7-15}$$

$$b_k = \frac{V_k^-}{\sqrt{Z_{0k}}} = \frac{1}{2}\left[\frac{V_k}{\sqrt{Z_{0k}}} - \sqrt{Z_{0k}}I_k\right] \tag{4-7-16}$$

If the characteristic impedance is also normalized so that $\sqrt{Z_{0k}} = 1$, then

$$V_k = a_k + b_k \tag{4-7-17}$$

$$I_k = a_k - b_k \tag{4-7-18}$$

$$a_k = \tfrac{1}{2}(V_k + I_k) \tag{4-7-19}$$

$$b_k = \tfrac{1}{2}(V_k - I_k) \tag{4-7-20}$$

Since from Eq. (4-7-6)

$$V_k = \sum_j^n Z_{kj}I_j \qquad \text{for } k = 1, 2, 3, \ldots, n \tag{4-7-21}$$

it follows that

$$a_k = \frac{1}{2}\sum_j^n (Z_{kj} + \delta_{kj})I_k \tag{4-7-22}$$

$$b_k = \frac{1}{2}\sum_j^n (Z_{kj} - \delta_{kj})I_k \tag{4-7-23}$$

where δ_{kj} is the *Kronecker delta* and is defined as

$$\delta_{kj} = 1 \qquad \text{if } k = j \tag{4-7-24}$$

$$\delta_{kj} = 0 \qquad \text{if } k \neq j \tag{4-7-25}$$

In matrix notation, Eqs. (4-7-22) and (4-7-23) can be written

$$\mathbf{a} = \tfrac{1}{2}(\mathbf{Z} + [\mathbf{I}])\mathbf{I} \tag{4-7-26}$$

$$\mathbf{b} = \tfrac{1}{2}(\mathbf{Z} - [\mathbf{I}])\mathbf{I} \tag{4-7-27}$$

where **a** and **b** are column matrices and [I] is the identity matrix. Since the impedance matrix **Z** and the identity matrix [I] are square matrices ($n \times n$), the matrix (**Z** − [I]) is surely $n \times n$ and may have an inverse. Thus

$$I = 2(Z + [I])^{-1}a \qquad (4\text{-}7\text{-}28)$$

Therefore Eq. (4-7-27) becomes

$$b = (Z - [I])(Z + [I])^{-1}a \qquad (4\text{-}7\text{-}29)$$

In comparing Eq. (4-7-29) with Eq. (4-7-3), the matrix **S** can be written

$$S = (Z - [I])(Z + [I])^{-1} \qquad (4\text{-}7\text{-}30)$$

Let the matrices **P** and **Q** be so defined that

$$P = Z - [I] \qquad (4\text{-}7\text{-}31)$$

$$Q = Z + [I] \qquad (4\text{-}7\text{-}32)$$

Since the impedance matrix **Z** is symmetric, the matrices **P** and **Q** are also symmetric and commutative; that is,

$$PQ = QP \qquad (4\text{-}7\text{-}33)$$

Multiplying the left- and right-hand sides of **PQ** and **QP** by Q^{-1} yields

$$Q^{-1}PQQ^{-1} = Q^{-1}QPQ^{-1} \qquad (4\text{-}7\text{-}34)$$

Then

$$Q^{-1}P = PQ^{-1} = S \qquad (4\text{-}7\text{-}35)$$

The transpose $\tilde{S}$ of **S** is

$$\tilde{S} = \widetilde{Q^{-1}P} = \widetilde{PQ^{-1}} = Q^{-1}P = PQ^{-1} = S \qquad (4\text{-}7\text{-}36)$$

This means that the terms S_{ij} and S_{ji} of the **S** matrix are equal and thus the matrix **S** has a symmetry.

PROPERTY 2: Unity Property. The sum of the products of each term of any one row or of any one column of the matrix **S** multiplied by its complex conjugate is unity; that is,

$$\sum_i^n S_{ij}S_{ij}^* = 1 \qquad \text{for } j = 1, 2, 3, \ldots, n \qquad (4\text{-}7\text{-}37)$$

From the principle of conservation of energy, if the microwave devices are lossless, the power input must be equal to the power output. The incident and reflected waves are related to the incident and reflected voltages by

$$a = \frac{V^+}{\sqrt{Z_0}} \qquad (4\text{-}7\text{-}38)$$

$$b = \frac{V^-}{\sqrt{Z_0}} \qquad (4\text{-}7\text{-}39)$$

It can be seen that

$$\text{Incident power} = P_+ = \tfrac{1}{2}aa^* = \tfrac{1}{2}|a|^2 \qquad (4\text{-}7\text{-}40)$$

$$\text{Reflected power} = P_- = \tfrac{1}{2}bb^* = \tfrac{1}{2}|b|^2 \qquad (4\text{-}7\text{-}41)$$

With no loss of generality, it is assumed that a wave of unit voltage is incident on port 1 of an n-port junction and that no voltage waves enter any other ports. Hence the power input is given by

$$P_{in} = a_1 a_1^* = |a_1|^2 \qquad (4\text{-}7\text{-}42)$$

which is equal to the power output leaving the ith port. That is,

$$P_{in} = a_1 a_1^* = P_{out} = \sum_i^n b_i b_i^* = b_1 b_1^* + b_2 b_2^* + \cdots + b_n b_n^* \qquad (4\text{-}7\text{-}43)$$

Since $b_i = S_{i1} a_1$, then

$$a_1 a_1^* = (S_{11}a_1)(S_{11}a_1)^* + (S_{21}a_1)(S_{21}a_1)^* + \cdots + (S_{n1}a_1)(S_{n1}a_1)^* \qquad (4\text{-}7\text{-}44)$$

Consequently,

$$1 = S_{11}S_{11}^* + S_{21}S_{21}^* + \cdots + S_{n1}S_{n1}^* \qquad (4\text{-}7\text{-}45)$$

or

$$1 = \sum_i^n S_{ij}S_{ij}^* = \sum_i^n |S_{ij}|^2 \qquad \text{for } j = 1, 2, 3, \ldots \qquad (4\text{-}7\text{-}46)$$

Since S_{ij} is symmetric,

$$1 = \sum_i^n S_{ij}S_{ij}^* = \sum_i^n |S_{ij}|^2 \qquad \text{for } i = 1, 2, 3, \ldots \qquad (4\text{-}7\text{-}47)$$

For a lossy junction, the power dissipated at the junction is

$$P_{diss} = \frac{1}{2} \sum_{n=1}^n (a_n a_n^* - b_n b_n^*) \qquad (4\text{-}7\text{-}48)$$

It can be shown that

$$\sum_{n=1}^n a_n a_n^* = \tilde{\mathbf{a}}\mathbf{a}^* \qquad (4\text{-}7\text{-}49)$$

and

$$\sum_{n=1}^n b_n b_n^* = \tilde{\mathbf{a}}\mathbf{S}\mathbf{S}^*\mathbf{a}^* \qquad (4\text{-}7\text{-}50)$$

It should be noted that the right-hand terms of Eqs. (4-7-49) and (4-7-50) are 1×1 matrices or simply numbers. Thus the power dissipated at the junction is given by

$$P_{diss} = \tfrac{1}{2}\tilde{\mathbf{a}}(1 - \mathbf{S}\mathbf{S}^*)\mathbf{a}^* \qquad (4\text{-}7\text{-}51)$$

PROPERTY 3: Zero Property. The sum of the products of each term of any row (or column) multiplied by the complex conjugate of the corresponding terms of any other row (or column) is zero.

$$\sum_i^n S_{ik}S_{ij}^* = 0 \qquad \text{for } k \neq j \qquad \begin{pmatrix} k = 1, 2, 3, \ldots, n \\ j = 1, 2, 3, \ldots, n \end{pmatrix} \qquad (4\text{-}7\text{-}52)$$

In general, the incident and reflected waves may exist at each of the n-ports. Then the incident and the reflected power for a lossless junction are

$$\sum_j^n a_j a_j^* = \sum_i^n b_i b_i^* \qquad (4\text{-}7\text{-}53)$$

Substitution of Eq. (4-7-1) and its complex conjugate in Eq. (4-7-52) yields

$$\sum_j^n a_j a_j^* = \sum_j^n \left(\sum_i^n S_{ij}S_{ij}^*\right)a_j a_j^* + \left[\sum_k' \sum_j \left(\sum_i S_{ik}S_{ij}^*\right)a_k a_j^*\right]$$
$$+ \left[\sum_k' \sum_j \left(\sum_i S_{ik}S_{ij}^*\right)a_k a_j^*\right]^* \qquad (4\text{-}7\text{-}54)$$

where the prime on $\sum'$ indicates that the terms of $k = j$ are not included in the sum. The first term on the right-hand side of Eq. (4-7-53) can be simplified by the use of Eq. (4-7-46). The last two terms on the right are of the form $(A + A^*)$, which is equal to twice the real part of A. Equation (4-7-53) can be simplified to

$$0 = 2 \operatorname{Re} \sum_k' \sum_j \left(\sum_i S_{ij}S_{ij}^*\right)a_k a_j^* \qquad (4\text{-}7\text{-}55)$$

Since the factor of $a_k a_j^*$ is nonzero,

$$\sum_i^n S_{ik}S_{ij}^* = 0 \qquad (4\text{-}7\text{-}56)$$

where $k \neq j$
$k = 1, 2, 3, \ldots, n$
$j = 1, 2, 3, \ldots, n$

For example, if only a_1 and a_2^* exist at port 1 and port 2, respectively, with all other ports terminated in their characteristic impedance, Eq. (4-7-56) becomes

$$S_{11}S_{12}^* + S_{21}S_{22}^* + S_{31}S_{32}^* + \cdots + S_{n1}S_{n2}^* = 0 \qquad (4\text{-}7\text{-}57)$$

Since S_{ij} is symmetric, it can be shown that

$$S_{11}S_{21}^* + S_{12}S_{22}^* + S_{13}S_{23}^* + \cdots + S_{1n}S_{2n}^* = 0 \qquad (4\text{-}7\text{-}58)$$

This has proved the statement of Property 3 of the S parameters.

PROPERTY 4: Phase Shift. If any of the terminal planes (or reference planes), say the kth port, is moved away from the junction by an electric distance $\beta_k \ell_k$, each of the coefficients S_{ij} involving k will be multiplied by the factor $e^{-j\beta_k \ell_k}$.

It is apparent that a change in the specified location of the terminal planes of an arbitrary junction will affect only the phase of the scattering coefficients of the junction. In matrix notation, the new S parameter S' can be written

$$S' = \phi S \phi \tag{4-7-59}$$

where S is the old scattering matrix and

$$\phi = \begin{bmatrix} \phi_{11} & 0 & 0 & \cdots & 0 \\ 0 & \phi_{22} & 0 & \cdots & 0 \\ \cdot & \cdot & \cdot & & \cdot \\ 0 & 0 & 0 & \cdots & \phi_{nn} \end{bmatrix} \tag{4-7-60}$$

where

$$\phi_{11} = \phi_{22} = \phi_{kk} = e^{-j\beta_k \ell_k} \qquad \text{for } k = 1, 2, 3, \ldots, n$$

It should be noted that whereas S_{ii} involves the product of two factors of $e^{-j\beta\ell}$, the new S' will be changed by a factor of $e^{-j2\beta\ell}$.

4-8   Waveguide Tees

As noted, waveguide Tees may consist of the E-plane T, H-plane T, magic T, hybrid rings, corners, bends, and twists. All such waveguide components are discussed in this section.

4-8-1. Tee Junctions

In microwave circuits a waveguide or coaxial-line junction with three independent ports is commonly referred to as a *Tee junction*. From the S parameter theory of a microwave junction it is evident that a T-junction should be characterized by a matrix of third order containing nine elements, six of which should be independent. The characteristics of a three-port junction can be explained by three theorems of the T-junction. These theorems are derived from the equivalent-circuit representation of the T-junction. Their statements follow [4].

1. A short circuit may always be placed in one of the arms of a three-port junction in such a way that no power can be transferred through the other two arms.
2. If the junction is symmetrical about one of its arms, a short circuit can always be placed in that arm so that no reflections occur in power transmission between the other two arms (i.e., the arms present matched impedances).
3. It is impossible for a general three-port junction of arbitrary symmetry to present matched impedances at all three arms.

The E-plane Tee and H-plane Tee are described below.

I. E-Plane T (Series Tee). An E-plane T is a waveguide T in which the axis of its side arm is parallel to the E field of the main guide (see Fig. 4-8-1). If the collinear arms are symmetric about the side arm, there are two

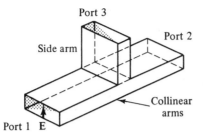

Fig. 4-8-1. *E*-plane T.　Port 1 E

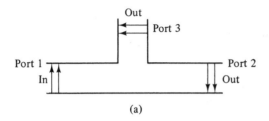

(a)

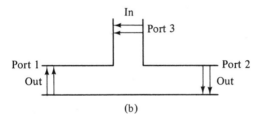

(b)

Fig. 4-8-2. Two-way transmission of *E*-plane T. (a) Input through main arm. (b) Input from side arm.

different transmission characteristics (see Fig. 4-8-2). It can be seen from Fig. 4-8-1 that if the *E*-plane *T* is perfectly matched with the aid of screw tuners or inductive or capacitive windows at the junction, the diagonal components of the scattering matrix, S_{11}, S_{22}, and S_{33} are zero because there will be no reflection. When the waves are fed into the side arm (port 3), the waves appearing at port 1 and port 2 of the collinear arm will be in opposite phase and in the same magnitude. Therefore

$$S_{13} = -S_{23} \qquad (4\text{-}8\text{-}1)$$

It should be noted that Eq. (4-8-1) does not mean that S_{13} is always positive and S_{23} is always negative. The negative sign merely means that S_{13} and S_{23} have opposite signs. For a matched junction, the **S** matrix is given by

$$\mathbf{S} = \begin{bmatrix} 0 & S_{12} & S_{13} \\ S_{21} & 0 & S_{23} \\ S_{31} & S_{32} & 0 \end{bmatrix} \qquad (4\text{-}8\text{-}2)$$

From Property 1 of the **S** matrix in Eq. (4-7-36), the symmetric terms in

Eq. (4-8-2) are equal and they are

$$S_{12} = S_{21}, \qquad S_{13} = S_{31}, \qquad S_{23} = S_{32} \qquad (4\text{-}8\text{-}3)$$

From Property 3 of the S matrix in Eq. (4-7-52), the sum of the products of each term of any column (or row) multiplied by the complex conjugate of the corresponding terms of any other column (or row) is zero and it is

$$S_{11}S_{12}^* + S_{21}S_{22}^* + S_{31}S_{32}^* = 0 \qquad (4\text{-}8\text{-}4)$$

Hence

$$S_{13}S_{23}^* = 0 \qquad (4\text{-}8\text{-}5)$$

This means that either S_{13} or S_{23}^*, or both, should be zero. However, from Property 2 of the S matrix in Eq. (4-7-37), the sum of the products of each term of any one row (or column) multiplied by its complex conjugate is unity; that is,

$$S_{21}S_{21}^* + S_{31}S_{31}^* = 1 \qquad (4\text{-}8\text{-}6)$$

$$S_{12}S_{21}^* + S_{32}S_{32}^* = 1 \qquad (4\text{-}8\text{-}7)$$

$$S_{13}S_{13}^* + S_{23}S_{23}^* = 1 \qquad (4\text{-}8\text{-}8)$$

Substitution of Eq. (4-8-3) in (4-8-6) results in

$$|S_{12}|^2 = 1 - |S_{13}|^2 = 1 - |S_{23}|^2 \qquad (4\text{-}8\text{-}9)$$

Equations (4-8-8) and (4-8-9) are contradictory, for if $S_{13} = 0$, S_{23} is also zero and thus Eq. (4-8-8) is false. In a similar fashion, if $S_{23} = 0$, S_{13} becomes zero and therefore Eq. (4-8-8) is not true. This inconsistency proves the statement of a T-junction in Section 4-8-1 that the junction cannot be matched to the three arms. In other words, the diagonal elements of the S matrix of a T-junction are not all zeros.

In general, when an E-plane T is constructed of an empty waveguide, it is poorly matched at the T-junction. Hence $S_{ij} \neq 0$ if $i = j$. However, since the collinear arm is usually symmetrical about the side arm, $|S_{13}| = |S_{23}|$ and $S_{11} = S_{22}$. Then the S matrix can be simplified to

$$\mathbf{S} = \begin{bmatrix} S_{11} & S_{12} & S_{13} \\ S_{12} & S_{11} & -S_{13} \\ S_{13} & -S_{13} & S_{33} \end{bmatrix} \qquad (4\text{-}8\text{-}10)$$

II. H-Plane T (Shunt Tee). An H-plane T is a waveguide T in which the axis of its side arm is "shunting" the E field or parallel to the H field of the main guide as shown in Fig. 4-8-3.

It can be seen that if two input waves are fed into port 1 and port 2 of the collinear arm, the output wave at port 3 will be in phase and additive. On the other hand, if the input is fed into port 3, the wave will split equally into port 1 and port 2 in phase and in the same magnitude. Therefore the

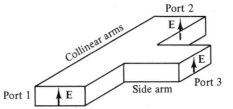

Fig. 4-8-3. *H*-plane T.

S matrix of the *H*-plant T is similar to Eqs. (4-8-2) and (4-8-10) except that

$$S_{13} = S_{23} \qquad (4\text{-}8\text{-}11)$$

4-8-2. Magic Tee (Hybrid T)

A magic T is a combination of the *E*-plane T and *H*-plane T (refer to Fig. 4-8-4). The magic T has several characteristics.

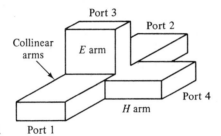

Fig. 4-8-4. Magic T.

1. If two waves of equal magnitude and the same phase are fed into port 1 and port 2, the output will be zero at port 3 and additive at port 4.
2. If a wave is fed into port 4 (the *H* arm), it will be divided equally between port 1 and port 2 of the collinear arms and will not appear at port 3 (the *E* arm).
3. If a wave is fed into port 3 (the *E* arm), it will produce an output of equal magnitude and opposite phase at port 1 and port 2. The output at port 4 is zero. That is, $S_{43} = S_{34} = 0$.
4. If a wave is fed into one of the collinear arms at port 1 or port 2, it will not appear in the other collinear arm at port 2, or port 1 because the *E* arm causes a phase delay while the *H* arm causes a phase advance. That is, $S_{12} = S_{21} = 0$.

Therefore the S matrix of a magic T can be expressed as

$$\mathbf{S} = \begin{bmatrix} 0 & 0 & S_{13} & S_{14} \\ 0 & 0 & S_{23} & S_{24} \\ S_{31} & S_{32} & 0 & 0 \\ S_{41} & S_{42} & 0 & 0 \end{bmatrix} \qquad (4\text{-}8\text{-}12)$$

The magic T is commonly used for mixing, duplexing, and impedance measurements. For example, there are two identical radar transmitters in equipment stock. A particular application requires twice more input power to an antenna than either transmitter can deliver. A magic T may be used to couple the two transmitters to the antenna in such a way that the transmitters do not load each other. The two transmitters should be connected to ports 3 and 4, respectively, as shown in Fig. 4-8-5. Transmitter No. 1, connected to port 3, causes a wave to emanate from port 1 and another to emanate from port 2; these waves are equal in magnitude but opposite in phase. Similarly, transmitter No. 2, connected to port 4, gives rise to a wave at port 1 and another at port 2, both equal in magnitude and in phase. At port 1 the two opposite waves cancel each other. At port 2 the two in-phase waves add together; so double output power at port 2 is obtained for the antenna as shown in Fig. 4-8-5.

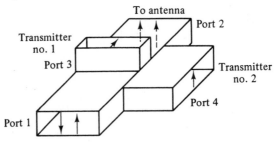

Fig. 4-8-5. Magic T-coupled transmitters to antenna.

4-8-3. Hybrid Rings (Rat-Race Circuits)

A hybrid ring consists of an annular line of proper electrical length to sustain standing waves, to which four arms are connected at proper intervals by means of series or parallel junctions. Figure 4-8-6 shows a hybrid ring with series junctions.

The hybrid ring has similar characteristics as those of the hybrid T. When a wave is fed into port 1, it will not appear at port 3 because the difference of phase shifts for the waves traveling in the clockwise and counterclockwise directions is 180°. Thus the waves are canceled at port 3. For the same reason, the waves fed into port 2 will not emerge at port 4 and so on.

The S matrix for an ideal hybrid ring can be expressed as

$$S = \begin{bmatrix} 0 & S_{12} & 0 & S_{14} \\ S_{21} & 0 & S_{23} & 0 \\ 0 & S_{32} & 0 & S_{34} \\ S_{41} & 0 & S_{43} & 0 \end{bmatrix} \tag{4-8-13}$$

It should be noted that the phase cancellation occurs only at a designed frequency for an ideal hybrid ring. In actual hybrid rings there are small

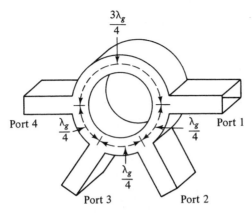

Fig. 4-8-6. Hybrid ring.

leakage couplings, and therefore the zero elements in the matrix Eq. (4-8-13) are not quite equal to zero.

4-8-4. Waveguide Corners, Bends, and Twists

The waveguide corner, bend, and twist are shown in Fig. 4-8-7. These waveguide components are normally used to change the direction of the guide through an arbitrary angle.

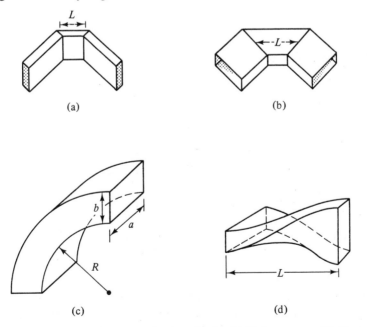

Fig. 4-8-7. Waveguide corner, bend, and twist. (a) *E*-plane corner. (b) *H*-plane corner. (c) Bend. (d) Continuous twist.

In order to minimize reflections from the discontinuities, it is desirable to have the mean length L between continuities equal to an odd number of quarter wavelengths. That is,

$$L = (2n + 1)\frac{\lambda_g}{4} \qquad (4\text{-}8\text{-}14)$$

where $n = 0, 1, 2, 3, \ldots,$
     $\lambda_g =$ wavelength in the waveguide

If the mean length L is an odd number of quarter wavelengths, the reflected waves from both ends of the waveguide section are completely canceled. For the waveguide bend, the minimum radius of curvature for a small reflection is given by Southworth [5] as

$$R = 1.5b \qquad \text{for an } E \text{ bend} \qquad (4\text{-}8\text{-}15)$$

and
$$R = 1.5a \qquad \text{for an } H \text{ bend} \qquad (4\text{-}8\text{-}16)$$

where a and b are the dimensions of the waveguide bend as illustrated in Fig. 4-8-7(c).

4-9 Directional Couplers

A directional coupler is a four-port waveguide junction as shown in Fig. 4-9-1. It consists of a primary waveguide 1–2 and a secondary waveguide 3–4. When all ports are terminated in their characteristic impedances, there is free transmission of power, without reflection, between port 1 and port 2, and there is no transmission of power between port 1 and port 3 or between port 2 and port 4 because no coupling exists between these two pairs of ports. The degree of coupling between port 1 and port 4 and between port 2 and port 3 depends on the structure of the coupler.

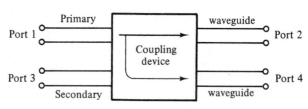

Fig. 4-9-1. Directional coupler.

The characteristics of a directional coupler can be expressed in terms of its coupling factor and its directivity. Assuming that the wave is propagating from port 1 to port 2 in the primary line, the coupling factor and the directivity are defined, respectively, by

$$\text{Coupling factor (dB)} = 10 \log_{10} \frac{P_1}{P_4} \qquad (4\text{-}9\text{-}1)$$

and
$$\text{Directivity (dB)} = 10 \log_{10} \frac{P_4}{P_3} \qquad (4\text{-}9\text{-}2)$$

where P_1 = the power input to port 1
P_3 = the power output from port 3
P_4 = the power output from port 4

It should be noted that port 2, port 3, and port 4 are terminated in their characteristic impedances. The coupling factor is a measure of the ratio of power levels in the primary and secondary lines. So if the coupling factor is known, a fraction of power measured at port 4 may be used to determine the power input at port 1. This significance is desirable for microwave power measurements because no disturbance, which may be caused by the power measurements, occurs in the primary line. The directivity is a measure of how well the forward traveling wave in the primary waveguide couples only to a specific port of the secondary waveguide. An ideal directional coupler should have infinite directivity. In other words, the power at port 3 must be zero because port 2 and port 4 are perfectly matched. Actually, well-designed directional couplers have a directivity of only 30 to 35 dB.

Several types of directional couplers exist, such as a two-hole directional coupler, four-hole directional coupler, reverse-coupling directional coupler (Schwinger coupler), and Bethe-hole directional coupler (refer to Fig. 4-9-2). Only the very commonly used two-hole directional coupler is described here.

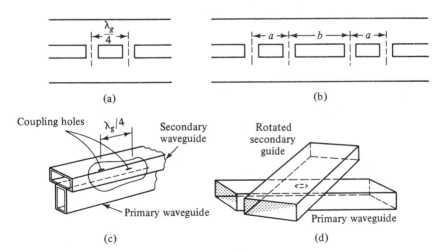

Fig. 4-9-2. Different directional couplers. (a) Two-hole directional coupler. (b) Four-hole directional coupler. (c) Schwinger coupler. (d) Bethe-hole directional coupler.

4-9-1. Two-Hole Directional Couplers

A two-hole directional coupler with traveling waves propagating in it is illustrated in Fig. 4-9-3. The spacing between the centers of two holes must be

$$L = (2n + 1)\frac{\lambda_g}{4} \tag{4-9-3}$$

where n is any positive integer.

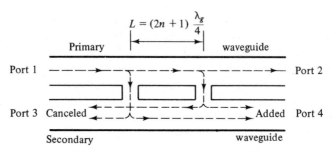

Fig. 4-9-3. Two-hole directional coupler.

A fraction of the wave energy entered into port 1 passes through the holes and is radiated into the secondary guide as the holes act as slot antennas. The forward waves in the secondary guide are in the same phase, regardless of the hole space, and are added at port 4. The backward waves in the secondary guide (waves are progressing from the right to left) are out of phase by $(2L/\lambda_g)2\pi$ radians and are canceled at port 3.

4-9-2. S Matrix of a Directional Coupler

In a directional coupler all four ports are completely matched. Thus the diagonal elements of the S matrix are zeros and

$$S_{11} = S_{22} = S_{33} = S_{44} = 0 \tag{4-9-4}$$

As noted, there is no coupling between port 1 and port 3 and between port 2 and port 4. So

$$S_{13} = S_{31} = S_{24} = S_{42} = 0 \tag{4-9-5}$$

Consequently, the S matrix of a directional coupler becomes

$$\mathbf{S} = \begin{bmatrix} 0 & S_{12} & 0 & S_{14} \\ S_{21} & 0 & S_{23} & 0 \\ 0 & S_{32} & 0 & S_{34} \\ S_{41} & 0 & S_{43} & 0 \end{bmatrix} \tag{4-9-6}$$

Equation (4-9-6) can be further reduced by means of the properties of the S parameters as described in Section 4-7. From Eq. (4-7-52) two relationships are given by

$$S_{12}S_{14}^* + S_{32}S_{34}^* = 0 \qquad (4\text{-}9\text{-}7)$$

and

$$S_{21}S_{23}^* + S_{41}S_{43}^* = 0 \qquad (4\text{-}9\text{-}8)$$

From Eq. (4-7-37) we can write

$$S_{12}S_{12}^* + S_{14}S_{14}^* = 1 \qquad (4\text{-}9\text{-}9)$$

Equations (4-9-7) and (4-9-8) can be also written

$$|S_{12}||S_{14}| = |S_{32}||S_{34}| \qquad (4\text{-}9\text{-}10)$$

and

$$|S_{21}||S_{23}| = |S_{41}||S_{43}| \qquad (4\text{-}9\text{-}11)$$

Since $S_{12} = S_{21}$, $S_{14} = S_{41}$, $S_{23} = S_{32}$, and $S_{34} = S_{43}$, then

$$|S_{12}| = |S_{34}| \qquad (4\text{-}9\text{-}12)$$

$$|S_{14}| = |S_{23}| \qquad (4\text{-}9\text{-}13)$$

Let

$$S_{12} = S_{34} = p \qquad (4\text{-}9\text{-}14)$$

where p is positive and real. Then from Eq. (4-9-8)

$$p(S_{23}^* + S_{41}) = 0 \qquad (4\text{-}9\text{-}15)$$

Let

$$S_{23} = S_{41} = jq \qquad (4\text{-}9\text{-}16)$$

where q is positive and real. Then from Eq. (4-9-9)

$$p^2 + q^2 = 1 \qquad (4\text{-}9\text{-}17)$$

The **S** matrix of a directional coupler is reduced to

$$\mathbf{S} = \begin{bmatrix} 0 & p & 0 & jq \\ p & 0 & jq & 0 \\ 0 & jq & 0 & p \\ jq & 0 & p & 0 \end{bmatrix} \qquad (4\text{-}9\text{-}18)$$

EXAMPLE 4-9-1: Directional Coupler. A symmetrical directional coupler with infinite directivity and a forward attenuation of 20 dB is used to monitor the power delivered to a load Z_ℓ (see Fig. 4-9-4). Bolometer 1 intro-

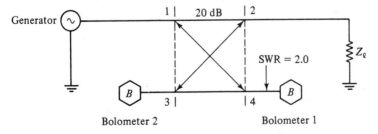

Fig. 4-9-4. Power measurements by directional coupler.

duces a VSWR of 2.0 on arm 4; bolometer 2 is matched to arm 3. If bolometer 1 reads 8 mW and bolometer 2 reads 2 mW, find

    (a) the amount of power dissipated in the load Z_ℓ, and

    (b) the VSWR on arm 2.

    Solution.   The wave propagation in the directional coupler is shown in Fig. 4-9-5.

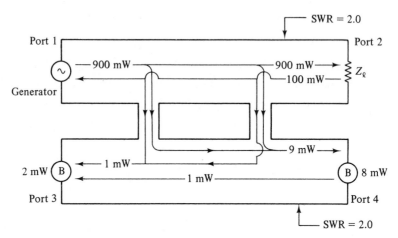

Fig. 4-9-5. Wave propagation in the directional coupler.

(a) Power dissipation at Z_ℓ.

1. The reflection coefficient at port 4 is

$$|\Gamma| = \frac{\rho - 1}{\rho + 1} = \frac{2 - 1}{2 + 1} = \frac{1}{3}$$

2. Since the incident power and reflected power are related by

$$P^- = P^+ |\Gamma|^2$$

where P^+ = incident power

        P^- = reflected power

Then

$$|\Gamma| = \frac{1}{3} = \sqrt{\frac{P^-}{P^+}} = \sqrt{\frac{P^-}{8 + P^-}}$$

The incident power to port 4 is $P_4^+ = 9$ mW, and the reflected power from port 4 is $P_4^- = 1$ mW.

3. Since port 3 is matched and the bolometer at port 3 reads 2 mW, 1 mW must be radiated through the holes.

4. Since 20 dB is equivalent to a power ratio of 100 to 1, the power input at port 1 is given by

$$P_1 = 100 P_4^+ = 900 \text{ mW}$$

and the power reflected from the load is

$$P_2^- = 100 \ (1 \text{ mW}) = 100 \text{ mW}$$

5. The power dissipated in the load is

$$P_\ell = P_2^+ - P_2^- = 900 - 100 = 800 \text{ mW}$$

(b) The reflection coefficient is calculated as

$$|\Gamma| = \sqrt{\frac{P^-}{P^+}} = \sqrt{\frac{100}{900}} = \frac{1}{3}$$

Then the VSWR on arm 2 is

$$\rho = \frac{1 + |\Gamma|}{1 - |\Gamma|} = \frac{1 + \frac{1}{3}}{1 - \frac{1}{3}} = 2.0$$

4-10 Circulators and Isolators

Both microwave circulators and microwave isolators are nonreciprocal transmission devices that use the property of Faraday rotation in the ferrite material. In order to understand the operating principles of circulators and isolators, let us describe the behavior of ferrites in the nonreciprocal phase shifter.

A nonreciprocal phase shifter consists of a thin slab of ferrite placed in a rectangular waveguide at a point where the dc magnetic field of the incident wave mode is circularly polarized. Ferrite is a family of $MeO \cdot Fe_2O_3$, where Me is a divalent metal iron. When a piece of ferrite is affected by a dc magnetic field, the ferrite exhibits a phenomenon of Faraday rotation. It does so because the ferrite is nonlinear material and its permeability is an asymmetrical tensor [6], as expressed by

$$\mathbf{B} = \hat{\mu}\mathbf{H} \tag{4-10-1}$$

where

$$\hat{\mu} = \mu_0(1 + \hat{\chi}) \tag{4-10-2}$$

and

$$\hat{\chi} = \begin{bmatrix} \chi & j\kappa & 0 \\ j\kappa & \chi & 0 \\ 0 & 0 & 0 \end{bmatrix} \tag{4-10-3}$$

which is the tensor magnetic susceptibility. Here χ is the diagonal susceptibility and κ is the off-diagonal susceptibility.

When a dc magnetic field is applied to a ferrite, the unpaired electrons in the ferrite material tend to line up with the dc field due to their magnetic dipole moment. However, the nonreciprocal precession of unpaired electrons in the ferrite causes their relative permeabilities (μ_r^+, μ_r^-) to be unequal and the wave in the ferrite is then circularly polarized. The propagation constant for a linearly polarized wave inside the ferrite can be expressed as [6]

$$\gamma_{\pm} = j\omega\sqrt{\epsilon\mu_0(\mu + \kappa)} \qquad (4\text{-}10\text{-}4)$$

where
$$\mu = 1 + \chi \qquad (4\text{-}10\text{-}5)$$

$$\mu_r^+ = \mu + \kappa \qquad (4\text{-}10\text{-}6)$$

$$\mu_r^- = \mu - \kappa \qquad (4\text{-}10\text{-}7)$$

The relative permeability μ_r changes with the applied dc magnetic field as given by

$$\mu_r^{\pm} = 1 + \frac{\gamma_e M_e}{|\gamma_e| H_{dc} \mp \omega} \qquad (4\text{-}10\text{-}8)$$

where γ_e = the gyromagnetic ratio of an electron
M_e = the saturation magnetization
ω = the angular frequency of a microwave field
H_{dc} = the dc magnetic field
μ_r^+ = the relative permeability in the clockwise direction (right or positive circular polarization)
μ_r^- = the relative permeability in the counterclockwise direction (left or negative circular polarization)

It can be seen from Eq. (4-10-8) that if $\omega = |\gamma_e| H_{dc}$, then μ_r^+ is infinite. This phenomenon is called the *gyromagnetic resonance* of the ferrite. A graph of μ_r is plotted as a function of H_{dc} for longitudinal propagation in Fig. 4-10-1.

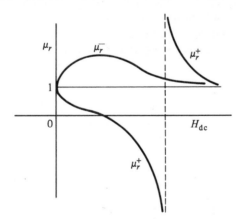

Fig. 4-10-1. Curves of μ_r versus H_{dc} for axial propagation.

If μ_r^+ is much larger than μ_r^- ($\mu_r^+ \gg \mu_r^-$), the wave in the ferrite is rotated in the clockwise direction. Consequently, the propagation phase constant β^+ for the forward direction differs from the propagation phase constant β^- for the backward direction. By choosing the length of the ferrite slab and the dc magnetic field so that

$$\omega = (\beta^+ - \beta^-)\ell = \frac{\pi}{2} \qquad (4\text{-}10\text{-}9)$$

A differential phase shift of 90° for the two directions of propagation can be obtained.

4-10-1. Microwave Circulators

A microwave circulator is a multiport waveguide junction in which the wave can flow only from the nth port to the $(n + 1)$th port in one direction (see Fig. 4-10-2). Although there is no restriction on the number of ports, the four-port microwave circulator is the most common. One type of four-port microwave circulator is a combination of two 3-dB side-hole directional couplers and a rectangular waveguide with two nonreciprocal phase shifters as shown in Fig. 4-10-3.

The operating principle of a typical microwave circulator can be analyzed with the aid of Fig. 4-10-3. Each of the two 3-dB couplers in the circulator introduces a phase shift of 90°, and each of the two phase shifters

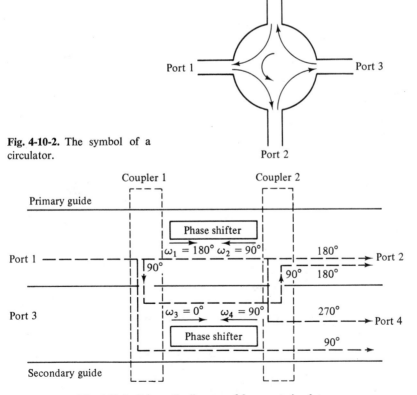

Fig. 4-10-2. The symbol of a circulator.

Fig. 4-10-3. Schematic diagram of four-port circulator.

produces a certain amount of phase change in a certain direction as indicated. When a wave is incident to port 1, the wave is split into two components by coupler 1. The wave in the primary guide arrives at port 2 with a relative phase change of 180°. The second wave propagates through the two couplers and the secondary guide and arrives at port 2 with a relative phase shift of 180°. Since the two waves reaching port 2 are in phase, the power transmission is obtained from port 1 to port 2. However, the wave propagates through the primary guide, phase shifter, and coupler 2 and arrives at port 4 with a phase change of 270°. The wave travels through coupler 1 and the secondary guide, and it arrives at port 4 with a phase shift of 90°. Since the two waves reaching port 4 are out of phase by 180°, the power transmission from port 1 to port 4 is zero. In general, the differential propagation constants in the two directions of propagation in a waveguide containing ferrite phase shifters should be

$$\omega_1 - \omega_3 = (2m + 1)\pi \qquad \text{radians/s} \qquad (4\text{-}10\text{-}10)$$

$$\omega_2 - \omega_4 = 2n\pi \qquad \text{radians/s} \qquad (4\text{-}10\text{-}11)$$

where m and n are any integers, including zeros.

A similar analysis shows that a wave incident to port 2 emerges at port 3 and so on. As a result, the sequence of power flow is designated as $1 \rightarrow 2 \rightarrow 3 \rightarrow 4 \rightarrow 1$.

Many types of microwave circulators are in use today. However, their principles of operation remain the same. Figure 4-10-4 shows a four-port circulator constructed of two magic Tees and a phase shifter. The phase shifter produces a phase shift of 180°. The explanation of how this circulator works is left as an exercise for the reader.

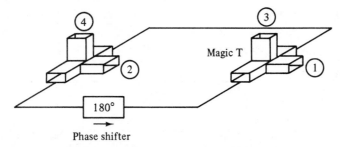

Fig. 4-10-4. A four-port circulator.

A perfectly matched, lossless, and nonreciprocal four-port circulator has an S matrix of the form

$$\mathbf{S} = \begin{bmatrix} 0 & S_{12} & S_{13} & S_{14} \\ S_{21} & 0 & S_{23} & S_{24} \\ S_{31} & S_{32} & 0 & S_{34} \\ S_{41} & S_{42} & S_{43} & 0 \end{bmatrix} \qquad (4\text{-}10\text{-}12)$$

Using the properties of S parameters as described in Section 3-1, the **S** matrix in Eq. (4-10-12) can be simplified to

$$\mathbf{S} = \begin{bmatrix} 0 & 0 & 0 & 1 \\ 1 & 0 & 0 & 0 \\ 0 & 1 & 0 & 0 \\ 0 & 0 & 1 & 0 \end{bmatrix} \tag{4-10-13}$$

4-10-2. Microwave Isolator

An isolator is a nonreciprocal transmission device that is used to isolate one component from reflections of other components in the transmission line. An ideal isolator completely absorbs the power for propagation in one direction and provides lossless transmission in the opposite direction. So the isolator is usually called *uniline*. Isolators are generally used to improve the frequency stability of microwave generators, such as klystrons and magnetrons, in which the reflection from the load affects the generating frequency. In such cases, the isolator placed between the generator and load prevents the reflected power from the unmatched load from returning to the generator. As a result, the isolator maintains the frequency stability of the generator.

Isolators can be constructed in many ways. They can be made by terminating ports 3 and 4 of a four-port circulator with matched loads. On the other hand, isolators can be made by inserting a ferrite rod along the axis of a rectangular waveguide as shown in Fig. 4-10-5. The isolator here is a Faraday-rotation isolator. Its operating principle can be explained as follows [7]. The input resistive card is in the y-z plane, and the output resistive card is displaced 45° with respect to the input card. The dc magnetic field, which is applied longitudinally to the ferrite rod, rotates the wave plane of polarization by 45°. The degrees of rotation depend on the length and diameter of the rod and on the applied dc magnetic field. An input TE_{01} dominant

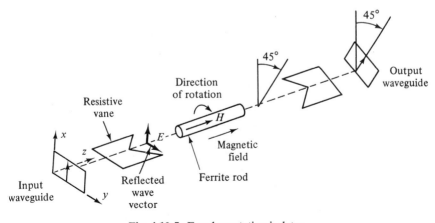

Fig. 4-10-5. Faraday-rotation isolator.

mode is incident to the left end of the isolator. Since the TE_{01} mode wave is perpendicular to the input resistive card, the wave passes through the ferrite rod without attenuation. The wave in the ferrite rod section is rotated clockwise by 45° and is normal to the output resistive card. As a result of rotation, the wave arrives at the output end without attenuation at all. On the contrary, a reflected wave from the output end is similarly rotated clockwise 45° by the ferrite rod. However, since the reflected wave is parallel to the input resistive card, the wave is thereby absorbed by the input card. The typical performance of these isolators is about 1-dB insertion loss in forward transmission and about 20-to 30-dB isolation in reverse attenuation.

4-11 Summary

Several resonators, reentrant cavities, and slow-wave structures in general use have been described. These passive devices play a major role in microwave source generation and amplification of microwave active devices, the subjects of the next chapter.

Resonant Frequency: The resonant frequency of a resonator is determined by solving Maxwell's field equations for a given boundary condition.

The resonant frequency of a rectangular resonator is given by

$$f_r = \frac{1}{2\sqrt{\mu\epsilon}} \sqrt{\left(\frac{m}{a}\right)^2 + \left(\frac{n}{b}\right)^2 + \left(\frac{p}{d}\right)^2} \qquad \text{(for TE and TM modes)}$$

The resonant frequency of a cylindrical resonator or a semicircular resonator is expressed by

$$f_r = \frac{1}{2\pi\sqrt{\mu\epsilon}} \sqrt{\left(\frac{X'_{np}}{a}\right)^2 + \left(\frac{q\pi}{d}\right)^2} \qquad \text{(for TE modes)}$$

$$f_r = \frac{1}{2\pi\sqrt{\mu\epsilon}} \sqrt{\left(\frac{X_{np}}{a}\right)^2 + \left(\frac{q\pi}{d}\right)^2} \qquad \text{(for TM modes)}$$

The Q of a Resonator: At resonance the peak energies stored in the electric and magnetic fields are equal. The Q is a measure of the frequency selectivity of a resonant or antiresonant circuit, and it is defined as

$$Q \equiv 2\pi \frac{\text{maximum energy stored}}{\text{energy dissipated per cycle}} = \frac{\omega W}{P}$$

where W is the maximum stored energy and P is the average power loss.

Reentrant Cavity: A reentrant cavity is one in which the metallic boundaries extend into the interior of the cavity. The commonly used radical reentrant cavity has the inductance and capacitance given by

$$L = \frac{\mu \ell}{2\pi} \ell n \frac{b}{a}$$

and
$$C = \epsilon_0 \left[\frac{\pi a^2}{d} - 4a \, \ell n \frac{0.765}{\sqrt{\ell^2 + (b-a)^2}} \right]$$

Slow-Wave Structure: The helix slow-wave structure is normally used in the helix traveling-wave tube (TWT) to reduce the wave velocity in the axial direction of the helix. The phase velocity in the axial direction is expressed by

$$v_{p\epsilon} = \frac{p}{\sqrt{\mu\epsilon[p^2 + (\pi d)^2]}}$$

where p is the helix pitch and d is the helix diameter.

In this chapter S-parameter theory, waveguide Tees, directional couplers, circulators, and isolators have been discussed in some detail.

S-parameter Theory: The S parameter, also called the S matrix or scattering-matrix, has four characteristic properties.

1. **Symmetry property:** If a microwave junction satisfies a reciprocity condition or if there are no active devices at the junction, the junction is a linear passive circuit, and the S parameters are equal to their corresponding transpose. That is,

$$\mathbf{S} = \tilde{\mathbf{S}}$$

2. **Unity property:** The sum of the products of each term of any one row or of any one column of the matrix **S** multiplied by its complex conjugate is unity. That is,

$$\sum_{i}^{n} S_{ij}S_{ij}^* = 1 \qquad \text{for } j = 1, 2, 3, \ldots, n$$

3. **Zero property:** The sum of the products of each term of any row (or column) multiplied by the complex conjugate of the corresponding term of any other row (or column) is zero. That is,

$$\sum_{i}^{n} S_{ij}S_{ij}^* = 0 \qquad \text{for } k \neq j \qquad \begin{pmatrix} k = 1, 2, 3, \ldots, n \\ j = 1, 2, 3, \ldots, n \end{pmatrix}$$

4. **Phase-shifting property:** If any of the terminal planes (or reference plane), say the kth port, is moved away from the junction by an electric distance $\beta_k \ell_k$, each of the coefficients S_{ij} involving k will be multiplied by the factor $e^{-j\beta_k \ell_k}$. In matrix notation, the new S parameter, S', can be expressed in terms of the old S parameter, **S**, and the phase matrix (ϕ matrix). That is,

$$\mathbf{S}' = \boldsymbol{\phi}\mathbf{S}\boldsymbol{\phi}$$

where **S** is the old scattering matrix and

$$\boldsymbol{\phi} = \begin{bmatrix} \phi_{11} & 0 & 0 & \cdots & 0 \\ 0 & \phi_{22} & 0 & \cdots & 0 \\ \cdot & \cdot & \cdot & \cdots & \cdot \\ 0 & 0 & 0 & \cdots & \phi_{nn} \end{bmatrix}$$

$$\phi_{11} = \phi_{22} = \phi_{kk} = e^{-j\beta_k \ell_k} \qquad \text{for } k = 1, 2, 3, \ldots, n$$

Directional Coupler: A directional coupler is a four-port waveguide junction consisting of a primary and a secondary guide. Its characteristics can be expressed in terms of its coupling factor and its directivity. Assuming that the wave is propagating from port 1 to port 2 in the primary guide and that port 3 and port 4 of the secondary guide are corresponding to the input side and the output side, the coupling factor and the directivity are defined, respectively, by

$$\text{Coupling factor (dB)} \equiv 10 \log_{10} \frac{P_1}{P_4}$$

$$\text{Directivity (dB)} \equiv 10 \log_{10} \frac{P_4}{P_3}$$

where P_1 = the power input to port 1
P_3 = the power output from port 3
P_4 = the power outpout from port 4

REFERENCES

[1,2] FUJISAWA, K., General treatment of klystron resonant cavities. *IRE Trans.* **MTT-6**, 344–358, October 1958.

[3] BECK, A. H. W., *Space-Charge Waves*. Pergamon Press, New York, 1958.

[4] MONTGOMERY, C. G., et al., *Principles of Microwave Circuits*. McGraw-Hill Book Company, New York, 1948.

[5] SOUTHWORTH, G. E., *Principles and Applications of Waveguide Transmission*, Chapters 8 and 9. D. Van Nostrand Co., Princeton, N.J., 1950.

[6] SOOHOO, R. F., *Theory and Application of Ferrites*. Prentice-Hall, Englewood Cliffs, N.J., 1960.

[7] BOWNESS, C., Microwave ferrites and their applications. *Microwave J.*, **1**, 13–21, July–August 1958.

SUGGESTED READINGS

1. ALTMAN, J. L., *Microwave Circuits*, Chapters 2, 3, and 4. D. Van Nostrand Company, Princeton, N.J., 1964.

2. ATWATER, H. A., *Introduction to Microwave Theory*, Chapter 4. McGraw-Hill Book Company, New York, 1962.

3. BRONWELL, A. B., and R. E. BEAM, *Theory and Application of Microwaves*, Chapters 17 and 18. McGraw-Hill Book Company, New York, 1947.

4. COLLIN, ROBERT E., *Foundations for Microwave Engineering*, Chapters 6 to 8. McGraw-Hill Book Company, New York, 1966.

5. GEWARTOWSKI, J. W., and H. A. WATSON, *Principles of Electron Tubes*, Chapter 8. D. Van Nostrand Company, Princeton, N.J., 1965.

6. GHOSE, R. N., *Microwave Circuit Theory and Analysis*, Chapter 8. McGraw-Hill Book Company, New York, 1963.

7. HEWLETT PACKARD. *S-Parameter Design.* Application Note 154, Hewlett Packard, Palo Alto, Calif., 1972.

8. HOGAN, C. L., The ferromagnetic Faraday effect at microwave frequencies and its applications. *Bell System Tech. J.*, **31**, 1. January 1952.

9. ISHII, T. K., *Microwave Engineering*, Chapter 4. Ronald Press, New York, 1966.

10. POLDER, D., On the theory of ferromagnetic research. *Phil. Mag.*, **40**, 99. 1949.

11. REICH, HERBERT J., et al., *Microwave Principles*, Chapters 4 and 5. D. Van Nostrand Company, Princeton, N.J., 1966.

Microwave Tubes

5-0 Introduction

chapter

5

We turn next to a quantitative or qualitative analysis of several conventional vacuum tubes and microwave tubes in common use. The conventional vacuum tubes, such as triodes, tetrodes, and pentodes, are still used as signal sources of low output power at low microwave frequencies. The most important microwave tubes at present are the linear-beam tubes (O-type) tabulated in Table 5-0-1. The paramount O-type tube is the two-cavity klystron, and it is followed by the reflex klystron. The helix traveling-wave tube (TWT), the coupled-cavity TWT, the forward-wave amplifier (FWA), and the backward-wave amplifier and oscillator (BWA and BWO) are also O-type tubes, but they have nonresonant periodic structures for electron interactions. The Twystron is a hybrid amplifier that uses combinations of klystron and TWT components.

TABLE 5-0-1. Linear-Beam Tubes (*O*-type).

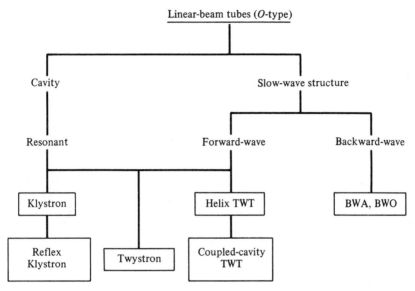

Several equally important microwave tubes include the crossed-field tubes (*M*-type) tabulated in Table 5-0-2. Here the magnetron is the oldest and still the most important of this family. Others are the forward-wave crossed-field amplifier (FWCFA), the dematron, and the carcinotron (*M*-

TABLE 5-0-2. Crossed-Field Tubes (*M*-type).

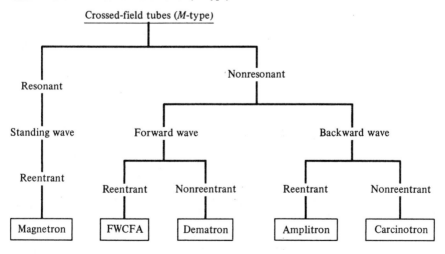

type backward-wave oscillator). Although it is impossible to discuss all such tubes in detail, the common operating principles of many will be described.

5-1   Conventional Vacuum Triodes, Tetrodes, and Pentodes

Conventional vacuum triodes, tetrodes, and pentodes are less useful signal sources at frequencies above 1 GHz because of lead-inductance and inter-elctrode-capacitance effects, transit-angle effects, and gain-bandwidth product limitations. These three effects are analyzed in detail in the following sections.

5-1-1.   Lead-Inductance and Interelectrode-Capacitance Effects

At frequencies above 1 GHz conventional vacuum tubes are impaired by parasitic-circuit reactances because the circuit capacitances between tube electrodes and the circuit inductance of the lead wire are too large for a microwave resonant circuit. Furthermore, as the frequency is increased up to the microwave range, the real part of the input admittance may be large enough to cause a serious overload of the input circuit and thereby reduce the operating efficiency of the tube. In order to gain a better understanding of the effects stated above, the triode circuit shown in Fig. 5-1-1 should be studied carefully.

Figure 5-1-1(b) shows the equivalent circuit of a triode circuit under the assumption that the interelectrode capacitances and cathode inductance are the only parasitic elements. Since $C_{gp} \ll C_{gk}$ and $\omega L_k \ll 1/(\omega C_{gk})$, the input voltage V_{in} can be written as

$$V_{in} = V_g + V_k = V_g + j\omega L_k g_m V_g \qquad (5\text{-}1\text{-}1)$$

and the input current as

$$I_{in} = j\omega C_{gk} V_g \qquad (5\text{-}1\text{-}2a)$$

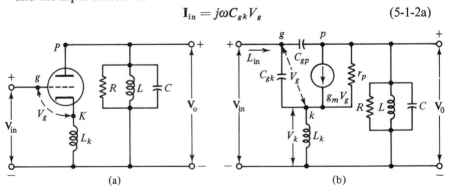

FIG. 5-1-1. Triode circuit (a) and its equivalent (b).

Substitution of Eq. (5-1-2a) in Eq. (5-1-1) yields

$$\mathbf{V}_{in} = \frac{\mathbf{I}_{in}(1 + j\omega L_k g_m)}{j\omega C_{gk}} \tag{5-1-2b}$$

The input admittance of the tube is approximately

$$\mathbf{Y}_{in} = \frac{\mathbf{I}_{in}}{\mathbf{V}_{in}} = \frac{j\omega C_{gk}}{1 + j\omega L_k g_m} = \omega^2 L_k C_{gk} g_m + j\omega C_{gk} \tag{5-1-3}$$

in which $\omega L_k g_m \ll 1$ has been replaced. The inequality is almost always true, since the cathode lead is usually short and is quite large in diameter, and the transconductance g_m is generally much less than one millimho.

The input impedance at very high frequencies is given by

$$\mathbf{Z}_{in} = \frac{1}{\omega^2 L_k C_{gk} g_m} - j \frac{1}{\omega^3 L_k^2 C_{gk} g_m^2} \tag{5-1-4}$$

The real part of the impedance is inversely proportional to the square of the frequency, and the imaginary part is inversely proportional to the third order of the frequency. When the frequencies are above 1 GHz, the real part of the impedance becomes small enough to nearly short the signal source. Consequently, the output power is decreased rapidly. Similarly, the input admittance of a pentode circuit is expressed by

$$\mathbf{Y}_{in} = \omega^2 L_k C_{gk} g_m + j\omega(C_{gk} + C_{gs}) \tag{5-1-5}$$

where C_{gs} is the capacitance between the grid and screen, and its input impedance is given by

$$\mathbf{Z}_{in} = \frac{1}{\omega^2 L_k C_{gk} g_m} - j \frac{C_{gk} + C_{gs}}{\omega^3 L_k^2 C_{gk}^2 g_m^2} \tag{5-1-6}$$

There are several ways to minimize the inductance and capacitance effects, such as a reduction in lead length and electrode area. This minimization, however, also limits the power-handling capacity.

5-1-2. Transit-Angle Effects

Another limitation in the application of conventional tubes at microwave frequencies is the electron transit angle between electrodes. The electron transit angle is defined as

$$\theta_g \equiv \omega \tau_g = \frac{\omega d}{v_0} \tag{5-1-7}$$

where $\tau_g = \dfrac{d}{v_0}$ is the transit time

$d =$ the separation between cathode and grid
$v_0 = 0.593 \times 10^6 \sqrt{V_0}$ is the velocity of the electron
$V_0 =$ the dc voltage

When frequencies are below microwave range, the transit angle is negligible. At microwave frequencies, however, the transit time (or angle) is large compared to the period of the microwave signal, and the potential between the cathode and the grid may alternate from 10 to 100 times during the electron transit. The grid potential during the negative half cycle thus removes energy that was given to the electron during the positive half cycle. Consequently, the electrons may oscillate back and forth in the cathode-grid space or return to the cathode. The overall result of transit-angle effects is to reduce the operating efficiency of the vacuum tube. The degenerate effect becomes more serious when frequencies are well above 1 GHz. Once electrons pass the grid, they are quickly accelerated to the anode by the high plate voltage.

When the frequency is below 1 GHz, the output delay is negligible in comparison with the phase of the grid voltage. This means that the trans-admittance is a real large quantity, which is the usual transconductance g_m. At microwave frequencies the transit angle is not negligible, and the trans-admittance becomes a complex number with a relatively small magnitude. This situation indicates that the output is decreased.

From the preceding analysis it is clear that the transit-angle effect can be minimized by first accelerating the electron beam with a very high dc voltage and then velocity-modulating it. This is indeed the principal operation of such microwave tubes as klystrons and magnetrons.

5-1-3. Gain-Bandwidth Product Limitation

In ordinary vacuum tubes the maximum gain is generally achieved by resonating the output circuit as shown in Fig. 5-1-2. In the above equivalent

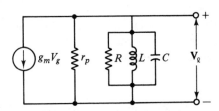

FIG. 5-1-2. Output-tuned circuit of a pentode.

circuit (see Fig. 5-1-2) it is assumed that $r_p \gg \omega L_k$. The load voltage is given by

$$V_\ell = \frac{g_m V_g}{G + j[\omega C - 1/(\omega L)]} \tag{5-1-8}$$

where $\quad G = \dfrac{1}{r_p} + \dfrac{1}{R}$

r_p = the plate resistance
R = the load resistance
L, C = the tuning elements

The resonant frequency is expressed by

$$f_r = \frac{1}{2\pi\sqrt{LC}} \tag{5-1-9}$$

and the maximum voltage gain $A_{\max}$ at resonance by

$$A_{\max} = \frac{g_m}{G} \tag{5-1-10}$$

Since the bandwidth is measured at the half-power point, the denominator of Eq. (5-1-8) must be related by

$$G = \omega C - \frac{1}{\omega L} \tag{5-1-11}$$

The roots of this quadratic equation are given by

$$\omega_1 = \frac{G}{2C} - \sqrt{\left(\frac{G}{2C}\right)^2 + \frac{1}{LC}} \tag{5-1-12}$$

$$\omega_2 = \frac{G}{2C} + \sqrt{\left(\frac{G}{2C}\right)^2 + \frac{1}{LC}} \tag{5-1-13}$$

Then the bandwidth can be expressed by

$$BW = \omega_2 - \omega_1 = \frac{G}{C} \quad \text{for} \quad \left(\frac{G}{2C}\right)^2 \gg \frac{1}{LC} \tag{5-1-14}$$

Hence, the gain-bandwidth product of the circuit of Fig. 5-1-2 is

$$A_m(BW) = \frac{g_m}{C} \tag{5-1-15}$$

It is important to note that the gain-bandwidth product is independent of frequency. For a given tube, a higher gain can be achieved only at the expense of a narrower bandwidth. This restriction is applicable to a resonant circuit only. In microwave devices either reentrant cavities or slow-wave structures are used to obtain a possible overall high gain over a broad bandwidth.

5-2 Linear-Beam Tubes (O-Type)

In the last section it was shown that the gain-bandwidth product of a conventional vacuum tube is independent of frequency. In other words, for a given tube, an increase in gain can be obtained only at the expense of a bandwidth. The remedy for this restriction is to use a nonresonant circuit, such as the reentrant cavity used in a reflex klystron and the helical slow-wave structure used in a traveling-wave tube. In Chapter 4 several types of cavities, resonators, and slow-wave structures were described. In this chapter we shall see how these passive devices are used in linear-beam tubes to increase power output and efficiency.

The advent of linear-beam tubes began with the Heil oscillators [1] in 1935 and the Varian brothers' klystron amplifier [1] in 1939. The work was advanced by the space-charge-wave propagation theory of Hahn and Ramo [1] in 1939 and continued with the invention of the helix-type traveling-wave tube (TWT) by R. Kompfner in 1944 [2]. From the early 1950s on, the low power output of linear-beam tubes made it possible to achieve high power levels, first rivaling and finally surpassing magnetrons, the early sources of microwave high power. Subsequently, several additional devices were developed, two of which have shown lasting importance. They are the extended interaction klystron [3] and the Twystron hybrid amplifier [4]. The present state of the art for U.S. high-power klystrons, gridded tubes, and traveling-wave tubes (TWTs) is shown in Figs. 5-2-1, 5-2-2, and 5-2-3, respectively.

In a linear-beam tube a magnetic field whose axis coincides with that of the electron beam is used to hold the beam together as it travels the length of the tube. *O*-type tubes derive their name from the French TPO (tubes à propagation des ondes) or from the word original (meaning the original type of tube). In these tubes electrons receive potential energy from the dc beam voltage before they arrive in the microwave interaction region, and this energy is converted into their kinetic energy. In the microwave interaction region the electrons are either accelerated or decelerated by the microwave field and then bunched as they drift down the tube. The bunched electrons,

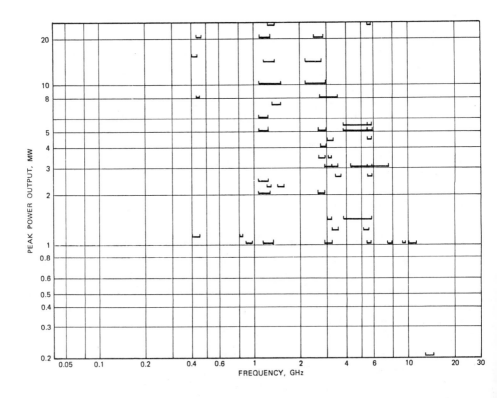

FIG. 5-2-1. State of the art for U.S. high-power klystrons.

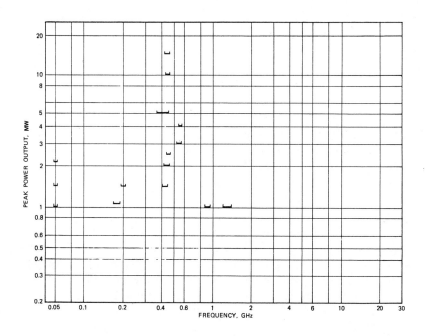

FIG. 5-2-2. State of the art for U.S. high-power gridded tubes.

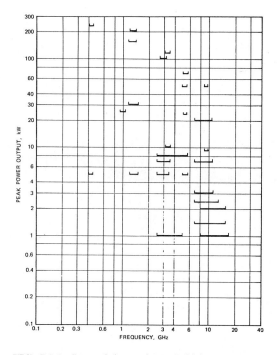

FIG. 5-2-3. State of the art for U.S. high-power TWTs.

179

in turn, induce current in the output structure. The electrons then give up their kinetic energy to the microwave fields and are collected by the collector.

O-type traveling-wave tubes are suitable for amplification. At present, klystron and TWT amplifiers can deliver a peak power output up to 30 MW (megawatts) with a beam voltage on the order of 100 kV at the frequency of 10 GHz. The average power outputs are up to 700 kW. The gain of these tubes is on the order of 30 to 70 dB, and the efficiency is from 15 to 60%. The bandwidth is from 1 to 8% for klystrons and 10 to 15% for TWTs.

5-2-1. Klystrons

The two-cavity klystron is a widely used microwave amplifier operated by the principles of velocity and current modulation. All electrons injected from the cathode arrive at the first cavity with uniform velocity. Those electrons passing the first cavity gap at zeros of the gap voltage (or signal voltage) pass through with unchanged velocity; those passing through the positive half cycles of the gap voltage undergo an increase in velocity; those passing through the negative swings of the gap voltage undergo a decrease in velocity. As a result of these actions, the electrons gradually bunch together as they travel down the drift space. The variation in electron velocity in the drift space is known as *velocity modulation*. The density of the electrons in the second cavity gap varies cyclically with time. The electron beam contains an ac component and is said to be current-modulated. The maximum bunching should occur approximately at midway between the second cavity grids during its retarding phase; thus the kinetic energy is transferred from the electrons to the field of the second cavity. The electrons then emerge from the second cavity with reduced velocity and finally terminate at the collector. The characteristics of a two-cavity klystron amplifier are as follows:

Efficiency:	About 40%
Power output:	Average power (CW power) is up to 500 kW and pulsed power is up to 30 MW at 10 GHz.
Power gain:	About 30 dB

The schematic diagram of a two-cavity klystron amplifier is shown in Fig. 5-2-1-1.

The cavity close to the cathode is known as the *buncher cavity* or input cavity, which velocity-modulates the electron beam. The other cavity is called the *catcher cavity* or output cavity; it catches energy from the bunched electron beam. The beam then passes through the catcher cavity and is terminated at the collector. The quantitative analysis of a two-cavity klystron can be described in four parts under the following assumptions.

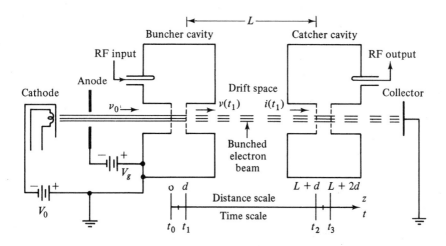

FIG. 5-2-1-1. Two-cavity klystron amplifier.

1. The electron beam is assumed to have a uniform density in the cross section of the beam.
2. Space-charge effects are negligible.
3. The magnitude of the microwave signal input is assumed to be much smaller than the dc accelerating voltage.

(1) Velocity-Modulation Process. When electrons are first accelerated by the high dc voltage V_0 before entering the buncher grids, their velocity is uniform.

$$v_0 = \sqrt{\frac{2eV_0}{m}} = 0.593 \times 10^6 \sqrt{V_0} \text{ m/s} \qquad (5\text{-}2\text{-}1)$$

In Eq. (5-2-1) it is assumed that electrons leave the cathode with zero velocity. When a microwave signal is applied to the input terminal, the gap voltage between the buncher grids appears as

$$V_s = V_1 \sin \omega t \qquad (5\text{-}2\text{-}2)$$

where V_1 is the amplitude of the signal and $V_1 \ll V_0$ is assumed.

In order to find the modulated velocity in the buncher cavity in terms of either the entering time t_0 or the exiting time t_1 and the gap transit angle θ_g as shown in Fig. 5-2-1-1, it is necessary to determine the average microwave voltage in the buncher gap as indicated in Fig. 5-2-1-2.

Since $V_1 \ll V_0$, the average transit time through the buncher gap d is

$$\tau \cong \frac{d}{v_0} = t_1 - t_0 \qquad (5\text{-}2\text{-}3)$$

The average gap transit angle can be expressed as

$$\theta_g = \omega\tau = \omega(t_1 - t_0) = \frac{\omega d}{v_0} \qquad (5\text{-}2\text{-}4)$$

181

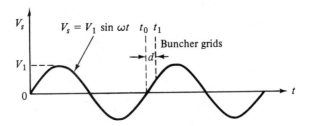

FIG. 5-2-1-2. Signal voltage in the buncher gap.

The average microwave voltage in the buncher gap can be found in the following way.

$$\langle V_s \rangle = \frac{1}{\tau} \int_{t_0}^{t_1} V_1 \sin \omega t \, dt = -\frac{V_1}{\omega \tau}[\cos \omega t_1 - \cos \omega t_0]$$

$$= \frac{V_1}{\omega \tau}\left[\cos \omega t_0 - \cos \omega \left(t_0 + \frac{d}{v_0}\right)\right] \qquad (5\text{-}2\text{-}5)$$

Let

$$\omega t_0 + \frac{\omega d}{2v_0} = \omega t_0 + \frac{\theta_g}{2} = A$$

and

$$\frac{\omega d}{2v_0} = \frac{\theta_g}{2} = B$$

Then using the trigonometric identity that $\cos (A - B) - \cos (A + B) = 2 \sin A \sin B$, Eq. (5-2-5) becomes

$$\langle V_s \rangle = V_1 \frac{\sin [\omega d/(2v_0)]}{\omega d/(2v_0)} \sin \omega \left(t_0 + \frac{d}{2v_0}\right)$$

$$= V_1 \frac{\sin (\theta_g/2)}{\theta_g/2} \sin (\omega t_0 + \theta_g/2) \qquad (5\text{-}2\text{-}6)$$

It is defined as

$$\beta_i \equiv \frac{\sin [\omega d/(2v_0)]}{\omega d/(2v_0)} = \frac{\sin (\theta_g/2)}{\theta_g/2} \qquad (5\text{-}2\text{-}7)$$

β_i is known as the *beam-coupling coefficient* of the input cavity gap (see Fig 5-2-1-3).

It can be seen that increasing the gap transit angle θ_g decreases the coupling between the electron beam and the buncher cavity; that is, the velocity modulation of the beam for a given microwave signal is decreased. Immediately after velocity modulation, the exit velocity from the buncher gap is given by

$$v(t_1) = \sqrt{\frac{2e}{m}\left[V_0 + \beta_i V_1 \sin \left(\omega t_0 + \frac{\theta_g}{2}\right)\right]}$$

$$= \sqrt{\frac{2e}{m}V_0\left[1 + \frac{\beta_i V_1}{V_0} \sin \left(\omega t_0 + \frac{\theta_g}{2}\right)\right]} \qquad (5\text{-}2\text{-}8)$$

where the factor $\beta_i V_1/V_0$ is called the *depth of velocity modulation*.

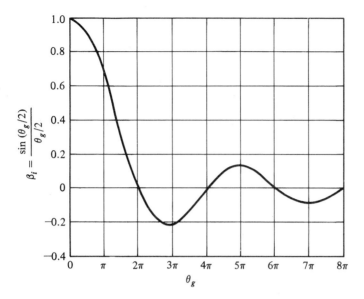

FIG. 5-2-1-3. Beam-coupling coefficient versus gap transit angle.

Using binomial expansion under the assumption of

$$\beta_i V_1 \ll V_0 \tag{5-2-9}$$

Eq. (5-2-8) becomes

$$v(t_1) = v_0 \left[1 + \frac{\beta_i V_1}{2V_0} \sin \left(\omega t_0 + \frac{\theta_g}{2} \right) \right] \tag{5-2-10}$$

Equation (5-2-10) is the equation of velocity modulation. Alternately, the equation of velocity modulation can be given by

$$v(t_1) = v_0 \left[1 + \frac{\beta_i V_1}{2V_0} \sin \left(\omega t_1 - \frac{\theta_g}{2} \right) \right] \tag{5-2-11}$$

(2) Bunching Process. Once the electrons leave the buncher cavity, they drift with a velocity given by Eq. (5-2-10) or (5-2-11) along in the field-free space between the two cavities. The effect of velocity modulation produces bunching of the electron beam—or current modulation. The electrons that pass the buncher at $V_s = 0$ travel through with unchanged velocity v_0 and become the bunching center. Those electrons that pass the buncher cavity during the positive half cycles of the microwave input voltage V_s travel faster than the electrons that passed the gap when $V_s = 0$. Those electrons that pass the buncher cavity during the negative half cycles of the voltage V_s travel slower than the electrons that passed the gap when $V_s = 0$. At a distance of ΔL along the beam from the buncher cavity, the beam electrons have drifted into dense clusters. Figure 5-2-1-4 shows the trajectories of minimum, zero, and maximum electron acceleration.

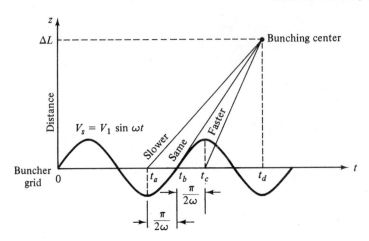

FIG. 5-2-1-4. Bunching distance.

The distance from the buncher grid to the location of dense electron bunching for the electron at t_b is

$$\Delta L = v_0(t_d - t_b) \qquad (5\text{-}2\text{-}12)$$

Similarly, the distances for the electrons at t_a and t_c are

$$\Delta L = v_{\min}(t_d - t_a) = v_{\min}\left(t_d - t_b + \frac{\pi}{2\omega}\right) \qquad (5\text{-}2\text{-}13)$$

and

$$\Delta L = v_{\max}(t_d - t_c) = v_{\max}\left(t_d - t_b - \frac{\pi}{2\omega}\right) \qquad (5\text{-}2\text{-}14)$$

From Eq. (5-2-10) or (5-2-11) the minimum and maximum velocities are

$$v_{\min} = v_0\left(1 - \frac{\beta_i V_1}{2V_0}\right) \qquad (5\text{-}2\text{-}15)$$

and

$$v_{\max} = v_0\left(1 + \frac{\beta_i V_1}{2V_0}\right) \qquad (5\text{-}2\text{-}16)$$

Substitution of Eqs. (5-2-15) and (5-2-16) in Eqs. (5-2-13) and (5-2-14), respectively, yields the distance

$$\Delta L = v_0(t_d - t_b) + \left[v_0\frac{\pi}{2\omega} - v_0\frac{\beta_i V_1}{2V_0}(t_d - t_b) - v_0\frac{\beta_i V_1}{2V_0}\frac{\pi}{2\omega}\right] \qquad (5\text{-}2\text{-}17)$$

and

$$\Delta L = v_0(t_d - t_b) + \left[-v_0\frac{\pi}{2\omega} + v_0\frac{\beta_i V_1}{2V_0}(t_d - t_b) + v_0\frac{\beta_i V_1}{2V_0}\frac{\pi}{2\omega}\right] \qquad (5\text{-}2\text{-}18)$$

The necessary condition for those electrons at t_a, t_b, and t_c to meet at the same distance ΔL is

$$v_0\frac{\pi}{2\omega} - v_0\frac{\beta_i V_1}{2V_0}(t_d - t_b) - v_0\frac{\beta_i V_1}{2V_0}\frac{\pi}{2\omega} = 0 \qquad (5\text{-}2\text{-}19)$$

and
$$-v_0\frac{\pi}{2\omega} + v_0\frac{\beta_i V_1}{2V_0}(t_d - t_b) + v_0\frac{\beta_i V_1}{2V_0}\frac{\pi}{2\omega} = 0 \qquad (5\text{-}2\text{-}20)$$

Consequently,
$$t_d - t_b \simeq \frac{\pi V_0}{\omega\beta_i V_1} \qquad (5\text{-}2\text{-}21)$$

and
$$\Delta L = v_0\frac{\pi V_0}{\omega\beta_i V_1} \qquad (5\text{-}2\text{-}22)$$

It should be noted that the mutual repulsion of the space charge is neglected, but the qualitative results are similar to the preceding representation when the effects of repulsion are included. Furthermore, the distance given by Eq. (5-2-22) is not the one for a maximum degree of bunching. Figure 5-2-1-5 shows the distance-time plot or Applegate diagram.

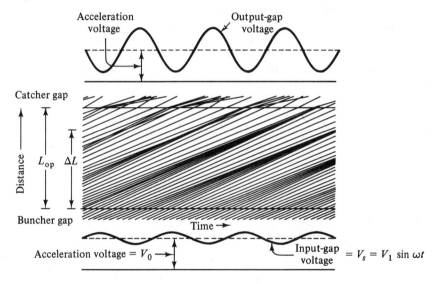

FIG. 5-2-1-5. Applegate diagram.

What should the spacing be between the buncher and catcher cavities in order to achieve a maximum degree of bunching? Since the drift region is field free, the transit time for an electron to travel a distance of L as shown in Fig. 5-2-1-1 is given by

$$T = t_2 - t_1 = \frac{L}{v(t_1)} = T_0\left[1 - \frac{\beta_i V_1}{2V_0}\sin\left(\omega t_1 - \frac{\theta_g}{2}\right)\right] \qquad (5\text{-}2\text{-}23)$$

where the binomial expansion of $(1 + x)^{-1}$ for $|x| \ll 1$ has been replaced. $T_0 = L/v_0$ is the dc transit time.
In terms of radians the above expression can be written

$$\omega T = \omega t_2 - \omega t_1 = \theta_0 - X\sin\left(\omega t_1 - \frac{\theta_g}{2}\right) \qquad (5\text{-}2\text{-}24a)$$

where $\qquad \theta_0 = \dfrac{\omega L}{v_0} = 2\pi N$ $\qquad\qquad\qquad$ (5-2-24b)

is the dc transit angle between cavities and N is the number of electron transit cycles in the drift space

$$X \equiv \frac{\beta_i V_1}{2 V_0} \theta_0 \qquad\qquad (5\text{-}2\text{-}24c)$$

is defined as the *bunching parameter* of a klystron

At the buncher gap a charge dQ_0 passing through at a time interval dt_0 is given by

$$dQ_0 = I_0 \, dt_0 \qquad\qquad (5\text{-}2\text{-}25)$$

where I_0 is the dc current. From the principle of conversation of charges this same amount of charge dQ_0 also passes the catcher at a later time interval dt_2. Hence

$$I_0 \, |dt_0| = i_2 \, |dt_2| \qquad\qquad (5\text{-}2\text{-}26)$$

where the absolute value signs are necessary because a negative value of the time ratio would indicate a negative current. i_2 is the current at the catcher gap. Rewriting Eq. (5-2-23) in terms of Eq. (5-2-10) yields

$$t_2 = t_0 + \tau + T_0 \left[1 - \frac{\beta_i V_1}{2 V_0} \sin \left(\omega t_0 + \frac{\theta_g}{2} \right) \right] \qquad (5\text{-}2\text{-}27)$$

Alternatively,

$$\omega t_2 - \left(\theta_0 + \frac{\theta_g}{2} \right) = \left(\omega t_0 + \frac{\theta_g}{2} \right) - X \sin \left(\omega t_0 + \frac{\theta_g}{2} \right) \qquad (5\text{-}2\text{-}28)$$

where $(\omega t_0 + \theta_g/2)$ is the buncher cavity departure angle and $\omega t_2 - (\theta_0 + \theta_g/2)$ is the catcher cavity arrival angle.

$\qquad$ Figure 5-2-1-6 shows the curves for the catcher cavity arrival angle as a function of the buncher cavity departure angle in terms of the bunching parameter X.

$\qquad$ Differentiation of Eq. (5-2-27) with respect to t_0 results in

$$dt_2 = dt_0 \left[1 - X \cos \left(\omega t_0 + \frac{\theta_g}{2} \right) \right] \qquad\qquad (5\text{-}2\text{-}29)$$

The current arriving at the catcher cavity is then given as

$$i_2(t_0) = \frac{I_0}{1 - X \cos (\omega t_0 + \theta_g/2)} \qquad\qquad (5\text{-}2\text{-}30)$$

In terms of t_2 the current is

$$i_2(t_2) = \frac{I_0}{1 - X \cos (\omega t_2 - \theta_0 - \theta_g/2)} \qquad\qquad (5\text{-}2\text{-}31)$$

In Eq. (5-2-31) the relationship of $t_2 = t_0 + \tau + T_0$ is used—namely, $\omega t_2 = \omega t_0 + \omega \tau + \omega T_0 = \omega t_0 + \theta_g + \theta_0$. Figure 5-2-1-7 shows curves of the beam

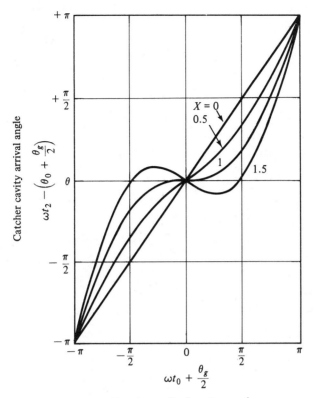

FIG. 5-2-1-6. Catcher arrival angle versus buncher departure angle.

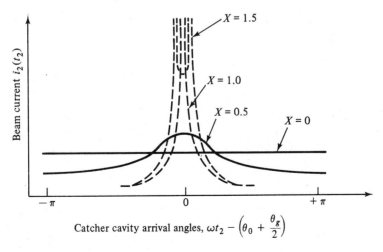

FIG. 5-2-1-7. Beam current i_2 versus catcher cavity arrival angle.

current $i_2(t_2)$ as a function of the catcher arrival angle in terms of the bunching parameter X.

The beam current at the catcher cavity is a periodic waveform of period $2\pi/\omega$ about dc current. Therefore the current i_2 can be expanded in a Fourier series and so

$$i_2 = a_0 + \sum_{n=1}^{\infty} (a_n \cos n\omega t_2 + b_n \sin n\omega t_2) \qquad (5\text{-}2\text{-}32)$$

where n is an integer, excluding zero. The series coefficients a_0, a_n, and b_n in Eq. (5-2-32) are given by the integrals,

$$a_0 = \frac{1}{2\pi} \int_{-\pi}^{\pi} i_2 \, d(\omega t_2)$$

$$a_n = \frac{1}{\pi} \int_{-\pi}^{\pi} i_2 \cos n\omega t_2 \, d(\omega t_2) \qquad (5\text{-}2\text{-}33)$$

$$b_n = \frac{1}{\pi} \int_{-\pi}^{\pi} i_2 \sin n\omega t_2 \, d(\omega t_2)$$

Substitution of Eqs. (5-2-26) and (5-2-27) in Eq. (5-2-33) yields

$$a_0 = \frac{1}{2\pi} \int_{-\pi}^{\pi} I_0 \, d(\omega t_0) = I_0 \qquad (5\text{-}2\text{-}34a)$$

$$a_n = \frac{1}{\pi} \int_{-\pi}^{\pi} I_0 \cos \left[(n\omega t_0 + n\theta_g + n\theta_0) - nX \sin \left(\omega t_0 + \frac{\theta_g}{2} \right) \right] d(\omega t_0)$$

$$(5\text{-}2\text{-}34b)$$

$$b_n = \frac{1}{\pi} \int_{-\pi}^{\pi} I_0 \sin \left[(n\omega t_0 + n\theta_g + n\theta_0) - nX \sin \left(\omega t_0 + \frac{\theta_g}{2} \right) \right] d(\omega t_0)$$

$$(5\text{-}2\text{-}34c)$$

By using the trigonometrical functions

$$\cos (A \pm B) = \cos A \cos B \mp \sin A \sin B$$

and

$$\sin (A \pm B) = \sin A \cos B \pm \cos A \sin B$$

the two integrals as shown in Eqs. (5-2-34b) and (5-2-34c) involve cosines and sines of a sine function. Each term of the integrand contains an infinite number of terms of Bessel functions. These are

$$\cos \left[nX \sin \left(\omega t_0 + \frac{\theta_g}{2} \right) \right] = J_0 \, (nX)$$

$$+ 2 \left[J_2(nX) \cos 2\left(\omega t_0 + \frac{\theta_g}{2} \right) \right]$$

$$+ 2 \left[J_4(nX) \cos 4\left(\omega t_0 + \frac{\theta_g}{2} \right) \right]$$

$$+ \cdots \qquad (5\text{-}2\text{-}34d)$$

and

$$\sin\left[nX\sin\left(\omega t_0 + \frac{\theta_g}{2}\right)\right] = 2\left[J_1(nX)\sin\left(\omega t_0 + \frac{\theta_g}{2}\right)\right]$$
$$+ 2\left[J_3(nX)\sin 3\left(\omega t_0 + \frac{\theta_g}{2}\right)\right]$$
$$+ \dots \hspace{3cm} (5\text{-}2\text{-}34e)$$

If the above series are substituted into the integrands of Eqs. (5-2-34b) and (5-2-34c), respectively, the integrals are readily evaluated term by term and the series coefficients are

$$a_n = 2I_0 J_n(nX) \cos(n\theta_g + n\theta_0) \hspace{2cm} (5\text{-}2\text{-}34f)$$
$$b_n = 2I_0 J_n(nX) \sin(n\theta_g + n\theta_0) \hspace{2cm} (5\text{-}2\text{-}34g)$$

where $J_n(nX)$ is the nth-order Bessel function of the first kind (see Fig. 5-2-1-8).

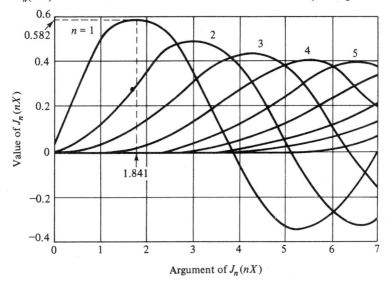

FIG. 5-2-1-8. Bessel functions $J_n(nX)$.

Substitution of Eqs. (5-2-34a, f, g) in (5-2-32) yields the beam current i_2 as

$$i_2 = I_0 + \sum_{n=1}^{\infty} 2I_0 J_n(nX) \cos n\omega(t_2 - \tau - T_0) \hspace{1cm} (5\text{-}2\text{-}35)$$

The fundamental component of the beam current at the catcher cavity has a magnitude

$$I_f = 2I_0 J_1(X) \hspace{3cm} (5\text{-}2\text{-}36)$$

This fundamental component I_f has its maximum amplitude at

$$X = 1.841 \hspace{3cm} (5\text{-}2\text{-}37)$$

The optimum distance L at which the maximum fundamental component of current occurs is computed from Eqs. (5-2-24b), (5-2-24c), and (5-2-37) as

$$L_{\text{optimum}} = \frac{3.682 v_0 V_0}{\omega \beta_i V_1} \qquad (5\text{-}2\text{-}38)$$

It is interesting to note that the distance given by Eq. (5-2-22) is approximately 15% less than the result of Eq. (5-2-38). The discrepancy is due in part to the approximations made in deriving Eq. (5-2-22) and to the fact that the maximum fundamental component of current will not coincide with the maximum electron density along the beam because the harmonic components exist in the beam.

(3) *Output Power and Beam Loading.* The maximum bunching should occur approximately at the midway between the catcher grids. The phase of the catcher gap voltage must be maintained in such a way that the bunched electrons, as they pass through the grids, encounter a retarding phase. When the bunched electron beam passes through the retarding phase, its kinetic energy is transferred to the field of the catcher cavity. When the electrons emerge from the catcher grids, they have reduced velocity and are finally collected by the collector.

(a) THE INDUCED CURRENT IN THE CATCHER CAVITY. Since the current induced by the electron beam in the walls of the catcher cavity is directly proportional to the amplitude of the microwave input voltage V_1, the fundamental component of the induced microwave current in the catcher is given by

$$i_{2\,\text{ind}} = \beta_0 i_2 = \beta_0 2 I_0 J_1(X) \cos \omega(t_2 - \tau - T_0) \qquad (5\text{-}2\text{-}39)$$

where β_0 is the beam coupling coefficient of the catcher gap. If the buncher and catcher cavities are identical, then $\beta_i = \beta_0$. The fundamental component of current induced in the catcher cavity then has a magnitude

$$I_{2\,\text{ind}} = \beta_0 I_2 = \beta_0 2 I_0 J_1(X) \qquad (5\text{-}2\text{-}40)$$

Figure 5-2-1-9 shows an output equivalent circuit in which R_{sho} represents the wall resistance of catcher cavity, R_B the beam loading resistance, R_L the external load resistance, and R_{sh} the effective shunt resistance.

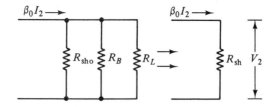

FIG. 5-2-1-9. Output equivalent circuit.

The output power delivered to the catcher cavity is given as

$$P_{\text{out}} = \frac{(\beta_0 I_2)^2}{2} R_{\text{sh}} = \frac{\beta_0 I_2 V_2}{2} \qquad (5\text{-}2\text{-}41)$$

where R_{sh} is the total equivalent shunt resistance of the catcher circuit, including the load, and V_2 is the fundamental component of the catcher gap voltage.

(b) EFFICIENCY OF KLYSTRON. The electronic efficiency of the klystron amplifier is defined as the ratio of the output power to the input power

$$\text{Efficiency} \equiv \frac{P_{\text{out}}}{P_{\text{in}}} = \frac{\beta_0 I_2 V_2}{2 I_0 V_0} \qquad (5\text{-}2\text{-}42)$$

in which the power losses to the beam loading and cavity walls are not included.

If the coupling is perfect, $\beta_0 = 1$, the maximum beam current approaches $I_{2\,\text{max}} = 2I_0(0.582)$, and the voltage V_2 is equal to V_0. Then the maximum electronic efficiency is about 58%. In practice, the electronic efficiency of a klystron amplifier is in the range of 15 to 30%. Since the efficiency is a function of the catcher gap transit angle θ_g, Fig. 5-2-1-10 shows the maximum efficiency of a klystron as a function of catcher transit angle.

(c) MUTUAL CONDUCTANCE OF A KLYSTRON AMPLIFIER. The equivalent mutual conductance of the klystron amplifier can be defined as the ratio of

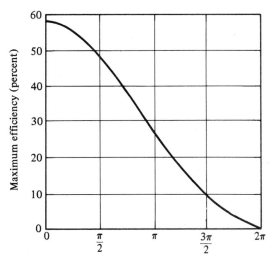

Catcher gap transit angle θ_g

FIG. 5-2-1-10. Maximum efficiency of klystron versus transit angle.

the induced output current to input voltage. That is,

$$|G_m| \equiv \frac{i_{2\,\text{ind}}}{V_1} = \frac{2\beta_0 I_0 J_1(X)}{V_1}$$ (5-2-43)

From Eq. (5-2-24c) the input voltage V_1 can be expressed in terms of the bunching parameter X as

$$V_1 = \frac{2V_0}{\beta_0 \theta_0} X$$ (5-2-44)

In Eq. (5-2-44) it is assumed that $\beta_0 = \beta_l$. Substitution of Eq. (5-2-44) in Eq. (5-2-43) yields the normalized mutual conductance as

$$\frac{|G_m|}{G_0} = \beta_0^2 \theta_0 \frac{J_1(X)}{X}$$ (5-2-45)

where $G_0 = I_0/V_0$ is the dc beam conductance. The mutual conductance is not a constant but decreases as the bunching parameter X increases. Figure 5-2-1-11 shows the curves of normalized transductance as a function of X.

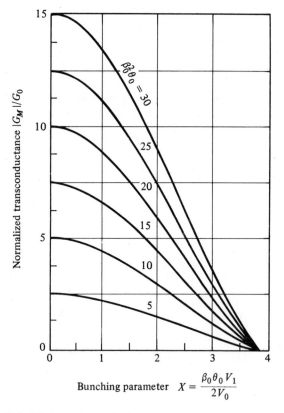

FIG. 5-2-1-11. Normalized transconductance versus bunching parameter.

It can be seen from the curves that, for a small signal, the normalized transconductance is maximum. That is,

$$\frac{|G_m|}{G_0} = \frac{\beta_0^2\theta_0}{2} \tag{5-2-46}$$

For a maximum output at $X = 1.841$, the normalized mutual conductance is

$$\frac{|G_m|}{G_0} = 0.316\beta_0^2\theta_0 \tag{5-2-47}$$

The voltage gain of a klystron amplifier is defined as

$$A_v \equiv \frac{|V_2|}{|V_1|} = \frac{\beta_0 I_2 R_{sh}}{V_1} = \frac{\beta_0^2\theta_0}{R_0}\frac{J_1(X)}{X}R_{sh} \tag{5-2-48}$$

where $R_0 = V_0/I_0$ is the dc beam resistance. Substitution of Eqs. (5-2-40) and (5-2-44) in Eq. (5-2-48) results in

$$A_v = G_m R_{sh} \tag{5-2-49}$$

(d) POWER REQUIRED TO BUNCH THE ELECTRON BEAM. As described earlier, the bunching action takes place in the buncher cavity. When the buncher gap transit angle is small, the average energy of electrons leaving the buncher cavity over a cycle is nearly equal to the energy with which they enter. However, when the buncher gap transit angle is large, the electrons that leave the buncher gap have greater average energy than when they enter. The difference between the average exit energy and the entrance energy must be supplied by the buncher cavity to bunch the electron beam. It is difficult to calculate the power required to produce the bunching action. Feenberg did some extensive work on beam loading [5]. The ratio of the power required to produce bunching action to the dc power required to perform the electron beam is given by Feenberg as

$$\frac{P_B}{P_0} = \frac{V_1^2}{2V_0^2}\left[\frac{1}{2}\beta_i^2 - \frac{1}{2}\beta_i\cos\left(\frac{\theta_g}{2}\right)\right] = \frac{V_1^2}{2V_0^2}F(\theta_g) \tag{5-2-50}$$

where

$$F(\theta_g) = \frac{1}{2}\left[\beta_i^2 - \beta_i\cos\left(\frac{\theta_g}{2}\right)\right]$$

Since the dc power is

$$P_0 = V_0^2 G_0 \tag{5-2-51}$$

where $G_0 = I_0/V_0$ is the equivalent electron beam conductance. The power given by the buncher cavity to produce beam bunching is

$$P_B = \frac{V_1^2}{2}G_B \tag{5-2-52}$$

where G_B is the equivalent bunching conductance. Substitution of Eqs. (5-2-51) and (5-2-52) in Eq. (5-2-50) yields the normalized electronic conductance as

$$\frac{G_B}{G_0} = \frac{R_0}{R_B} = F(\theta_g) \tag{5-2-53}$$

Figure 5-2-1-12 shows the normalized electronic conductance as a function of the buncher gap transit angle. It can be seen that there is a critical buncher

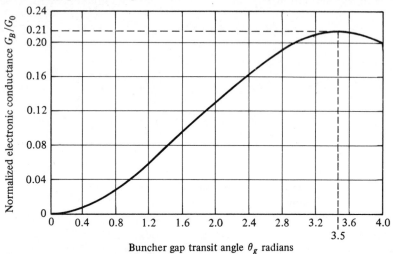

FIG. 5-2-1-12. Normalized electronic conductance versus buncher gap transit angle.

gap transit angle for a minimum equivalent bunching resistance. When the transit angle θ_g is 3.5 radians, the equivalent bunching resistance is about five times the electron beam resistance. The power delivered by the electron beam to the catcher cavity can be expressed as

$$\frac{V_2^2}{2R_{sh}} = \frac{V_2^2}{2R_{sho}} + \frac{V_2^2}{2R_B} + \frac{V_2^2}{2R_L} \tag{5-2-54}$$

As a result, the effective impedance of the catcher cavity is

$$\frac{1}{R_{sh}} = \frac{1}{R_{sho}} + \frac{1}{R_B} + \frac{1}{R_L} \tag{5-2-55}$$

Finally, the loaded quality factor of the catcher cavity circuit at the resonant frequency can be written

$$\frac{1}{Q_L} = \frac{1}{Q_0} + \frac{1}{Q_B} + \frac{1}{Q_{ext}} \tag{5-2-56}$$

where Q_L = the loaded quality of the whole catcher circuit
Q_0 = the quality of the catcher walls
Q_B = the quality of the beam loading
Q_{ext} = the quality of the external load

EXAMPLE 5-2-1: Klystron Amplifier. A two-cavity klystron amplifier has the following parameters:

$V_0 = 1000$ volts $\quad R_0 = 40$ kΩ

$I_0 = 25$ mA $\quad\quad f = 3$ GHz

Gap spacing in either cavity: $d = 1$ mm

Spacing between the two cavities: $L = 4$ cm

Effective shunt impedance, excluding beam loading: $R_{sh} = 30$ kΩ

(a) Find the input gap voltage to give maximum voltage V_2.
(b) Find the voltage gain, neglecting the beam loading in the output
cavity.
(c) Find the efficiency of the amplifier, neglecting beam loading.
(d) Calculate the beam loading conductance and show that neglecting it
was justified in the above calculations.

Solution.

(a) For maximum V_2, $J_1(X)$ must be maximum. This means $J_1(X) = 0.582$ at $X = 1.841$.

The electron velocity just leaving the cathode is

$$v_0 = (0.593 \times 10^6)\sqrt{V_0} = (0.593 \times 10^6)\sqrt{10^3} = 1.88 \times 10^7 \text{ m/s}$$

The gap transit angle is

$$\theta_g = \omega \frac{d}{v_0} = 2\pi(3 \times 10^9)\frac{10^{-3}}{1.88 \times 10^7} = 1 \text{ radian}$$

The beam-coupling coefficient is

$$\beta_i = \beta_0 = \frac{\sin(\theta_g/2)}{\theta_g/2} = \frac{\sin(1/2)}{1/2} = 0.952$$

The dc transit angle between the cavities is

$$\theta_0 = \omega T_0 = \omega \frac{L}{v_0} = 2\pi(3 \times 10^9)\frac{4 \times 10^{-2}}{1.88 \times 10^7} = 40 \text{ radians}$$

The maximum input voltage V_1 is then given by

$$V_{1 \text{ max}} = \frac{2V_0 X}{\beta_i \theta_0} = \frac{2(10^3)(1.841)}{(0.952)(40)} = 96.5 \text{ volts}$$

(b) The voltage gain is found as

$$A_v = \frac{\beta_0^2 \theta_0}{R_0} \frac{J_1(X)}{X} R_{sh} = \frac{(0.952)^2(40)(0.582)(30 \times 10^3)}{4 \times 10^4 \times 1.841} = 8.595$$

(c) The efficiency can be found as follows.

$$I_2 = 2I_0 J_1(X) = 2 \times 25 \times 10^{-3} \times 0.582 = 29.1 \times 10^{-3} \text{ A}$$

$$V_2 = \beta_0 I_2 R_{sh} = (0.952)(29.1 \times 10^{-3})(30 \times 10^3) = 831 \text{ volts}$$

$$\text{Efficiency} = \frac{\beta_0 I_2 V_2}{2I_0 V_0} = \frac{(0.952)(29.1 \times 10^{-3})(831)}{2(25 \times 10^{-3})(10^3)} = 46.2\%$$

(d) Calculate the beam loading conductance (refer to Fig. 5-2-1-9). The beam loading conductance G_B is

$$G_B = \frac{G_0}{2}\left(\beta_0^2 - \beta_0 \cos \frac{\theta_g}{2}\right) = \frac{25 \times 10^{-6}}{2}[(0.952)^2 - (0.952) \cos (28.6°)]$$

$$= 8.8 \times 10^{-7} \text{ mho}$$

Then the beam loading resistance R_B is

$$R_B = \frac{1}{G_B} = 1.14 \times 10^6 \qquad \text{ohms}$$

In comparison with R_L and R_{sho} or the effective shunt resistance R_{sh}, the beam loading resistance is like an open circuit and thus can be neglected in the preceding calculations.

(4) State of the Art

(a) EXTENDED INTERACTION. The most common form of extended interaction has been attained recently by coupling two or more adjacent klystron cavities. Figure 5-2-1-13 shows schematically a five-section extended-interaction cavity as compared to a single-gap klystron cavity.

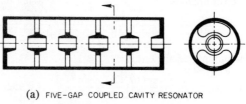

(a) FIVE-GAP COUPLED CAVITY RESONATOR

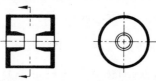

(b) SINGLE-GAP KLYSTRON CAVITY

FIG. 5-2-1-13. Comparison of a five-gap extended interaction cavity with a single-gap klystron cavity. (a) Five-gap coupled cavity resonator. (b) Single-gap klystron cavity. (*After A. Staprans et al. [18]; reprinted with permission from the IEEE, Inc.*)

(b) HIGH EFFICIENCY AND LARGE POWER. In the 1960s much effort was devoted to improving the efficiency of klystrons. For instance, a 50-kW experimental tube has demonstrated 75% efficiency in the industrial heating band [6]. The VA-884D klystron is a five-cavity amplifier whose operating characteristics are listed in Table 5-2-1.

TABLE 5-2-1. VA-884D Operating Characteristics.

Frequency	5.9–6.45	GHz
Power output	14	kW
Gain	52	dB
Efficiency	36	%
Electronic bandwidth (1 dB)	75	MHz
Beam voltage	16.5 ·	kV
Beam current	2.4	A

One of the better-known high-peak power klystrons is the tube developed specifically for use in the 2-mile Stanford Linear Accelerator [7] at Palo Alto, California. A cutaway view of the tube is shown in Fig. 5-2-1-14. The operating characteristics of this tube are listed in Table 5-2-2.

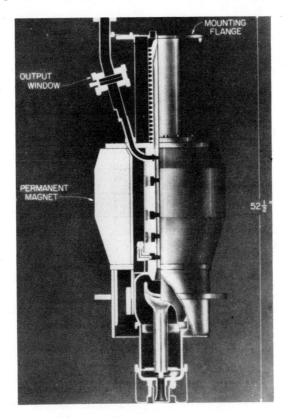

FIG. 5-2-1-14. Cutaway of 24-MW S-band permanent-magnet-focused klystron. (*Courtesy Stanford Linear Accelerator Center.*)

TABLE 5-2-2.   Operating Characteristics of the Stanford Linear Accelerator Center High-Power Klystron.

Frequency	2.856	GHz
RF pulse width	2.5	μsec
Pulse repetition rate	60–360	pps
Peak power output	24	MW
Beam voltage	250	kV
Beam current	250	A
Gain	50 to 55	dB
Efficiency	about 36	%
Weight of permanent (focusing) magnet	363	kg
Electronic bandwidth (1 dB)	20	MHz

5-2-2. Reflex Klystrons

If a fraction of the output power is fed back to the input cavity and if the loop gain has a magnitude of unity with a phase shift of multiple 2π, the klystron will oscillate. However, a two-cavity klystron oscillator is usually not constructed because, when the oscillation frequency is varied, the resonant frequency of each cavity and the feedback path phase shift must be readjusted for a positive feedback. The reflex klystron is a single-cavity klystron that overcomes the disadvantages of the two-cavity klystron oscillator. It is a low-power generator of 10 to 500-mW output at a frequency range of 1 to 25 GHz. The efficiency is about 20 to 30%. This type is widely used in the laboratory for microwave measurements and in microwave receivers as local oscillators in commercial, military, and airborne Doppler radars as well as missiles. The theory of the two-cavity klystron can be applied to the analysis of the reflex klystron with slight modification. A schematic diagram of the reflex klystron is shown in Fig. 5-2-2-1.

The electron beam injected from the cathode is first velocity-modulated by the cavity-gap voltage. Some electrons accelerated by the accelerating field enter the repeller space with greater velocity than those with unchanged velocity. Some electrons decelerated by the retarding field enter the repeller region with less velocity. All electrons turned around by the repeller voltage then pass through the cavity gap in bunches that occur once per cycle. On their return journey the bunched electrons pass through the gap during the retarding phase of the alternating field and give up their kinetic energy to the electromagnetic energy of the field in the cavity. Oscillator output energy is then taken from the cavity. The electrons are finally collected by the walls of the cavity or other grounded metal parts of the tube. Figure 5-2-2-2 shows an Applegate diagram for the $1\frac{3}{4}$ mode of a reflex klystron.

(1) *Velocity Modulation.* The analysis of a reflex klystron is similar to that of a two-cavity klystron. For simplicity, the effect of space-charge forces on the electron motion will again be neglected. The electron entering the

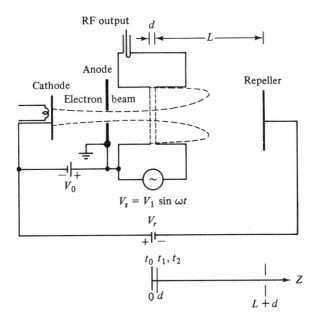

t_0 = the time for the electron entering the gavity gap at $z = 0$

t_1 = the time for the same electron leaving the cavity gap at $z = d$

t_2 = the time for the same electron returned by the retarding field at $z = d$ and collected on the walls of the cavity

FIG. 5-2-2-1. Schematic diagram of a reflex klystron.

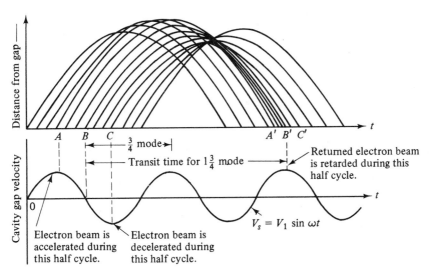

FIG. 5-2-2-2. Applegate diagram with gap voltage for a reflex klystron.

199

cavity gap from the cathode at $z = 0$ and time t_0 is assumed to have uniform velocity

$$v_0 = 0.593 \times 10^6 \sqrt{V_0} \qquad (5\text{-}2\text{-}57)$$

The same electron leaves the cavity gap at $z = d$ at time t_1 with velocity

$$v(t_1) = v_0 \left[1 + \frac{\beta_i V_1}{2V_0} \sin \left(\omega t_1 - \frac{\theta_g}{2} \right) \right] \qquad (5\text{-}2\text{-}58)$$

This expression is identical to Eq. (5-2-11), for the problems up to this point are identical to those of a two-cavity klystron amplifier. The same electron is forced back to the cavity at $z = d$ and time t_2 by the retarding electric field E, which is given by

$$E = \frac{V_r + (V_0 + V_1 \sin \omega t)}{L} \qquad (5\text{-}2\text{-}59)$$

This retarding field E is assumed to be constant in the z direction. The force equation for one electron in the repeller region is

$$M \frac{d^2 z}{dt^2} = -eE = -e \frac{(V_r + V_0)}{L} \qquad (5\text{-}2\text{-}60\text{a})$$

where $\mathbf{E} = -\nabla V$ is used in the z direction only, V_r is the magnitude of the repeller voltage, and $|V_1 \sin \omega t| \ll (V_r + V_0)$ is assumed.

Integration of Eq. (5-2-60a) twice yields

$$\frac{dz}{dt} = \frac{-e(V_r + V_0)}{mL} \int_{t_1}^{t} dt = \frac{-e(V_r + V_0)}{mL}(t - t_1) + K_1 \qquad (5\text{-}2\text{-}60\text{b})$$

at $t = t_1$, $\dfrac{dz}{dt} = v(t_1) = K_1$

$$z = \frac{-e(V_r + V_0)}{mL} \int_{t_1}^{t} (t - t_1)\, dt + v(t_1) \int_{t_1}^{t} dt$$

$$z = \frac{-e(V_r + V_0)}{2mL}(t - t_1)^2 + v(t_1)(t - t_1) + K_2$$

at $t = t_1$, $z = d = K_2$, then

$$z = \frac{-e(V_r + V_0)}{2mL}(t - t_1)^2 + v(t_1)(t - t_1) + d \qquad (5\text{-}2\text{-}61)$$

On the assumption that the electron leaves the cavity gap at $z = d$ and time t_1 with a velocity of $v(t_1)$ and returns to the gap at $z = d$ and time t_2, then

at $t = t_2$, $z = d$,

$$0 = \frac{-e(V_r + V_0)}{2mL}(t_2 - t_1)^2 + v(t_1)(t_2 - t_1)$$

The round-trip transit time in the repeller region is given by

$$T' = t_2 - t_1 = \frac{2mL}{e(V_r + V_0)}v(t_1) = T'_0\left[1 + \frac{\beta_i V_1}{2V_0} \sin\left(\omega t_1 - \frac{\theta_g}{2}\right)\right] \quad (5\text{-}2\text{-}62\text{a})$$

$$\text{where} \quad T'_0 = \frac{2mLv_0}{e(V_r + V_0)} \quad (5\text{-}2\text{-}62\text{b})$$

is the round-trip dc transit time of the center-of-the-bunch election.

Multiplication of Eq. (5-2-62a) through by a radian frequency results in

$$\omega(t_2 - t_1) = \theta'_0 + X' \sin\left(\omega t_1 - \frac{\theta_g}{2}\right) \quad (5\text{-}2\text{-}63\text{a})$$

$$\text{where} \quad \theta'_0 = \omega T'_0 \quad (5\text{-}2\text{-}63\text{b})$$

is the round trip dc transit angle of the center-of-the-bunch electron

$$X' \equiv \frac{\beta_i V_1}{2V_0}\theta'_0 \quad (5\text{-}2\text{-}63\text{c})$$

is the bunching parameter of the reflex klystron oscillator.

(2) **Power Output and Efficiency.** In order for the electron beam to generate a maximum amount of energy to the oscillation, the returning electron beam must cross the cavity gap when the gap field is maximum retarding. In this way, a maximum amount of kinetic energy can be transferred from the returning electrons to the cavity walls. It can be seen from Fig. 5-2-2-2 that for a maximum energy transfer, the round-trip transit angle, referring to the center of the bunch, must be given by

$$\omega(t_2 - t_1) = \omega T'_0 = \left(n - \frac{1}{4}\right)2\pi = N2\pi = 2\pi n - \frac{\pi}{2} \quad (5\text{-}2\text{-}64)$$

where $V_1 \ll V_0$ is assumed
     $n =$ any positive integer for cycle number
     $N = n - \frac{1}{4}$ is the number of modes

The current modulation on the electron beam as it reenters the cavity from the repeller region can be determined in the same manner as in Section 5-2-1 for a two-cavity klystron amplifier. It can be seen from Eqs. (5-2-24a) and (5-2-63a) that the bunching parameter X' of a reflex klystron oscillator has a negative sign with respect to the bunching parameter X of a two-cavity klystron amplifier. Furthermore, the beam current injected into the cavity gap from the repeller region flows in the negative z direction. Consequently, the beam current of a reflex klystron oscillator can be written

$$i_{2t} = -I_0 - \sum_{n=1}^{\infty} 2I_0 J_n(nX') \cos n(\omega t_2 - \theta'_0 - \theta_g) \quad (5\text{-}2\text{-}65)$$

The fundamental component of the current induced in the cavity by the modulated electron beam is given by

$$i_2 = -\beta_i I_2 = 2I_0\beta_i J_1(X')\cos(\omega t_2 - \theta'_0) \qquad (5\text{-}2\text{-}66)$$

in which θ_g has been neglected as a small quantity compared with θ'_0. The magnitude of the fundamental component is

$$I_2 = 2I_0\beta_i J_1(X') \qquad (5\text{-}2\text{-}67)$$

The dc power supplied by the beam voltage V_0 is

$$P_{dc} = V_0 I_0 \qquad (5\text{-}2\text{-}68)$$

and the ac power delivered to the load is given by

$$P_{ac} = \frac{V_1 I_2}{2} = V_1 I_0 \beta_i J_1(X') \qquad (5\text{-}2\text{-}69)$$

From Eqs. (5-2-63b), (5-2-63c), and (5-2-64) the ratio of V_1 over V_0 is expressed by

$$\frac{V_1}{V_0} = \frac{2X'}{\beta_i(2\pi n - \pi/2)} \qquad (5\text{-}2\text{-}70)$$

Substitution of Eq. (5-2-70) in Eq. (5-2-69) yields the power output as

$$P_{ac} = \frac{2V_0 I_0 X' J_1(X')}{2\pi n - \pi/2} \qquad (5\text{-}2\text{-}71)$$

Therefore the electronic efficiency of a reflex klystron oscillator is defined as

$$\text{Efficiency} \equiv \frac{P_{ac}}{P_{dc}} = \frac{2X' J_1(X')}{2\pi n - \pi/2} \qquad (5\text{-}2\text{-}72)$$

The factor $X'J_1(X')$ reaches a maximum value of 1.25 at $X' = 2.408$ and $J_1(X') = 0.52$, In practice, the $n = 1$ or $n = 2$ modes have the most power output. If $n = 2$ or $1\frac{3}{4}$ mode, the maximum electronic efficiency becomes

$$\text{Efficiency}_{max} = \frac{2(2.408)J_1(2.408)}{2\pi(2) - \pi/2} = 22.7\% \qquad (5\text{-}2\text{-}73)$$

The maximum theoretical efficiency of a reflex klystron oscillator ranges from 20 to 30%. Figure 5-2-2-3 shows a curve of $X'J_1(X')$ vs. X'.

For a given beam voltage V_0, the relationship between the repeller voltage and cycle number n required for oscillation is found by inserting Eqs. (5-2-64) and (5-2-57) into Eq. (5-2-62b).

$$\frac{V_0}{(V_r + V_0)^2} = \frac{(2\pi n - \pi/2)^2}{8\omega^2 L^2}\frac{e}{m} \qquad (5\text{-}2\text{-}74)$$

The power output can be expressed in terms of the repeller voltage V_r. That is,

$$P_{ac} = \frac{V_0 I_0 X' J_1(X')(V_r + V_0)}{\omega L}\sqrt{\frac{e}{2mV_0}} \qquad (5\text{-}2\text{-}75)$$

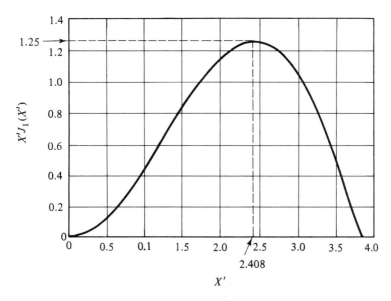

FIG. 5-2-2-3. $X'J_1(X')$ versus X'.

It can be seen from Eq. (5-2-74) that, for a given beam voltage V_0 and cycle number n or mode number N, the center repeller voltage V_r can be determined in terms of the center frequency. Then the power output at the center frequency can be calculated from Eq. (5-2-75). When the frequency varies from the center frequency and the repeller voltage about the center voltage, the power output will vary accordingly, assuming a bellshape (see Fig. 5-2-2-4).

(3) Electronic Admittance. From Eq. (5-2-66) the induced current can be written in phasor form as

$$i_2 = 2I_0\beta_i J_1(X')e^{-j\theta_0'} \qquad (5\text{-}2\text{-}76a)$$

The voltage across the gap at time t_2 can also be written in phasor form.

$$V_2 = V_1 e^{-j\pi/2} \qquad (5\text{-}2\text{-}76b)$$

The ratio of i_2 to V_2 is defined as the electronic admittance of the reflex klystron. That is,

$$Y_e = \frac{I_0}{V_0}\frac{\beta_i^2\theta_0'}{2}\frac{2J_1(X')}{X'}e^{j(\pi/2-\theta_0')} \qquad (5\text{-}2\text{-}77)$$

The amplitude of the phasor admittance indicates that the electronic admittance is a function of the dc beam admittance, the dc transit angle, and the second transit of the electron beam through the cavity gap. It is evident that the electronic admittance is nonlinear, since it is proportional to the factor $2J_1(X')/X'$, and X' is proportional to the signal voltage. This factor of proportionality is shown in Fig. 5-2-2-5. When the signal voltage goes to zero, the factor approaches unity.

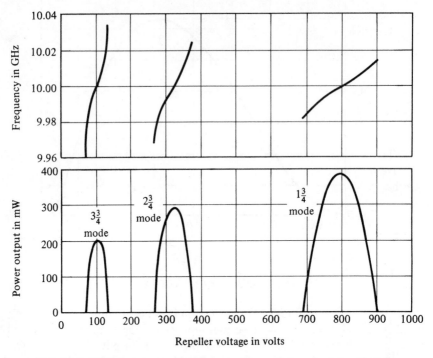

FIG. 5-2-2-4. Power output and frequency characteristics of a reflex klystron.

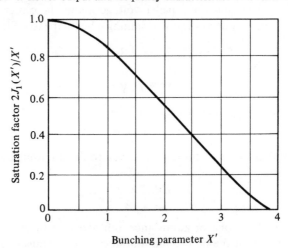

FIG. 5-2-2-5. Reflex klystron saturation factor.

The equivalent circuit of a reflex klystron is shown in Fig. 5-2-2-6. In this circuit L and C are the energy storage elements of the cavity, G_c the copper losses of the cavity, G_b the beam loading conductance, and G_ℓ the load conductance.

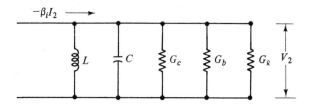

$-\beta_i I_2 \longrightarrow$

FIG. 5-2-2-6. Equivalent circuit of a reflex klystron.

The necessary condition for oscillations is that the magnitude of the negative real part of the electronic admittance as given by Eq. (5-2-77) not be less than the total conductance of the cavity circuit. That is,

$$|-G_e| \geq G \qquad (5\text{-}2\text{-}78)$$

where $G = G_c + G_b + G_\ell = 1/R_{sh}$ and R_{sh} is the effective shunt resistance.

Equation (5-2-77) can be rewritten in rectangular form.

$$Y_e = G_e + jB_e \qquad (5\text{-}2\text{-}79)$$

Since the electronic admittance as shown in Eq. (5-2-77) is in exponential form, its phase is $\pi/2$ when θ_0' is zero. The rectangular plot of the electron admittance Y_e is a spiral (see Fig. 5-2-2-7). Any value of θ_0' for which the spiral lies in the area to the left of line $(-G - jB)$ will yield oscillation. Since the spiral cuts the negative real axis at

$$\theta_0' = \left(n - \frac{1}{4}\right)2\pi = N2\pi \qquad (5\text{-}2\text{-}80)$$

where N is the mode number as indicated in the plot, the phenomenon verifies the early analysis.

EXAMPLE 5-2-2: Reflex Klystron. A reflex klystron operates under the following conditions:

$V_0 = 600$ volts
$R_{sh} = 15$ kΩ
$f_r = 9$ GHz
$L = 1$ mm
$\dfrac{e}{m} = 1.759 \times 10^{11}$ (MKS system)

The tube is oscillating at f_r at the peak of the $n = 2$ mode or $1\frac{3}{4}$ mode. Assume that the transit time through the gap and beam loading can be neglected.

(a) Find the value of the repeller voltage V_r.
(b) Find the direct current necessary to give a microwave gap voltage of 200 volts.
(c) What is the electronic efficiency under this condition?

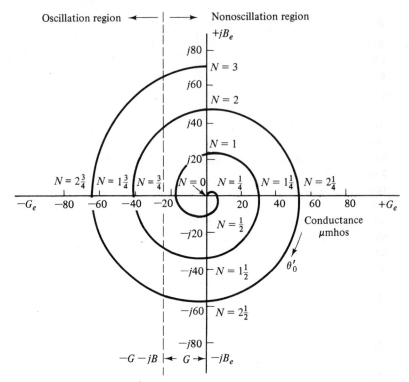

FIG. 5-2-2-7. Electronic admittance spiral of a reflex klystron.

Solution.

(a) From Eq. (5-2-74)

$$\frac{V_0}{(V_r + V_0)^2} = \left(\frac{e}{m}\right)\frac{(2\pi n - \pi/2)^2}{8\omega^2 L^2}$$

$$= (1.759 \times 10^{11})\frac{(2\pi 2 - \pi/2)^2}{8(2\pi \times 9 \times 10^9)^2(10^{-3})^2} = 0.832 \times 10^{-3}$$

$$(V_r + V_0)^2 = \frac{600}{0.832 \times 10^{-3}} = 0.721 \times 10^6$$

$$V_r = 250 \quad \text{volts}$$

(b) Assume that $\beta_0 = 1$. Since

$$V_1 = I_2 R_{sh} = 2I_0 J_1(X')R_{sh}$$

the direct current I_0 is

$$I_0 = \frac{V_1}{2J_1(X')R_{sh}} = \frac{200}{2(0.582)(15 \times 10^3)} = 11.45 \quad \text{mA}$$

(c) From Eq. (5-2-72) the electronic efficiency is

$$\text{Efficiency} = \frac{2X'J_1(X')}{2\pi n - \pi/2} = \frac{2(1.841)(0.582)}{2\pi(2) - \pi/2} = 19.5\%$$

5-2-3. Helix Traveling-Wave Tubes (TWT)

Since Kompfner invented the helix traveling-wave tube (TWT) in 1944 [9], its basic circuit has changed little. For broad-band applications, the helix TWTs are almost exclusively used, whereas for high-average-power purposes, such as radar transmitters, the coupled-cavity TWTs are commonly used.

In previous sections klystrons and reflex klystrons were analyzed in some detail. Before starting to describe the TWT, it seems appropriate to compare the basic operating principles of both the TWT and the klystron. In the case of the TWT, the microwave circuit is nonresonant and the wave propagates with the same speed as the electrons in the beam. The initial effect on the beam is a small amount of velocity modulation caused by the weak electric fields associated with the traveling wave. Just as in the klystron, this velocity modulation later translates to current modulation, which then induces an RF current in the circuit, causing amplification. However, there are some major differences between the TWT and the klystron.

1. The interaction of electron beam and RF field in the TWT is continuous over the entire length of the circuit, but the interaction in the klystron occurs only at the gaps of a few resonant cavities.
2. The wave in the TWT is a propagating wave; the wave in the klystron is not.
3. In the coupled-cavity TWT there is a coupling effect between the cavities, whereas each cavity in the klystron operates independently.

A helix traveling-wave tube consists of an electron beam and a slow-wave structure. The electron beam is focused by a constant magnetic field along the electron beam and the slow-wave structure. This is termed an O-type traveling-wave tube. The slow-wave structure is either the helical type or folded-back line. The applied signal propagates around the turns of the helix and produces an electric field at the center of the helix, directed along the helix axis. The axial electric field progresses with a velocity that is very close to the velocity of light multiplied by the ratio of helix pitch to helix circumference. When the electrons enter the helix tube, an interaction takes place between the moving axial electric field and the moving electrons. On the average, the electrons transfer energy to the wave on the helix. This interaction causes the signal wave on the helix to become larger. The electrons entering the helix at zero field are not affected by the signal wave; those electrons entering the helix at the accelerating field are accelerated, and those at the retarding field are decelerated. As the electrons travel further along the helix, they bunch at the collector end. The bunching shifts the phase by $\pi/2$. Each electron in the bunch encounters a stronger retarding field. Then the microwave energy of the electrons is delivered by the electron bunch to the wave on the helix. The amplification of the signal wave is accomplished.

The characteristics of the traveling-wave tube are

Frequency range: 3 GHz and higher
Bandwidth: about 0.8 GHz
Efficiency: 20 to 40%
Power output: up to 10 kW average
Power gain: up to 60 dB

(1) *Amplification Process.* A schematic diagram of a helix-type traveling-wave tube is shown in Fig. 5-2-3-1.

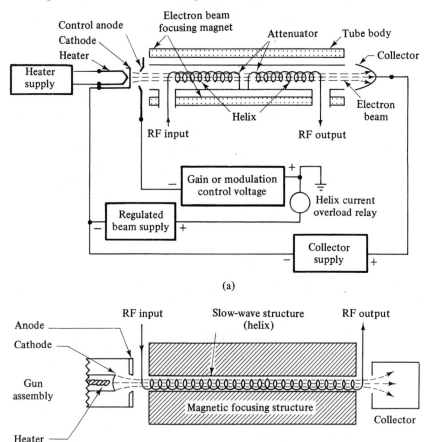

(a)

(b)

FIG. 5-2-3-1. Diagram of helix traveling-wave tube. (a) Schematic diagram of helix traveling wave tube. (b) Simplified circuit.

The slow-wave structure of the helix is characterized by the Brillouin diagram shown in Fig. 4-5-3. The phase shift per period of the fundamental wave on the structure is given by

$$\theta_1 = \beta_0 L \tag{5-2-81}$$

where $\beta_0 = \omega/v_0$ is the phase constant of the average beam velocity and L is the period or pitch.

Since the dc transit time of an electron is given by

$$T_0 = \frac{L}{v_0} \tag{5-2-82}$$

the phase constant of the nth space-harmonic is

$$\beta_n = \frac{\omega}{v_0} = \frac{\theta_1 + 2\pi n}{v_0 T_0} = \beta_0 + \frac{2\pi n}{L} \tag{5-2-83}$$

In Eq. (5-2-83) the axial space-harmonic phase velocity is assumed to be synchronized with the beam velocity for possible interactions between the electron beam and electric field. That is,

$$v_{np} = v_0 \tag{5-2-84}$$

Equation (5-2-83) is identical to Eq. (4-5-13). In practice, the dc velocity of the electrons is adjusted to be slightly greater than the axial velocity of the electromagnetic wave for energy transfer. When a signal voltage is coupled into the helix, the axial electric field exerts a force on the electrons as a result of the following relationships:

$$\mathbf{F} = -e\mathbf{E} \quad \text{and} \quad \mathbf{E} = -\nabla V \tag{5-2-85}$$

The electrons entering the retarding field are decelerated and those in the accelerating field are accelerated. They begin forming a bunch centered about those electrons that enter the helix during the zero field. This process is shown in Fig. 5-2-3-2.

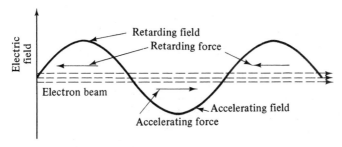

FIG. 5-2-3-2. Interactions between electron beam and electric field.

Since the dc velocity of the electrons is slightly greater than the axial wave velocity, more electrons are in the retarding field than in the accelerating field, and a great amount of energy is transferred from the beam to the electromagnetic field. The microwave signal voltage is, in turn, amplified by the amplified field. The bunch continues to become more compact, and a larger amplification of the signal voltage occurs at the end of the helix. The magnet produces an axial magnetic field to prevent spreading of the electron beam as it travels down the tube. An attenuator placed near the center of the helix reduces all the waves traveling along the helix to nearly zero so that the reflected waves from the mismatched loads can be prevented from reaching the input and causing oscillation. The bunched electrons emerging from the attenuator induce a new electric field with the same frequency. This field, in turn, induces a new amplified microwave signal on the helix.

The motion of electrons in the helix-type traveling-wave tube can be quantitatively analyzed in terms of the axial electric field. If the traveling wave is propagating in the z direction, the z component of the electric field can be expressed as

$$E_z = E_1 \sin (\omega t - \beta_p z) \qquad (5\text{-}2\text{-}86)$$

where E_1 is the magnitude of the electric field in the z direction. If $t = t_0$ at $z = 0$, the electric field is assumed maximum. $\beta_p = \omega/v_p$ is the axial phase constant of the microwave, and v_p is the axial phase velocity of the wave.

The equation of motion of the electron is given by

$$m \frac{dv}{dt} = -eE_1 \sin (\omega t - \beta_p z) \qquad (5\text{-}2\text{-}87)$$

Assume that the velocity of the electron is

$$v = v_0 + v_e \cos (\omega_e t + \theta_e) \qquad (5\text{-}2\text{-}88)$$

Then

$$\frac{dv}{dt} = -v_e \omega_e \sin (\omega_e t + \theta_e) \qquad (5\text{-}2\text{-}89)$$

where $v_0 =$ the dc electron velocity
$v_e =$ the magnitude of velocity fluctuation in the velocity-modulated electron beam
$\omega_e =$ the angular frequency of velocity fluctuation
$\theta_e =$ the phase angle of the fluctuation

Substitution of Eq. (5-2-89) in Eq. (5-2-87) yields

$$mv_e \omega_e \sin (\omega_e t + \theta_e) = eE_1 \sin (\omega t - \beta_p z) \qquad (5\text{-}2\text{-}90)$$

For interactions between the electrons and the electric field, the velocity of the velocity-modulated electron beam must be approximately equal to the dc electron velocity. This is

$$v \simeq v_0 \qquad (5\text{-}2\text{-}91)$$

Hence the distance z traveled by the electrons is

$$z = v_0(t - t_0) \qquad (5\text{-}2\text{-}92)$$

and

$$mv_e \omega_e \sin (\omega_e t + \theta_e) = eE_1 \sin [\omega t - \beta_p v_0(t - t_0)] \qquad (5\text{-}2\text{-}93)$$

Comparison of the left and right-hand sides of Eq. (5-2-93) shows that

$$v_e = \frac{eE_1}{m\omega_e} \qquad (5\text{-}2\text{-}94)$$

$$\omega_e = \beta_p(v_p - v_0)$$

$$\text{and} \quad \theta_e = \beta_p v_0 t_0$$

It can be seen that the magnitude of the velocity fluctuation of the electron beam is directly proportional to the magnitude of the axial electric field.

(2) Convection Current. In order to determine the relationship between the circuit and electron beam quantities, the convection current induced in the electron beam by the axial electric field and the microwave axial field produced by the beam can first be developed. When the space-charge effect is considered, the electron velocity, the charge density, the current density, and the axial electric field will perturbate about their averages or dc values. Mathematically, these quantities can be expressed as

$$v = v_0 + v_1 e^{j\omega t - \gamma z} \qquad (5\text{-}2\text{-}95)$$

$$\rho = \rho_0 + \rho_1 e^{j\omega t - \gamma z} \qquad (5\text{-}2\text{-}96)$$

$$J = -J_0 + J_1 e^{j\omega t - \gamma z} \qquad (5\text{-}2\text{-}97)$$

$$E_z = E_1 e^{j\omega t - \gamma z} \qquad (5\text{-}2\text{-}98)$$

where $\gamma = \alpha_e + j\beta_e$ is the propagation constant of the axial waves. The minus sign is attached to J_0 so that J_0 may be a positive in the negative z direction. For a small signal, the electron beam-current density can be written

$$J = \rho v \simeq -J_0 + J_1 e^{j\omega t - \gamma z} \qquad (5\text{-}2\text{-}99)$$

In Eq. (5-2-99), $-J_0 = \rho_0 v_0$, $J_1 = \rho_1 v_0 + \rho_0 v_1$, and $\rho_1 v_1 \simeq 0$ have been replaced. If an axial electric field exists in the structure, it will perturbate the electron velocity according to the force equation. Hence the force equation can be written

$$\frac{dv}{dt} = -\frac{e}{m} E_1 e^{j\omega t - \gamma z} = \left(\frac{\partial}{\partial t} + \frac{dz}{dt}\frac{\partial}{\partial z}\right)v = (j\omega - \gamma v_0)v_1 e^{j\omega t - \gamma z} \qquad (5\text{-}2\text{-}100)$$

where dz/dt has been replaced by v_0. Thus

$$v_1 = \frac{-e/m}{j\omega - \gamma v_0} E_1 \qquad (5\text{-}2\text{-}101)$$

In accordance with the law of conservation of electric charge, the continuity equation can be written

$$\nabla \cdot J + \frac{\partial \rho}{\partial t} = (-\gamma J_1 + j\omega \rho_1)e^{j\omega t - \gamma z} = 0 \qquad (5\text{-}2\text{-}102)$$

It follows that

$$\rho_1 = -\frac{j\gamma J_1}{\omega} \qquad (5\text{-}2\text{-}103)$$

Substituting Eqs. (5-2-101) and (5-2-103) in

$$J_1 = \rho_1 v_0 + \rho_0 v_1 \tag{5-2-104}$$

gives

$$J_1 = j\frac{\omega}{v_0}\frac{e}{m}\frac{J_0}{(j\omega - \gamma v_0)^2}E_1 \tag{5-2-105}$$

In Eq. (5-2-105), $-J_0 = \rho_0 v_0$ has been replaced. If the magntiude of the axial electric field is uniform over the cross-sectional area of the electron beam, the spatial ac current i will be proportional to the dc current I_0 with the same proportionality constant for J_1 and J_0. Therefore the convection current in the electron beam is given by

$$i = j\frac{\beta_e I_0}{2V_0(j\beta_e - \gamma)^2}E_1 \tag{5-2-106}$$

where $\beta_e \equiv \dfrac{\omega}{v_0}$ is defined as the phase constant of the velocity-modulated electron beam

$v_0 = \sqrt{\dfrac{2e}{m}V_0}$ has been used

This equation is called the *electronic equation,* for it determines the convection current induced by the axial electric field. If the axial field and all parameters are known, the convection current can be found by means of Eq. (5-2-106)

(3) Axial Electric Field. The convection current in the electron beam induces an electric field in the slow-wave circuit. This induced field adds to the field already present in the circuit and causes the circuit power to increase with distance. The coupling relationship between the electron beam and the slow-wave helix is shown in Fig. 5-2-3-3.

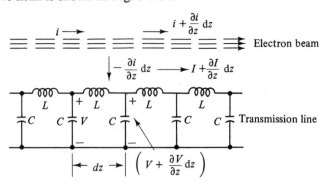

FIG. 5-2-3-3. Electron beam coupled to equivalent circuit of a slow-wave helix.

For simplicity, the slow-wave helix is represented by a distributed loss-less transmission line. The parameters are defined as follows:

$L =$ inductance per unit length
$C =$ capacitance per unit length
$I =$ alternating current in transmission line
$V =$ alternating voltage in transmission line
$i =$ convection current

Since the transmission line is coupled to a convection-electron beam current, a current is then induced in the line. The current flowing into the left-end portion of the line of length dz is i, and the current flowing out of the right end of dz is $[i + (\partial i/\partial z)\, dz]$. However, since the net change of current in the length dz must be zero the current flowing out of the electron beam into the line must be $[-(\partial i/\partial z)\, dz]$. Application of transmission-line theory and Kirchhoff's current law to the electron beam results, after simplification, in

$$\frac{\partial I}{\partial z} = -C\frac{\partial V}{\partial t} - \frac{\partial i}{\partial z} \qquad (5\text{-}2\text{-}107)$$

Then

$$-\gamma I = -j\omega C V + \gamma i \qquad (5\text{-}2\text{-}108)$$

in which $\partial/\partial z = -\gamma$ and $\partial/\partial t = j\omega$ are replaced.

From Kirchhoff's voltage law the voltage equation, after simplification, is

$$\frac{\partial V}{\partial z} = -L\frac{\partial I}{\partial t} \qquad (5\text{-}2\text{-}109)$$

Similarly,

$$-\gamma V = -j\omega L I \qquad (5\text{-}2\text{-}110)$$

Elimination of the circuit current I from Eqs. (5-2-108) and (5-2-109) yields

$$\gamma^2 V = -V\omega^2 LC - \gamma i j\omega L \qquad (5\text{-}2\text{-}111)$$

If the convection-electron beam current is not present, Eq. (5-2-111) reduces to a typical wave equation of a transmission line. When $i = 0$, the propagation constant is defined from Eq. (5-2-111) as

$$\gamma_0 \equiv j\omega\sqrt{LC} \qquad (5\text{-}2\text{-}112)$$

and the characteristic impedance of the line can be determined from Eqs. (5-2-108) and (5-2-110).

$$Z_0 = \sqrt{\frac{L}{C}} \qquad (5\text{-}2\text{-}113)$$

When the electron beam current is present, Eq. (5-2-111) can be written in terms of Eqs. (5-2-112) and (5-2-113).

$$V = -\frac{\gamma\gamma_0 Z_0}{\gamma^2 - \gamma_0^2} i \qquad (5\text{-}2\text{-}114)$$

Since $E_z = -\nabla V = -(\partial V / \partial z) = \gamma V$, the axial electric field is given by

$$E_1 = -\frac{\gamma^2 \gamma_0 Z_0}{\gamma^2 - \gamma_0^2} i \qquad (5\text{-}2\text{-}115)$$

This equation is called the *circuit equation* because it determines how the axial electric field of the slow-wave helix is affected by the spatial ac electron beam current.

(4) Wave Modes. The wave modes of a helix-type traveling-wave tube can be determined by solving the electronic and circuit equations simultaneously for the propagation constants. Each solution for the propagation constants represents a mode of traveling wave in the tube. It can be seen from Eqs. (5-2-106) and (5-2-115) that there are four distinct solutions for the propagation constants. This means that there are four modes of traveling wave in the O-type traveling-wave tube. Subsitution of Eq. (5-2-106) in Eq. (5-2-115) yields

$$(\gamma^2 - \gamma_0^2)(j\beta_e - \gamma)^2 = -j\frac{\gamma^2 \gamma_0 Z_0 \beta_e I_0}{2V_0} \qquad (5\text{-}2\text{-}116)$$

Equation (5-2-116) is of fourth order in γ and thus will have four roots. Its exact solutions can be obtained with numerical methods and a digital computer. However, the approximate solutions may be found by equating the dc electron beam velocity to the axial phase velocity of the traveling wave, which is equivalent to setting

$$\gamma_0 = j\beta_e \qquad (5\text{-}2\text{-}117)$$

Then Eq. (5-2-116) reduces to

$$(\gamma - j\beta_e)^3 (\gamma + j\beta_e) = 2C^3 \beta_e^2 \gamma^2 \qquad (5\text{-}2\text{-}118)$$

where C is the traveling-wave tube gain parameter and is defined as

$$C \equiv \left(\frac{I_0 Z_0}{4V_0}\right)^{1/3} \qquad (5\text{-}2\text{-}119)$$

It can be seen from Eq. (5-2-118) that there are three forward traveling waves corresponding to $e^{-j\beta_e z}$, and one backward traveling wave corresponding to $e^{+j\beta_e z}$. Let the propagation constant of the three forward traveling waves be

$$\gamma = j\beta_e - \beta_e C \delta \qquad (5\text{-}2\text{-}120)$$

where it is assumed that $C\delta \ll 1$.

Substitution of Eq. (5-2-120) in Eq. (5-2-118) results in

$$(-\beta_e C \delta)^3 (j2\beta_e - \beta_e C \delta) = 2C^3 \beta_e^2 (-\beta_e^2 - 2j\beta_e^2 C \delta + \beta_e^2 C^2 \delta^2) \qquad (5\text{-}2\text{-}121)$$

Since $C\delta \ll 1$, Eq. (5-2-121) is reduced to

$$\delta = (-j)^{1/3} \qquad (5\text{-}2\text{-}122)$$

From the theory of complex variables the three roots of $(-j)$ can be plotted in Fig. 5-2-3-4.

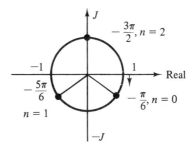

FIG. 5-2-3-4. Three roots of $(-j)$.

Equation (5-2-122) can be written in exponential form as

$$\delta = (-j)^{1/3} = e^{-j[(\pi/2 + 2n\pi)/3]} \qquad (n = 0, 1, 2) \qquad (5\text{-}2\text{-}123)$$

The first root δ_1 at $n = 0$ is

$$\delta = e^{-j(\pi/6)} = \frac{\sqrt{3}}{2} - j\frac{1}{2} \qquad (5\text{-}2\text{-}124)$$

The second root δ_2 at $n = 1$ is

$$\delta_2 = e^{-j(5\pi/6)} = -\frac{\sqrt{3}}{2} - j\frac{1}{2} \qquad (5\text{-}2\text{-}125)$$

The third root δ_3 at $n = 2$ is

$$\delta_3 = e^{-j(3\pi/2)} = j \qquad (5\text{-}2\text{-}126)$$

The fourth root δ_4 corresponding to the backward traveling wave can be obtained by setting

$$\gamma = -j\beta_e - \beta_e C \,\delta_4 \qquad (5\text{-}2\text{-}127)$$

Similarly,

$$\delta_4 = -j\frac{C^2}{4} \qquad (5\text{-}2\text{-}128)$$

Thus the values of the four propagation constants γ are given by

$$\gamma_1 = -\beta_e C \frac{\sqrt{3}}{2} + j\beta_e\left(1 + \frac{C}{2}\right) \qquad (5\text{-}2\text{-}129)$$

$$\gamma_2 = \beta_e C \frac{\sqrt{3}}{2} + j\beta_e\left(1 + \frac{C}{2}\right) \qquad (5\text{-}2\text{-}130)$$

$$\gamma_3 = j\beta_e(1 - C) \qquad (5\text{-}2\text{-}131)$$

$$\gamma_4 = -j\beta_e\left(1 - \frac{C^3}{4}\right) \qquad (5\text{-}2\text{-}132)$$

These four propagation constants represent four different modes of wave propagation in the O-type helical traveling-wave tube. It is concluded that the wave corresponding to γ_1 is a forward wave and that its amplitude grows exponentially with distance; the wave corresponding to γ_2 is also a forward wave, but its amplitude decays exponentially with distance; the wave corresponding to γ_3 is also a forward wave, but its amplitude remains constant;

and the fourth wave corresponding to γ_4 is a backward wave, and there is no change in amplitude. The growing wave propagates at a phase velocity slightly lower than the electron beam velocity, and the energy flows from the electron beam to the wave. The decaying wave propagates the same velocity as that of the growing wave, but the energy flows from the wave to the electron beam. The constant-amplitude wave travels at a velocity slightly higher than the electron beam velocity, but no net energy exchange occurs between the wave and the electron beam. The backward wave progresses in the negative z direction with a velocity slightly higher than the velocity of the electron beam inasmuch as the typical value of C is about 0.02.

(5) Gain Consideration. For simplicity, it is assumed that the structure is perfectly matched so that there is no backward traveling wave. Such is usually the case. Even though there is a reflected wave from the output end of the tube traveling backward toward the input end, the attenuator placed around the center of the tube will subdue the reflected wave to a minimum or zero level. So the total circuit voltage will be the sum of three forward voltages corresponding to the three forward traveling waves. This is equivalent to

$$V(z) = V_1 e^{-\gamma_1 z} + V_2 e^{-\gamma_2 z} + V_3 e^{-\gamma_3 z} = \sum_{n=1}^{3} V_n e^{-\gamma_n z} \qquad (5\text{-}2\text{-}133)$$

The input current can be found from Eq. (5-2-106) as

$$i(z) = - \sum_{n=1}^{3} \frac{I_0}{2V_0 C^2} \frac{V_n}{\delta_n^2} e^{-\gamma_n z} \qquad (5\text{-}2\text{-}134)$$

in which $C\delta \ll 1$, $E_1 = \gamma V$, and $\gamma = j\beta_e(1 - C\delta)$ have been used.

The input fluctuating component of velocity of the total wave may be found from Eq. (5-2-101) as

$$v_1(z) = \sum_{n=1}^{3} j \frac{v_0}{2V_0 C} \frac{V_n}{\delta_n} e^{-\gamma_n z} \qquad (5\text{-}2\text{-}135)$$

In Eq. (5-2-135) $E_1 = \gamma V$, $C\delta \ll 1$, $\beta_e v_0 = \omega$, and $v_0 = \sqrt{(2e/m)V_0}$ have been used.

To determine the amplification of the growing wave, the input reference point is set at $z = 0$ and the output reference point is taken at $z = \ell$. It follows that at $z = 0$ the voltage, current, and velocity at the input point are given by

$$V(0) = V_1 + V_2 + V_3 \qquad (5\text{-}2\text{-}136)$$

$$i(0) = -\frac{I_0}{2V_0 C^2}\left(\frac{V_1}{\delta_1^2} + \frac{V_2}{\delta_2^2} + \frac{V_3}{\delta_3^2}\right) \qquad (5\text{-}2\text{-}137)$$

$$v_1(0) = -j\frac{v_0}{2V_0 C}\left(\frac{V_1}{\delta_1} + \frac{V_2}{\delta_2} + \frac{V_3}{\delta_3}\right) \qquad (5\text{-}2\text{-}138)$$

The simultaneous solution of Eqs. (5-2-136), (5-2-137), and (5-2-138) with $i(0) = 0$ and $v_1(0) = 0$ is

$$V_1 = V_2 = V_3 = \frac{V(0)}{3} \qquad (5\text{-}2\text{-}139)$$

Since the growing wave is increasing exponentially with distance, it will predominate over the total voltage along the circuit. When the length ℓ of the slow-wave structure is sufficiently large, the output voltage will be almost equal to the voltage of the growing wave. Substitution of Eqs. (5-2-130) and (5-2-139) in Eq. (5-2-133) yields the output voltage as

$$V(\ell) \simeq \frac{V(0)}{3} \exp\left(\frac{\sqrt{3}}{2}\beta_e C\ell\right) \exp\left[-j\beta_e\left(1 + \frac{C}{2}\right)\ell\right] \quad (5\text{-}2\text{-}140)$$

The factor $\beta_e\ell$ is conventionally written $2\pi N$, where N is the circuit length in electronic wavelength—that is,

$$N = \frac{\ell}{\lambda_e} \quad \text{and} \quad \beta_e = \frac{2\pi}{\lambda_e} \quad (5\text{-}2\text{-}141)$$

The amplitude of the output voltage is then given by

$$V(\ell) = \frac{V(0)}{3} \exp\left(\sqrt{3}\,\pi NC\right) \quad (5\text{-}2\text{-}142)$$

The output power gain in decibels is defined as

$$A_p \equiv 10 \log\left|\frac{V(\ell)}{V(0)}\right|^2 = -9.54 + 47.3 NC \quad \text{dB} \quad (5\text{-}2\text{-}143)$$

where NC is a numerical number.

The output power gain as shown in Eq. (5-2-143) indicates an initial loss at the circuit input of 9.54 dB. This loss is due to the fact that the input voltage splits into three waves of equal magnitude and the growing wave voltage is only one-third the total input voltage. It can also be seen that the power gain is proportional to the length N in electronic wavelength of the slow-wave structure and the gain parameter C of the circuit.

5-2-4. Coupled-Cavity Traveling-Wave Tubes

Helix traveling-wave tubes (TWTs) produce at most up to several kilowatts of average power output. Here we describe coupled-cavity traveling-wave tubes, which are used for high-power applications.

(1) Physical description. The term *coupled cavity* means that a coupling is provided by a long slot that strongly couples the magnetic component of the field in adjacent cavities in such a manner that the passband of the circuit is mainly a function of this one variable. Figure 5-2-4-1 shows two coupled-cavity circuits that are principally used in traveling-wave tubes.

As far as the coupling effect is concerned, there are two types of coupled-cavity circuits in traveling-wave tubes. The first type consists of the fundamentally forward-wave circuits that are normally used for pulse applications requiring at least half a megawatt of peak power. These coupled-cavity circuits exhibit negative mutual inductive coupling between the cavities and operate with the fundamental space-harmonic. The clover leaf [12] and

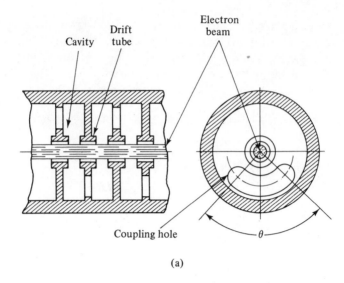

(a)

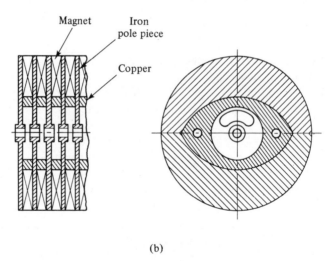

(b)

FIG. 5-2-4-1. Coupled-cavity circuits in the TWTs. (a) Basic coupled-cavity circuit. (b) Coupled-cavity circuit with integral periodic-permanent-magnet (PPM) focusing. (*After J. T. Mendel [17]; reprinted by permission of the IEEE, Inc.*)

centipede circuits [13] (see Fig. 5-2-4-2) belong to this type. The second type is the first space-harmonic circuit, which has positive mutual coupling between the cavities. These circuits operate with the first spatial harmonic and are commonly used for pulse or continuous wave (CW) applications from one to several hundred kilowatts of power output [14]. In addition, the long-slot

FIG. 5-2-4-2. Centipede and clover leaf coupled-cavity circuits. (*After A. Staprans et al. [18]; reprinted by permission of the IEEE, Inc.*)

circuit of the positive mutual coupling-cavity circuit operates at the fundamental spatial harmonic with a higher frequency mode. This circuit is suitable for megawatt power output. Figure 5-2-4-3 shows several space-harmonic coupled-cavity circuits.

FIG. 5-2-4-3. Space-harmonic coupled-cavity circuits. (*After A. Staprans et al. [18]; reprinted by permission of the IEEE, Inc.*)

(2) *Principles of Operation.* Any repetitive series of lumped *LC* elements constitute a propagating filter-type circuit. The coupled cavities in the traveling-wave tube are usually highly overcoupled, resulting in a bandpass-filter-type characteristic. When the slot angle (θ) as shown in Fig. 5-2-4-1(a) is larger than 180°, the passband is close to its practical limits. The drift tube is formed by the reentrant part of the cavity, just as in the case of a klystron. During the interaction of the RF field and the electron beam in the traveling-wave tube a phase change occurs between the cavities as a function of frequency. A decreasing phase characteristic is reached if the mutual inductance of the coupling slot is positive, whereas an increasing phase characteristic is obtained if the mutual inductive coupling of the slot is negative [12].

The amplification of the traveling-wave tube interaction requires that the electron beam interact with a component of the circuit field that has an increasing phase characteristic with frequency. The circuit periodicity can give rise to field components that have phase characteristics [15] as shown in Fig. 5-2-4-4. In Fig. 5-2-4-4 the angular frequency (ω) is plotted as a function

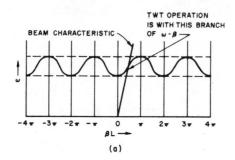

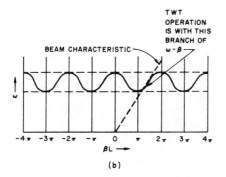

FIG. 5-2-4-4. ω-β diagrams for coupled-cavity circuits. (a) Fundamental forward-wave circuit (negative mutual inductance coupling between cavities). (b) Fundamental backward-wave circuit (positive mutual inductive coupling between cavities). (*After A. Staprans et al. [18]; reprinted by permission of the IEEE, Inc.*)

of the phase shift ($\beta\ell$) per cavity. The ratio of ω to β is equal to the phase velocity. For a circuit having positive mutual inductive coupling between the cavities, the electron beam velocity is adjusted to be approximately equal to the phase velocity of the first forward-wave spatial harmonic. For the circuits with negative mutual inductive coupling, the fundamental branch component of the circuit wave is suitable for synchronism with the electron beam and is normally used by the traveling-wave tube. The coupled-cavity equivalent circuit has been developed by Curnow [16] as shown in Fig. 5-2-4-5.

In Fig. 5-2-4-5 inductances are used to represent current flow and capacitors to represent the electric fields of the cavities. The circuit can be evolved into a fairly simple configuration. Loss in the cavities can be approximately calculated by adding resistance in series with the circuit inductance.

(3) *Microwave Characteristics.* When discussing the power capability of traveling-wave tubes, it is important to make a clear distinction between the average and the peak power because these two figures are limited by

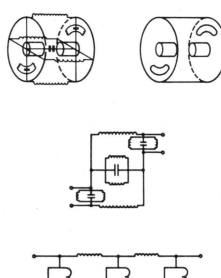

FIG. 5-2-4-5. Equivalent circuits for a slot coupled cavity. (*From A. Staprans et al. [18]; reprinted by permission of the IEEE, Inc.*)

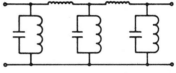

totally different factors. The average power at a given frequency is almost always limited by thermal considerations relative to the RF propagating circuit. However, the peak RF power capability depends on the voltage for which the tube can be designed. The beam current varies as the three-half power of the voltage, and the product of the beam current and the voltage determines the total beam power. That is,

$$I_{\text{beam}} = KV_0^{3/2} \qquad (5\text{-}2\text{-}143a)$$

and

$$P_{\text{beam}} = KV_0^{5/2} \qquad (5\text{-}2\text{-}143b)$$

where V_0 is the beam voltage and K is the electron-gun perveance.

For a solid-beam electron gun with good optics, the perveance is generally considered to be between 1 to 2×10^{-6}. Once the perveance is fixed, the required voltage for a given peak beam power is then uniquely determined. Figures 5-2-4-6 and 5-2-4-7 demonstrate the difference between peak and average power capability and the difference between periodic-permanent-magnet (PPM) and solenoid-focused designs [17].

Coupled-cavity traveling-wave tubes are constructed with a limited amount of gain per section of cavities to ensure stability. Each cavity section is terminated by either a matched load or an input or output line in order to reduce gain variations with frequency. Cavity sections are cascaded to achieve higher tube gain than can be tolerated in one section of cavities. Stable gain greater than 60 dB can be obtained over about 30% bandwidth by this method.

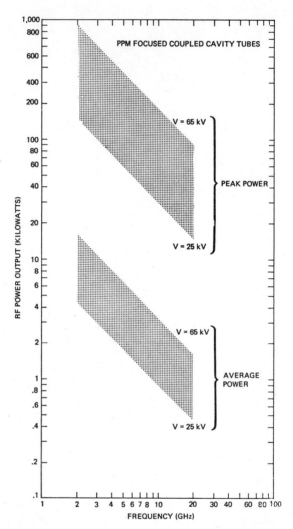

FIG. 5-2-4-6. Peak and average power capability of typical TWTs in field use. (*After J. T. Mendel [17]; reprinted by permission of the IEEE, Inc.*)

The overall efficiency of coupled-cavity traveling-wave tubes is determined by the amount of energy converted to RF energy and the energy dissipated by the collector. Interaction efficiencies from 10 to 40% have been achieved from coupled-cavity traveling-wave tubes. Overall efficiencies of 20 to 55% have been obtained [18].

(4) High Power and Gridded Control. Coupled-cavity traveling-wave tubes are the most versatile devices used for amplification at microwave frequencies with high gain, high power, high efficiency, and wide bandwidth.

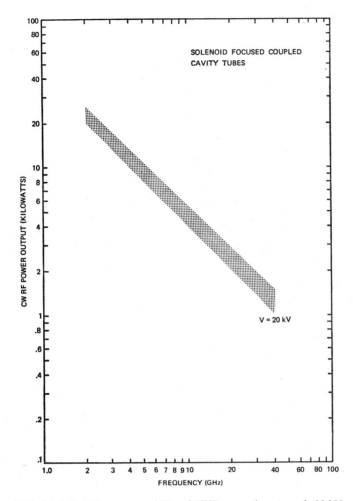

FIG. 5-2-4-7. CW power capability of TWTs operating at nearly 20 kV. (*From J. T. Mendel [17]; reprinted by permission of the IEEE, Inc.*)

High-power TWTs have four main sections: electron gun for electron emission, slow-wave structure for effective beam interaction, magnetic circuit for beam focusing, and collector structure for collecting electron beams and dissipating heat energy. Specifically, the physical components of a coupled-cavity traveling-wave tube consist of an electron emitter, a shadow grid, a control grid, a modulating anode, a coupled-cavity circuit, a solenoid magnetic circuit, and a collector depression structure (see Fig. 5-2-4-8).

After electrons are emitted from the cathode, the electron beam has a tendency to spread out due to the electron-repelling force. On the other hand, the electron beam must be small enough for effective interaction with the slow-wave circuit. Usually the diameter of the electron beam is smaller than

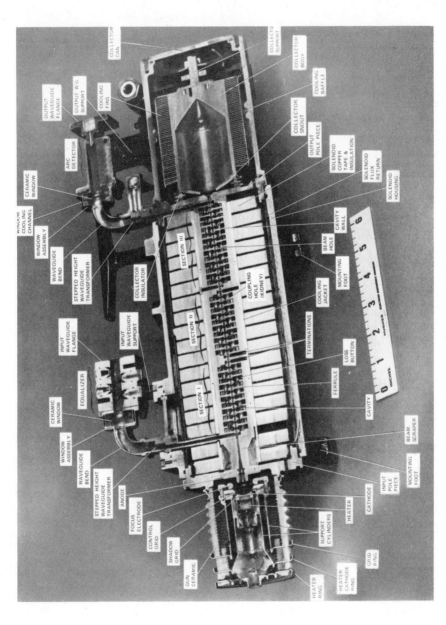

FIG. 5-2-4-8. Cutaway of a coupled-cavity traveling-wave tube. (*Courtesy Hughes Aircraft Company, Electron Dynamics Division.*)

one-tenth wavelength of the signal. Coupled-cavity traveling-wave high-power tubes utilize a shadow-grid technique to control the electron beam; so the device is called a *gridded traveling-wave tube* (GTWT). As shown in Fig. 5-2-4-9a [22], the electron emitter of a gridded traveling-wave tube has two control electrodes: one shadow grid near the cathode and one control grid slightly away from the cathode. The shadow grid, which is at cathode potential and interposed between the cathode and the control grid, suppresses electron emission from those portions of the cathode that would give rise to interception at the control grid. The control grid, which is at a positive potential, controls the electron beam. These grids can control far greater beam power than would otherwise be possible.

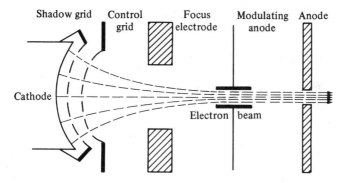

FIG. 5-2-4-9a. Electron emitter with control electrodes for gridded high-power TWT.

In general, an anode modulation technique is frequently used in traveling-wave tubes to eliminate voltage pulsing through the lower unstable beam voltage and to reduce modulator power requirements for high-power pulse output. In gridded traveling-wave tubes the modulator applies a highly regulated positive grid drive voltage with respect to the cathode to turn the electron beam on for RF amplification. A not well-regulated negative grid bias voltage with respect to the cathode is applied to cut the electron beam off. Thus the anode modulator acts as a pulse switch for the electron beam of the gridded traveling-wave tube.

The anode of the electron gun is operated at a voltage higher than that of the slow-wave structure to prevent positive ions formed by the electron beam in the region of the slow-wave structure from draining toward and bombarding the cathode.

(5) High Efficiency and Collector Voltage Depression. After passing through the output cavity, the electron beam strikes a collector electrode. The function of the collector electrode could be performed by replacing the second grid of the output cavity with a solid piece of metal. However, using a separate electrode may have two advantages. First, the collector can be made as large as desired in order to collect the electron beam at a lower density, thus minimizing localized heating. If the collector were a part of the slow-wave circuit, its size would be limited by the maximum gap capacitance consistent with good high-frequency performance. Secondly, using a separate collector can reduce its potential considerably below the beam voltage in the RF interaction region, thereby reducing the power dissipated in the collector and increasing the overall efficiency of the device. Gridded traveling-wave high-power tubes have a separate collector that dissipates the electrons in the form of heat. A cooling mechanism absorbs the heat by thermal conduction to a cooler surface.

The efficiency of a gridded traveling-wave high-power tube is the ratio of the RF power output to the product of cathode voltage (beam voltage) and cathode current (beam current). It may be expressed in terms of the product of the electronic efficiency and the circuit efficiency. The electronic efficiency expresses the percentage of the dc or pulsed input power that is converted into RF power on the slow-wave structure. The circuit efficiency, on the other hand, determines the percentage of dc input power that is delivered to the load exterior to the tube. The electron beam does not extract energy from any dc power supply unless the electrons are actually collected by an electrode connected to that power supply. If a separate power supply is connected between cathode and collector and if the cavity grids intercept a negligible part of the electron beam, the power supply between the cathode and collector will be the only one supplying any power to the tube. For a gridded traveling-wave tube, the collector voltage is normally operated at about 40% of the cathode voltage. Thus the overall efficiency of conversion of dc to RF power is almost twice the electronic efficiency. Under this condition the tube is operating with collector voltage depression.

(6) *Normal Depression and Overdepression of Collector Voltage.* Most gridded traveling-wave tubes are very sensitive to variations of collector depression voltages below normal depression level, since the tubes operate closer to the knee of the electron spent beam curves. Figure 5-2-4-9b [22] shows the spent beam curves for a typical gridded traveling-wave tube.

Under normal collector depression voltage V_c at -7.5 kV with full saturated power output, the spent beam electrons are collected by the collector and returned to the cathode. Thus the collector current I_c is about 2.09 A. A small amount of electrons intercepted by the beam scraper or slow-wave circuit contributes the tube body current for about 0.178 A. Very few electrons with lower kinetic energy reverse the direction of their velocity inside the collector and fall back onto the output pole piece. These returning electrons yield a current I_r of 0.041 A, which is only a small fraction of the body current I_b. These values are shown in Fig. 5-2-4-9b.

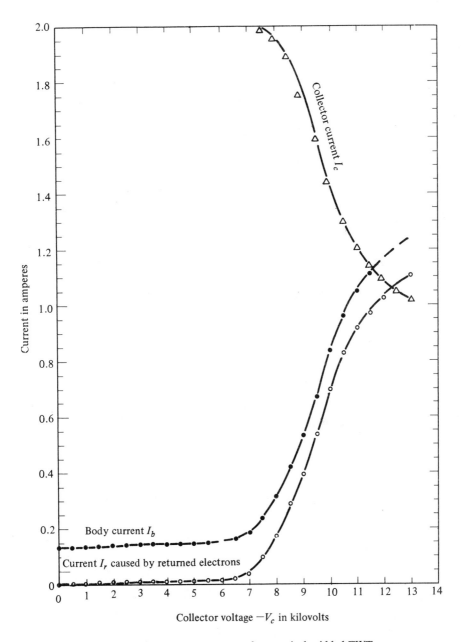

Collector current I_c

Body current I_b

Current I_r caused by returned electrons

Current in amperes

Collector voltage $-V_c$ in kilovolts

FIG. 5-2-4-9b. Spent beam curves for a typical gridded TWT.

When the collector voltage is overdepressed from the normal level of −7.5 kV to the worst case of about −11.5 kV, a greater number of the spent electrons inside the collector reverse the direction of their velocity by a highly negative collector voltage and fall back onto the grounded output pole piece because the potential of the pole piece is 11.5 kV higher than the collector voltage. It can be seen from Fig. 5-2-4-9b that when the collector voltage is overdepressed from −7.5 to −11.5 kV, the collector current is decreased sharply from 2.01 to 1.14 A and the body current is increased rapidly from 0.237 to 1.110 A. The body current consists of two parts: one part is the current due to the electrons intercepted by the circuit or the beam scrapers and another part is the current due to the electrons returned by the overdepressed collector voltage. Figure 5-2-4-9c [22] shows the impact probability of returned electrons by certain overdepressed collector voltage.

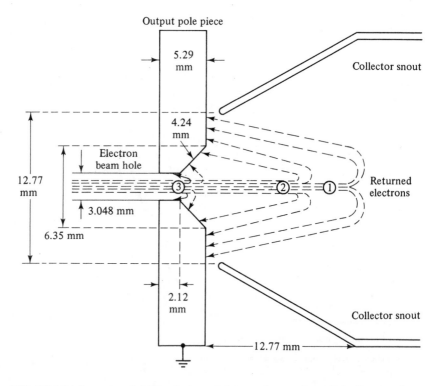

FIG. 5-2-4-9c. Impact probability of returned electrons by overdepressed collector voltage.

(7) *Two-Stage Collector Voltage Depression Technique.* When the spent electron beam arrives in the collector, the kinetic energies of each electron are different. Under the normal operation at a collector voltage of about 40%

of the cathode voltage, very few electrons will be returned by the negative collector voltage. Consequently, the tube body current is very small and negligible because the returned electrons are the only ones intercepted by the cavity grids and the slow-wave circuit. However, when the collector is more negative, more electrons with lower energy will reverse their direction of velocity and fall onto the output pole piece. Thus the tube body current will increase sharply. Since electrons of various energy classes exist inside the collector, two-stage collector voltage depression may be utilized. Each stage is biased at a different voltage. Specifically, the main collector may be biased at 40% depression of the cathode voltage for normal operation, but the collector snout may be grounded to the output pole piece for overdepression operation. As a result, the returned electrons will be collected by the collector snout and returned to the cathode even though the collector voltage is over-depressed to be more negative. Since the collector is cooled by a cooling mechanism, the overheating problem for overdepression is eased. Figure 5-2-4-10 shows a structure of two-stage collector voltage depression and Fig. 5-2-4-11 a basic interconnection of a gridded traveling-wave tube with its power supplies [22].

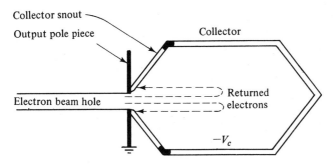

FIG. 5-2-4-10. Diagram for two-stage collector depression.

(8) Stabilization of Cathode and Collector Voltages. The cathode voltage of a gridded traveling-wave tube is negative with respect to ground so that the electrons can be emitted from the cathode. In order to maintain a constant beam power for a uniform gain, the cathode voltage must be constant. In addition, the phase shift through the tube is directly related to the beam velocity; thus high resolution and low ripple are required in the cathode voltage power supply to avoid undesirable phase-shift variations. Consequently, the cathode power supply of the gridded traveling-wave high-power tube is usually regulated for better than 1% over line and load changes and is also well filtered because of the critical requirements on the cathode voltage with respect to ground. The cathode power supply provides the tube body current. Under normal operation the body current is very small in com-

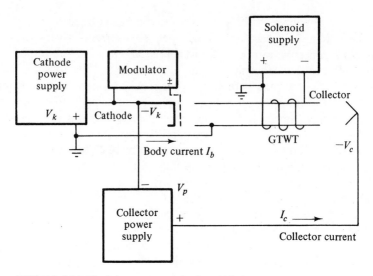

FIG. 5-2-4-11. Basic interconnection of a gridded traveling-wave tube with its power supplies.

parison with the collector current. Figure 5-2-4-12 shows a basic interconnection for a gridded traveling-wave tube with two regulated power supplies.

Figure 5-2-4-13 illustrates a voltage-regulator circuit for a cathode power supply. The voltage regulator indicated in the circuit consists of two devices: one differential amplifier and one tetrode tube. The solid-state

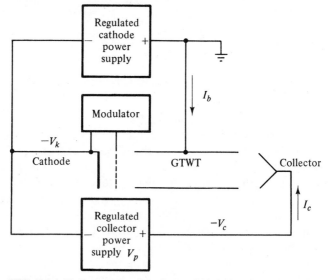

FIG. 5-2-4-12. Interconnections for a gridded traveling-wave high-power tube.

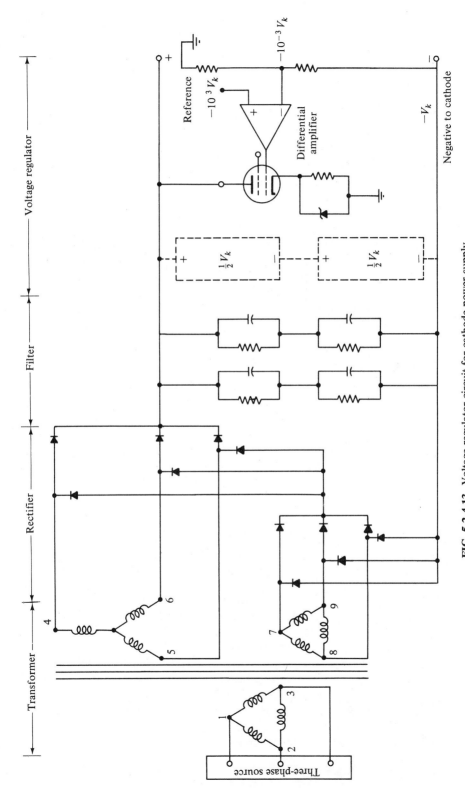

FIG. 5-2-4-13. Voltage-regulator circuit for cathode power supply.

differential amplifier amplifies the difference between the preset reference voltage and the voltage that is one-thousandth of the output voltage. The reference voltage is adjustable to a preset level. The output voltage of the differential amplifier drives the control grid of the regulator tube between cutoff and saturation in order to null the difference voltage.

As shown in Fig. 5-2-4-12, the negative terminal of the collector power supply is connected to the cathode and the positive terminal to the collector electrode. The collector depression voltage is the difference of the two regulated supply voltages. The cathode supply provides the tube body current and the collector supply yields the collector current. The ratio of the collector current over the body current is about 10 for an operation of 40% voltage depression. Thus the power delivered to the tube by the collector supply is about four times larger than the power furnished by the cathode supply. An electrical transient may occur in the circuit when power supplies are just being switched on or off. The reasons for an electrical transient may be due to two factors.

FIRST.   LOAD CHANGES. When the load of a generator is suddenly increased, a larger current is needed. Since the generator cannot meet the demand instantaneously, the output voltage of the generator drops momentarily. Conversely, when the load of a generator is suddenly decreased, the output voltage drops accordingly.

SECOND.   SWITCHING ON OR OFF. When the switch of a generator is just turned on or off, the armature current in the armature conductors produces armature reaction. The nature of armature reaction decreases the terminal voltage for lagging loads.

When an electrical transient is created in the circuit, the collector voltage is overdepressed. As a result, the spent electrons inside the collector reverse the direction of their velocity by the highly negative collector voltage and fall back onto the grounded output pole piece. The tube body current is sharply increased and the collector current is greatly decreased. When the returned electrons impact the output pole piece, the pole piece will be damaged by high heat. The damage of the output pole piece creates a mismatch in the interaction circuit and degrades the performance of the tube. In particular, the tube gain, efficiency, bandwidth, and power output are affected accordingly by the circuit mismatch when the collector voltage is overdepressed below the normal depression level. Furthermore, the large body current may burn out the solid-state differential amplifier of the cathode voltage regulator and vary the electron beam of the gridded tube. If the damage is beyond the tolerability of the gridded traveling-wave high-power tube, the tube may cease to function.

In order to maintain a constant collector depression voltage, the collector voltage must be regulated. There are three possible ways to do so [22].

FIRST. REGULATOR IN SERIES WITH THE COLLECTOR POWER SUPPLY. In this method a voltage regulator is incorporated in series with the collector supply as shown in Fig. 5-2-4-14 so that the output voltage of the collector supply may be regulated at a certain level with respect to ground. Since the output voltage of the cathode supply is highly regulated at a certain level, the difference between the two regulated voltages will produce a well-regulated voltage with respect to ground at the collector electrode.

SECOND. REGULATOR IN PARALLEL WITH THE COLLECTOR SUPPLY. In this method a voltage regulator is inserted in parallel with the collector supply as shown in Fig. 5-2-4-15 so that a regulated voltage with respect to ground at normal depression may be achieved at the collector terminal.

THIRD. REGULATOR BETWEEN THE CATHODE VOLTAGE AND THE COLLECTOR VOLTAGE. In this method the collector depression voltage is regulated with respect to the cathode voltage as shown in Fig. 5-2-4-16. If the collector voltage is overdepressed above the normal depression value (absolute value), the differential amplifier 2 tends to adjust the cathode voltage below its fixed level (absolute value). When the cathode voltage is dropped, the collector voltage will be readjusted to its normal depression level with respect to ground.

(9) Wide Bandwidth and Stagger-Tuned Circuit. The characteristics of a gridded traveling-wave tube are wide bandwidth, high gain, and high power. High power is accomplished by high cathode voltage with shadow-grid control, and high gain is achieved by electron interaction in multicoupled cavities. Coupled-multicavity gridded traveling-wave tubes are operated with their cavities stagger-tuned so as to obtain greater bandwidth at some reduction in gain. This situation is analogous to the well-known design of wideband IF amplifiers in which each stage is tuned to a slightly different frequency in order to improve the overall gain-bandwidth product.

(10) Beam Focusing and Magnetic Circuit. All traveling-wave tubes require some means of holding the electron beam together as it travels through the slow-wave circuit of the tube, for the beam tends to spread out as a result of the mutual repulsive forces between electrons. The magnetic circuit of a gridded traveling-wave tube consists of a solenoid structure and two soft iron pole pieces. The magnetic lines are parallel to the direction of propagation of the electron beam. The input and output pole pieces function as

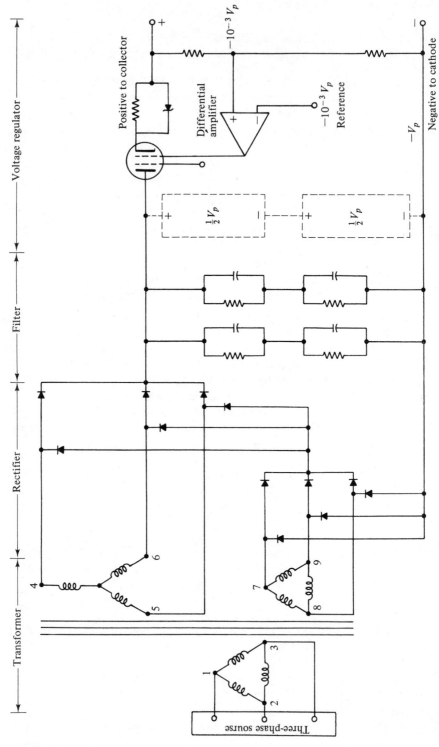

FIG. 5-2-4-14. Regulator in series with collector supply.

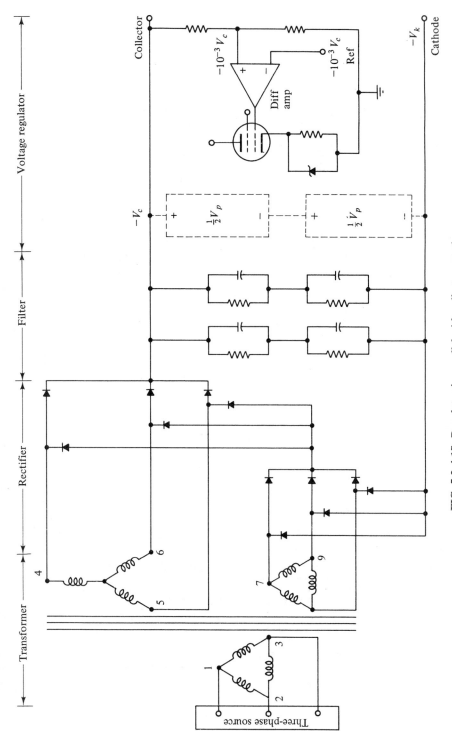

FIG. 5-2-4-15. Regulator in parallel with collector supply.

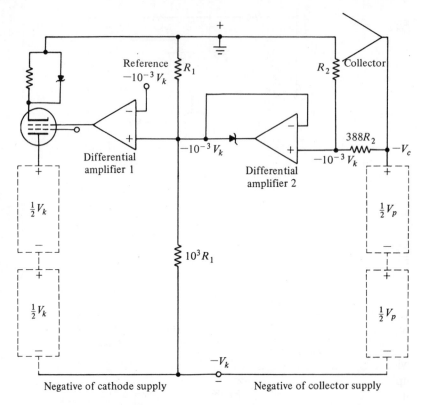

FIG. 5-2-4-16. Regulator between cathode voltage and collector voltage.

magnetic shields. In addition, a number of permalloy "field straightener" disks are mounted perpendicular to the axis of the tube. Since these disks act as equipotential planes with respect to the magnetic field, they force the magnetic field to be axially symmetric with respect to the axis of the tube. The magnetic circuit is surrounded by an external magnetic shield that reduces the leakage field ouside the shield to a negligible amount. In certain tube structures in which the interaction structure is short enough, permanent magnet focusing is often used. In some high-power gridded tubes for airborne and space applications, periodic-permanent-magnet (PPM) focusing is utilized. Figure 5-2-4-17 illustrates four common methods of magnetic focusing.

5-2-5. Twystron Hybrid Amplifier

Several hybrid devices using combinations of klystron and traveling-wave tube components have been devised in order to achieve a better performance than each tube can obtain separately. The only widely used hybrid

MAGNETIC
FIELD LINES

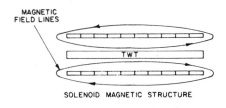

SOLENOID MAGNETIC STRUCTURE

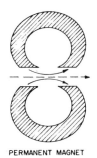

PERMANENT MAGNET

ALNICO RADIAL SEGMENTS

ALNICO RINGS

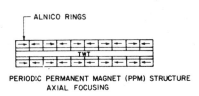

PERIODIC PERMANENT MAGNET (PPM) STRUCTURE
AXIAL FOCUSING

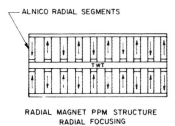

RADIAL MAGNET PPM STRUCTURE
RADIAL FOCUSING

FIG. 5-2-4-17. Methods of magnetic focusing. (*Courtesy Hughes Aircraft Company, Electron Dynamics Division.*)

device is known under the trade name Twystron amplifier [4]. The Twystron amplifier consists of a multicavity klystron input section and a traveling-wave output section. Figure 5-2-5-1 is a schematic circuit diagram of a Twystron amplifier compared to the klystron, coupled-cavity traveling-wave tube, and extended interaction klystron, as a function of interaction impedance. The chief feature of the Twystron amplifier is that it combines the advantages of klystrons and traveling-wave tubes. Figure 5-2-5-2 shows a cutaway of a typical S-band Twystron amplifier.

The four-cavity stagger-tuned klystron driver section provides higher gain at the end bands than its midband because the driver section can be heavily loaded without excessive power loss. The traveling-wave output circuit (clover leaf) is designed principally for high efficiency over the desired bandwidth. Its gain at the band edges is 6 to 10 dB below the values of the midband. The combination of klystron and traveling-wave tube yields a relatively flat gain characteristic over the entire frequency range as shown in Fig. 5-2-5-3. Gain flatness and high efficiency are achieved in the Twystron amplifier because of the more efficient bunching in the klystron at the band edges where the output slow-wave circuit gain is low.

The principal applications of Twystron amplifiers are in land-based or shipboard high-power radar transmitters. Typical power output levels at S

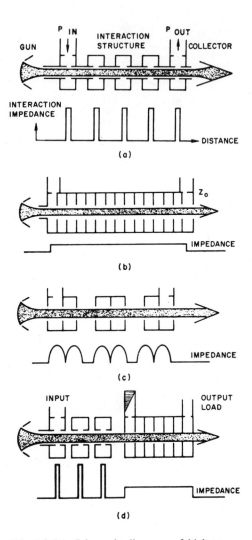

FIG. 5-2-5-1. Schematic diagrams of high-power linear-beam tubes. (a) Klystron. (b) Coupled cavity TWT. (c) Extended interaction klystron. (d) Hybrid Twystron amplifier. (*After A. Staprans et al. [18]; reprinted by permission of the IEEE, Inc.*)

and C bands are from 1 to 10 MW (megawatts) for peak power and from 1 to 30 KW for average power [19]. Table 5-2-3 lists typical tube characteristics for several Twystron amplifiers. Figure 5-2-5-4 shows typical power output versus frequency at a C-band Twystron amplifier.

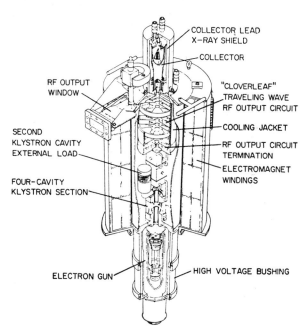

FIG. 5-2-5-2. Cutaway of a Twystron amplifier at S-band.
(*After A. Staprans et al. [18]; reprinted with permission of the IEEE, Inc.*)

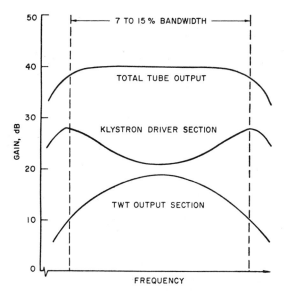

FIG. 5-2-5-3. Relative gain of a Twystron amplifier.
(*After A. Staprans et al. [18]; reprinted by permission of the IEEE, Inc.*)

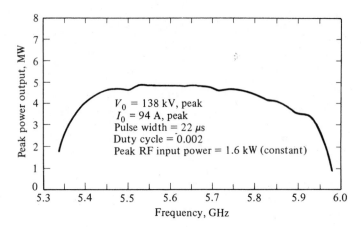

FIG. 5-2-5-4. Power output versus frequency for C-band Twystron amplifier, VA-146. (*After A. Staprans et al. [18]; reprinted with permission of the IEEE, Inc.*)

TABLE 5-2-3. Typical Twystron Amplifier Characteristics.

Tube type	VA-145	VA-915	VA-146	VA-145LV
Frequency bandwidth, GHz	2.7–2.9 2.9–3.1 3.0–3.2	3.1–3.6	5.4–5.9	3.1–3.5
Peak power output, MW	3.5	7.0	4.0	1.0
Average power output, kW	7.0	28.0	10.0	1.0
Pulse width, μs	10.0	40.0	20.0	50.0
Efficiency, %	35.0	30.0	30.0	30.0
Beam voltage, kV	117.0	180.0	140.0	80.0
Beam current, A	80.0	150.0	95.0	45.0
Drive power, kW	0.3	3.0	2.0	1.0

5-2-6. Backward-Wave Amplifiers (BWA)

The O-type backward-wave amplifier has an RF signal impressed on the slow-wave structure near the collector, and the signal output at the gun end as shown in Fig. 5-2-6-1.

Separate power supplies are provided for the anode and the helix. This arrangement makes the adjustment of the beam current independent of the helix voltage. The electron beam is assumed to be confined in the center of helix by an axial magnetic field. The electrons near the cathode are velocity-modulated as in the ordinary traveling-wave tube.

While the signal waves are traveling toward the cathode, the electrons are bunched somewhere near the collector. If the bunched electrons are decelerated by the microwave fields, the bunched electron beams transfer

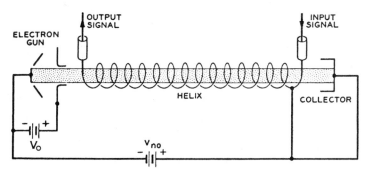

FIG. 5-2-6-1. Schematic diagram of backward-wave amplifier.

energy to the input microwave fields. The energy propagates on the slow-wave structure toward the output, and the amplification of a microwave signal in the backward-wave tube is accomplished.

The electronic equation as shown in Eq. (5-2-106) is derived under the assumption that the traveling waves move forward in the positive z direction. However, as described in Section 5-2-3 the direct current density J_0 is assumed to be a positive number in the negative z dirention. Hence the convection-beam current in the electron induced by a spatial harmonic of the circuit field in the backward-wave amplifier is given by

$$i = -j\frac{\beta_e I_0}{2V_0(j\beta_e - \gamma)^2}E_1 \qquad (5\text{-}2\text{-}144)$$

It should be noted that the electronic equation as shown in Eq. (5-2-144) for a backward-wave amplifier differs from the corresponding equation given by Eq. (5-2-106) for a forward-wave amplifier by a minus sign.

The circuit equation shown in Eq. (5-2-115) is applicable to backward waves as well as forward waves. The equation is repeated here.

$$E_1 = -\frac{\gamma^2\gamma_0 Z_0}{\gamma^2 - \gamma_0^2}i \qquad (5\text{-}2\text{-}145)$$

As analyzed in previous sections, a simultaneous solution of Eqs. (5-2-144) and (5-2-145) yields four propagation constants.

$$\gamma_1 = -\beta_e C\frac{\sqrt{3}}{2} + j\beta_e\left(1 - \frac{C}{2}\right) \qquad (5\text{-}2\text{-}146)$$

$$\gamma_2 = \beta_e C\frac{\sqrt{3}}{2} + j\beta_e\left(1 - \frac{C}{2}\right) \qquad (5\text{-}2\text{-}147)$$

$$\gamma_3 = j\beta_e(1 + C) \qquad (5\text{-}2\text{-}148)$$

$$\gamma_4 = -j\beta_e\left(1 + \frac{C^3}{4}\right) \qquad (5\text{-}2\text{-}149)$$

where $C = [I_0 Z_0/(4V_0)]^{1/3}$ is the traveling-wave tube gain parameter. These four propagation constants represent four different modes of wave propagation in the O-type backward-wave tube. The predominant wave is the one that is represented by

$$V(z) = V_1 e^{-\gamma_1(\ell-z)} = V_1 \exp\left[\frac{\sqrt{3}}{2}\beta_e C(\ell - z)\right] \exp\left[-j\beta_e\left(1 - \frac{C}{2}\right)(\ell - z)\right]$$

$$(5\text{-}2\text{-}150)$$

5-3   Crossed-Field Tubes (M-Type)

In Section 5-2 several linear-beam tubes in general use were described in detail. In these tubes the dc magnetic field that is in parallel with the dc electric field is used merely to focus the electron beam. In crossed-field devices, however, the dc magnetic field and the dc electric field are perpendicular to each other. In all crossed-field tubes the dc magnetic field plays a direct role in the RF interaction process. In this section several commonly used crossed-field tubes, such as magnetron, forward-wave crossed-field amplifier (FWCFA), and M-carcinotron, are discussed.

Crossed-field tubes derive their name from the fact that the dc electric field and the dc magnetic field are perpendicular to each other. They are also called M-type tubes after the French TPOM (tubes à propagation des ondes à champs magnétique; tubes for propagation of waves in a magnetic field). In a crossed-field tube the electrons emitted by the cathode are accelerated by the electric field and gain velocity; but the greater their velocity, the more their path is bent by the magnetic field. If an RF field is applied to the anode circuit, those electrons entering the circuit during the retarding field are decelerated and give up some of their energy to the RF field. Consequently, their velocity is decreased, and these slower electrons will then travel the dc electric field far enough to regain essentially the same velocity as before. Because of the crossed-field interactions, only those electrons that have given up sufficient energy to the RF field can travel all the way to the anode. This phenomenon would make the M-type devices relatively efficient. Those electrons entering the circuit during the accelerating field are accelerated by means of receiving enough energy from the RF field and returned back toward the cathode. This back bombardment of the cathode produces heat in the cathode and decreases operational efficiency.

Crossed-field tubes (M-type) as tabulated in Table 5-0-2 are the magnetron, forward-wave crossed-field amplifier (FWCFA), dematron, amplitron, and carcinotron. The present state of the art for U.S. high-power magnetrons and crossed-field amplifiers is depicted in Figs. 5-3-1-1 and 5-3-1-2, respectively.

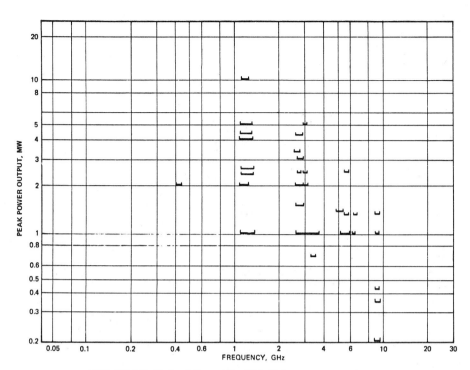

FIG. 5-3-1-1. State of the art for U.S. high-power magnetrons.

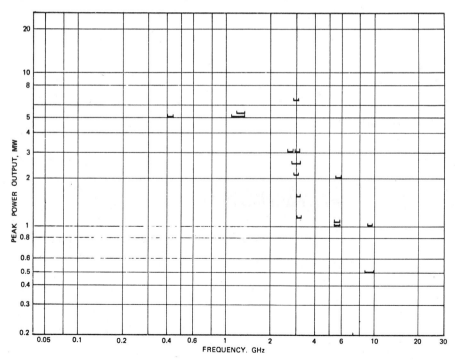

FIG. 5-3-1-2. State of the art for U.S. high-power CFAS.

5-3-1. Magnetrons

Although Hull invented the magnetron in 1921 [20], it remained primarily an interesting laboratory device until about 1940. During World War II an urgent need for high-power microwave generators for radar transmitters led to the rapid development of the magnetron to its present state.

All magnetrons consist of some form of anode and cathode operated in a dc magnetic field normal to a dc electric field between the cathode and anode. Due to the crossed field between the cathode and anode, the electrons emitted from the cathode are influenced by the crossed field to move in curved paths. If the dc magnetic field is strong enough, the electrons will not arrive in the anode but return instead to the cathode. Consequently, the anode current is cut off.

Magnetrons can be classified into three types.

Type 1   Split-Anode Magnetron. This type of magnetron uses a static negative resistance between two anode segments.

Type 2   Cyclotron-Frequency Magnetrons. This type operates under the influence of synchronism between an alternating component of electric field and a periodic oscillation of electrons in a direction parallel to the field.

Type 3   Traveling-Wave Magnetrons. This type depends on the interaction of electrons with a traveling electromagnetic field of linear velocity. They are customarily referred simply as "magnetrons."

Negative-resistance magnetrons ordinarily operate at frequencies below the microwave region. Although cyclotron-frequency magnetrons operate at frequencies above microwave range, but its power output is very small (about 1 watt at 3 GHz) and its efficiency is very low (about 10% in the split-anode type and 1% in the single-anode type). So the first two types of magnetrons will not be discussed in this text. In this section only the traveling-wave magnetrons are discussed.

(1) Physical Description. Schematic diagrams of two typical models of magnetron oscillators are shown in Fig. 5-3-1-3. In a cylindrical magnetron several reentrant cavities are connected to the gaps. The dc voltage V_0 is applied between the cathode and the anode. The magnetic flux B_0 is into the page. When the dc voltage and the magnetic flux density are adjusted properly, the electrons will follow cycloidal paths in the cathode-anode space under the influence of both the electric field E and the magnetic flux density B as shown in Fig. 5-3-1-4.

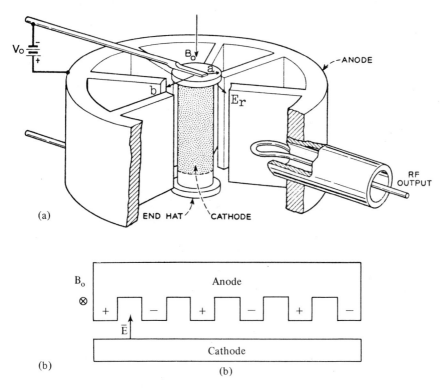

(a)

(b)

FIG. 5-3-1-3. Schematic diagrams of magnetrons. (a) Cylindrical model of a magnetron. (b) Linear model of a magnetron.

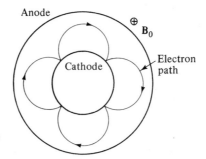

FIG. 5-3-1-4. Electron path in circular magnetrons.

(2) Principles of Operation. As described previously, the electrons emitted by the cathode move in a curved path between cathode and anode under the combined force of electric and magnetic fields. Hence both the electric and magnetic forces control the behavior of the electrons. This fact can be seen from the results of solving the field equations.

The equations of motion for electrons in the cylindrical magnetron can be written with the aid of Eqs. (1-3-3) as

$$\frac{d^2r}{dt^2} - r\left(\frac{d\phi}{dt}\right)^2 = \frac{e}{m}E_r - \frac{e}{m}rB_z\frac{d\phi}{dt} \tag{5-3-1}$$

$$\frac{1}{r}\frac{d}{dt}\left(r^2\frac{d\phi}{dt}\right) = \frac{e}{m}B_z\frac{d\phi}{dt} \tag{5-3-2}$$

where $B_0 = B_z$ is assumed in the positive z direction.

The electron will acquire a tangential as well as a radial velocity. Whether the electron will just graze at the anode and return back toward the cathode depends on the relative magnitudes of V_0 and B_0. This condition is called the *cutoff condition* of the magnetron. The solution of Eqs. (5-3-1) and (5-3-2) with the boundary conditions applied yields the cutoff magnetic flux density as

$$B_{oc} = \frac{[8V_0(e/m)]^{1/2}}{b[1 - (a^2/b^2)]} \tag{5-3-3}$$

This means that if $B_0 > B_{oc}$ for a given V_0, the electrons will not reach the anode. Conversely, the cutoff voltage is given by

$$V_{oc} = \frac{e}{8m}B_0^2b^2\left(1 - \frac{a^2}{b^2}\right)^2 \tag{5-3-4}$$

This means that if $V_0 < V_{oc}$ for a given B_0, the electrons will not reach the anode. Equation (5-3-4) is often called the *Hull cutoff voltage equation*.

Since the magnetic field is normal to the motion of electrons that travel in a cycloidal path, the outward centrifugal force is equal to the pulling force. Thus

$$\frac{mv^2}{R} = evB \tag{5-3-5}$$

where R = the radius of the cycloidal path
v = the tangential velocity of the electron

The cyclotron angular frequency of the circular motion of the electron is then given by

$$\omega_c = \frac{v}{R} = \frac{eB}{m} \tag{5-3-6}$$

The period for one complete revolution can be expressed as

$$T = \frac{2\pi}{\omega} = \frac{2\pi m}{eB} \tag{5-3-7}$$

Since the slow-wave structure is closed on itself, or "reentrant," oscillations are possible only if the total phase shift around the structure is an integral multiple of 2π radians. So if there are N reentrant cavities in the anode structure, the phase shift between two adjacent cavities can be written

$$\phi_n = \frac{2\pi n}{N} \tag{5-3-8}$$

where n is an integer indicating the nth mode of oscillation.

In order for oscillations to be produced in the structure, the anode dc voltage must be adjusted so that the average rotational velocity of the electrons corresponds to the phase velocity of the field in the slow-wave structure. Magnetron oscillators are ordinarily operated in the π mode. That is,

$$\phi_n = \pi \qquad (\pi \text{ mode}) \tag{5-3-9}$$

Figure 5-3-1-5 shows the lines of force in the π mode of an eight-cavity magnetron. It is clear that in π mode the excitation is largely in the cavities,

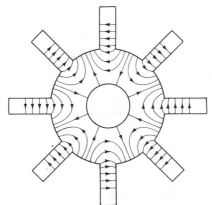

FIG. 5-3-1-5. Lines of force in π mode of eight-cavity magnetron.

having opposite phase in successive cavities. The successive rise and fall of adjacent anode-cavity fields may be regarded as a traveling wave along the surface of the slow-wave structure. In order for the energy to be transferred from the moving electrons to the traveling field, the electrons must be decelerated by a retarding field when they pass through each anode cavity. If L is the mean separation between cavities, the phase constant of the fundamental-mode field is given by

$$\beta_0 = \frac{2\pi n}{NL} \tag{5-3-10}$$

The traveling-wave field of the slow-wave structure can be obtained by solving Maxwell's equations subject to the boundary conditions. The solution for the fundamental ϕ component of the electric field has the form [8]

$$E_{\phi 0} = jE_1 e_{j(\omega t - \beta_0 \phi)} \tag{5-3-11}$$

where E_1 is a constant and β_0 is given in Eq. (5-3-10).

Thus the traveling field of the fundamental mode will travel around the structure with angular velocity

$$\frac{d\phi}{dt} = \frac{\omega}{\beta_0} \tag{5-3-12}$$

where $d\phi/dt$ can be found from Eq. (5-3-11).

When the cyclotron frequency of the electrons is equal to the angular frequency of the field, the interaction between the field and the electron occurs and the energy is transferred. That is,

$$\omega_c = \beta_0 \frac{d\phi}{dt} \tag{5-3-13}$$

The analysis of the linear M-type magnetron of Fig. 5-3-1-3(b) is similar to that of the cylindrical magnetron oscillator.

(3) Microwave Characteristics. For many years magnetrons have been the high-power sources in operating frequencies as high as 70 GHz. Military radar relies on conventional traveling-wave magnetrons to generate high peak power RF pulses. No other microwave devices could perform the same function with the same size, weight, voltage, and efficiency range as the conventional magnetrons. At the present state of the art magnetrons can deliver a peak power output of up to 40 MW with the dc voltage on the order of 50 kV at the frequency of 10 GHz. The average power outputs are up to 800 kW. Its efficiency is very high, ranging from 40 to 70%.

The beacon magnetrons are miniature conventional magnetrons that deliver peak outputs as high as 3.5 kW and yet weigh less than 1kg (2 pounds). These devices are ideal when a very compact, low-voltage source of pulsed power is required, such as in airborne, missile, satellite, or Doppler systems. Most of the beacon magnetrons exhibit negligible frequency shift and provide long-life performance under the most severe environmental and temperature conditions.

CW (continuous wave) magnetrons can produce output power from 100 to 300 W at X band. These devices are available with a variety of tuning mechanisms and cooling systems to fit specific applications. Some CW magnetrons are designed to meet airborne environments and are also ideal as laboratory CW power sources.

5-3-2. Forward-Wave Crossed-Field Amplifier (FWCFA)

(1) Physical Description. The crossed-field amplifier (CFA) is an outgrowth of the magnetron. CFAs can be grouped by their mode of operation as forward-wave or backward-wave types and by their electron stream source as emitting sole or injected beam types. The first group concerns the direction of the phase and group velocity of the energy on the microwave circuit. This can be seen from the ω-β diagrams of Fig. 5-2-4-4. Since the electron stream reacts to the RF electric field forces, the behavior of the phase velocity with frequency is of prime concern. The second group emphasizes the method by which electrons reach the interaction region and how they are controlled. This can be seen in the schematic diagrams of Fig. 5-3-2-1.

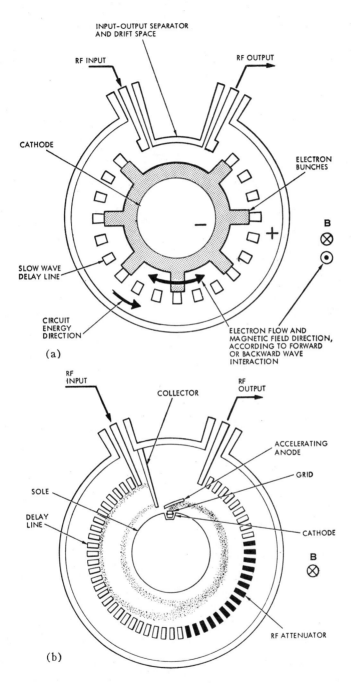

FIG. 5-3-2-1. Schematic diagrams of CFAs. (a) Continuous-cathode emitting-sole CFA, forward wave or backward wave. (b) Injected-beam CFA. (*After J. F. Skowron [21]; reprinted by permission of the IEEE, Inc.*)

In the forward-wave mode the helix-type slow-wave structure is often selected as the microwave circuit for the crossed-field amplifier, whereas in the backward-wave mode the strapped bar line represents a satisfactory choice. A structure of strapped crossed-field amplifier is shown in Fig. 5-3-2-2.

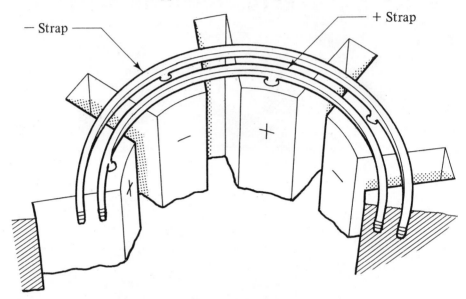

FIG. 5-3-2-2. Diagram of a strapped CFA. (*After J. F. Skowron [21]; reprinted by permission of the IEEE, Inc.*)

(2) Principles of Operation. In the emitting-sole tube current emanated from the cathode is in response to the electric field forces in the space between the cathode and anode. The amount of current is a function of the dimension, the applied voltage, and the emission properties of the cathode. The perveance of the interaction geometry tends to be quite high, about 5×10^{-10} to 10×10^{-10}, which results in a high-current and high-power capability at relative low voltage. In the injected-beam tube the electron beam is produced in a separate gun assembly and is injected into the interaction region.

The beam-circuit interaction features are similar in both the emitting-sole and the injected-beam tubes. Favorably phased electrons continue toward the positively polarized anode and are ultimately collected, whereas unfavorably phased electrons are directed toward the negative polarized electrode.

In linear-beam interactions as discussed in Section 5-2-3, the electron stream is first accelerated by an electric gun to the full dc velocity; the dc velocity is approximately equal to the axial phase velocity of the RF field in a slow-wave structure. After interaction occurs, the spent electron beam leaves the interaction region with a low average velocity. The difference in velocity is accounted for by the RF energy created on the microwave circuit. In the crossed-field amplifier the electron is exposed to the dc electric field force, the magnetic field force, the electric field force of the RF field, and even the space-charge force from other electrons. The last force is normally not considered in analytic approaches because of its complexity. Inder the influence of the three forces the electrons travel in spiral trajectories in a direction tending along equipotentials. The exact motion has been subject to much analysis by means of a computer. Figure 5-3-2-3 shows the pattern of the electron flow in the crossed-field amplifier by computerized techniques [21]. It can be seen that when the spoke is positively polarized or the RF field is in the positive half cycle, the electrons speed up toward the anode; while the spoke is negatively polarized or the RF field is in the negative half cycle, the electrons are returned toward the cathode. Consequently, the electron beam moves in a spiral path in the interaction region.

(*3*) *Microwave Characteristics.* The crossed-field amplifier is characterized by its low or moderate gain, moderate bandwidth, high efficiency, saturated amplification, small size, low weight, and high perveance. These features have allowed the crossed-field amplifier to be used in a variety of electronic systems ranging from low-power high-reliability space communications to multimegawatt high-average-power coherent pulsed radar.

5-3-3. M-Carcinotron Oscillators

The *M*-carcinotron oscillator is an *M*-type backward-wave oscillator. The interaction between the electrons and the slow-wave structure takes place in a space of crossed fields. A linear model of an *M*-carcinotron is shown in Fig. 5-3-3-1. The slow-wave structure is in parallel with an electrode known as the sole. A dc electric field is maintained between the grounded slow-wave structure and the negative sole. A dc magnetic field is directed into the page. The electrons emitted from the cathode are bent through a 90° angle by the magnetic field. The electrons interact with a backward-wave space harmonic of the circuit, and the energy in the circuit flows opposite to the direction of the electron motion. The slow-wave structure is terminated at the collector end, and the RF signal output is removed at the electron gun end. Since the *M*-carcinotron is a crossed-field device, its efficiency is very high, ranging from 30 to 60%.

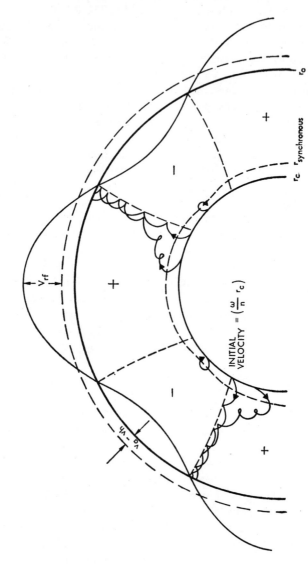

FIG. 5-3-2-3. Motion of electrons in CFA. *(After J. F. Skowron [21]; reprinted with permission of the IEEE, Inc.)*

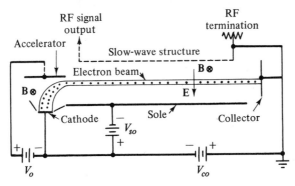

FIG. 5-3-3-1. Linear model of an *M*-carcinotron oscillator. (*After Gewartowski and Watson [10]; reprinted by permission of McGraw-Hill Book Company.*)

5-4 Summary

Most of the microwave tubes in common use have been described. These tubes include such types as conventional vacuum tubes, klystrons, reflex klystrons, traveling-wave tubes, and magnetrons. The essential points in the discussion of each device emphasize the principles of operation of the device and its power output. In summary, Fig. 5-4-1 shows the CW (continuous wave) power output in watts for several commonly used microwave tubes against their operating frequencies.

Conventional Vacuum Tubes: The fact that the conventional vacuum triode has power-frequency limitation was discussed in Section 5-1. It is quite clear from Fig. 5-4-1 that the power output of the vacuum triode decreases rapidly when the operating frequency is about 3 GHz. The power-frequency limitations are due to the following three effects:

1. Lead-inductance and interelectrode capacitance effect
2. Transit-angle effect
3. Gain-bandwidth product limitation

Linear-Beam Tubes (*O*-type): All linear-beam tubes are operated by the velocity-modulation process in multicavity resonator or slow-wave structure. In general, their power outputs are high. For example, the CW power output of an extended interaction oscillator (multicavity klystron) is from 1 to 10 W at frequencies of 40 to 160 GHz. The characteristics of CW power outputs for two-cavity klystron, reflex klystron, and backward-wave oscillator (BWO) are shown in Fig. 5-4-1.

Crossed-Field Tubes (*M*-type): All crossed-field tubes are operated under the combined force of electric and magnetic fields. In crossed-field devices, the dc magnetic field and the dc electric field are perpendicular to each other; so the dc magnetic field plays a direct role in the RF interaction process. The power outputs of the *M*-type devices are very high. A conventional magnetron can deliver a peak power output of up to 40 MW with

253

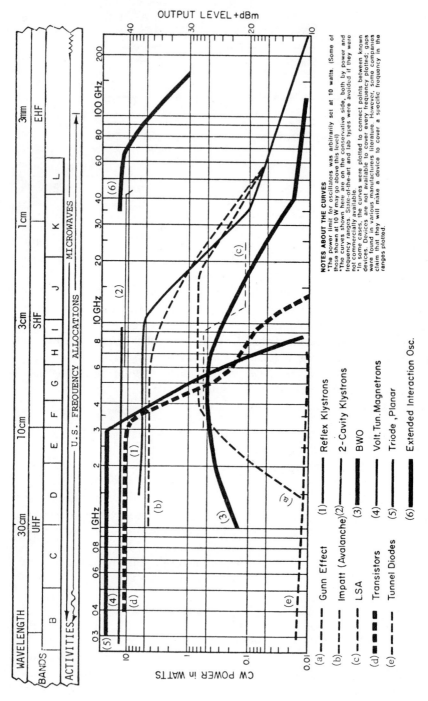

FIG. 5-4.1. CW power of microwave devices.

a dc voltage of 50 kV at the frequency of 10 GHz. A CW magnetron can produce power outputs from 100 to 300 W at *X* band.

REFERENCES

[1] WARNECKE, R. R., et al., Velocity Modulated Tubes in *Advances in Electrons*, Vol. 3. Academic Press, New York, 1951.

[2] CHODOROW, M., and C. SUSSKIND, *Fundamentals of Microwave Electrons*. McGraw-Hill Book Company, New York, 1964.

[3] CHODOROW, M., and T. WESSEL-BERG, A high-efficiency klystron with distributed interaction. *IRE Trans. Electron Devices*, ED-8, 44–55, January 1961.

[4] LaRUE, A. D., and R. R. RUBERT, Multi-Megawatt Hybrid TWT's at *S*-band and *C*-band, presented to the IEEE Electron Devices Meeting, Washington, D.C., October 1964.

[5] FEENBERG, E., Notes on velocity modulation. *Sperry Gyroscope Laboratories Report. 5521–1043,* Chapter I, pp. 41–44.

[6] LIEN, E. L., High Efficiency Klystron Amplifier in *Conv. Rec. MOGA 70* (8th Int. Conf., Amsterdam, The Netherlands, September 1970).

[7] MERDINIAN, G., and J. V. LEBACQZ, High power, permanent magnet focused, *S*-band klystron for linear accelerator use. *Proc. 5th Int. Conf. on Hyperfrequency Tubes* (Paris, France, September 1964).

[8] HUTTER, R. G. E., *Beam and Wave Electronics in Microwave Tubes*. D. Van Nostrand Company, Princeton, N.J., 1960.

[9] KOMPFNER, R., The traveling-wave tube as amplifier at microwaves. *Proc. IRE*, 35, 124–127, February 1947.

[10] GEWARTOWSKY, J. V., and H. A. WATSON, *Principles of Electron Tubes*, p. 391. D. Van Nostrand Company, Princeton, N.J., 1965.

[11] REICH, H. J., et al., *Microwave Principles*, p. 259. D. Van Nostrand Company, Princeton, N.J., 1966.

[12] CHODOROW, M., and R. A. CRAIG, Some new circuits for high-power traveling-wave tubes. *Proc. IRE*, 45, 1106–1118, August 1957.

[13] ROUMBANIS, T., et al., A megawatt *X*-band TWT amplifier with 18% bandwidth. *Proc. High-Power Microwave Tubes Symp*, Vol. 1 (The Hexagon, Fort Monmouth, N.J., September 25–26, 1962).

[14] RUETZ, A. J., and W. H. YOCOM, High-power traveling-wave tubes for radar systems. *IRE Trans. Mil. Electron.*, MIL-5, 39–45, April 1961.

[15] BRILLOUIN, L., *Wave Propagation in Periodic Structures*, 2nd ed. Dover, New York, 1953.

[16] CURNOW, H. J., A general equivalent circuit for coupled-cavity slow-wave structures. *IEEE Trans. Microwave Theory and Tech.*, **MTT-13**, 671–675, September 1965.

[17] MENDEL, J. T., Helix and coupled-cavity traveling-wave tubes. *Proc. IEEE*, **61**, No. 3, 280–298, March 1973.

[18] STAPRANS, A., et al., High-power linear-beam tubes. *Proc. IEEE*, **61**, No. 3, 299–330, March 1973.

[19] ROUMBANIS, T., Centipede Twystron Amplifiers and Traveling-Wave Tubes for Broadband, High-Efficiency, Super-Power Amplification. *Proc. 7th Int. Conf. on Microwave and Optical Generation and Amplification* (Hamburg, Germany, September 16–20, 1968).

[20] HULL, A. W., *Physical Review*. **18**, 31, 1921.

[21] SKOWRAN, JOHN F., The continuous-cathode (emitting-sole) crossed-field amplifier. *Proc. IEEE*, **61**, No. 3, 330–356, March 1973.

[22] LIAO, SAMUEL Y., The Effect of Collector Voltage Overdepression on Tube Performance of the Gridded Traveling-Wave Tubes. Hughes Aircraft Company, El Segundo, Calif., August 1977.

SUGGESTED READINGS

1. ANGELAKOS, D. J., and T. E. EVERHART, *Microwave Communications*, Chapters 3 and 4. McGraw-Hill Book Company, New York, 1968.

2. BRONWELL, A. B., and R. E. BEAM, *Theory and Application of Microwaves*, Chapters 6 and 8. McGraw-Hill Book Company, New York, 1947.

3. COLLIN, ROBERT E., *Foundations for Microwave Engineering*, Chapter 9. McGraw-Hill Book Company, New York, 1966.

4. GEWARTOWSKI, J. W., and H. A. WATSON, *Principles of Electron Tubes*, Chapters 5, 6, and 10 to 13. D. Van Nostrand Company, Princeton, N.J., 1965.

5. *IEEE. Proceedings*. Special issue, High-Power Microwave Tubes, Vol. 61, No. 3, March 1973.

6. MILLMAN, JACOB, and C. C. HALKIAS, *Electronic Devices and Circuits*, Chapters 7 and 8. McGraw-Hill Book Company, New York, 1967.

7. REICH, HERBERT J., et al., *Microwave Theory and Techniques*, Chapters 10 to 15. D. Van Nostrand Company, Princeton, N.J., 1953.

8. REICH, HERBERT J., et al., *Microwave Principles*, Chapters 8 to 12. D. Van Nostrand Company, Princeton, N.J., 1966.

9. SPANGENBERG, KARL R., *Vacuum Tubes*, Chapters 17 and 18. McGraw-Hill Book Company, New York, 1948.

Microwave
Solid-State Devices

6-0   Introduction

Microwave tubes in common use were analyzed in Chapter 5. Another class of microwave-source devices that is becoming increasingly important at microwave frequencies is the solid-state device. These devices can be broken down into four groups. In the first group are the microwave transistors [3, 4], microwave tunnel diodes [5], and microwave field-effect transistors (FETs) [25 to 39]. This group is discussed in Section 6-1. The second group, which is characterized by the junction effect of the semiconductor, is called the transferred electron device (TED) [1, 2]. These devices include the Gunn diode, limited space-charge accumulation diode (LSA diode), indium phosphide diode (InP diode), and cadmium telluride diode (CdTe diode). This group

6

257

is analyzed in Section 6-2. The devices of the third group, which are operated by the bulk effect of the semiconductor, are referred to as the avalanche diodes: the impact ionization avalanche transit-time diodes (IMPATT diodes) [6], the trapped plasma avalanche triggered transit-time diodes (TRAPATT diodes) [7], and the barrier injected transit-time diodes (BARITT diodes) [130]. The avalanche diodes are described in Section 6-3. The last group includes the quantum-electronic solid-state devices, such as the ruby masers [8] and the semiconductor lasers [9], which are able to function by the stimulated emission of radiation, the parametric devices, which utilize a nonlinear reactance for amplication [156–163], and the infrared devices, which use a number of sensitive thermal detectors to recognize or identify the target at night [164–181]. Sections 6-4 through 6-6 deal with this group. All the microwave power-source solid-state devices analyzed in this chapter are tabulated in Table 6-0-1.

TABLE 6-0-1.   Microwave Solid-State Devices.

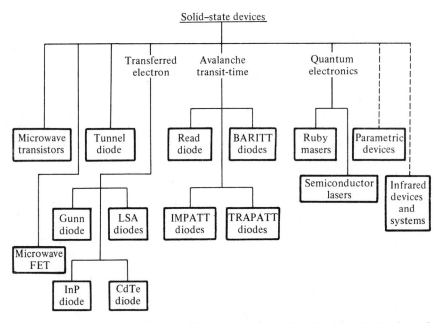

In studying microwave solid-state devices, the electrical behavior of solids is the first item to be investigated. In the following sections it will be seen that the transport of charge through a semiconductor depends not only on the properties of the electron but also on the arrangement of atoms in the solids. Semiconductors are a group of substances having electrical conductivities that are intermediate between metals and insulators. Since the conductivity of the semiconductors can be varied over wide ranges by changes in their temperature, optical excitation, and impurity content, they are the natural choices for electronic devices. The properties of important semiconductors are tabulated in Table 6-0-2.

TABLE 6-0-2. Properties of Important Semiconductors.

| Semiconductor | Band gap (eV) | | Mobility at 300°K (cm²/volt-s) | | Dielectric constant |
	$0°K$	$300°K$	Holes	Electrons	
C	5.51	5.47	1600	1800	5.5
Ge	0.89	0.803	1900	3900	16
Si	1.16	1.12	600	1500	11.8
AlSb	1.75	1.63	420	200	11
GaSb	0.80	0.67	1400	4000	15
GaAs	1.52	1.43	400	8500	10.9
GaP	2.40	2.24	75	110	10
InSb	0.26	0.16	750	78,000	17
InAs	0.46	0.33	460	33,000	14.5
InP	1.34	1.29	150	4600	14
CdS	2.56	2.42	50	300	10
CdSe	1.85	1.70		800	10
ZnO		3.2		200	9
ZnS	3.70	3.60		165	8

The energy bands of a semiconductor play a major role in their electrical behavior. For any semiconductor, there is a forbidden energy region in which no allowable states can exist. The energy band above the forbidden region is called the *conduction band*, and the bottom of the conduction band is designated by E_c. The energy band below the forbidden region is called the *valence band*, and the top of the valence band is designated by E_v. The separation between the energy of the lowest conduction band and that of the highest valence band is called the *energy band gap* E_g, which is the most important parameter in semiconductors.

Electron energy is conventionally defined as positive when measured upward whereas the hole energy as positive when measured downwards. A simplified band diagram is shown in Fig. 6-0-1.

Only a few years ago it seemed that microwave transistors would be useful for generating power up to about 5 GHz. Since their inception, avalanche diodes have produced in excess of 4 watts CW (continuous wave) at 5 GHz. Gunn diodes had been considered only for local oscillators or low power transmitter applications, but recent results indicate that a single Gunn diode can generate an output power of 1 watt at *X* band. At higher microwave frequencies, and even well into the millimeter range, LSA diodes can provide the highest peak power of any solid-state device, up to 250 watts in *C* band,

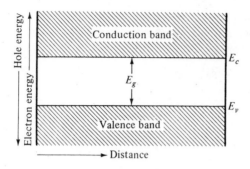

FIG. 6-0-1. Energy-band diagram.

100 watts in *X* band, and 50 watts in *Ku* band. Since the pulsed Gunn and TRAPATT diodes are essentially transit-time devices, their operating frequency is approximately determined by the thickness of the active layer in the diode. An operating frequency of 10 GHz requires an active layer thickness on the order of 10 μm (microns). Thus only a limited voltage can be applied to such a thin layer because of breakdown limitations. Consequently, the peak power capability of both the pulsed Gunn diodes and the TRAPATT diodes is greatly limited at higher frequencies. On the other hand, the peak power capability of an LSA diode is approximately proportional to the square of the thickness of the active layer because its operating frequency is independent of the thickness of the active layer. Thus the LSA diode is capable of producing higher peak power than either the pulsed Gunn diodes or the TRAPATT diodes. Figure 6-0-2 shows peak power versus frequency for these three devices.

Solid-state microwave power sources are widely used in radar, communications, navigational and industrial electronics, and medical and

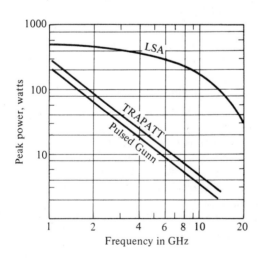

FIG. 6-0-2. Peak-power levels achieved by microwave diode [11]. (*Courtesy of Horizon House, Microwave Journal.*)

biological equipment. A list of several representative applications for microwave solid-state devices is shown in Table 6-0-3.

TABLE 6-0-3. Applications of Microwave Solid-State Devices [12]. (After Sobol and Sterzer, Reprinted by Permission from the IEEE, Inc.)

Devices	Applications	Advantages
Transistor	L-band transmitters for telemetry systems and phased array radar systems L- and S-band transmitters for communications systems	Low cost, low power supply, reliable, high CW power output, light weight
TED	C-, X-, and Ku-band ECM amplifiers for wideband systems X- and Ku-band transmitters for radar systems, such as traffic control	Low power supply (12 volt), low cost, light weight, reliable, low noise, high gain
IMPATT	Transmitters for millimeter-wave communications systems	Low power supply, low cost, reliable, high CW power output, light weight
TRAPATT	S-band pulsed transmitters for phased array radar systems	High peak and average power, reliable, low power supply, low cost
BARITT	Local oscillators in communications and radar receivers	Low cost, low power supply, reliable, low noise

6-1 Microwave Transistors, Tunnel Diodes, and Microwave Field-Effect Transistors

In Section 5-1 the power-frequency limitations on conventional vacuum power tubes were analyzed in some detail. The question naturally arises as to whether microwave power transistors, microwave tunnel diodes, and microwave field-effect transistors also have power-frequency limitations. The answer is yes. In this section the physical structures, operating principles, and, microwave characteristics, as well as the power-frequency limitations of microwave power transistors, microwave tunnel diodes, and microwave field-effect transistors are described in detail. Before describing microwave transistors, however, it seems appropriate to demonstrate the concentrations of a typical *PNP* transistor (see Fig. 6-1-1) in 6-1-1 for comparison with microwave solid-state devices in later sections.

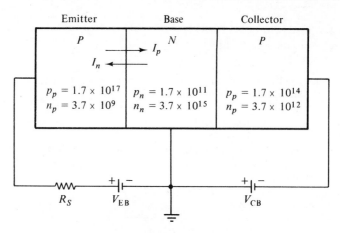

FIG. 6-1-1. Typical *p-n-p* transistor.

6-1-1. Microwave Transistors

The invention of the transistor (contraction for transfer resistor) by William Shockley [13] and coworkers at Bell Laboratory in 1948 had a revolutionary impact on electronic technology in general and solid-state devices in particular. Since then transistors and related semiconductor devices have replaced vacuum tubes for lower power sources. Microwave power transistor technology has advanced significantly during the past decade. The microwave transistor is a nonlinear device, and its principle of operation is similar to that of the low-frequency device, but requirements for dimensions, process control, heat sinking, and packaging are much more severe [3, 12, 14].

(1) Physical Structures. All microwave transistors are now planar in form and almost all are of the silicon *n-p-n* type. The geometry can be characterized as follows: (a) interdigitated, (b) overlay, and (c) matrix (also called mesh or emitter grid) as shown in Fig. 6-1-2. The interdigitated type is for a small signal and power, but the overlay type and matrix type are for small power only. The figure of merit for the three surface geometries shown in Fig. 6-1-2 is listed in Table 6-1-1 [12].

The state of the art of transistor fabrication today limits the emitter width w to about 1 μm and the base thickness t to about 0.2 μm. Emitter length ℓ is of the order of 25 μm for the overlay and matrix. Table 6-1-2 [12] lists some typical performance figures.

(2) Principles of Operation. As noted, operation of a microwave power transistor is similar to that of a low-frequency device, but class C operation is much better than operation in class A or AB mode. Figure 6-1-3 shows a simple model of a microwave transistor.

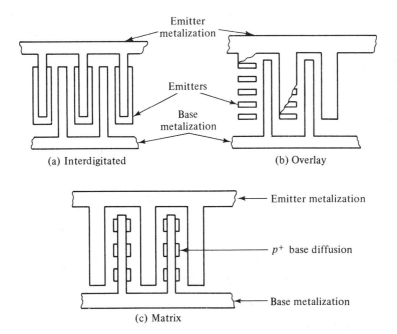

FIG. 6-1-2. Surface geometrics of microwave power transistor [12]. (*After Sobol et al.; reprinted with permission from the IEEE, Inc.*)

TABLE 6-1-1. Figure of Merit (*M*) of Various Surface Geometries. (From Sobol and Sterzer [12]; reprinted by permission of the IEEE, Inc.)

Surface geometry and unit cell	$M = \dfrac{EP}{BA}$
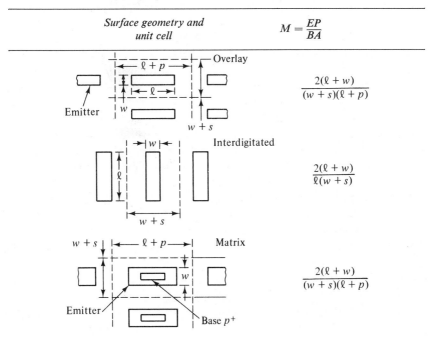 Overlay	$\dfrac{2(\ell + w)}{(w + s)(\ell + p)}$
Interdigitated	$\dfrac{2(\ell + w)}{\ell(w + s)}$
Matrix	$\dfrac{2(\ell + w)}{(w + s)(\ell + p)}$

TABLE 6-1-2. Typical Performance of Microwave Power Transistors (28-volt bias, CW, common-base circuit). (From Sobol and Sterzer [12]; reprinted by permission of the IEEE, Inc.)

Frequency (GHz)	Power (watts)	Gain (dB)	Efficiency (%)
1.0	20	10	60
2.0	10	7	50
3.0	5	5	30

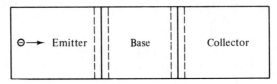

FIG. 6-1-3. Transit-time model of a microwave transistor.

When the transistor is biased for class C operation, both the emitter-base and collector-base junctions are reverse-biased, and no current flows in the absence of an applied signal. The depletion layer at the emitter-base junction is much smaller than that at the collector-base junction. When an RF voltage of sufficient magnitude is applied to the emitter-base junction, the junction is forward-biased for a fraction of an RF cycle. Thus the electrons are injected into the base. The injected carriers transit the base by a combined diffusion and drift flow and are then accelerated in the collector-base depletion region. The electric field in the collector-base depletion region, even at the minimum swing of the RF signal, is sufficiently large to accelerate electrons to their saturation velocity. The flow of energetic electrons injected during the time of forward emitter-base bias represents a pulse of current in the collector circuit. The peak of the current pulse occurs when the electrons traverse the collector-base depletion region with a saturation velocity. If the collector circuit is a real load at the driving-signal frequency, an output power can be taken from the load at the fundamental frequency.

(3) Microwave Characteristics. There are several approaches in analyzing the characteristics of microwave transistors. The most common is a combination of internal-parameter and two-port analyses. Figure 6-1-4 shows a double-diffused *n-p-n* microwave transistor with its impurity profile [10], and Fig. 6-1-5 shows the current-equivalent circuits. It can be seen from the equivalent circuits that

$r_e =$ the emitter resistance

$r_b = \dfrac{w}{\ell} r_0$ is the base resistance, where w is the emitter strip width

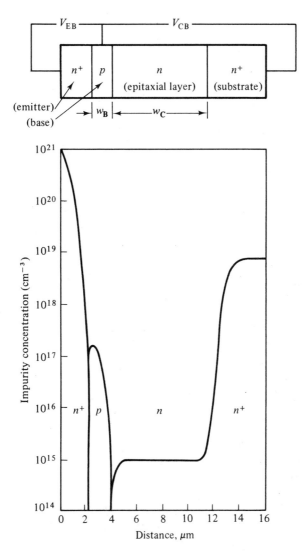

FIG. 6-1-4. Double-diffused *n-p-n* microwave transistor. (*After S. M. Sze [10]; reprinted by permission of John Wiley and Sons, Inc.*)

and ℓ is the emitter strip length

$r_0 = \dfrac{\rho_b}{w}$ and ρ_b is the average resistivity of the base layer

$C_e =$ the emitter depletion-layer capacitance,

$C_c = C_0 w \ell$ is the collector depletion-layer capacitance, where C_0 is the collector capacitance per unit area

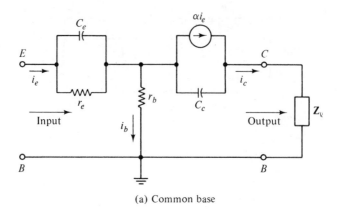

(a) Common base

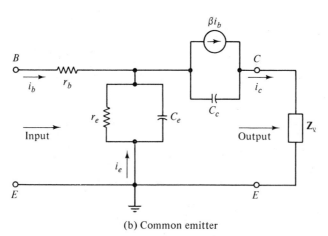

(b) Common emitter

FIG. 6-1-5. Current-equivalent circuits of microwave transistors.

(A) CURRENT GAIN. The dc common-base current gain α_0, also referred to as h_{FB} from the four-terminal hybrid parameters (where the subscripts F and B refer to forward and common base, respectively), is defined by

$$\alpha_0 = h_{FB} = \frac{\Delta I_C}{\Delta I_E} \tag{6-1-1}$$

Similarly, the dc common-emitter current gain β_0, also referred to as h_{FE}, is defined by

$$\beta_0 = h_{FE} = \frac{\Delta I_C}{\Delta I_B} \tag{6-1-2}$$

Since $I_E = I_B + I_C$, then α_0 and β_0 are related to each other by

$$\beta_0 = \frac{\alpha_0}{1 - \alpha_0} \tag{6-1-3}$$

For a small signal, the common-base current gain α is defined as

$$\alpha = h_{fb} = \frac{\partial I_C}{\partial I_E}\Big|_{\Delta V_{CB}=0} \tag{6-1-4}$$

Similarly, the small-signal common-emitter current gain β is defined as

$$\beta = h_{fe} = \frac{\partial I_C}{\partial I_B}\Big|_{\Delta V_{CE}=0} \tag{6-1-5}$$

Substitution of Eqs. (6-1-1) and (6-1-2) in Eqs. (6-1-4) and (6-1-5) yields

$$\alpha = \alpha_0 + I_E \frac{\partial \alpha_0}{\partial I_E} \tag{6-1-6}$$

$$\beta = \beta_0 + I_B \frac{\partial \beta_0}{\partial I_B} \tag{6-1-7}$$

and

$$\beta = \frac{\alpha}{1 - \alpha} \tag{6-1-8}$$

(B) CUTOFF FREQUENCY. The charge-carrier transit-time cutoff frequency f_T of a charge-control type of microwave transistor is defined [4] by the relation

$$f_T = \frac{1}{2\pi\tau} \tag{6-1-9}$$

where τ is the average time for a charge carrier moving at an average velocity v to traverse the emitter-collector distance L. This delay time represents the sum of four delays encountered sequentially by the minority charge-carrier. That is,

$$\tau = \tau_E + \tau_B + \tau_X + \tau_C \tag{6-1-10}$$

where τ_E = emitter-base junction charging time
τ_B = base transit time
τ_X = base-collector depletion-layer transit time
τ_C = collector-base junction charging time

For typical microwave transistors, the base and base-collector depletion-layer transit times are much greater than the junction charging times. Therefore the average time for a carrier is approximately equal to

$$\tau = \tau_B + \tau_X \tag{6-1-11}$$

The base-collector depletion-layer transit time is dependent on the collector-base bias and the epitaxial doping level. In general, a CW class C microwave power transistor would have a dc breakdown voltage about two times higher than the bias voltage. The typical 28-volt microwave power transistor requires an epitaxial layer with an impurity level of $N_d = 5 \times 10^{15}$ per cubic centimeter or 1-ohm·cm resistivity. The depletion-layer transit time corresponding to 28-volt operation and 1-ohm·cm resistivity is

$$\tau_X = 1.7 \times 10^{-11} \text{ second} \tag{6-1-12}$$

The base transit time is a function of the base width δ and is given by

$$\tau_B = 8.3 \times 10^{-10} \, \delta^2 \text{ second} \qquad (6\text{-}1\text{-}13)$$

Therefore the charge-carrier transit-time cutoff frequency can be simplified [12] and expressed as

$$f_T = \frac{1}{2\pi(\tau_X + \tau_B)} \qquad (6\text{-}1\text{-}14)$$

(C) POWER GAIN G_p. As described in Section 4-7, when frequencies are in the microwave region, it is difficult to achieve short and open circuits for measuring Z, Y, and H parameters. An alternative method is to use S parameters in traveling waves for solving power gain at microwave frequencies. The two-port network of a microwave transistor amplifier is shown in Fig. 6-1-6 and the two S parameter equations are expressed as

$$\mathbf{b_1} = \mathbf{S_{11}a_1} + \mathbf{S_{12}a_2}$$
$$\mathbf{b_2} = \mathbf{S_{21}a_1} + \mathbf{S_{22}a_2} \qquad (6\text{-}1\text{-}15)$$

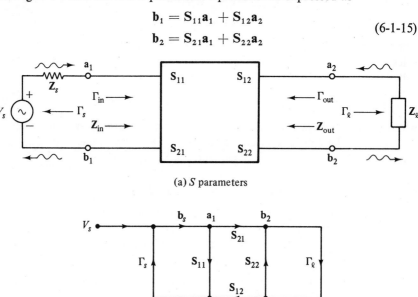

(a) S parameters

(b) Signal flow

FIG. 6-1-6. Two-port network of a microwave transistor amplifier.

The power gain G_P of a microwave transistor amplifier is defined as the ratio of the output power P_ℓ delivered to the load $\mathbf{Z}_\ell$ over the input power P_{avs} available from the source to the network. That is,

$$G_p = \frac{P_\ell}{P_{avs}} \qquad (6\text{-}1\text{-}16)$$

The power delivered to the load is the resultant of the power incident on the load minus the power reflected from the load.

$$P_\ell = |\mathbf{b}_2|^2 - |\mathbf{a}_2|^2 = |\mathbf{b}_2|^2(1 - |\Gamma_\ell|^2) \qquad (6\text{-}1\text{-}16\text{a})$$

where

$$\Gamma_\ell = \frac{Z_\ell - Z_0}{Z_\ell + Z_0}$$

is the reflection coefficient of the load.

The power available from the source is given by

$$P_{\mathrm{avs}} = \frac{|\mathbf{b}_s|^2}{(1 - |\Gamma_s|^2)} \qquad (6\text{-}1\text{-}16\text{b})$$

where $\Gamma_s = \dfrac{Z_s - Z_0}{Z_s + Z_0}$ is the reflection coefficient of the source

$\mathbf{b}_s = $ a function of $\mathbf{b}_2$ that is to be determined

Then the power gain is expressed by

$$G_p = \frac{|\mathbf{b}_2|^2}{|\mathbf{b}_s|^2}(1 - |\Gamma_s|^2)(1 - |\Gamma_\ell|^2) \qquad (6\text{-}1\text{-}16\text{c})$$

The transfer function from $\mathbf{b}_s$ to $\mathbf{b}_2$ can be derived by the nontouching loop rules of signal flow theory. The nontouching loop rules, which are often called *Mason's rules*, have the following terms.

1. **Path.** A path is a series of directed lines followed in sequence and in the same direction in such a way that no node is touched more than once. The value of the path is the product of all coefficients encountered enroute. In Fig. 6-1-6(b) there is only one path from $\mathbf{b}_s$ to $\mathbf{b}_2$, and the value of the path is S_{21}. There are two paths from $\mathbf{b}_s$ to $\mathbf{b}_1$, and the values are S_{11} and $S_{21}\Gamma_\ell S_{12}$.
2. **First-order loop.** A first-order loop is defined as the product of all coefficients along the paths starting from a node and moving in the direction of the arrows back to that original node without passing the same node twice. In Fig. 6-1-6(b) there are three first-order loops, and the values are $S_{11}\Gamma_s$, $S_{22}\Gamma_\ell$, and $S_{21}\Gamma_\ell S_{12}\Gamma_s$.
3. **Second-order loop.** A second-order loop is defined as the product of any two nontouching first-order loops. In Fig. 6-1-6(b) there is only one second-order loop and its value is $S_{11}\Gamma_s S_{22}\Gamma_\ell$.
4. **Third-order loop.** A third-order loop is the product of any three nontouching first-order loops. In Fig. 6-1-6(b) there is no third-order loop.

Then the transfer function for the ratio of the dependent variable in question to the independent variable of the source is expressed by

$$\mathbf{T} = \frac{P_1[1 - \sum L(1)^1 + \sum L(2)^1 - \sum L(3)^1 + \cdots] + P_2[1 - \sum L(1)^2 + \sum L(2)^2 \cdots] + P_3[1 \cdots]}{1 - \sum L(1) + \sum L(2) - \sum L(3) + \cdots}$$

$$(6\text{-}1\text{-}17)$$

where

$$P_1, P_2, P_3 \ldots = \text{the various paths connecting these variables}$$

$\sum L(1), \sum L(2), \sum L(3) \ldots = $ the sums of all first-order, second-order, third-order loops ..., respectively

$\sum L(1)^1, \sum L(2)^1, \sum L(3)^1 \ldots = $ the sums of all first-order, second-order, third-order loops that do not touch the first path between the variables

$\sum L(1)^2, \sum L(2)^2, \sum L(3)^2 \ldots = $ the sums of all first-order, second-order, third-order loops that do not touch the second paht

From Fig. 6-1-6(b) the transfer function of $\mathbf{b}_2$ over $\mathbf{b}_s$ is given by

$$\frac{\mathbf{b}_2}{\mathbf{b}_s} = \frac{S_{21}}{1 - S_{11}\Gamma_s - S_{22}\Gamma_\ell - S_{21}\Gamma_\ell S_{12}\Gamma_s + S_{11}\Gamma_s S_{22}\Gamma_\ell} \qquad (6\text{-}1\text{-}18)$$

Substitution of Eq. (6-1-18) in (6-1-16b) yields the power gain

$$G_p = \frac{|S_{21}|^2(1 - |\Gamma_s|^2)(1 - |\Gamma_\ell|^2)}{|(1 - S_{11}\Gamma_s)(1 - S_{22}\Gamma_\ell) - S_{21}S_{12}\Gamma_s\Gamma_\ell|^2} \qquad (6\text{-}1\text{-}19)$$

(D) MAXIMUM AVAILABLE POWER GAIN $G_{\max}$. In order to maximize the forward power gain $G_{\max}$ of a microwave transistor amplifier, the input and output networks must be conjugately matched. First, however, the stability of a transistor amplifier should be described. Stability or resistance to oscillation is a most important factor in amplifier design and is determined by the S parameters, the synthesized source, and the load impedances. Oscillations are only possible if either the input or output port or both have negative resistance. This situation occurs if S_{11} or S_{22} is greater than unity. Yet even with negative resistance the amplifier might still be stable.

There are two traditional expressions for stability—conditional stability and unconditional stability. A network is conditionally stable if the real part of the input impedance Z_{in} and the output impedance Z_{out} is greater than zero for some positive real source and load impedances at a specific frequency. A network is unconditionally stable if the real part of the input impedance Z_{in} and the output impedance Z_{out} is greater than zero for all positive real source and load impedances at a specific frequency.

The maximum power gain $G_{\max}$ that can be realized for a microwave transistor amplifier without external feedback is defined as the forward power gain when the input and output are simultaneously and conjugately matched. Conjugately matched conditions mean that the reflection coefficient Γ_s of the source is equal to the conjugate of the input reflection coefficient Γ_{in} and that the reflection coefficient Γ_ℓ of the load is equal to the conjugate of the output reflection coefficient Γ_{out}. These are

$$\Gamma_s = \Gamma_{in}^* \quad \text{and} \quad \Gamma_\ell = \Gamma_{out}^*$$

In order for a transistor amplifier to be unconditionally stable, the magnitudes of $S_{11}, S_{22}, \Gamma_{in}$, and Γ_{out} must be smaller than unity and the transistor's inherent stability factor K must be greater than unity and positive. K is computed from

$$K = \frac{1 + |\Delta|^2 - |S_{11}|^2 - |S_{22}|^2}{2|S_{12}S_{21}|} > 1 \qquad (6\text{-}1\text{-}20)$$

where

$$\Delta \equiv S_{11}S_{22} - S_{12}S_{21}$$

The input and output reflection coefficients are given by

$$\Gamma_{in} = S_{11} + \frac{S_{21}S_{12}\Gamma_\ell}{1 - S_{22}\Gamma_\ell} \qquad (6\text{-}1\text{-}20a)$$

and

$$\Gamma_{out} = S_{22} + \frac{S_{21}S_{12}\Gamma_s}{1 - S_{11}\Gamma_s} \qquad (6\text{-}1\text{-}20b)$$

The boundary conditions for stability are given by

$$|\Gamma_{in}| = 1 = \left| S_{11} + \frac{S_{21}S_{12}\Gamma_\ell}{1 - S_{22}\Gamma_\ell} \right| \qquad (6\text{-}1\text{-}21a)$$

and

$$|\Gamma_{out}| = 1 = \left| S_{22} + \frac{S_{21}S_{12}\Gamma_s}{1 - S_{11}\Gamma_s} \right| \qquad (6\text{-}1\text{-}21b)$$

Substitution of the real and imaginary values for the S parameters in Eqs. (6-1-21a) and (6-1-21b) yields the solutions of Γ_s and Γ_ℓ as

$$R_s(\text{radius of } \Gamma_s \text{ circle}) = \frac{|S_{12}S_{21}|}{|S_{11}|^2 - |\Delta|^2} \qquad (6\text{-}1\text{-}21c)$$

$$C_s(\text{center of } \Gamma_s \text{ circle}) = \frac{C_s^*}{|S_{11}|^2 - |\Delta|^2} \qquad (6\text{-}1\text{-}21d)$$

$$R_\ell(\text{radius of } \Gamma_\ell \text{ circle}) = \frac{|S_{12}S_{21}|}{|S_{22}|^2 - |\Delta|^2} \qquad (6\text{-}1\text{-}21e)$$

$$C_\ell(\text{center of } \Gamma_\ell \text{ circle}) = \frac{C_\ell^*}{|S_{22}|^2 - |\Delta|^2} \qquad (6\text{-}1\text{-}21f)$$

where

$$\Delta = S_{11}S_{22} - S_{12}S_{21}$$
$$C_s = S_{11} - \Delta S_{22}^*$$
$$C_\ell = S_{22} - \Delta S_{11}^*$$

The reflection coefficient of the source impedance required to conjugately match the input of the transistor for maximum power gain is

$$\Gamma_{sm} = C_s^* \left[\frac{B_s \pm \sqrt{B_s^2 - 4|C_s|^2}}{2|C_s|^2} \right] \qquad (6\text{-}1\text{-}21g)$$

where

$$B_s = 1 + |S_{11}|^2 - |S_{22}| - |\Delta|^2 \qquad (6\text{-}1\text{-}21h)$$

The reflection coefficient of the load impedance required to conjugately match the output of the transistor for maximum power gain is

$$\Gamma_{\ell m} = \mathbf{C}_\ell^* \left[\frac{B_\ell \pm \sqrt{B_\ell^2 - 4 | \mathbf{C}_\ell |^2}}{2 | \mathbf{C}_\ell |^2} \right] \tag{6-1-21i}$$

where $B_\ell = 1 + | \mathbf{S}_{22} |^2 - | \mathbf{S}_{11} |^2 - | \Delta |^2$ (6-1-21j)

If the computed values of B_s and B_ℓ are negative, then the plus sign should be used in front of the radical in Eqs. (6-1-21g) and (6-1-21i). Conversely, if B_s and B_ℓ are positive, then the negative sign should be used.

Stability circles can be plotted directly on a Smith chart. These circles separate the output or input planes into stable and potentially unstable regions. A stability circle plotted on the output plane indicates the values of all loads that provide negative real input impedance, thereby causing the circuit to oscillate. A similar circle can be plotted on the input plane to indicate the values of all loads that provide negative real output impedance and again cause oscillation. A negative real impedance is defined as a reflection coefficient that has a magnitude that is greater than unity. The regions of instability occur within the circles whose centers and radii are expressed by Eqs. (6-1-21c) through (6-1-21f).

By using an appropriate sign, only one answer is possible in either equation of Eqs. (6-1-21g) and (6-1-21i) and a value of less than unity is obtained. The maximum available power gain possible is expressed as

$$G_{\max} = \frac{| \mathbf{S}_{21} |}{| \mathbf{S}_{12} |} | K \pm (K^2 - 1)^{1/2} | \tag{6-1-22}$$

So maximum available power gain is obtained only if the microwave transistor amplifier is loaded with Γ_{sm} and $\Gamma_{\ell m}$ as reflection coefficients. The maximum frequency of oscillation is determined after the maximum available power gain is achieved.

(E) UNILATERAL POWER GAIN G_u.   The unilateral power gain G_u is the forward power gain in a feedback amplifier having its reverse power gain set to zero ($| S_{12} |^2 = 0$) by adjusting a lossless reciprocal feedback network connected around the microwave transistor amplifier. That is,

$$G_u = \frac{| \mathbf{S}_{21} |^2 (1 - | \Gamma_s |^2)(1 - | \Gamma_\ell |^2)}{| 1 - \mathbf{S}_{11} \Gamma_s |^2 | 1 - \mathbf{S}_{22} \Gamma_\ell |^2} \tag{6-1-23}$$

The maximum unilateral power gain is obtained when $\Gamma_s = \mathbf{S}_{11}^*$ and $\Gamma_\ell = \mathbf{S}_{22}^*$. Then

$$G_{u\,\max} = \frac{| \mathbf{S}_{21} |^2}{(1 - | \mathbf{S}_{11} |^2)(1 - | \mathbf{S}_{22} |^2)} \tag{6-1-24}$$

(F) AMPLIFIER DESIGN WITH S PARAMETERS.   Figure 6-1-7(a) shows a microwave transistor-amplifier circuit with an input and an output matching networks. When $\mathbf{S}_{12}$ is very small or zero, maximum gain can be achieved if

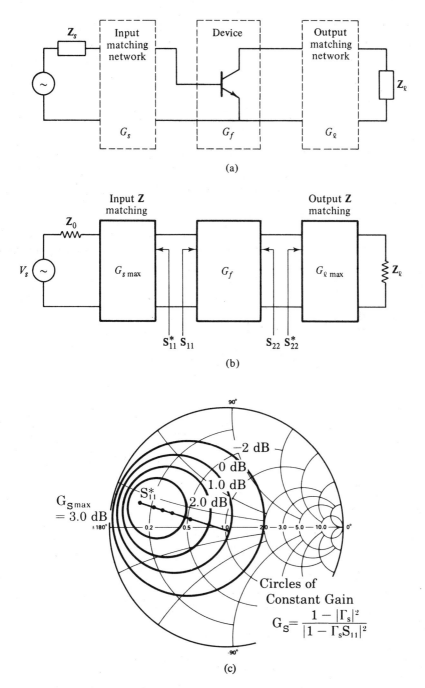

FIG. 6-1-7. Microwave transistor circuits. (a) Amplifier matching circuit. (b) Amplifier circuit for maximum gain. (c) Constant power gain circles.

273

the input and output of the transistor are terminated in impedances that are
the complex conjugates of S_{11} and S_{22}, respectively.

The unilateral power gain G_u as indicated by Eq. (6-1-23) can be written

$$G_u = \frac{1 - |\Gamma_s|^2}{|1 - S_{11}\Gamma_s|^2} \cdot |S_{21}|^2 \cdot \frac{1 - |\Gamma_\ell|^2}{|1 - S_{22}\Gamma_\ell|^2}$$

$$= g_s \cdot g_f \cdot g_\ell \qquad \text{(numerical values)}$$

$$= G_s + G_f + G_\ell \qquad \text{(dB)} \qquad\qquad (6\text{-}1\text{-}23a)$$

where $\quad g_s = \dfrac{1 - |\Gamma_s|^2}{|1 - S_{11}\Gamma_s|^2}$ is the additional power gain or loss resulting
from the input impedance matching network
between the device and source

$g_f = |S_{21}|^2$ is the forward power gain of the device with the
input and output terminated in matching loads

$g_\ell = \dfrac{1 - |\Gamma_\ell|^2}{|1 - S_{22}\Gamma_\ell|^2}$ is the additional power gain or loss due to
the impedance matching network between the
output of the device and the load

Maximum unilateral transducer gain can be accomplished by choosing
impedance matching networks such that $\Gamma_s = S_{11}^*$, $Z_s = Z_0$, $\Gamma_\ell = S_{22}^*$ and
$Z_\ell = Z_0$ as shown in Fig. 6-1-7(b). Then Eq. (6-1-24) can be written

$$G_{u\,\max} = \frac{1}{1 - |S_{11}|^2} \cdot |S_{21}|^2 \cdot \frac{1}{1 - |S_{22}|^2}$$

$$= g_{s\,\max} \cdot g_f \cdot g_{\ell\,\max} \qquad \text{(numerical values)}$$

$$= G_{s\,\max} + G_f + G_{\ell\,\max} \qquad \text{(dB)} \qquad\qquad (6\text{-}1\text{-}24a)$$

where $\quad g_{s\,\max} = \dfrac{1}{1 - |S_{11}|^2}$

$g_f = |S_{21}|^2$

$g_{\ell\,\max} = \dfrac{1}{1 - |S_{22}|^2}$

Constant Gain Circles: It is obvious that for $\Gamma_s = S_{11}^*$ or $\Gamma_\ell = S_{22}^*$, the
power gain G_s or G_ℓ is equal to maximum, respectively. It is also clear that
for $|\Gamma_s| = 1$ or $|\Gamma_\ell| = 1$, the power gain G_s or G_ℓ has a value of zero. For
any arbitrary value of G_s or G_ℓ between these extremes of zero and $G_{s\,\max}$
or $G_{\ell\,\max}$, solutions for Γ_s or Γ_ℓ lie on a circle.

For $0 < G_s < G_{s\,\max}$ dB,

$$g_s = \frac{1 - |\Gamma_s|^2}{|1 - S_{11}\Gamma_s|^2} \qquad \text{(numerical values)} \qquad\qquad (6\text{-}1\text{-}24b)$$

It is convenient to plot these circles on a Smith chart. The circles have their
centers located on the vector drawn from the center of the Smith chart to

the point S_{11}^* or S_{22}^*. The distance from the center of the Smith chart to the center of the constant gain circle along the vector S_{11}^* or S_{22}^* is given by

$$d = \frac{g_n |S_{11}|}{1 - |S_{11}|^2(1 - g_n)} \qquad (6\text{-}1\text{-}24c)$$

The radius of the constant gain circle is expressed by

$$r = \frac{\sqrt{1 - g_n}(1 - |S_{11}|^2)}{1 - |S_{11}|^2(1 - g_n)} \qquad (6\text{-}1\text{-}24d)$$

where g_n is the normalized gain value for the gain circle g_s or g_t, respectively. That is,

$$g_n = \frac{g_s}{g_{s\,max}} = g_s(1 - |S_{11}|^2) \qquad (6\text{-}1\text{-}24e)$$

For example, let $\Gamma_s = S_{11}^* = 0.707\underline{/155°}$ as shown on the Smith chart; the constant gain circles are drawn on a Smith chart as shown in Fig. 6-1-7(c). Any value of Γ_s along a 1-dB circle would result in a power gain G_s of 1 dB and so on. The maximum power $G_{s\,max}$ is 3 dB.

Design Example for $G_{u\,max}$: The S parameters of a microwave transistor can be measured by using the HP-8410S network analyzer system. For frequency at 1 GHz, a certain transistor measured with a 50-Ω resistance matching the input and output has the following parameters:

$$S_{11} = 0.707\underline{/25°}$$
$$S_{22} = 0.51\underline{/-20°}$$
$$S_{21} = 5\underline{/180°}$$

The desired amplifier circuit is shown in Fig. 6-1-7(d).

Step 1. Calculation and Plot of the Input and Output Constant Power Gain Circles. From Eq. (6-1-24a),

$$G_{u\,max} = 10 \log \left(\frac{1}{1 - |S_{11}|^2}\right) + 10 \log |S_{21}|^2 + 10 \log \left(\frac{1}{1 - |S_{22}|^2}\right)$$

$$= 10 \log \left(\frac{1}{1 - |0.707|^2}\right) + 10 \log |5|^2 + 10 \log \left(\frac{1}{1 - |0.51|^2}\right)$$

$$= 3\,dB + 14\,dB + 1.33\,dB$$

$$= 18.33\,dB$$

For this example, only the two circles representing maximum gain are needed. These circles have zero radius and are located at S_{11}^* and S_{22}^* [see Fig. 6-1-7(e)].

Step 2. Overlap of Two Smith Charts. To facilitate the design of the matching networks, two Smith charts 180° out of phase are overlapped with each other. The original chart can then be used to read impedances and the overlaid chart to read admittances as shown in Fig. 6-1-7(e).

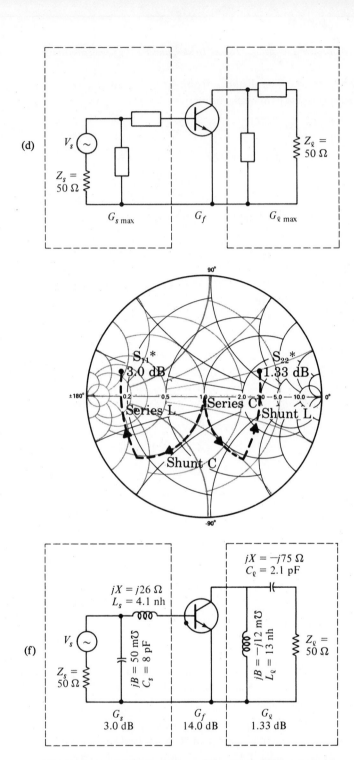

FIG. 6-1-7. (cont.) (d) Desired amplifier circuit. (e) Constant gain circles. (f) Transistor amplifier circuit. (*c, d, e, and f are through the courtesy of Hewlett-Packard Co.*)

Step 3. Determination of the Output Matching Network. To determine the matching network for the output, we start the load impedance of 50 Ω at the center of the Smith chart and proceed along the constant unit resistance circle in the counterclockwise direction until we arrive at the constant conductance circle that intersects the point representing S_{22}^*.

$$jX = -j1.4 \times 50 = -j75 \text{ ohms}$$
$$C_\ell = 2.1 \text{ pF}$$

Then we add an inductive susceptance along the constant conductance circle so that the impedance looking into the matching network will be equal to S_{22}^*.

$$jB = -j0.60\left(\frac{1}{50}\right) = -j12 \text{ millimhos}$$
$$L_\ell = 13 \text{ nH}$$

Step 4. Determination of the Input Matching Network. The same procedures can be applied at the input side, and the results are a shunt capacitor and a series inductor.

$$jX = j26 \text{ ohms}$$
$$L_s = 4.1 \text{ nH}$$
$$jB = j50 \text{ millimhos}$$
$$C_s = 8 \text{ pF}$$

Step 5. Final Amplifier Network. Figure 6-1-7(f) shows the final microwave transistor-amplifier circuit.

(4) Power-Frequency Limitations. In Section 5-1 power-frequency limitations on conventional vacuum tubes were analyzed in detail. When the operating frequency is above 1 GHz, the output power of tubes is considerably decreased. The question then arises as to whether microwave power transistors have any limitations on their frequency and output power. The answer is positive. Several authors have discussed this subject. Early [17] first introduced the power-frequency limitations inherent in (a) the limiting velocity of carriers in semiconductors and (b) the maximum fields attainable in semiconductors without the onset of avalanche multiplication. These basic ideas were later developed and discussed in detail by Johnson [4], who made three assumptions.

    I. There is a maximum possible velocity of carriers in a semiconductor. This is the "saturated drift velocity v_s," and is on the order of 6×10^6 cm/s for electrons and holes in silicon and germanium.

    II. There is a maximum electric field E_m that can be sustained in a semiconductor without having dielectric breakdown. This field is about 10^5 volts/cm in germanium and 2×10^5 volts/cm in silicon.

III. The maximum current that a microwave power transistor can carry is limited by the base width.

With these three postulates Johnson derived four basic equations for the power-frequency limitations on microwave power transistors.

FIRST EQUATION.   Voltage-Frequency Limitation.

$$V_m f_T = \frac{E_m v_s}{2\pi} = \begin{cases} 2 \times 10^{11} \text{ volts/s for silicon} \\ 1 \times 10^{11} \text{ volts/s for germanium} \end{cases} \qquad (6\text{-}1\text{-}25)$$

where $f_T = \dfrac{1}{2\pi\tau}$ is the charge-carrier transit-time cutoff frequency

$\tau = \dfrac{L}{v}$ is the average time for a charge carrier moving at an average velocity v to traverse the emitter-collector distance L

$V_m = E_m L_{\min}$ is the maximum allowable applied voltage

v_s = the maximum possible saturated drift velocity

E_m = the maximum electric field

With the carriers moving at a velocity v_s of 6×10^6 cm/s, the transit time can be reduced even further by decreasing the distance L. The lower limit on L can be reached when the electric field becomes equal to the dielectric breakdown field. However, the present state of the art of microwave transistor fabrication limits the emitter-collector length L to about 25 μm for overlay and matrix devices and to nearly 250 μm for interdigitated devices [12]. Consequently, there is an upper limit on cutoff frequency. In practice, the attainable cutoff frequency is considerably less than the maximum possible frequency indicated by Eq. (6-1-25) because the saturated velocity v_s and the electric field intensity will not be uniform.

SECOND EQUATION.   Current-Frequency Limitation.

$$(I_m X_c) f_T = \frac{E_m v_s}{2\pi} \qquad (6\text{-}1\text{-}26)$$

where I_m = the maximum current of the device

$X_c = \dfrac{1}{\omega_T C_0} = \dfrac{1}{2\pi f_T C_0}$ is the reactive impedance

C_0 = the collector-base capacitance

It should be noted that the relationship $2\pi f_T \tau_0 \simeq 2\pi f_T \tau = 1$ was used in deriving Eq. (6-1-26) from Eq. (6-1-25). In practice, no maximum current exists because the area of the device cannot be increased without bound. If the impedance level is zero, the maximum current through a velocity-saturated sample might be infinite. However, the limited impedance will limit the maximum current for a maximum attainable power.

THIRD EQUATION.   Power-Frequency Limitation.

$$(P_m X_c)^{1/2} f_T = \frac{E_m v_s}{2\pi} \tag{6-1-27}$$

This equation was obtained by multiplying Eq. (6-1-25) by Eq. (6-1-26) and replacing $V_m I_m$ by P_m. It is significant that, for a given device impedance, the power capacity of a device must be decreased as the device cutoff frequency is increased. For a given product of $E_m v_s$ (i.e., a given material), the maximum power that can be delivered to the carriers traversing the transistor is infinite if the cross section of the transistor can be made as large as possible. In other words, the value of the reactance X_c must approach zero. Thus Eq. (6-1-27) allows the results to be predicted. Figure 6-1-8 shows a graph of Eq. (6-1-27) and the experimental results reported from manufacturers [18].

FOURTH EQUATION.   Power Gain-Frequency Limitation.

$$(G_m V_{th} V_m)^{1/2} f = \frac{E_m v_s}{2\pi} \tag{6-1-28}$$

where $G_m =$ the maximum available power gain

$V_{th} = \dfrac{KT}{e}$ is the thermal voltage

$K =$ Boltzmann's constant, 1.38×10^{-23} joule/°K

$T =$ the absolute temperature in degrees Kelvin

$e =$ the electron charge (1.60×10^{-19} coul)

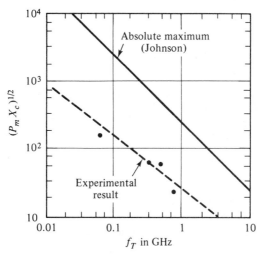

FIG. 6-1-8. $(P_m X_c)^{1/2}$ versus f_T. (*After B. C. De Loach* [18]; *reprinted by permission of Academic Press.*)

The maximum available power gain of a transistor was derived by Johnson [14] as

$$G_m = \left(\frac{f_T}{f}\right)^2 \frac{Z_{out}}{Z_{in}}$$ (6-1-29)

where Z_{out} and Z_{in} are the output and input impedances, respectively. If the electrode series resistances are assumed to be zero, the ratio of the output impedance to the input impedance can be written

$$\frac{Z_{out}}{Z_{in}} = \frac{C_{in}}{C_{out}}$$ (6-1-30)

where C_{in} is the input capacitance and C_{out} is the output (base-collector) capacitance. When the maximum total carrier charges Q_m move to the collector in a carrier-base transit time τ_b and with a thermal voltage V_{th} the input capacitance C_{in} and the emitter diffusion capacitance C_d are related by

$$C_{in} = C_d \cong \frac{Q_m}{V_{th}} = \frac{I_m \tau_b}{V_{th}}$$ (6-1-31)

The output capacitance is given by

$$C_{out} = \frac{I_m \tau_0}{V_m}$$ (6-1-32)

Substitution of Eqs. (6-1-25), (6-1-31), and (6-1-32) in Eq. (6-1-29) yields Eq. (6-1-28). The actual performance of a microwave transistor will fall far short of that predicted by Eq. (6-1-28). At present the high-frequency limit of a 28-volt silicon n-p-n transistor operating at the 1-watt level is approximately 10 GHz. Typical power gains of microwave transistors lie in the 6-to 10-dB range.

6-1-2. Microwave Tunnel Diodes

Since the publication of Esaki's classic paper on tunnel diodes in 1958 [19], the potential of tunnel diodes for microwave applications was quickly established. Prior to 1958 the anomalous characteristics of some p-n junctions was observed by many scientists, but the irregularities were rejected immediately because they did not follow the "classic" diode equation. Esaki, however, described this anomalous phenomenon by applying a quantum tunneling theory. The tunneling phenomenon is a majority carrier effect. The tunneling time of carriers through the potential energy barrier is not governed by the classic transit time concept—that the transit time is equal to the barrier width divided by the carrier velocity—but rather by the quantum transition probability per unit time. Tunnel diodes are useful in many circuit applications in microwave amplification, microwave oscillation, and binary memory because of their low cost, light weight, high speed, low-power operation, low noise, and high peak-current to valley-current ratio.

(*1*) **Principles of Operation.** The tunnel diode is a negative-resistance semiconductor *p-n* junction diode. The negative resistance is created by the tunnel effect of electrons in the *p-n* junction. The doping of both the *p* and *n* regions of the tunnel diode is very high, impurity concentrations of 10^{19} to 10^{20} atoms/cm^3 being used; and the depletion-layer barrier at the junction is very thin, on the order of 100 Å or 10^{-6} cm. Classically, it is only possible for those particles to pass over the barrier if and only if they have an energy equal to or greater than the height of the potential barrier. However, quantum mechanically, if the barrier is less than 3 Å, there is an appreciable probability that particles will tunnel through the potential barrier even though they do not have enough kinetic energy to pass over the same barrier. In addition to the barrier thinness, there must also be filled energy states on the side from which particles will tunnel and allowed empty states on the other side into which particles penetrate through at the same energy level. In order to understand the tunnel effects fully, let us analyze the energy-band pictures of a heavily doped *p-n* diode. Figure 6-1-9 shows energy-band diagrams of a tunnel diode.

Under open-circuit conditions or at zero-bias equilibrium, the upper levels of electron energy of both the *p* type and *n* type are lined up at the same Fermi level as shown in Fig. 6-1-9(a). Since there are no filled states on one side of the junction that are at the same energy level as empty allowed states on the other side, there is no flow of charge in either direction across the junction and the current is zero, as shown at point (a) of the volt-ampere characteristic curve of a tunnel diode in Fig. 6-1-10.

In ordinary diodes the Fermi level exists in the forbidden band. Since the tunnel diode is heavily doped, the Fermi level exists in the valence band in *p*-type and in the conduction band in *n*-type semiconductors. When the tunnel diode is forward-biased by a voltage between zero and the value that would produce peak tunneling current I_p ($0 < V < V_p$), the energy diagram is as shown in part (1) of Fig. 6-1-9(b). Accordingly, the potential barrier is decreased by the magnitude of the applied forward-bias voltage. A difference in Fermi levels in both sides is created. Since there are filled states in the conduction band of the *n* type at the same energy level as allowed empty states in the valence band of the *p* type, the electrons tunnel through the barrier from the *n* type to the *p* type, giving rise to a forward tunneling current from the *p* type to the *n* type as shown in sector (1) of Fig. 6-1-10(a). As the forward bias is increased to V_p, the picture of the energy band is as shown in part (2) of Fig. 6-1-9(b). A maximum number of electrons can tunnel through the barrier from the filled states in the *n* type to the empty states in the *p* type, giving rise to the peak current I_p in Fig. 6-1-10(a). If the bias voltage is further increased, the condition shown in part (3) of Fig. 6-1-9(b) is reached. The tunneling current decreases as shown in sector (3) of Fig. 6-1-10(a).

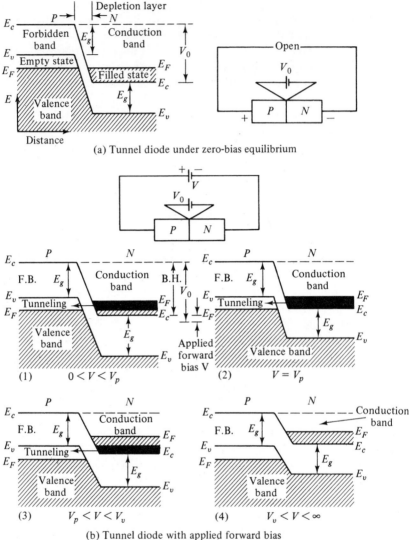

(a) Tunnel diode under zero-bias equilibrium

(b) Tunnel diode with applied forward bias

E_F is the Fermi level representing the energy state with 50%
  probability of being filled if no forbidden band exists
V_0 is the potential barrier of the junction
E_g is the energy required to break a covalent bond, which is
  0.72 eV for germanium and 1.10 eV for silicon
E_c is the lowest energy in the conduction band
E_v is the maximum energy in the valence band
V is the applied forward bias
F.B. stands for the forbidden band
B.H. represents the barrier height

FIG. 6-1-9. Energy-band diagrams of tunnel diode.

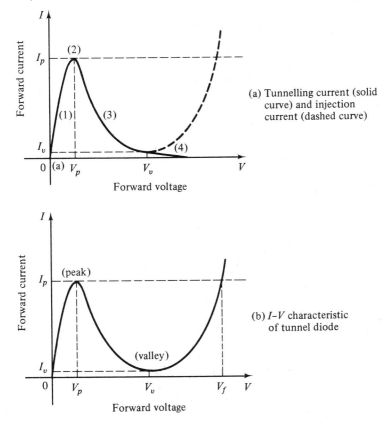

FIG. 6-1-10. Voltage-ampere characteristics of tunnel diode.

Finally, at a very large bias voltage, the band structure of part (4) of Fig. 6-1-9(b) is obtained. Since there are now no allowed empty states in the *p* type at the same energy level as filled states in the *n* type, no electrons can tunnel through the barrier and the tunneling current drops to zero as shown at point (4) of Fig. 6-1-10(a).

When the forward-bias voltage *V* is increased above the valley voltage V_v, the ordinary injection current *I* at the *p-n* junction starts to flow. This injection current is increased exponentially with the forward voltage as indicated by the dashed curve of Fig. 6-1-10(a). The total current, given by the sum of the tunneling current and the injection current, results in the volt-ampere characteristic of the tunnel diode as shown in Fig. 6-1-10(b). It can be seen from the figure that the total current reaches a minimum value I_v (or valley current), somewhere in the region where the tunnel-diode characteristic meets the ordinary *p-n* diode characteristic. The ratio of peak current to valley current (I_p/I_v) can theoretically reach 50 to 100. However, practically, this ratio is about 15.

(2) Microwave Characteristics. The tunnel diode is useful in microwave oscillators and amplifiers because the diode exhibits a negative-resistance characteristic in the region between peak current I_p and valley current I_v. The V-I characteristic of a tunnel diode with the load line is shown in Fig. 6-1-11.

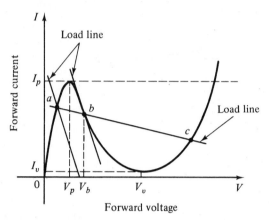

FIG. 6-1-11. V-I characteristic of tunnel diode with load line.

Here the abc load line intersects the characteristic curve in three points. Points a and c are stable points and point b is unstable. If the voltage and current vary about b, the final values of I and V would be given by point a or c, but not by b. Since the tunnel diode has two stable states for this load line, the circuit is called *bistable*, and it can be utilized as a binary device in switching circuits. However, microwave oscillation or amplification generated by the tunnel diode is our major concern in this section. The second load line intersects the V-I curve at point b only. This point is stable and shows a dynamic negative conductance that enables the tunnel diode to function as a microwave amplifier or oscillator. The circuit with a load line crossing point b in the negative-resistance region is called *astable*. Another load line crossing point a in the positive-resistance region indicates a *monostable* circuit. The negative conductance in Fig. 6-1-11 is given by

$$-g = \left.\frac{\partial i}{\partial v}\right|_{V_b} = \frac{1}{-R_n} \tag{6-1-33}$$

where R_n is the magnitude of negative resistance.

For a small variation of the forward voltage about V_b, the negative resistance is constant and the diode circuit behavior is stable. A small-signal equivalent circuit for the tunnel diode operated in the negative-resistance region is shown in Fig. 6-1-12. Here R_s and L_s denote the inductance and resistance of the packaging circuit of a tunnel diode. The junction capacitance

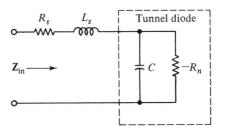

FIG. 6-1-12. Equivalent circuit of tunnel diode.

C of the diode is usually measured at the valley point. R_n is the negative resistance of the diode. Typical values of these parameters for a tunnel diode having a peak current I_p of 10 mA are

$$-R_n = -30 \ \Omega, \quad R_s = 1 \ \Omega \quad L_s = 5 \ \text{nH}, \quad C = 20 \ \text{pF}$$

The input impedance Z_{in} of the equivalent circuit as shown in Fig. 6-1-12 is given by

$$Z_{\text{in}} = R_s + j\omega L_s + \frac{R_n[j/(\omega C)]}{-R_n - j/(\omega C)}$$

$$= R_s - \frac{R_n}{1 + (\omega R_n C)^2} + j\left[\omega L_s - \frac{\omega R_n^2 C}{1 + (\omega R C)^2}\right] \quad (6\text{-}1\text{-}34)$$

For the resistive cutoff frequency, the real part of the input impedance Z_{in} must be zero. Consequently, from Eq. (6-1-34) the resistive cutoff frequency is given by

$$f_c = \frac{1}{2\pi R_n C} \sqrt{\frac{R_n}{R_s} - 1} \quad (6\text{-}1\text{-}35)$$

For the self-resonance frequency, the imaginary part of the input impedance must be zero. Thus

$$f_r = \frac{1}{2\pi R_n C} \sqrt{\frac{R_n^2 C}{L_s} - 1} \quad (6\text{-}1\text{-}36)$$

The tunnel diode can be connected either in parallel or in series with a resistive load as an amplifier, and its equivalent circuits are shown in Fig. 6-1-13.

(A) PARALLEL LOADING. It can be seen from Fig. 6-1-13(a) that the output power in the load resistance is given by

$$P_{\text{out}} = \frac{V^2}{2R_\ell} \quad (6\text{-}1\text{-}37)$$

One part of this output power is generated by the small input power through the tunnel diode amplifier with a gain of A, and this part can be written

$$P_{\text{in}} = \frac{V^2}{2AR_\ell} \quad (6\text{-}1\text{-}38)$$

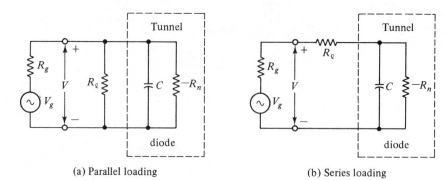

(a) Parallel loading   (b) Series loading

FIG. 6-1-13. Equivalent circuits of tunnel diodes.

Another part of the output power is generated by the negative resistance, and it is expressed as

$$P_n = \frac{V^2}{2R_n} \tag{6-1-39}$$

Therefore

$$\frac{V^2}{2AR_\ell} + \frac{V^2}{2R_n} = \frac{V^2}{2R_\ell} \tag{6-1-40}$$

and the gain equation of a tunnel diode amplifier is given by

$$A = \frac{R_n}{R_n - R_\ell} \tag{6-1-41}$$

When the negative resistance R_n of the tunnel diode approaches the load resistance R_ℓ, the gain A approaches infinity and the system goes into oscillation.

(B) SERIES LOADING.   In the series circuit shown in Fig. 6-1-13(b) the power gain A is given by

$$A = \frac{R_\ell}{R_\ell - R_n} = \frac{1}{1 - R_n/R_\ell} \tag{6-1-41a}$$

The device remains stable in the negative-resistance region without switching if $R_\ell < R_n$.

A tunnel diode can be connected to a microwave circulator to make a negative-resistance amplifier as shown in Fig. 6-1-14. A microwave circulator is a multiport junction in which the power may flow only from port 1 to port 2, port 2 to port 3, and so on in the direction shown. Although the number of ports is not restricted, microwave circulators with four ports are most commonly used. If the circulator is perfect and has a positive real characteristic impedance R_0, an amplifier with infinite gain can be built by selecting a negative-resistance tunnel diode whose input impedance has a real part equal to $-R_0$ and an imaginary part equal to zero. The reflection coefficient

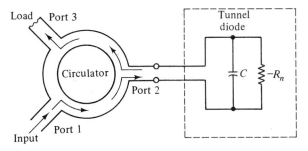

FIG. 6-1-14. Tunnel diode connected to circulator.

from Fig. 6-1-13 is infinite. In general, the reflection coefficient is given by

$$\Gamma = \frac{-R_n - R_0}{-R_n + R_0} \qquad (6\text{-}1\text{-}42)$$

(3) State of the Art. Since the first tunnel diode was reported by Esaki in 1958, this device has been intensively investigated as a microwave power source. Its ability to operate at high frequencies of about 100 GHz has been demonstrated [20], but its power output in the high-frequency range has been very low. Table 6-1-3 shows the state of the art for tunnel diode oscillators.

TABLE 6-1-3.   State of the Art for Tunnel Diode Oscillators.

Frequency (GHz)	*Material*	*CW Power (mW)*	*Efficiency (%)*	*Reference*
5	GaAs	9.1	–	[21]
7.9	GaAs	2.0	–	[22]
9.0	GaAs	2.5	2	[23]
10.5	GaAs	1.5	2	[23]
49.0	GaAs	0.2	–	[24]

(4) Power-Frequency Limitations. The fact that the tunnel diode uses a quantum mechanism for the electrons to tunnel through potential barriers in order to exhibit a negative resistance means that there are severe limitations on power output and oscillation frequency. The negative resistance occurs only in the forward-bias direction at a voltage intermediate between zero voltage and the diode's cut-in voltage. It is apparent that the maximum voltage that can be applied to a tunnel diode in a negative-resistance region is less than the bandgap voltage. For Si, Ge, GaAs, and GaSb, the bandgap voltage is 1.10, 0.72, 1.40, and 0.68 volt, respectively. Consequently, tunnel diodes have a low power output.

6-1-3. Microwave Field-Effect Transistors (FETs)

After Shockley and his coworkers invented the transistor in 1948, he proposed in 1952 a new type of field-effect transistor (FET) in which the conductivity of a layer of a semiconductor is modulated by a transverse electric field [25]. In a conventional transistor both the majority and the minority carriers are involved; so this type of transistor is customarily referred to as a *bipolar* transistor. In a field-effect transistor the current flow is carried by one type of carrier only. This type is referred to as a *unipolar* transistor. Our purpose here is to describe the physical structures, principles of operation, microwave characteristics, and power-frequency limitations of unipolar field-effect transistors. Since the microwave Schottky barrier-gate field-effect transistor has the capability of amplifying small signals up to the frequency range of X band with low-noise figures, it has lately replaced the parametric amplifier in airborne radar systems because the latter is complicated in fabrication and expensive in production.

The unipolar field-effect transistor has several advantages over the bipolar junction transistor.

1. It may have voltage gain in addition to current gain.
2. Its efficiency is higher than that of a bipolar transistor.
3. Its noise figure is lower.
4. Its operating frequency is up to X band.
5. Its input resistance is very high, up to several megohms.

(*1*) **Physical Structures.** Unipolar field-effect transistors may consist of two types—*p-n* junction gate or Schottky barrier gate. In 1938 Schottky suggested that the potential barrier could arise from stable space charges in the semiconductor alone without the presence of a chemical layer [57]. The model derived from his theory is known as the *Schottky barrier*. The material may be either silicon or gallium arsenide (GaAs), and the channel type may be either *n* channel or *p* channel.

(A) JUNCTION FIELD-EFFECT TRANSISTOR (JFET). The junction field-effect transistor (JFET) is the one originally proposed by Shockley [25]. Figure 6-1-15 shows the schematic diagram and circuit symbol for an *n*-channel junction field-effect transistor (JFET). The *n*-type material is sandwiched between two highly doped layers of *p*-type material that is designated P^+. This type of device is called an *n-channel JFET*. If the middle part is a *p*-type semiconductor, the device is called a *p-channel JFET*. The two *p*-type regions in the *n*-channel JFET as shown in Fig. 6-1-15 are referred to as *gates*. Each end of the *n* channel is joined by a metallic contact. In accordance with the directions of the biasing voltages shown in Fig. 6-1-15, the left-hand contact that supplies the source of the flowing electrons is referred to as the

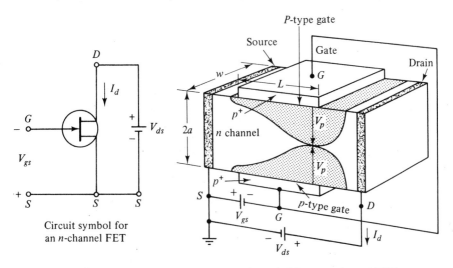

FIG. 6-1-15. Schematic diagram and circuit symbol for an *n*-channel FET.

source, whereas the right-hand contact that drains the electrons out of the material is called the *drain.* The circuit symbol for an *n*-channel JFET is also shown in Fig. 6-1-15. The direction of the drain current I_d is flowing from the drain to the device. For a *p*-channel JFET, the polarities of the two biasing voltages V_{gs} and V_{ds} are interchanged, the head of the arrow points away from the device, and the drain current I_d flows away from the device.

(B) SCHOTTKY-BARRIER FIELD-EFFECT TRANSISTOR (MESFET). The Schottky-barrier field-effect transistor was developed by many scientists and engineers, such as Mead [26] and Hooper [27], and it is sometimes called a *metal-semiconductor field-effect transistor* (MESFET). Figure 6-1-16 shows

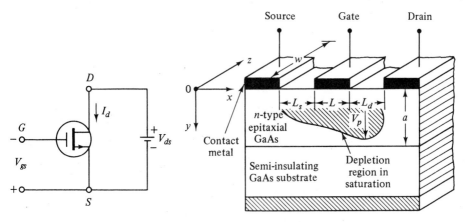

FIG. 6-1-16. Schematic diagram and symbol of a MESFET.

the schematic diagram and circuit symbol for a Schottky barrier-gate GaAs field-effect transistor. The device is of an interdigitated structure, fabricated by using an n-type GaAs expitaxial film about 0.15 to 0.35 μm thick on a semi-insulating substrate. The n-channel layer is doped with either sulfur or tin in a doping concentration N between 8×10^{16} and 2×10^{17} per cubic centimeter. The electron mobility in the layer is in the range of 3000 to 4500 cm^2/V·s. The Schottky barrier gate is evaporated aluminum. The source and drain contacts are Au—Ge, Au—Te, or Au—Te—Ge alloys. A contact metallization pattern of gold is used to bring the source, drain, and gate contacts out to bonding pads over the semi-insulating substrate.

(2) *Principles of Operation.* In the FETs of Figs. 6-1-15 and 6-1-16 a voltage is applied in the direction to reverse-bias the p-n junction between the source and the gate, while the source and the drain electrodes are forward-biased. Under this bias condition the majority carriers (electrons) flow in the n-type expitaxial layer from the source electrode, through the channel beneath the gate, to the drain electrode. The current in the channel causes a voltage drop along its length so that the Schottky barrier-gate electrode becomes progressively more reverse-biased toward the drain electrode. As a result, a charge-depletion region is set up in the channel and gradually pinches off the channel against the semi-insulating substrate toward the drain end. As the reverse bias between the source and the gate increases, so does the height of the charge-depletion region. The decrease of the channel height in the nonpinched-off region will increase the channel resistance. Consequently, the drain current I_d will be modulated by the gate voltage V_{gs}. This phenomenon is analogous to the characteristics of the collector current I_c versus the collector voltage V_{ce} with the base current I_b as parameter in a bipolar transistor. In other words, a family of curves of the drain current I_d versus the voltage V_{ds} between the source and drain with the gate voltage V_{gs} as parameter will be generated in a unipolar Schottky barrier-gate gallium arsenide field-effect transistor as shown in Fig. 6-1-17.

The transconductance of a FET is expressed as

$$g_m = \frac{dI_d}{dV_{gs}}\bigg|_{V_{ds}=\text{constant}} \quad \text{mhos} \tag{6-1-43}$$

For a fixed drain-to-source voltage V_{ds}, the drain current I_d is a function of the reverse-biasing gate voltage V_{gs}. Because the drain current I_d is controlled by the field effect of the gate voltage V_{gs}, this device is therefore referred to as the field-effect transistor (FET). When the drain current I_d is continuously increasing, the ohmic voltage drop between the source and the channel reverse-biases the p-n junction further. As a result, the channel is eventually pinched off. When the channel is pinched off, the drain current I_d will remain almost constant even though the drain-to-source voltage V_{ds} is continuously increased.

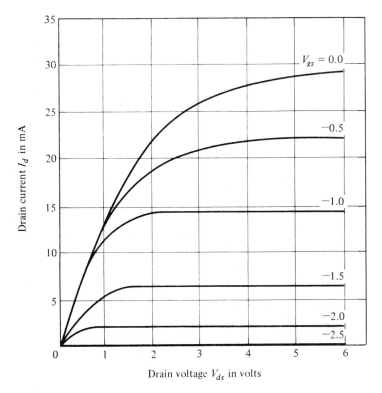

FIG. 6-1-17. Voltage-current characteristics of a typical *n*-channel GaAs Schottky barrier-gate FET.

PINCH-OFF VOLTAGE V_p. The pinch-off voltage is the gate reverse voltage that removes all the free charge from the channel. From Fig. 6-1-16 Poisson's equation for the voltage in the *n* channel, in terms of volume charge density, is given by

$$\frac{d^2V}{dy^2} = -\frac{\rho}{\epsilon} = -\frac{qN}{\epsilon} = -\frac{qN}{\epsilon_r\epsilon_0} \qquad (6\text{-}1\text{-}44)$$

where $\rho =$ the volume charge density in coulombs per cubic meter
$q =$ the charge in coulombs
$N =$ the electron concentration in electrons per cubic meter
$\epsilon =$ the permittivity of the material in farads per meter
$\epsilon = \epsilon_r\epsilon_0$, ϵ_r is the relative dielectric constant
$\epsilon_0 = 8.854 \times 10^{-12}$ F/m is the permittivity of free space

Integration of Eq. (6-1-44) once and application of the boundary condition of the electric field $E = -(dV/dy) = 0$ at $y = a$ yield

$$\frac{dV}{dy} = -\frac{qN}{\epsilon}(y - a) \qquad \text{volts per meter} \qquad (6\text{-}1\text{-}45)$$

Integration of Eq. (6-1-45) once and application of the boundary condition $V = 0$ at $y = 0$ result in

$$V = -\frac{qN}{2\epsilon}(y^2 - 2ay) \qquad \text{volts} \qquad (6\text{-}1\text{-}46)$$

Then the pinch-off voltage V_p at $y = a$ is written

$$V_p = \frac{qNa^2}{2\epsilon} \qquad \text{volts} \qquad (6\text{-}1\text{-}47)$$

where a is the height of the channel in meters.

Equation (6-1-47) indicates that the pinch-off voltage is a function of the doping concentration N and the channel height a. Doping may be increased to the limit set by the gate breakdown voltage and the pinch-off voltage may be made large enough so that drift saturation effects just become dominant.

(3) **Microwave Characteristics.** For microwave frequencies, the Schottky barrier-gate field-effect transistor has a very short channel length, and its velocity saturation occurs in the channel before reaching the pinched path. The microwave characteristics of a Schottky barrier-gate field-effect transistor depend not only on the intrinsic parameters, such as g_m, G_d, R_i, C_{sg} and C_{dg}, but also on the extrinsic parameters R_g, R_s, C_{sd}, R_p, and C_p. Figure 6-1-18 shows the cross section of a Schottky barrier-gate field-effect transistor and its equivalent circuit.

The microwave properties of Schottky barrier field-effect transistors were investigated and analyzed by many scientists and engineers [28–34]. The noise behavior of Schottky barrier-gate field-effect transistors at microwave frequencies has been investigated and measured by van der Ziel and others [35–39].

Intrinsic Elements:
  g_m is the transconductance of the FET.
  G_d is the drain conductance.
  R_i is the input resistance.
  C_{sg} is the source-gate capacitance.
  C_{dg} is the drain-gate (or feedback) capacitance.

Extrinsic Elements:
  R_g is the gate metallization resistance.
  R_s is the source-gate resistance.
  C_{sd} is the source-drain capacitance.
  R_p is the gate bonding-pad parasitic resistance.
  C_p is the gate bonding-pad parasitic capacitance.
  Z_ℓ is the load impedance.

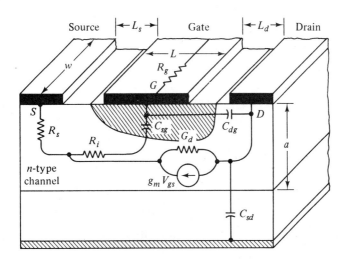

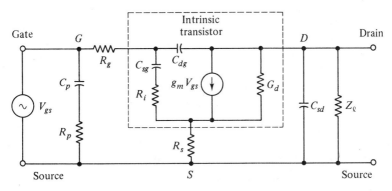

FIG. 6-1-18. Cross-section and equivalent circuit of a Schottky-barrier FET.

The values of these intrinsic and extrinsic elements depend on the channel type, material, structure, and dimensions of the Schottky barrier-gate FET. The large values of the extrinsic resistances will seriously decrease the power gain and efficiency and increase the noise figure of the MESFET. It is advantageous to increase the channel doping N as high as possible in order to decrease the relative influence of the feedback capacitance C_{dg} and to increase the transconductance g_m and the dc open-circuit voltage gain. However, an increase in concentration N decreases the breakdown voltage of the gate. A doping of 10^{18} per cubic centimeter might be an upper limit.

(A) DRAIN CURRENT I_d.   The drain current I_d of a Schottky barrier-gate FET is expressed [29] as

$$I_d = I_p \frac{3(u^2 - p^2) - 2(u^3 - p^3)}{1 + \eta(u^2 - p^2)} \qquad \text{amperes} \qquad (6\text{-}1\text{-}48)$$

where $I_p = \dfrac{qN\mu aWV_p}{3L}$ is the saturation current for the Shockley case
at $V_{gs} = 0$

$\mu =$ low-field mobility in square meters per volt-second

$q = 1.6 \times 10^{-19}$ coul is the electron charge

$N =$ doping concentration in electrons per cubic meter

$a =$ channel height

$W =$ gate width

$L =$ gate length

$V_p =$ pinch-off voltage as defined in Eq. (6-1-47)

$u = \left(\dfrac{V_{ds} + |V_{gs}|}{V_p}\right)^{1/2}$ is the normalized sum of the drain and gate voltages with respect to the pinch-off voltage

$p = \left(\dfrac{|V_{gs}|}{V_p}\right)^{1/2}$ is the normalized gate voltage with respect to the pinch-off voltage

$\eta = \dfrac{\mu V_p}{v_s L} = \dfrac{v}{v_s}$ is the normalized drift velocity with respect to the saturation drift velocity

$v_s =$ saturation drift velocity

$v = \dfrac{\mu E_x}{1 + \mu E_x/v_s}$ is the drift velocity in the channel

$E_x =$ absolute value of the electric field in the channel

Figure 6-1-17 shows a plot of the drain current I_d versus the drain voltage V_{ds} with the gate voltage V_{gs} as parameter for a typical n-channel GaAs Schottky barrier-gate field-effect transistor.

(B) CUTOFF FREQUENCY f_{co}. The cutoff frequency of a Schottky barrier-gate FET in a circuit depends on the way in which the transistor is being made. In a wideband lumped circuit the cutoff frequency is expressed [31] as

$$f_{co} = \frac{g_m}{2\pi C_g} = \frac{v_s}{4\pi L} \qquad \text{Hz} \qquad (6\text{-}1\text{-}49)$$

where $g_m =$ transconductance

$C_g =$ gate capacitance

$L =$ gate length

$v_s =$ saturation drift velocity

It is interesting to note that the cutoff frequency of the lumped circuit analysis as shown in Eq. (6-1-49) is different from the charge-carrier transit-time cutoff frequency as shown in Eq. (6-1-25) by a factor of one-half.

(C) MAXIMUM FREQUENCY OF OSCILLATION f_{max}. The maximum frequency of oscillation depends on the device transconductance and the drain resistance in a distributed circuit. It is expressed [29] as

$$f_{\max} = \frac{f_{co}}{2}(g_m R_d)^{1/2} = \frac{f_{co}}{2}\left[\frac{\mu E_p(u_m - p)}{v_s(1 - u_m)}\right]^{1/2} \quad \text{Hz} \qquad (6\text{-}1\text{-}50)$$

where R_d = drain resistance

g_m = device tranconductance

E_p = electric field at the pinch-off region in the channel

u_m = saturation normalization of u,

v_s, μ, and p are defined previously

For $p = 0$ (i.e., $V_{gs} = 0$) and $\eta \gg 1$, so that $f_{co} = v_s/(4\pi L)$ and $u_m \ll 1$,

$$f_{\max} = \gamma\frac{v_s}{L}\left(\frac{3}{\eta}\right)^{1/3} \quad \text{Hz} \qquad (6\text{-}1\text{-}51)$$

where $\gamma = 0.14$ for $\mu E_p/v_s = 13$ and $\gamma = 0.18$ for $\mu E_p/v_s = 20$ in case of GaAs.

It has been found experimentally [29] that the maximum frequency of oscillation for a gallium arsenide FET with gate length less than 10 μm is

$$f_{\max} = \frac{33 \times 10^3}{L} \quad \text{Hz} \qquad (6\text{-}1\text{-}52)$$

where L = gate length in meters.

The maximum frequency of oscillation $f_{\max}$ is similar to the cutoff frequency f_{co} determined by the transit time. From Eq. (6-1-25) the charge-carrier transit-time cutoff frequency is

$$f_{co} = \frac{1}{2\pi\tau} = \frac{v_s}{2\pi L} \quad \text{Hz} \qquad (6\text{-}1\text{-}53)$$

where $\tau = \dfrac{L}{v_s}$ is the transit time in seconds

L = gate length in meters

v_s = saturation drift velocity in meters per second

It is evident that the gallium arsenide FET has a better figure of merit than the silicon FET for an X-band amplifier because the saturation drift velocity v_s is 2×10^7 cm/s for GaAs at an electric field of 3 kV/cm and 8×10^6 cm/s for silicon at 15 kV/cm. In comparing Eq. (6-1-53) with (6-1-49), the difference is a factor of one-half.

The highest frequency of oscillation for maximum power gain with input and output networks matched is given [32] as

$$f_{\max} = \frac{f_{co}}{2}\left(\frac{R_d}{R_s + R_g}\right)^{1/2} \quad \text{Hz} \qquad (6\text{-}1\text{-}54)$$

where R_d = drain resistance

R_s = source-to-gate resistance

R_g = gate metallization resistance

(D) POWER GAIN.   In the microwave frequency range the measurable circuit parameters for a two-port Schottky barrier-gate field-effect transistor are the scattering parameters (S parameters) as described in Section 4-7 and Section 6-1-1 for bipolar junction transistors. After the values of S parameters are obtained, the power gain G_p, the maximum available power gain G_{max}, and the unilateral power gain G_u can readily be calculated by using the equations derived in Section 6-1-1.

Power Gain G_p: The power gain G_p of a microwave Schottky barrier-gate FET amplifier is defined as the ratio of the output power P_ℓ delivered to the load Z_ℓ over the input power P_{avs} available from the source to the FET as shown in Eq. (6-1-16). By application of Mason's rules as described in Section 6-1-1, the power gain for a Schottky barrier-gate FET amplifier as shown in Eq. (6-1-19) is given by

$$G_p = \frac{|S_{21}|^2(1 - |\Gamma_s|^2)(1 - |\Gamma_\ell|^2)}{|(1 - S_{11}\Gamma_s)(1 - S_{22}\Gamma_\ell) - S_{21}S_{12}\Gamma_s\Gamma_\ell|^2} \qquad (6\text{-}1\text{-}55)$$

Maximum available power gain G_{max}: In order to maximize the forward power gain G_{max} for a microwave FET amplifier, the input and output networks must be conjugately matched. The maximum available power as shown in Eq. (6-1-22) is

$$G_{max} = \frac{|S_{21}|}{|S_{12}|} | K \pm (K^2 - 1)^{1/2} | \qquad (6\text{-}1\text{-}56)$$

where K is the MESFET's inherent stability factor. It is expressed in Eq. (6-1-20) as

$$K = \frac{1 + |\Delta|^2 - |S_{11}|^2 - |S_{22}|^2}{2|S_{12}S_{21}|} > 1 \qquad (6\text{-}1\text{-}57)$$

where $\Delta \equiv S_{11}S_{22} - S_{12}S_{21}$

In order for a MESFET amplifier to be unconditionally stable, the stability factor must be greater than positive unity.

Unilateral Power Gain G_u: The unilateral power gain G_u is the forward power gain in a feedback amplifier having its reverse power gain set to zero (i.e., $|S_{12}|^2 = 0$) by adjusting a lossless reciprocal feedback network connected around the microwave MESFET amplifier as shown in Eq. (6-1-23). That is,

$$G_u = \frac{|S_{21}|^2(1 - |\Gamma_s|^2)(1 - |\Gamma_\ell|^2)}{|1 - S_{11}\Gamma_s|^2 |1 - S_{22}\Gamma_\ell|^2} \qquad (6\text{-}1\text{-}58)$$

The maximum unilateral power gain is achieved with $\Gamma_s = S_{11}^*$ and $\Gamma_\ell = S_{22}^*$. Then

$$G_{u\,max} = \frac{|S_{21}|^2}{(1 - |S_{11}|^2)(1 - |S_{22}|^2)} \qquad (6\text{-}1\text{-}59)$$

(E) NOISE FIGURE F. The overall effect of many noise sources in an electronic circuit is frequently specified by means of the noise figure of the circuit. The noise figure F of any linear two-port network can be defined in terms of its performance with a standard noise source connected to its input terminals. That is,

$$F \equiv \frac{\text{available noise power at output}}{\text{available noise power at input}} = \frac{N_0}{GkTB} \qquad (6\text{-}1\text{-}60)$$

where kTB = the available noise power of the standard source in a bandwidth B at temperature $T = 290°K$, and Boltzmann constant $k = 1.381 \times 10^{-23} J/°K$.

           G = the available power gain of the network at the frequency of the band considered

           N_0 = the available noise power at outputs

The noise N_0 at the output of the network within the same frequency band arises from the amplification of the input noise and from the noise N_n generated within the network. Then the output noise N_0 can be expressed as

$$N_0 = N_n + GkTB \qquad (6\text{-}1\text{-}61)$$

Thus the single-frequency noise figure results in

$$F = 1 + \frac{N_n}{GkTB} \qquad (6\text{-}1\text{-}62)$$

In general, when the noise powers are expressed by their noise temperatures, Eq. (6-1-62) becomes

$$F = 1 + \frac{T_n}{T_0} \qquad (6\text{-}1\text{-}63)$$

where T_n = noise temperature of the network in degrees Kelvin

           T_0 = ambient noise temperature at $290°K$

There are three types of noise figures for a MESFET amplifier.

1. *Intrinsic Noise Figure:* The intrinsic noise figure of a GaAs MESFET is given [35, 37] as

$$F = 2 + \gamma\left(\frac{E}{E_{sat}}\right)^3 \qquad (6\text{-}1\text{-}64)$$

where E = electric field in volts per meter

           E_{sat} = saturation electric field at $300 \, kV/m$

           $\gamma = 6$

At a frequency of 10 GHz, a noise figure of 6.6 dB for GaAs FET has been measured [36], which is much better than the noise figure of any other FET or bipolar transistor. At 5 GHz the noise figure for GaAs FET is approximately 3 dB. For silicon MESFET, the noise figure could be approximately expressed [35, 37] as

$$F = 2 + \gamma\left(\frac{E}{E_{sat}}\right)^2 \qquad (6\text{-}1\text{-}65)$$

where $\gamma = 2.3$ and E_{sat} = saturation electric field at $1500 \, kV/m$.

2. *Intervalley Scattering Noise Figure in GaAs:* According to the energy-band theory of the *n*-type gallium arsenide GaAs, the low-mobility upper valley is separated by an energy gap of 0.36 eV from the high-mobility lower valley, and the lower valley is separated by an energy gap of 1.43 eV from the valence band. The saturation drift velocity of electrons is at the electric field of 300 kV/m. The intervalley scattering noise figure is given [36] as

$$F = \frac{T_{nv}}{T_0}(1 - p) + \frac{T_{ni}}{T_0} + 1 \qquad (6\text{-}1\text{-}66)$$

where $T_0 = 290°K$ is the ambient noise temperature

T_{nv} = noise temperature of electrons in the valence band

T_{ni} = noise temperature of intervalley scattering

$1 - p$ = ratio of the number of electrons in the valence band

p = population probability of electrons in the lower and upper valleys (satellite valleys)

3. *Extrinsic Noise Figure:* Several sources of extrinsic noise for a Schottky barrier-gate field-effect transistor (MESFET) are

(a) Gate metallization resistance R_g—If the gate is made of relatively long, narrow, and thin aluminum film, the resistance between the gate bonding pad and the active transistor region is quite large.

(b) Source resistance R_s—The region between the source and gate contributes the source resistance.

(c) Drain resistance R_d—The region between the drain contact and the drain end of the channel causes resistance.

(d) Bonding pad resistance R_p—The gate bonding pads lie on the *n*-channel expitaxial end, and it may be represented by an impedance of a resistance R_p in series with a capacitance C_p.

The extrinsic noise figure of a MESFET contributed by these noise-source resistances is expressed [35] as

$$F = F_0 + \frac{R_n}{G_s}[(G_s - G_{on})^2 + (B_s - B_{on})^2] \qquad (6\text{-}1\text{-}67)$$

where F_0 = optimum noise figure

R_n = noise resistance

$Y_s = G_s + jB_s$ is the source admittance

$Y_{on} = G_{on} + jB_{on}$ is the optimum source admittance with respect to noise

The noise parameters F_0, G_{on}, B_{on}, and R_n are determined by calculating the noise figure F from Eq. (6-1-67) for four different source admittances $Y_s = G_s + jB_s$. It was found that the optimum extrinsic noise figure does not depend on the drain and gate voltages, and it is about 4 dB at 2 GHz and 8 dB at 8 GHz.

If the values of R_g, R_p, and R_s are reduced to minimum, the noise figure could be very much improved. This can be achieved by using heavier metallization of the gate, removal of the low-resistivity epitaxial layer below the gate bonding pad, and reduction of the distance between the source and gate. The best noise figure and power gain may be obtained for 0-volt gate voltage and 4-volt drain voltage for a GaAs MESFET.

(4) Power-Frequency Limitations. In Section 6-1-1-1 power-frequency limitations for microwave bipolar transistors were described. The question then arises as to whether microwave unipolar Schottky barrier-gate field-effect transistors have limitations on their frequency and power gain as well. The answer is yes. The basic principle of constant gain-bandwidth product is still valid. When the frequency is increased, the power gain is decreased. Figure 6-1-19 shows the measured power gain for a silicon Schottky barrier-gate FET [28].

Here the power gain near the X band is very low. However, the gallium arsenide (GaAs) microwave FET has about 4 dB power gain higher than the counterpart of silicon MESFET. So the best figure of a GaAs MESFET may

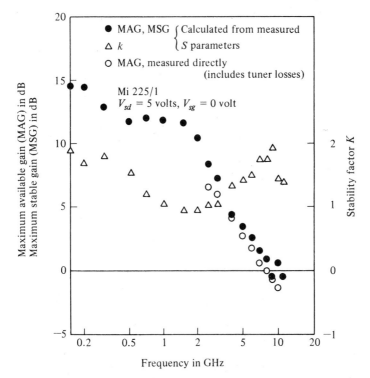

FIG. 6-1-19. Measured power gain for silicon MESFET. (*After P. Wolf [28]; reprinted with permission from IBM, Inc.*)

be about 6 dB at the X-band frequency range. The output power may reach a range of 1 watt in X band.

6-2 Transferred Electron Devices (TEDs)

The application of two-terminal semiconductor devices at microwave frequencies has seen an increased usage during the past decade. The CW, average, and peak power outputs of these devices at higher microwave frequencies are much larger than those obtainable with the best power transistor. The common characteristic of all active two-terminal solid-state devices is their negative resistance. The real part of their impedance is negative over a range of frequencies. In a positive resistance the current through the resistance and the voltage across it are in phase. The voltage drop across a positive resistance will be positive and a power of (I^2R) will be dissipated in the resistance. In a negative resistance, however, the current and voltage are out of phase by 180°. The voltage drop across a negative resistance will be negative, and a power of $(-I^2R)$ will be generated by the power supply associated with the negative resistance. In other words, positive resistances absorb power (passive devices), whereas negative resistances generate power (active devices). In this section the transferred electron devices (TEDs) are analyzed.

The differences between microwave transistors and transferred electron devices (TEDs) are fundamental. Transistors operate with either junctions or gates, but TEDs are bulk devices having no junctions or gates. The majority of transistors are fabricated from elemental semiconductors, such as silicon or germanium, whereas TEDs are fabricated from compound semiconductors, such as gallium arsenide (GaAs), indium phosphide (InP) or cadmium telluride (CdTe). Transistors operate with "warm" electrons whose energy is not much larger than the thermal energy (0.026 eV at room temperature) of electrons in the semiconductor, whereas TEDs operate with "hot" electrons whose energy is very much greater than the thermal energy. Because of their fundamental differences, the theory and technology of transistors cannot be applied to TEDs.

6-2-1. Gunn-Effect Diodes—GaAs Diode

Gunn-effect diodes are named after J. B. Gunn, who in 1963 discovered periodic fluctuations of current passing through the n-type gallium arsenide (GaAs) specimen when the applied voltage exceeded a certain critical value. Two years later, in 1965, B. C. DeLoach, R. C. Johnston, and B. G. Cohen discovered the impact ionization avalanche transit-time (IMPATT) mechanism in silicon, which employs the avalanching and transit-time properties of the diode to generate microwave frequencies. In later years the limited space-charge-accumulation diode (LSA diode) and the indium phosphide diode (InP diode) were also successfully developed. These are bulk devices in the sense that microwave amplification and oscillation are derived from the bulk negative-resistance property of uniform semiconductors rather than from the junction negative-resistance property between two different-type semiconductors such as the tunnel diode.

(1) Background. After inventing the transistor, Shockley suggested in 1954 that two-terminal negative-resistance devices using semiconductors may have advantages over transistors at high frequencies [40]. In 1961 Ridley and Watkins described a new method for obtaining negative differential mobility in semiconductors [41]. The principle involved is to heat carriers in a light-mass, high-mobility subband with an electric field so that the carriers can transfer to a heavy-mass, low-mobility, higher-energy subband when they have a high enough temperature. Ridley and Watkins also mentioned that Ge-Si alloys and some III-V compounds may have suitable subband structures in the conduction bands. Their theory for achieving negative differential mobility in bulk semiconductors by transferring electrons from high-mobility energy bands to low-mobility energy bands was taken a step further by Hilsum in 1962 [42]. Hilsum carefully calculated the transferred electron effect in several III-V compounds and was the first to use the terms of transferred electron amplifiers (TEAs) and oscillators (TEOs). He predicted accurately that a TEA-bar of semi-insulating GaAs would be operated at 373°K at a field of 3200 volts/cm. Unfortunately, Hilsum's attempts to verify his theory experimentally failed because the GaAs diode available to him at that time was not of sufficiently high quality.

It was not until 1963 that J. B. Gunn of IBM discovered the so-called Gunn effect from thin disks of *n*-type GaAs and *n*-type InP specimens while studying the noise properties of semiconductors [43]. Gunn did not connect —and even immediately rejected—his discoveries with the theories of Ridley, Watkins, and Hilsum. In 1963 Ridley predicted [44] that the field domain is continually moving down through the crystal, disappearing at the anode and then reappearing at a favoured nucleating center, and starting the whole cycle once more. Finally, Krömer stated [45] that the origin of the negative differential mobility is Ridley–Watkins–Hilsum's mechanism of electron transfer into the satellite valleys that occur in the conduction bands of both the *n*-type GaAs and the *n*-type InP and that the properties of the Gunn effect are the current oscillations due to the periodic nucleation and disappearance of traveling space-charge instability domains. Thus the correlation of theoretical predictions and experimental discoveries completed the theory of transferred electron devices (TEDs).

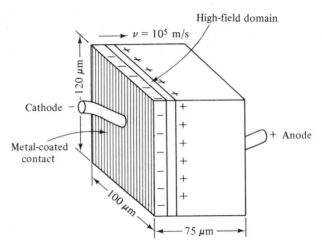

FIG. 6-2-1. Schematic diagram for *n*-type GaAs diode.

(2) Gunn Effect. A schematic diagram of a uniform *n*-type GaAs diode with ohmic contacts at the end surfaces is shown in Fig. 6-2-1.

J. B. Gunn observed the Gunn effect in the *n*-type GaAs bulk diode in 1963, an effect best explained by Gunn himself, who published several papers about his observations [46–49]. He stated in his first paper [46] that:

> Above some critical voltage, corresponding to an electric field of 2000–4000 volts/cm, the current in every specimen became a fluctuating function of time. In the GaAs specimens, this fluctuation took the form of a periodic oscillation superimposed upon the pulse current. . . . The frequency of oscillation was determined mainly by the specimen, and not by the external circuit. . . . The period of oscillation was usually inversely proportional to the specimen length and closely equal to the transit time of electrons between the electrodes, calculated from their estimated velocity of slightly over 10^7 cm/s. . . . The peak pulse microwave power delivered by the GaAs specimens to a matched load was measured. Values as high as 0.5 W at 1 Gc/s, and 0.15 W at 3 Gc/s, were found, corresponding to 1–2% of the pulse input power.*

From Gunn's observation the carrier drift velocity is linearly increased from zero to a maximum when the electric field is varied from zero to a threshold value. When the electric field is beyond the threshold value of 3000 volts/cm for the *n*-type GaAs, the drift velocity is decreased and the diode exhibits a phenomenon of negative resistance. This situation is shown in Fig. 6-2-2.

The current fluctuations are shown in Fig. 6-2-3. The current waveform was produced by applying a voltage pulse of 16-volt amplitude and 10-

---

*After J. B. Gunn [46]; represented with permission of IBM, Inc.

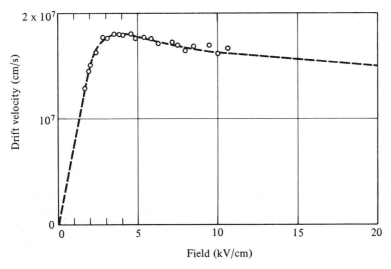

FIG. 6-2-2. Drift velocity of electrons in *n*-type GaAs versus electric field. (*After Gunn and Elliott, [47].*)

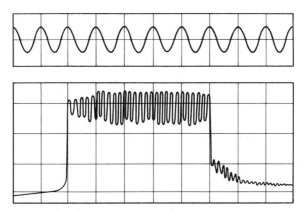

FIG. 6-2-3. Current waveform of *n*-type GaAs reported by Gunn. (*After J. B. Gunn [48]; reprinted by permission from IBM, Inc.*)

nanosecond duration to a specimen of *n*-type GaAs 2.5×10^{-3} cm in length. The oscillation frequency was 4.5 GHz. The lower trace had 2 nanoseconds per centimeter in the horizontal axis and 0.23 ampere per centimeter in the vertical axis. The upper trace was the expanded view of the lower trace. Gunn found that the period of these oscillations was equal to the transit time of the electrons through the specimen calculated from the threshold current.

Gunn also discovered that the threshold electric field E_{th} varied with the length and type of material. He developed an elaborate capacitive probe for plotting the electric field distribution within a specimen of *n*-type GaAs of length $L = 210$ μm and cross-sectional area 3.5×10^{-3} cm^2 with a low-field resistance of 16 Ω. Instabilities of current occurred at specimen voltages above 59 volts. This means that the threshold field is

$$E_{th} = \frac{V}{L} = \frac{59}{210 \times 10^{-6} \times 10^3} = 2810 \qquad \text{volts/cm} \qquad (6\text{-}2\text{-}1)$$

For a specimen voltage less than the threshold value the waveform $V(t)$ is shown in Fig. 6-2-4(a) measured at equal intervals of X in the vertical position and time t in the horizontal position. The values of X are 0, 40, 80, 120, and 160 μm reading upward and the time scale is 0.5 ns per major division. It is apparent that under these circumstances the time variation of potential at a point X merely reproduces that at $X = L$, multiplied by the fraction X/L, as would be expected for a homogeneous conductor. If the initial voltage exceeds the threshold, however, very different waveforms are shown in Fig. 6-2-4(b). When the instability begins, at about the middle of the pulse, the potential at $X = L$ rises sharply, remains high until almost the end of the pulse, and then drops rapidly again. This variation does not occur at other points in the specimen. A roughly equal rise in potential, of about 55 volts, occurs simultaneously at all points, but the drop takes place earlier at small values of X. At a given value of X the values of V before the rise and after the fall are approximately equal [48].

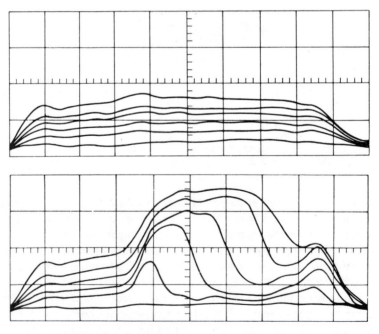

FIG. 6-2-4. Potential waveforms of n-type GaAs. (*After J. B. Gunn [49]; reprinted by permission of IBM, Inc.*)

Additional information can be obtained from a display of $\partial V/\partial t$ as a function of X, with t as a parameter, as shown in Fig. 6-2-5. Since V is slightly larger than it is in Fig. 6-2-4(b), the instabilities start at the beginning rather than at about the middle of the current pulse. In trace 1 the variation of $\partial V/\partial t$ with t fixed is approximately near X, showing that the electric field $(\mathbf{E} = -\nabla V = -\partial V/\partial x)$ is building up uniformly in the specimen. In trace 2, since it is delayed from trace 1 by 6.6×10^{-11} s, this linear distribution is beginning to undergo a distortion. By trace 11 this distortion has taken the form of a well-defined negative spatial pulse extending over a

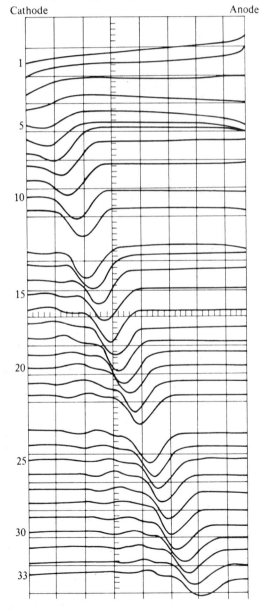

FIG. 6-2-5. Derivative of potential V with respect to time t. (*After J. B. Gunn [49]; reprinted with permission from IBM, Inc.*)

distance of about 30 μm. Elsewhere in the crystal $\partial V/\partial t = 0$. From this instant on, the negative maximum of the negative pulse propagates in the positive X direction (which is also the direction of electron flow) with no change in slope until it reaches the anode shortly after trace 33. The horizontal line indicates the distance scale of 26.5 μm per division.

The constant vertical separation between successive traces corresponds to a fixed time interval of 6.6×10^{-11} s, and the horizontal coordinate is proportional to X. Thus the inverse slope of a line joining the negative pulse maxima is equal to the velocity of the pulse, 8×10^6 cm/s. This value is about equal to that estimated for the drift velocity of electrons at the threshold. The average electric field within the "shock," taking the ratio of the specimen voltage and the pulse width, is found to be about 2×10^4 volts/cm. Since the $\partial V/\partial t$ pulse travels at about the same velocity as the drift velocity of electrons in the n-type GaAs and since $\partial V/\partial t$ is independent of the tranverse coordinate Y, it may be concluded that the Gunn effect is related to a high-field domain drifting in the axial direction of the specimen with about the same velocity as that of the electron beam.

(3) *Principles of Operation.* Many explanations have been offered for the Gunn effect. In 1964 Kroemer [45] suggested that Gunn's observations were in complete agreement with the two-valley model of the Ridley–Watkins–Hilsum (RHW) theory.

(A) TWO-VALLEY MODEL THEORY. Prior to the discovery of the Gunn effect, Kroemer proposed a negative-mass microwave amplifier in 1958 [50] and 1959 [51]. According to the energy-band theory of the n-type GaAs, a high-mobility lower valley is separated by an energy of 0.36 eV from a low-mobility upper valley as shown in Fig. 6-2-6. Table 6-2-1 lists the data for the two valleys in the n-type GaAs.

TABLE 6-2-1.   Data for Two Valleys in GaAs.

Valley	Effective mass M_e	Mobility μ	Separation Δ
Lower	$M_{e\ell} = 0.068$	$\mu_\ell = 8000\dfrac{cm^2}{V \cdot s}$	$\Delta = 0.36$ eV
Upper	$M_{eu} = 1.2$	$\mu_u = 180\dfrac{cm^2}{V \cdot s}$	$\Delta = 0.36$ eV

In Fig. 6-2-6 K indicates the wave number that is synonymous with the phase constant β, and E is equal to $\hbar\omega$ in quantum mechanics, where $\hbar$ is Planck's constant h divided by 2π. Therefore the E-K diagram is equivalent to the ω-β diagram of microwave slow-wave structure discussed in Section 4-2.

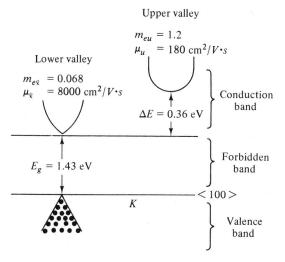

FIG. 6-2-6. Two-valley-model electron energy versus wave number for *n*-type GaAs.

If the electron densities in the lower and upper valleys are n_ℓ and n_u, the conductivity of the *n*-type GaAs is

$$\sigma = e(\mu_\ell n_\ell + \mu_u n_u) \qquad (6\text{-}2\text{-}2)$$

where $e =$ the electron charge

$\mu =$ the electron mobility in square centimeter per volt-second $(V \cdot s)$

$n =$ the electron density

When a sufficiently high field E is applied to the specimen, the electrons are accelerated and their effective temperature rises above the lattice temperature. Furthermore, the lattice temperature is also increased. Thus the electron density n and mobility μ are both functions of the electric field E. Differentiation of Eq. (6-2-2) with respect to E yields

$$\frac{d\sigma}{dE} = e\left(\mu_\ell \frac{dn_\ell}{dE} + \mu_u \frac{dn_u}{dE}\right) + e\left(n_\ell \frac{d\mu_\ell}{dE} + n_u \frac{d\mu_u}{dE}\right) \qquad (6\text{-}2\text{-}3)$$

If the total electron density is given by $n = n_\ell + n_u$ and it is assumed that μ_ℓ and μ_u are proportional to E^p, where p is a constant, then

$$\frac{d}{dE}(n_\ell + n_u) = \frac{dn}{dE} = 0 \qquad (6\text{-}2\text{-}4)$$

$$\frac{dn_\ell}{dE} = -\frac{dn_u}{dE} \qquad (6\text{-}2\text{-}5)$$

and

$$\frac{d\mu}{dE} \propto \frac{dE^p}{dE} = pE^{p-1} = p\frac{E^p}{E} \propto p\frac{\mu}{E} = \mu\frac{p}{E} \qquad (6\text{-}2\text{-}6)$$

Substitution of Eqs. (6-2-4) to (6-2-6) in Eq. (6-2-3) results in

$$\frac{d\sigma}{dE} = e(\mu_\ell - \mu_u)\frac{dn_\ell}{dE} + e(n_\ell\mu_\ell + n_u\mu_u)\frac{p}{E} \qquad (6\text{-}2\text{-}7)$$

Differentiation of Ohm's law $J = \sigma E$ with respect to E yields

$$\frac{dJ}{dE} = \sigma + \frac{d\sigma}{dE}E \qquad (6\text{-}2\text{-}8)$$

Equation (6-2-8) can be rewritten as

$$\frac{1}{\sigma}\frac{dJ}{dE} = 1 + \frac{d\sigma/dE}{\sigma/E} \qquad (6\text{-}2\text{-}9)$$

It is apparent that for negative resistance the current density J must decrease with increasing field E or the ratio of dJ/dE must be negative. Such would be the case only if the right-hand term of Eq. (6-2-9) is less than zero. In other words, the condition for negative resistance is

$$-\frac{d\sigma/dE}{\sigma/E} > 1 \qquad (6\text{-}2\text{-}10)$$

Substitution of Eqs. (6-2-2) and (6-2-7) in Eq. (6-2-10) with $f = n_u/n_\ell$ results in [41]

$$\left[\left(\frac{\mu_\ell - \mu_u}{\mu_\ell + f\mu_u}\right)\left(-\frac{E}{n_\ell}\frac{dn_\ell}{dE}\right) - p\right] > 1 \qquad (6\text{-}2\text{-}11)$$

It should be noted that the field exponent p is a function of the lattice scattering mechanism and should be negative and large. The lattice scattering makes impurity scattering quite undesirable, since when this is dominant, the mobility rises with increasing field and thus p is positive. When lattice scattering is dominant, however, p is negative and will depend on the lattice and carrier temperature. The first (parenthetical term) in Eq. (6-2-11) must be positive for satisfying unequality. This means that $\mu_\ell > \mu_u$. Electrons must begin in a low-mass valley and transfer to a high-mass valley when they are heated by the electric field. The maximum value of this term is unity—that is, when $\mu_\ell \gg \mu_u$, the factor dn_ℓ/dE in the second parentheses must be negative. This quantity represents a rate at which electrons transfer to the upper valley with respect to electric field and the rate will depend on differences between the electron densities, electron temperature, and energy gaps in the two valleys.

On the basis of the Ridley–Watkins–Hilsum theory described earlier, the band structure of a semiconductor must satisfy three criteria in order to exhibit negative resistance [52].

1. The energy difference between the bottom of the lower valley and the bottom of the upper valley must be several times larger than the thermal energy (about 0.026 eV) at room temperature. This means that $\Delta E > KT$, or $\Delta E > 0.026$ eV.

2. The energy difference between the valleys must be smaller than the energy difference between the conduction and valence bands. This means that $\Delta E < E_g$. Otherwise the semiconductor will break down and become highly conductive before the electrons begin to transfer to the upper valleys because hole-electron pair formation is created.
3. The electron velocities (dE/dk) must be much smaller in the upper valleys than in the lower valleys.

The two most useful semiconductors, silicon and germanium, do not meet all these criteria. However, some compound semiconductors, such as gallium arsenide (GaAs), indium phosphide (InP), and cadmium telluride (CdTe), do satisfy these criteria; but others, such as indium arsenide (InAs), gallium phosphide (GaP), and indium antimonide (InSb), do not. Figure 6-2-7(a) shows a possible current density versus field characteristic of a two-valley semiconductor. Under the equilibrium condition the electron densities in the lower and upper valleys remain the same. When the applied electric field is lower than the electric field of the lower valley $(E < E_\ell)$ no electrons will transfer to the upper valley as shown in Fig. 6-2-7(b). When the applied

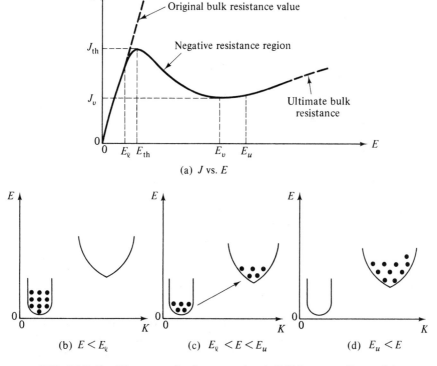

(a) J vs. E

(b) $E < E_\varrho$ (c) $E_\varrho < E < E_u$ (d) $E_u < E$

FIG. 6-2-7. Possible current density versus electric field for two-valley-model semiconductor.

electric field is higher than that of the lower valley and lower than that of the upper valley ($E_\ell < E < E_u$) electrons will begin to transfer to the upper valley as shown in Fig. 6-2-7(c). When the applied electric field is higher than that of the upper valley ($E_u < E$) all electrons will transfer to the upper valley as shown in Fig. 6-2-7(d).

A mathematical analysis of differential negative resistance requires a detailed analysis of high-field carrier transports [53, 54]. From electric field theory the magnitude of the current density in a semiconductor is given by

$$J = qnv \tag{6-2-12}$$

where q = electric charge
n = electron density
v = average electron velocity

Differentiation of Eq. (6-2-12) with respect to electric field E yields

$$\frac{dJ}{dE} = qn\frac{dv}{dE} \tag{6-2-13}$$

The condition for negative differential conductance can then be written

$$\frac{dv_d}{dE} = \mu_n < 0 \tag{6-2-14}$$

where μ_n denotes the negative mobility (see Fig. 6-2-8).

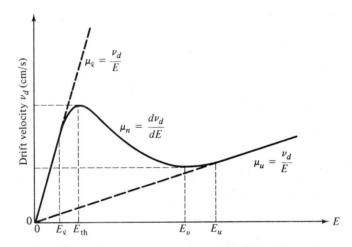

FIG. 6-2-8. Electron drift velocity versus electric field.

The direct measurement of the dependence of drift velocity on the electric field and direct evidence for the existence of the negative differential mobility were made by Ruch and Kino [55]. The experimental results, along with the theoretical results of analysis by Butcher and Fawcett [53], are shown in Fig. 6-2-9.

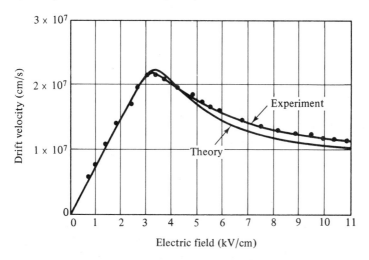

FIG. 6-2-9. Theoretical velocity-field characteristics of GaAs.

(B) HIGH-FIELD DOMAIN. In the last section we described how differential resistance can occur when an electric field of a certain range is applied to a multivalley semiconductor compound, such as the *n*-type GaAs. In this section we demonstrate how a decrease in drift velocity with increasing electric field can lead to the formation of a high-field domain for microwave generation and amplification.

In the *n*-type GaAs diode the majority carriers are electrons. When a small voltage is applied to the diode, the electric field and conduction current density are uniform throughout the diode. At low voltage the GaAs is ohmic, since the drift velocity of the electrons is proportional to the electric field. This situation was shown in Fig. 6-2-2. The conduction current density in the diode is given by

$$\mathbf{J} = \sigma\mathbf{E}_x = \frac{\sigma V}{L}\mathbf{U}_x = \rho v_x\mathbf{U}_x \qquad (6\text{-}2\text{-}15)$$

where $\mathbf{J}$ = conduction current density
σ = conductivity
$\mathbf{E}_x$ = electric field in the x direction
L = length of the diode
V = applied voltage
ρ = charge density
v = drift velocity
$\mathbf{U}$ = unit vector

The current is carried by free electrons that are drifting through a background of fixed positive charge. The positive charge, which is due to impurity atoms that have donated an electron (donors), is sometimes reduced by impurity atoms that have accepted an electron (acceptors). As long as the fixed charge is positive, the semiconductor is *n* type, since the principal carriers are the negative charges. The density of donors less the density of acceptors is referred as *doping*. When the space charge is zero, the carrier density is equal to the doping.

When the applied voltage is above the threshold value, which was measured at about 3000 volts/cm times the thickness of the GaAs diode, a high-field domain is formed near the cathode that reduces the electric field in the rest of the material and causes the current to drop to about two-thirds of its maximum value. This situation occurs because the applied voltage given by

$$V = -\int_0^L E_x \, dx \qquad (6\text{-}2\text{-}16)$$

For a constant voltage V an increase in the electric field within the specimen must be accompanied by a decrease in the electric field in the rest of the diode. The high-field domain then drifts with the carrier stream across the electrodes and disappears at the anode contact. When the electric field increases, the electron drift velocity decreases and the GaAs exhibits negative resistance.

More specifically, it is assumed that at point A on the J-E plot as shown in Fig. 6-2-10(b) there exists an excess (or accumulation) of negative charge that could be due to a random noise fluctuation or possibly a permanent nonuniformity in doping in the *n*-type GaAs diode. An electric field is then created by the accumulated charges as shown in Fig. 6-2-10(d). The field to the left of point A is lower than that to the right. If the diode is biased at point E_A on the J-E curve, this situation would imply that the carriers (or current) flowing into point A are greater than those flowing out of point A, thereby increasing the excess negative space charge at A. Furthermore, when the electric field to the left of point A is lower than it was before, the field to the right is then greater than the original one, resulting in an even greater space-charge accumulation. This process continues until the low and high fields both reach values outside the differential negative-resistance region and settle at points 1 and 2 in Fig. 6-2-10(a) where the currents in the two field regions are equal. As a result of this process, a traveling space-charge accumulation is formed. This process, of course, depends on the condition that the number of electrons inside the crystal is large enough to allow the necessary amount of space charge to be built up during the transit time of the space-charge layer.

The pure accumulation layer discussed above is the simplest form of space-charge instability. When positive and negative charges are separated by a small distance, then a dipole domain is formed as shown in Fig. 6-2-11. The electric field inside the dipole domain would be greater than the fields on either side of the dipole in Fig. 6-2-11(c). Because of the negative dif-

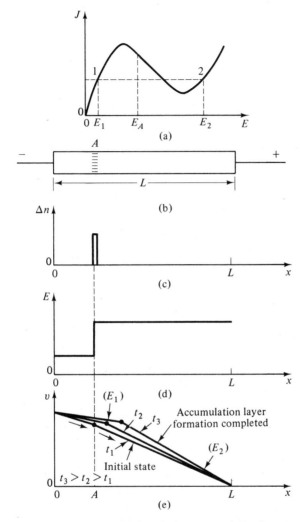

FIG. 6-2-10. Formation of an electron accumulation layer in GaAs. (*After H. Kroemer [56]; reprinted by permission of the IEEE, Inc.*)

ferential resistance, the current in the low-field side would be greater than that in the high-field side. The two field values will tend toward equilibrium conditions outside the differential negative-resistance region, where the low and high currents are the same as described in the previous section. Then the dipole field reaches a stable condition and moves through the specimen toward the anode. When the high-field domain disappears at the anode, a new dipole field starts forming at the cathode and the process is repeated.

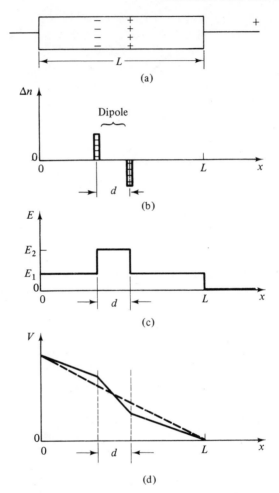

FIG. 6-2-11. Formation of an electron dipole layer in GaAs. (*After H. Kroemer [56]; reprinted with permission from the IEEE, Inc.*)

In general, the high-field domain has the following properties [52]:

1. A domain will start to form whenever the electric field in a region of the sample increases above the threshold electric field and will drift with the carrier stream through the device. When the electric field increases, the electron drift velocity decreases and the GaAs diode exhibits negative resistance.

2. If additional voltage is applied to a device containing a domain, the domain will increase in size and absorb more voltage than was added and the current will decrease.

3. A domain will not disappear before reaching the anode unless the voltage is dropped appreciably below threshold (for a diode with uniform doping and area).

4. The formation of a new domain can be prevented by decreasing the voltage slightly below threshold (in a nonresonant circuit).

5. A domain will modulate the current through a device as the domain passes through regions of different doping and cross-sectional area, or the domain may disappear. The effective doping may be varied in regions along the drift path by additional contacts.

6. The domain's length is generally inversely proportional to the doping; so devices with the same product of doping multiplied by length will behave similarly in terms of frequency multiplied by length, voltage/length, and efficiency.

7. As a domain passes a point in the device, the domain can be detected by a capacitive contact, since the voltage changes suddenly as the domain passes. The presence of a domain anywhere in a device can be detected by a decreased current or by a change in differential impedance.

It should be noted that properties (3) and (6) are valid only when the length of the domain is much longer than the thermal diffusion length for carriers, which for GaAs is about 1 μm for a doping of 10^{16} per cubic centimeter and about 10 μm for a doping of 10^{14} per cubic centimeter.

(*4*) *Modes of Operation.* Since Gunn first announced his observation of microwave oscillation in the *n*-type GaAs and *n*-type InP diodes in 1963, various modes of operation have been developed, depending on the material parameters and operating conditions. As noted, the formation of a strong space-charge instability depends on the conditions that enough charge is available in the crystal and that the specimen is long enough so that the necessary amount of space charge can be built up within the transit time of the electrons. This requirement sets up a criterion for the various modes of operation of bulk negative-differential-resistance devices. Copeland proposed four basic modes of operation of uniformly doped bulk diodes with low-resistance contacts [60] as shown in Fig. 6-2-12.

First: Gunn oscillation mode. This mode is defined in the region where the product of frequency multiplied by length is about 10^7 cm/s and the product of doping multiplied by length is greater than 10^{12} per centimeter squared. In this region the device is unstable because of the cyclic formation of either the accumulation layer or the high-field domain. In a circuit with relatively low impedance the device operates in the high-field

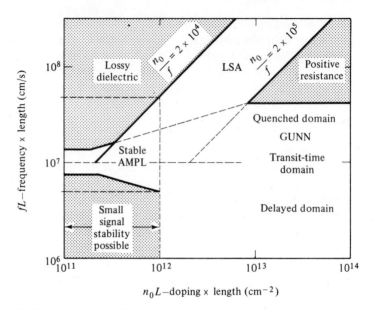

FIG. 6-2-12. Modes of operation for Gunn diodes. (*After J. A. Copeland [60]; reprinted by permission of the IEEE, Inc.*)

domain mode and the frequency of oscillation is near the intrinsic frequency. When the device is operated in a relatively high-Q cavity and coupled properly to the load, the domain is quenched and/or delayed before nucleating. In this case, the oscillation frequency is almost entirely determined by the resonant frequency of the cavity and has a value of several times the intrinsic frequency.

Second: Stable amplification mode. This mode is defined in the region where the product of frequency times length is about 10^7 cm/s and the product of doping times length is between 10^{11} and 10^{12} per centimeter squared.

Third: LSA oscillation mode. This mode is defined in the region where the product of frequency times length is above 10^7 cm/s and the quotient of doping divided by frequency is between 2×10^4 and 2×10^5.

Fourth: Bias-circuit oscillation mode. This mode occurs only when there is either Gunn or LSA oscillation, and it is usually at the region where the product of frequency times length is too small to appear in the figure. When a bulk diode is biased to threshold, the average current suddenly drops as Gunn oscillation begins. The drop in current at the threshold can lead to oscillations in the bias circuit that are typically 1 kHz to 100 MHz [61].

The first three modes are discussed in detail in this section. Before doing so, however, let us consider the criterion for classifying the modes of operation.

(I) CRITERION FOR CLASSIFYING THE MODES OF OPERATION. The Gunn-effect diodes are basically made from an *n*-type GaAs, with the concentrations of free electrons ranging from 10^{14} to 10^{17} per cubic centimeter at room temperature. Its typical dimensions are 150×150 μm in cross section and 30 μm long. During the early stages of space-charge accumulation the time rate of growth of the space-charge layers is given by

$$Q(X, t) = Q(X - vt, 0) \exp\left(\frac{t}{\tau_d}\right) \qquad (6\text{-}2\text{-}17)$$

where $\tau_d = \dfrac{\epsilon}{\sigma} = \dfrac{\epsilon}{en_0 |\mu_n|}$ is the magnitude of the negative dielectric
relaxation time
ϵ = the permittivity
n_0 = the doping concentration
μ_n = the negative mobility
e = the electron charge
σ = the conductivity

Figure 6-2-13 clarifies Eq. (6-2-17).

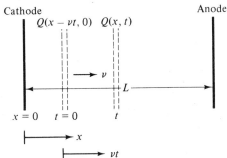

FIG. 6-2-13. Space-charge accumulation with a velocity of *v*.

If Eq. (6-2-17) remains valid throughout the entire transit time of the space-charge layer, the factor of maximum growth is given by

$$\text{Growth factor} = \frac{Q(L, L/v)}{Q(0, 0)} = \exp\left(\frac{L}{v\tau_d}\right) = \exp\left(\frac{Ln_0 e |\mu_n|}{\epsilon v}\right) \qquad (6\text{-}2\text{-}18)$$

In Eq. (6-2-18) the layer is assumed to start at the cathode at $t = 0$, $X = 0$ and arrive at the anode at $t = L/v$ and $X = L$. For a large space-charge growth, this factor must be larger than unity. This means that

$$n_0 L > \frac{\epsilon v}{e |\mu_n|} \qquad (6\text{-}2\text{-}19)$$

This is the criterion for classifying the modes of operation for the Gunn-effect diodes. For *n*-type GaAs, the value of $\epsilon v/(e|\mu_n|)$ is about 10^{12} per centimeter squared, where $|\mu_n|$ is assumed to be 100 cm²/$V \cdot s$.

(II) GUNN OSCILLATION MODES ($10^{12}/cm^2 \lesssim (n_0 L) < 10^{14}/cm^2$). Most Gunn-effect diodes have the product of doping and length ($n_0 L$) greater than $10^{12}/cm^2$. However, the mode that Gunn himself observed had a product $n_0 L$ that is much less. When the product of $n_0 L$ is greater than $10^{12}/cm^2$ in GaAs, the space-charge perturbations in the specimen increase exponentially in space and time in accordance with Eq. (6-2-17). Thus a high-field domain is formed and moves from the cathode to the anode as described earlier. The frequency of oscillation is given by the relation [62]

$$f = \frac{v_{dom}}{L_{eff}} \qquad (6\text{-}2\text{-}20)$$

where v_{dom} is the domain velocity and L_{eff} is the effective length that the domain travels from the time it is formed until the time that a new domain begins to form.

Gunn described the behavior of Gunn oscillators under several circuit configurations [63]. When the circuit is mainly resistive or the voltage across the diode is constant, the period of oscillation is the time required for the domain to drift from the cathode to the anode. This mode is not actually typical of microwave applications. Negative conductivity devices are usually operated in resonant circuits, such as high-Q resonant microwave cavities. When the diode is in a resonant circuit, the frequency can be tuned to a range of about an octave without loss of efficiency [64].

As described previously, the normal Gunn domain mode (or Gunn oscillation mode) is operated with the electric field greater than the threshold field ($E > E_{th}$). The high-field domain drifts along the specimen until it reaches the anode or until the low-field value drops below the sustaining field E_s required to maintain v_s as shown in Fig. 6-2-14. The sustaining drift velocity for GaAs is $v_s = 10^7$ cm/s.

Since the electron drift velocity v varies with the electric field, three possible domain modes for the Gunn domain mode exist.

(a) *Transit-time domain mode* ($fL \simeq 10^7$ cm/s): When the electron drift velocity v_d is equal to the sustaining velocity v_s, the high-field domain is

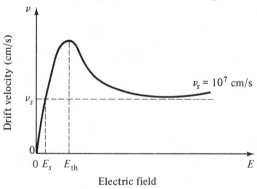

FIG. 6-2-14. Electron drift velocity versus electric field.

stable. In other words, the electron drift velocity is given by

$$v_d = v_s = fL \simeq 10^7 \text{ cm/s} \tag{6-2-21}$$

Then the oscillation period is equal to the transit time—that is, $\tau_0 = \tau_t$. This situation is shown in Fig. 6-2-15(a). The efficiency is below 10% because the current is collected only when the domain arrives at the anode.

(*b*) *Delayed domain mode* (10^6 cm/s $< fL < 10^7$ cm/s): When the transit time is chosen so that the domain is collected while $E < E_{th}$ as shown in Fig. 6-2-15(b), a new domain cannot form until the field rises above threshold again. In this case, the oscillation period is greater than the transit

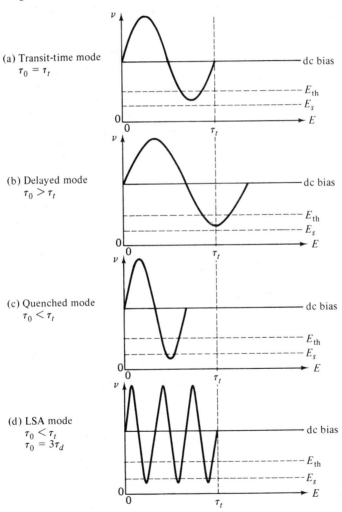

FIG. 6-2-15. Gunn domain modes.

time—that is, $\tau_0 > \tau_t$. This delayed mode is also called inhibited mode. The efficiency of this mode is about 20%.

(c) *Quenched domain mode* ($fL > 2 \times 10^7$ cm/s): If the bias field drops below the sustaining field E_s during the negative half cycle as shown in Fig. 6-2-15(c), the domain collapses before it reaches the anode. When the bias field swings back above threshold, a new domain is nucleated and the process repeats. Therefore the oscillations occur at the frequency of the resonant circuit rather than at the transit-time frequency. It has been found that the resonant frequency of the circuit is several times the transit-time frequency, since one dipole does not have enough time to readjust and absorb the voltage of the other dipoles [65, 66]. Theoretically, the efficiency of quenched domain oscillators can reach 13% [65].

(III) LIMITED-SPACE-CHARGE ACCUMULATION (LSA) MODE ($fL > 2 \times 10^7$ cm/s). When the frequency is very high, the domains do not have sufficient time to form while the field is above threshold. As a result, most of the domains are maintained in the negative conductance state during a large fraction of the voltage cycle. Any accumulation of electrons near the cathode has time to collapse while the signal is below threshold. Thus the LSA mode is the simplest mode of operation, and it consists of a uniformly doped semiconductor without any internal space charges. In this instance, the internal electric field would be uniform and proportional to the applied voltage. The current in the device is then proportional to the drift velocity at this field level. The efficiency of the LSA mode can reach 20%.

The oscillation period τ_0 should be no more than several times larger than the magnitude of the dielectric relaxation time in the negative conductance region τ_d. The oscillation indicated in Fig. 6-2-15(d) is $\tau_0 = 3\tau_d$. It is appropriate here to define the LSA boundaries. As described earlier, the sustaining drift velocity is 10^7 cm/s as shown in Eq. (6-2-21) and Fig. 6-2-14. For the n-type GaAs, the product of doping and length ($n_0 L$) is about 10^{12} per centimeter squared. At the low-frequency limit, the drift velocity is taken to be

$$v_\ell = fL = 5 \times 10^6 \text{ cm/s} \qquad (6\text{-}2\text{-}22)$$

The ratio of $n_0 L$ to fL yields

$$\frac{n_0}{f} = \frac{10^{12}}{5 \times 10^6} = 2 \times 10^5 \qquad (6\text{-}2\text{-}23)$$

At the upper-frequency limit it is assumed that the drift velocity is

$$v_u = fL = 5 \times 10^7 \text{ cm/s} \qquad (6\text{-}2\text{-}24)$$

and the ratio of $n_0 L$ to fL is

$$\frac{n_0}{f} = 2 \times 10^4 \qquad (6\text{-}2\text{-}25)$$

Both the upper and lower boundaries of the LSA mode are indicated in Fig. 6-2-12. The LSA mode is discussed further in Section 6-2-2.

(IV) STABLE AMPLIFICATION MODE ($n_0 L < 10^{12}/\text{cm}^2$). When the $n_0 L$ product of the device is less than about 10^{12} per centimeter squared, the device will exhibit amplification at the transit-time frequency rather than spontaneous oscillation. This situation occurs because the negative conductance is utilized without domain formation. There are too few carriers for domain formation within the transit time. Therefore amplification of signals near the transit-time frequency can be accomplished. This type of mode was first observed by Thim and Barber [66]. Furthermore, Uenohara showed that there are types of amplification depending on the *fL* product of the device [67] as shown in Fig. 6-2-16. However, no explanations were given.

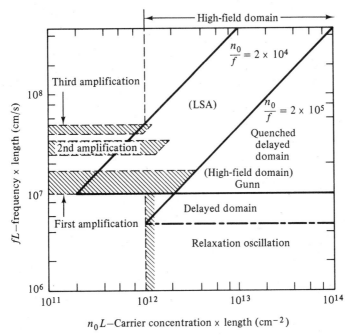

FIG. 6-2-16. Mode chart. (*After Uenohara [67]; reprinted by permission of McGraw-Hill Book Company.*)

The various modes of operation of Gunn diodes can be classified on the basis of the times in which various processes occur. These times are defined as follows.

τ_t = domain transit time
τ_d = the dielectric relaxation time at low field
τ_g = domain growth time

$\tau_0 =$ the natural period of oscillation of a high-Q external electric circuit

The modes described previously are summarized in Table 6-2-2.

TABLE 6-2-2. Modes of Operation of Gunn Oscillators.

Modes	Time relationships	Doping level	Nature of circuit
Stable amplifier	$\tau_0 \gtrsim \tau_t$	$n_0 L < 10^{12}$	Nonresonant
Gunn domain	$\tau_g \lesssim \tau_t$ $\tau_0 = \tau_t$	$n_0 L > 10^{12}$	Nonresonant; constant voltage
Quenched domain	$\tau_g \lesssim \tau_t$ $\tau_0 < \tau_t$	$n_0 L > 10^{12}$	Resonant; finite impedance
Delayed domain	$\tau_g \lesssim \tau_t$ $\tau_0 > \tau_t$	$n_0 L > 10^{12}$	Resonant; finite impedance
LSA	$\tau_0 < \tau_g$ $\tau_0 > \tau_d$	$2 \times 10^4 < \left(\dfrac{n_0}{f}\right) < 2 \times 10^5$	Multiply resonant; high impedance; high dc bias

(5) *Microwave Generation and Amplification*

(A) MICROWAVE GENERATION. As described earlier in Section 6-2-1(3)(B) if the applied field is less than threshold, the specimen is stable. If, however, the field is greater than threshold, the sample is unstable and divides up into two domains of different conductivity and different electric field but the same drift velocity. Figure 6-2-17 shows the stable and unstable regions.

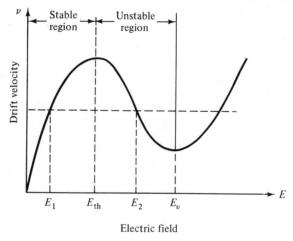

FIG. 6-2-17. Electric field versus drift velocity.

At the initial formation of the accumulation layer, the field behind the layer decreases and the field in the front of it increases. This process continues as the layer travels from the cathode toward the anode. As the layer approaches the anode, the field behind it begins to increase again; and after the layer is collected by the anode, the field in the whole sample is higher than threshold. When the high-field domain disappears at the anode, a new dipole field starts forming again at the cathode and the process repeats itself. Since current density is proportional to the drift velocity of the electrons, a pulsed current output is obtained. The oscillation frequency of the pulsed current is given by

$$f = \frac{v_d}{L_{eff}} \qquad (6\text{-}2\text{-}26)$$

where v_d is the velocity of the domain or approximately the drift velocity of the electrons and L_{eff} is the effective length that the domain travels.

The highest reported Gunn oscillator power-frequency performance for both pulsed and CW operation is listed in Table 6-2-3. Experiments have

TABLE 6-2-3. Gunn Oscillator: State of the Art.

Frequency (GHz)	Material	Operation	Power (W)	Efficiency (%)	Reference
1.16	GaAs	pulsed	100	6	[70]
3.0	GaAs	CW	0.06	5	[71]
3.05	GaAs	pulsed	200	29	[72]
3.2	GaAs	pulsed	2.5	7	[73]
5.0	GaAs	pulsed	2	–	[74]
7.0	GaAs	CW	0.032	3	[75]
8.3	GaAs	pulsed	19	22	[76]
8.7	GaAs	CW	0.78	2.5	[77]
9.0	GaAs	CW	0.02	3	[75]
13.3	GaAs	CW	0.17	3	[78]
15.0	GaAs	pulsed	10.00	10	[78]
24.8	GaAs	CW	0.12	5.2	[79]

shown that the n-type GaAs diodes have yielded 200-W pulses at 3.05 GHz and 780-mW CW power at 8.7 GHz. Efficiencies of 29% have been obtained in pulsed operation at 3.05 GHz and 5.2% in CW operation at 24.8 GHz. Predictions have been made that 250-kW pulses from a single block of n-type GaAs are theoretically possible up to 100 GHz.

The source generation of solid-state microwave devices has many advantages over vacuum tube devices that they are beginning to replace. However, at present they also have serious drawbacks that could prevent more widespread application. The most important disadvantages are

1. low efficiency at frequencies above 10 GHz

2. small tuning range

3. large dependence of frequency on temperature

4. high noise

These problems are common to both avalanche diodes and transferred electron devices [80].

(B) MICROWAVE AMPLIFICATION. When an RF signal is applied to a Gunn oscillator, amplification of the signal occurs, provided that the signal frequency is low enough to allow the space charge in the domain to readjust itself. There is a critical value of fL above which the device will not amplify. Below this frequency limit the sample presents an impedance with a negative real part that can be utilized for amplification. If n_0L becomes less than 10^{12} per square centimeter, domain formation is inhibited and the device exhibits a nonuniform field distribution that is stable with respect to time and space. Such a diode can amplify signals in the vicinity of the transit-time frequency and its harmonics without oscillation. If this device is used in a circuit with enough positive feedback, it will oscillate. Hakki has shown that the oscillation diode can amplify at nearby frequencies or can be used simultaneously as an amplifier and local oscillator [81]. However, the output power of a stable amplifier is quite low because of the limitation imposed by the value of n_0L.

In contrast to the stable amplifier, the Gunn-effect diode must oscillate at the transit-time frequency while it is amplifying at some other frequency. The value of n_0L must be larger than 10^{12} per square centimeter in order to establish traveling domain oscillations; hence substantially larger output power can be obtained. Because of the presence of high-field domains, this amplifier is called a *traveling domain amplifier* (TDA).

Although a large number of possible amplifier circuits exist, the essential feature of each is to provide both a broadband circuit at the signal frequency and a short circuit at the Gunn oscillation frequency. In order to maintain stability with respect to the signal frequency, the Gunn diode must see a source admittance whose real part is larger than the negative conductance of the diode. The simplest circuit satisfying this condition is shown in Fig. 6-2-18 [82]. An average gain of 3 dB was exhibited between 5.5 and 6.5 GHz.

Gunn diodes have been used in conjunction with circulator-coupled networks in the design of high-level wideband transferred electron amplifiers that have a voltage gain-bandwidth product in excess of 10 GHz for frequencies from 4 to about 16 GHz. Linear gains of 6 to 12 dB per stage and saturated-output-power levels in excess of 0.5 W have been realized [83]. Table 6-2-4 lists the performance of several amplifiers that have been designed in the past ten years.

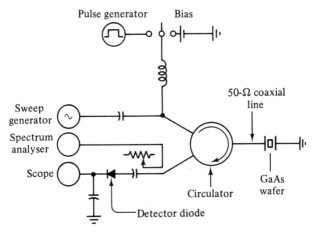

FIG. 6-2-18. Gunn-diode amplifier circuit. (*After H. W. Thim [82]; reprinted by permission of the IEEE, Inc.*)

TABLE 6-2-4. CW Gunn-Diode Amplifier Performance.*

Frequency band	3-dB bandwidth (GHz)	Small signal gain (dB)	Power gain (dB)	Efficiency (%)
C	4.5– 8.0	8	3	3
X	7.5–10.75	12	1.65	2.3
	8.0–12.0	6	1.8	2.5
Ku	12.0–16.0	6	1.5	2.5
	13.0–15.0	8	0.36	2

*After B. S. Perlman et al. [83]; reprinted by permission of the IEEE, Inc.

6-2-2. LSA Diodes

The abbreviation LSA stands for the Limited Space-charge Accumulation mode of the Gunn diode. As described previously, if the product $n_0 L$ is larger than 10^{12} per square centimeter and if the ratio of doping n_0 to frequency f is within 2×10^5 to 2×10^4 cm^3/s, the high-field domains and the space-charge layers do not have sufficient time to build up. The magnitude of the RF voltage must be large enough to drive the diode below threshold during each cycle in order to dissipate space charge. Also, the portion of each cycle during which the RF voltage is above threshold must be short enough to prevent the domain formation and the space-charge accumulation. Only the primary accumulation layer forms near the cathode; the rest of the sample remains fairly homogeneous. So with limited space-charge formation the remainder of the sample appears as a series negative resistance that increases the frequency of the oscillations in the resonant circuit. Copeland discovered the LSA mode of the Gunn diode in 1966 [84]. In the LSA mode the diode is placed in a resonator tuned to an oscillation frequency of

$$f_0 = \frac{1}{\tau_0} \tag{6-2-27}$$

The device is biased to several times the threshold voltage (see Fig. 6-2-19).

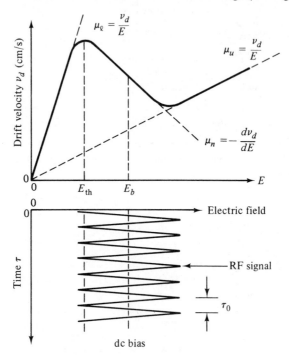

FIG. 6-2-19. LSA mode operation.

As the RF voltage swings beyond the threshold, the space charge starts building up at the cathode. Since the oscillation period τ_0 of the RF signal is less than the domain-growth time constant τ_g, the total voltage swings below the threshold before the domain can form. Furthermore, since τ_0 is much greater than the dielectric relaxation time τ_d, the accumulated space charge is drained in a very small fraction of the RF cycle. Therefore the device spends most of the RF cycle in the negative-resistance region, and the space charge is not allowed to build up. The frequency of oscillation in the LSA mode is independent of the transit time of the carriers and is determined solely by the circuit external to the device. Also, the power-impedance product does not fall off as $1/f_0^2$; thus the output power in the LSA mode can be greater than that in the other modes.

The LSA mode does have limitations. It is very sensitive to load conditions, temperatures, and doping fluctuations [85]. In addition, the RF circuit must allow the field to build up quickly in order to prevent domain formation. The power output of an LSA oscillator can be simply written

$$P = \eta V_0 I_0 = \eta (M E_{\text{th}} L)(n_0 e v_0 A) \qquad (6\text{-}2\text{-}28)$$

where $\eta =$ the dc-to-RF conversion efficiency (primarily a function of material and circuit considerations)

$V_0 =$ the operating voltage

$I_0 =$ the operating current

$M =$ the multiple of the operating voltage above negative-resistance threshold voltage

$E_{\text{th}} =$ the threshold field (about 3400 volts/cm)

$L =$ the device length (about 50 to 200 μm)

$n_0 =$ the donor concentration (about $10^{15} e/\text{cm}^3$)

$e =$ the electron charge (1.6×10^{-19} coul)

$v_0 =$ the average carrier drift velocity (about 10^7 cm/s)

$A =$ the device area (about $4 \times 10^{-4} - 20 \times 10^{-4}$ cm^2)

For an LSA oscillator, n_0 is primarily determined by the desired operating frequency f_0 so that, for a properly designed circuit, peak power output is directly proportional to the volume (LA) of the device length L multiplied by the area A of the active layer. Active volume cannot be increased indefinitely. In the theoretical limit, this is due to electrical wavelength and skin-depth considerations. In the practical limit, however, available bias, thermal dissipation capability, or technological problems associated with material uniformity limit device length. The best results are listed in Table 6-2-5.

TABLE 6-2-5.   LSA Oscillator Capabilities.

Frequency (GHz)	Power (W)	Operation	Duty cycle	Efficiency (%)	Reference
1.75	6000	pulsed	6×10^{-6}	15	[86]
5.0	500	pulsed	10^{-3}	10	[85]
7.0	2000	pulsed	6×10^{-6}	5	[86]
10.0	250	pulsed	10^{-3}	8	[85]
15.0	150	pulsed	10^{-3}	6	[85]
16.0	150	pulsed	6×10^{-6}	6	[86]
18.0	1.5	pulsed	10^{-5}	23	[86]
51.0	400	pulsed	10^{-5}	9	[84]
88.0	0.02	CW		2	[84]

6-2-3. InP Diodes

When Gunn first announced his Gunn effect in 1963, the diodes he investigated were of gallium arsenide (GaAs) and was indium phosphide (InP). The GaAs diode was described in detail earlier. In this section the *n*-type InP diode is discussed. Both the GaAs diode and the InP diode operate basically the same way in a circuit with dc voltage applied at the electrodes. In the ordinary Gunn effect in the *n*-type GaAs, the two-valley model theory is the foundation for explaining the electrical behavior of the Gunn effect. However, Hilsum proposed that indium phosphide and some alloys of indium gallium antimonide should work as three-level devices [90]. Figure 6-2-20 shows the three-valley model for indium phosphide.

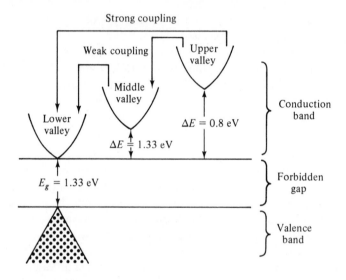

FIG. 6-2-20. Three-valley-model energy level for InP diode.

·It can be seen that InP, besides having an upper-valley energy level and a lower-valley energy level similar to the model shown in Fig. 6-2-6 for *n*-type GaAs, also has a third middle-valley energy level. In GaAs the electron transfer process from the lower valley to the upper valley is comparatively slow. At a particular voltage above threshold current flow consists of a larger contribution of electrons from the lower valley rather than from the upper valley. Because of this larger contribution from the lower energy level, a relatively low peak-to-valley current ratio results, which is shown in Fig. 6-2-21(a).

The InP diode has a larger peak-to-valley current ratio as shown in Fig. 6-2-21(b) because an electron transfer proceeds rapidly as the field increases. This situation occurs because the coupling between the lower valley and upper valley in InP is weaker than in the GaAs. The middle-valley energy level provides the additional energy loss mechanism required to avoid breakdown due to the high energies acquired by the lower-valley electrons from the weak coupling. It can be seen from Fig. 6-2-20 that the lower valley is weakly coupled to the middle valley but strongly coupled to the

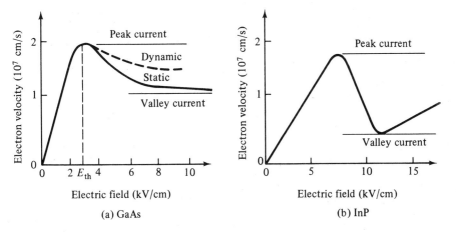

FIG. 6-2-21. Peak-to-valley current ratio.

upper valley to prevent breakdown. This situation ensures that under normal operating conditions electrons concentrate in the middle valley. Because InP has a greater energy separation between the lower valley and the nearest energy levels, the thermal excitation of electrons has less effect, and the degradation of its peak-to-valley current ratio is about four times less than in the GaAs [91].

The mode of operation of InP is unlike the domain oscillating mode in which a high-field domain is formed that propagates with a velocity of about 10^7 cm/s. The result is an output current waveform that is transit-time dependent. This mode reduces the peak-to-valley current ratio so that efficiency is reduced. For this reason, an operating mode is usually sought where charge domains are not formed. The three-valley model of InP inhibits the formation of domains because the electron diffusion coefficient is increased by the stronger coupling [91]. From experiments performed by Taylor and Colliver [92] it was determined that epitaxial InP oscillators operate through a transit-time phenomenon and do not oscillate in a bulk mode of the LSA type. From their findings it was determined that it is not appropriate to attempt to describe the space-charge oscillations in InP in terms of modes known to exist in GaAs devices. Taylor and Colliver also determined that the frequencies obtained from a device were dependent on the active-layer thickness. The InP oscillator could be tuned over a large frequency range, bounded only by the thickness, by adjusting the cavity size. Figure 6-2-22 shows the frequency ranges for the different active-layer thicknesses and lines of constant electron velocity [91, 92]. It can be seen from the figure that only a few InP devices operate in the domain formation area. In each case on the graph, the maximum efficiency occurs at about midband [92]. Table 6-2-6 summarizes the highest power and efficiencies for InP diodes.

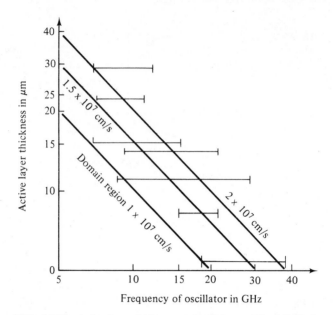

FIG. 6-2-22. Active layer thickness versus frequency for InP diode. (*From B. C. Taylor and D. J. Colliver [92]; reprinted by permission of the IEEE, Inc.*)

TABLE 6-2-6. Best Performance of InP Diode.

Frequency (GHz)	Thickness (μm)	Power (W)	Operation	Efficiency (%)	Reference
5.5	–	3.05	pulsed	14.7	[91]
8.5	28	0.95	pulsed	7.0	[92]
10.75	–	1.33	pulsed	12.0	[91]
13.8	11	0.5	pulsed	6.3	[92]
15	–	1.13	pulsed	15	[91]
18	11	1.05	pulsed	4.2	[92]
18	–	0.2	CW	10.2	[91]
25	5.4	0.65	pulsed	2.6	[92]
26	–	0.15	CW	6	[91]
29.4	5.4	0.23	pulsed	2	[92]
33	5.4	0.1	pulsed	0.4	[92]
37	–	0.01	CW	1	[91]

6-2-4. CdTe Diodes

The Gunn effect, first observed by Gunn as a time variation in the current through samples of n-type GaAs when the voltage across the sample exceeded a critical value, has since been observed in n-type InP, n-type CdTe, alloys of n-GaAs and n-GaP, and in InAs. In n-type cadmium telluride (CdTe), the Gunn effect was first seen by Foyt and McWhorter [93], who observed a time variation of the current through samples 250 to 300 μm long with a carrier concentration of 5 $\times$ 10^{14} per cubic centimeter and a room temperature mobility of 1000 cm²/$V \cdot s$. Ludwig, Halsted, and Aven [94] confirmed the existence of current oscillations in *n*-CdTe, and Ludwig has further reported studies of the Gunn effect in CdTe over a wider range of sample doping levels and lengths [95]. It has been confirmed that the same mechanism—the field-induced transfer of electrons to a higher conduction band minimum (Gunn effect)—applies in CdTe just as it does in GaAs. From the two-valley model theory in CdTe, as in GaAs, the $\langle 000 \rangle$ minimum is the lowest in energy. The effective mass $m_{\text{eff}} = 0.11$m (electron mass) and the intrinsic mobility $\mu \simeq 1100$ cm²/$V \cdot s$ at room temperature. Hilsum has estimated that $\langle 111 \rangle$ minima are the next lowest in energy, being 0.51 eV higher than $\langle 000 \rangle$ minimum [96]. In comparing the Gunn effect in CdTe to that in GaAs, a major difference is the substantially higher threshold field, about 13 kV/cm for CdTe compared with about 3 kV/cm for GaAs [97]. Qualitatively, the higher threshold can be thought of as associated with the relatively strong coupling of the electrons to longitudinal optical phonons, which limits the mobility—and hence the rate of energy acquisition from the applied field—and also provides an efficient mechanism for transferring energy to the lattice, thereby minimizing the kinetic energy in the electron distribution.

The ratio of peak-to-valley current is another parameter of interest. In CdTe, as in GaAs, the spike amplitude can be as large as 50 % of the maximum total current. A similar maximum efficiency for CdTe and GaAs can be expected. Since the domain velocities in CdTe and GaAs are approximately equal, samples of the same length will operate at about the same frequency in the transit-time mode. The high threshold field of CdTe combined with its poor thermal conductivity creates a heating problem. However, if sufficiently short pulses are used so that the heat can be dissipated, the high operating field of the sample can be an advantage.

6-3 Avalanche Transit-Time Devices

Avalanche transit-time diode oscillators rely on the effect of voltage breakdown across a reverse-biased *p-n* junction to produce a supply of holes and electrons. Ever since the development of modern semiconductor device theory scientists have speculated on whether it was possible to make a two-terminal negative-resistance device. The tunnel diode was the first such device to be realized in practice. Its operation depends on the properties of a forward-biased *p-n* junction in which both the *p* and *n* regions are heavily doped. The other two devices are the transferred electron devices and the avalanche transit-time devices. In this section the latter type is discussed.

The transferred electron devices or the Gunn oscillators operate simply by the application of a dc voltage to a bulk semiconductor. There are no *p-n* junctions in this device. Its frequency is a function of the load and of the natural frequency of the circuit. The avalanche diode oscillator uses carrier impact ionization and drift in the high-field region of a semiconductor junction to produce a negative resistance at microwave frequencies. The device was originally proposed in a theoretical paper by Read [100] in which he analyzed the negative-resistance properties of an idealized *n⁺-p-i-p⁺* diode. Two distinct modes of avalanche oscillator have been observed. One is the IMPATT mode, which stands for *IMPact ionization Avalanche Transit Time* operation. In this mode the typical dc-to-RF conversion efficiency is 5 to 10%, and frequencies are as high as 100 GHz with Si diodes. The other mode is the TRAPATT mode, which represents *TRApped Plasma Avalanche Triggered Transit* operation. Its typical conversion efficiency is from 20 to 60%.

Another type of active microwave devices is the BARITT diodes (*BARrier Injected Transit-Time diodes*) [12, 101]. It has long drift regions similar to those of IMPATT diodes. The carriers traversing the drift regions of BARITT diodes, however, are generated by minority carrier injection from forward-biased junctions rather than being extracted from the plasma of an avalanche region. Several different types of device structure have been operated as BARITT diodes, such a *p-n-p*, *p-n-v-p*, *p-n*-metal, and metal-*n*-metal. BARITT diodes have low noise figures of 15 dB, but their bandwidth is relatively narrow with low output power.

6-3-1. Read Diode

(1) Physical Description. The basic operating principle of IMPATT diodes can be most easily understood by reference to the first proposed avalanche diode, the Read diode [100]. The theory of this device was presented by Read in 1958, but the first real experimental Read diode was reported by Lee et. al in 1965 [102]. A mode of the original Read diode with a doping profile and a dc electric field distribution that exists when a large reverse bias is applied across the diode is shown in Fig. 6-3-1.

The Read diode is an *n⁺-p-i-p⁺* structure, where the superscript plus sign denotes very high doping and the *i* or *v* refers to intrinsic material. The device consists essentially of two regions. One is the *n⁺-p* junction region at which avalanche multiplication occurs. This region is also called the high-field region or the avalanche region. The other is the *i* or *v* region through which the generated holes must drift in moving to the *p⁺* contact. This region is also called the intrinsic region or the drift region. The *p* region is very thin. The space between the *n⁺-p* junction and the *i-p⁺* junction is called the space-charge region. Similar devices can be built in the *p⁺-n-i-n⁺* structure, in which electrons generated from avalanche multiplication drift through the *i* region.

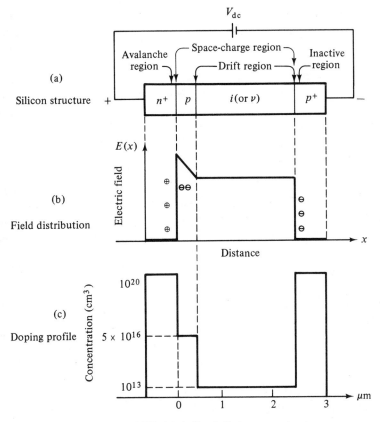

FIG. 6-3-1. Read diode.

The Read diode oscillator consisted of an n^+-p-i-p^+ diode biased in reverse and mounted in a microwave cavity. The impedance of the cavity is mainly inductive and is matched to the mainly capacitive impedance of the diode so as to form a resonant circuit. The device can produce a negative ac resistance that, in turn, delivers power from the dc bias to the oscillation.

(2) Avalanche Multiplication. When the reverse-biased voltage is well above the punchthrough or breakdown voltage, the space-charge region always extends from the n^+-p junction through the p and i regions to the i-p^+ junction. The fixed charges in the various regions are shown in Fig. 6-3-1(b). A positive charge gives a rising field in moving from left to right. The maximum field, which occurs at the n^+-p junction, is about several hundred kilovolts per centimeter. Carriers (holes) moving in the high field near the n^+-p junction acquire energy to knock valence electrons into the conduction band, thus producing hole-electron pairs. The rate of pair production, or

avalanche multiplication, is a sensitive nonlinear function of the field. By proper doping, the field can be given a relatively sharp peak so that avalanche multiplication is confined to a very narrow region at the n^+-p junction. The electrons move into the n^+ region and the holes drift through the space-charge region to the p^+ region with a constant velocity v_d of about 10^7 cm/s for silicon. The field throughout the space-charge region is above about 5 kV/cm. The transit time of a hole across the space charge region L is given by

$$\tau = \frac{L}{v_d} \qquad\qquad (6\text{-}3\text{-}1)$$

(3) *Carrier Current $I_0(t)$ and External Current $I_e(t)$.* As described previously, the Read diode is mounted in a microwave resonant circuit. An ac voltage can be maintained at a given frequency in the circuit, and the total field across the diode is the sum of the dc and ac fields. This total field causes breakdown at the n^+-p junction during the positive half of the ac voltage cycle if the field is above the breakdown voltage, and the carrier current (or the hole current in this case) $I_0(t)$ generated at the n^+-p junction by the avalanche multiplication grows exponentially with time while the field is above the critical value. During the negative half cycle, when the field is below the breakdown voltage, the carrier current $I_0(t)$ decays exponentially to a small steady-state value. The carrier current $I_0(t)$ is the current at the junction only and is in the form of a pulse of very short duration as shown in Fig. 6-3-2(d). Therefore the carrier current $I_0(t)$ reaches its maximum in the middle of the ac voltage cycle, or one-quarter of a cycle later than the voltage. Under the influence of the electric field the generated holes are injected into the space-charge region toward the negative terminal. As the injected holes traverse the drift space, they induce a current $I_e(t)$ in the external circuit as shown in Fig. 6-3-2(d).

When the holes generated at the n^+-p junction drift through the space-charge region, they cause a decrease of the field in accordance with Poisson's equation.

$$\frac{dE}{dx} = -\frac{\rho}{\epsilon} \qquad\qquad (6\text{-}3\text{-}2)$$

where ρ is the volume charge density and ϵ is the permittivity.

Since the drift velocity of the holes in the space-charge region is constant, the induced current $I_e(t)$ in the external circuit is simply equal to

$$I_e(t) = \frac{Q}{\tau} = \frac{v_d Q}{L} \qquad\qquad (6\text{-}3\text{-}3)$$

where Q = the total charge of the moving holes
            v_d = the hole drift velocity
            L = the length of the space-charge region

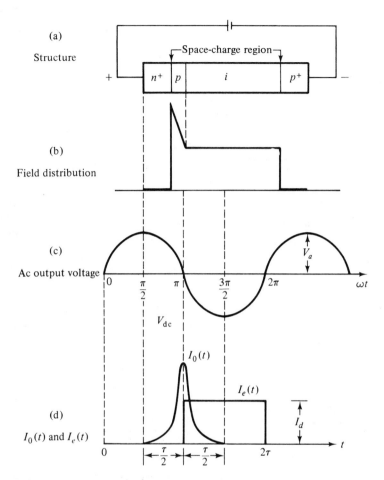

FIG. 6-3-2. Field, voltage and currents in Read diode. (*After Read [100]; reprinted by permission of the Bell System, AT&T Co.*)

It can be seen that the induced current $I_e(t)$ in the external circuit is equal to the average current in the space-charge region. When the pulse of hole current $I_0(t)$ is suddenly generated at the n^+-p junction, a constant current $I_e(t)$ starts flowing in the external circuit and continues to flow during the time τ in which the holes are moving across the space-charge region. Thus, on the average, the external current $I_e(t)$ due to the moving holes is delayed by $\tau/2$ or 90° relative to the pulsed carrier current $I_0(t)$ generated at the n^+-p junction. Since the carrier $I_0(t)$ is delayed by one-quarter of a cycle or 90° relative to the ac voltage, the external current $I_e(t)$ is then delayed by 180° relative to the voltage as shown in Fig. 6-3-2(d). Therefore the cavity should be tuned to give a resonant frequency as

$$2\pi f = \frac{\pi}{\tau}$$

then

$$f = \frac{1}{2\tau} = \frac{v_d}{2L} \qquad (6\text{-}3\text{-}4)$$

Since the output ac voltage and the external current $I_e(t)$ are out of phase by 180°, negative conductance occurs and the Read diode can be used for microwave oscillation and amplification. For example, taking $v_d = 10^7$ cm/s for silicon, the optimum operating frequency for a Read diode with an i-region length of 2.5 μm is 20 GHz.

(4) **Output Power and Quality Factor Q.** The external current $I_e(t)$ approaches a square wave, being very small during the positive half cycle of the ac voltage and almost constant during the negative half cycle. Since the direct current I_d supplied by the dc bias is the average external current or conductive current, it follows that the amplitude of variation of $I_e(t)$ is approximately equal to I_d. If V_a is the amplitude of the ac voltage, the ac power delivered is found to be

$$P = 0.707 V_a I_d \qquad \text{watts/unit area} \qquad (6\text{-}3\text{-}5)$$

The quality factor Q of a circuit is defined as

$$Q = \omega \frac{\text{average stored energy}}{\text{average dissipated energy}} \qquad (6\text{-}3\text{-}6)$$

Since the Read diode supplies ac energy, it has a negative Q in contrast to the positive Q of the cavity. At the stable operating point, the negative Q of the diode is equal to the positive Q of the cavity circuit. If the amplitude of the ac voltage increases, the stored energy, or energy of oscillation, increases faster than the energy delivered per cycle. This is the condition required in order for a stable oscillation to be possible.

6-3-2. IMPATT Diodes

(1) **Characteristics.** A theoretical Read diode made of a n^+-p-i-p^+ or p^+-n-i-n^+ structure has been analyzed. Its basic physical mechanism is the interaction of the impact ionization avalanche and the transit time of charge carriers. Hence the Read-type diodes are called IMPATT diodes. These diodes exhibit a differential negative resistance by two effects.

1. The impact ionization avalanche effect, which causes the carrier current $I_0(t)$ and the ac voltage to be out of phase by 90°.
2. The transit-time effect, which further delays the external current $I_e(t)$ relative to the ac voltage by 90°.

The first IMPATT operation as reported by Johnston et al. [103] in 1965, however, was obtained from a simple *p-n* junction. The first real Read-type IMPATT diode was reported by Lee et al. [102], as described previously. From the small-signal theory developed by Misawa [104] it has been confirmed that a negative resistance of the IMPATT diode can be obtained from a junction diode with any doping profile. Many IMPATT diodes consist of a high doping avalanching region followed by a drift region where the field is low enough that the carriers can traverse through it without avalanching. The Read diode is the basic type in the IMPATT diode family. The others are the one-sided abrupt *p-n* junction, the linearly graded *p-n* junction, and the *p-i-n* diode, all of which are shown in Fig. 6-3-3. The principle of operation of these devises, however, is essentially similar to the mechanism described for the Read diode.

(2) *Negative Resistance.* Small-signal analysis of a Read diode results in the following expression for the real part of the diode terminal impedance [104]:

$$R = R_s + \frac{2L_d^2}{v_d \epsilon A} \frac{1}{1 - (\omega^2/\omega_r^2)} \frac{1 - \cos\theta}{\theta^2} \qquad (6\text{-}3\text{-}7)$$

where R_s = the passive resistance of the inactive region
v_d = the carrier drift velocity
L_d = the length of the drift space-charge region
A = the diode cross section
ϵ = dielectric permittivity
θ = the transit angle, given by

$$\theta = \omega\tau = \omega\frac{L_d}{v_d} \qquad (6\text{-}3\text{-}8)$$

and ω_r is the avalanche resonant frequency, defined by

$$\omega_r \equiv \frac{(2\alpha' v_d I_0)^{1/2}}{\epsilon A} \qquad (6\text{-}3\text{-}9)$$

In Eq. (6-3-9) the quantity α' is the derivative of the ionization coefficient with respect to the electric field. This coefficient, the number of ionizations per centimeter produced by a single carrier, is a sharply increasing function of the electric field. The variation of the negative resistance with the transit angle when $\omega > \omega_r$ is plotted in Fig. 6-3-4. The peak value of the negative resistance occurs near $\theta = \pi$. For transit angles larger than π and approaching $3\pi/2$, the negative resistance of the diode increases rapidly. For practical purposes, the Read-type IMPATT diodes work well only in a frequency range around the π transit angle. That is,

$$f = \frac{1}{2\tau} = \frac{v_d}{2L} \qquad (6\text{-}3\text{-}10)$$

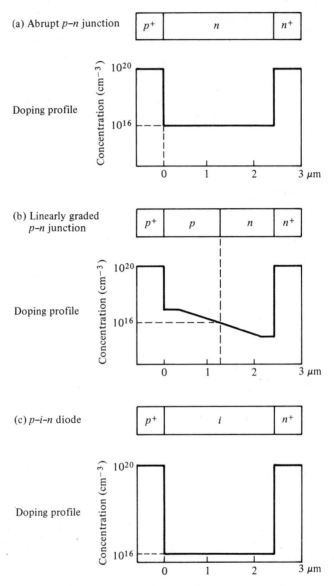

FIG. 6-3-3. Three typical silicon IMPATT diodes. (*After R. L. Johnston et al. [103]; reprinted by permission of the Bell System, AT&T Co.*)

(3) *Power Output and Efficiency.* At a given frequency the maximum output power of a single diode is limited by semiconductor materials and the attainable impedance levels in microwave circuitry. For a uniform avalanche, the maximum voltage that can be applied across the diode is given by

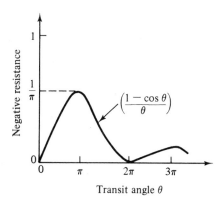

FIG. 6-3-4. Negative resistance versus transit angle.

$$V_m = E_m L \qquad (6\text{-}3\text{-}11)$$

where L is the depletion length and E_m is the maximum electric field. This maximum applied voltage is limited by the breakdown voltage. Furthermore, the maximum current that can be carried by the diode is also limited by the avalanche breakdown process, for the current in the space-charge region causes an increase in the electric field. The maximum current is given by

$$I_m = J_m A = \sigma E_m A = \frac{\epsilon}{\tau} E_m A = \frac{v_d \epsilon E_m A}{L} \qquad (6\text{-}3\text{-}12)$$

Therefore the upper limit of the power input is given by

$$P_m = I_m V_m = E_m^2 \epsilon v_d A \qquad (6\text{-}3\text{-}13)$$

The capacitance across the space-charge region is defined as

$$C = \frac{\epsilon A}{L} \qquad (6\text{-}3\text{-}14)$$

Substitution of Eq. (6-3-14) in Eq. (6-3-13) and application of $2\pi f \tau \doteq 1$ yield

$$P_m f^2 = \frac{E_m^2 v_d^2}{4\pi^2 X_c} \qquad (6\text{-}3\text{-}15)$$

It is interesting to note that this equation is identical to Eq. (6-1-27) of the power-frequency limitation for the microwave power transistor. The maximum power that can be given to the mobile carriers decreases as $1/f^2$. For silicon, this electronic limit is dominant at frequencies as high as 100 GHz. The efficiency of the IMPATT diodes is given by

$$\eta = \frac{P_{ac}}{P_{dc}} = \left(\frac{V_a}{V_d}\right)\left(\frac{I_a}{I_d}\right) \qquad (6\text{-}3\text{-}16)$$

For an ideal Read-type IMPATT diode, the ratio of the ac voltage to the applied voltage is about 0.5 and the ratio of the ac current to the dc current is about $2/\pi$, so that the efficiency would be about $1/\pi$ or over 30%. However, for practical IMPATT diodes, the efficiency is usually less than 30% due to the space-charge effect, the reverse-saturation-current effect, the high-frequency-skin effect, and the ionization-saturation effect.

IMPATT diodes are at present the most powerful CW solid-state microwave power sources. The diodes have been fabricated from Ge, Si, and GaAs and can probably be constructed from other semiconductors as well. IMPATT diodes provide potentially reliable, compact, inexpensive, and moderately efficient microwave power sources. Table 6-3-1 shows the present performance of IMPATT diodes.

TABLE 6-3-1.   Best Performance of IMPATT Diodes.

Frequency (GHz)	Power output (W)	Efficiency (%)	Operation	Material	Reference
0.425	435.0	22.0	pulsed	Si	[105]
0.450	450.0	43.0	CW	Ge	[106]
0.775	180.0	60.0	pulsed	Si	[105]
1.05	420.0	33.0	pulsed	Si	[105]
3.00	7.5	40.0	pulsed	Ge	[107]
6.00	0.62	12.1	CW	Ge	[108]
6.50	1.30	3.6	pulse	GaAs	[109]
6.80	0.38	7.8	pulse	GaAs	[109]
8.00	30.00	4.5	pulse	Si	[110]
13.8	2.70	10.2	CW	Si	[111]
16.5	2.10	5.0	pulsed	GaAs	[112]
21.0	8.80	4.50	pulsed	Si	[110]
27.0	0.40	2.00	pulsed	GaAs	[113]
50.0	0.64	10.00	CW	Si	[114]
107.0	0.74	3.20	CW	Si	[114]

6-3-3. TRAPATT Diodes

(1) Introduction. The abbreviation TRAPATT stands for *TRApped Plasma Avalanche Triggered Transit* mode, a mode first reported by Prager et. al [120]. It is a high-efficiency microwave generator capable of operating from several hundred megahertz to several gigahertz. The basic operation of the oscillator is a semiconductor *p-n* junction diode reverse-biased to current densities well in excess of those encountered in normal avalanche operation. High-peak-power diodes are typically silicon n^+-p-p^+ (or p^+-n-n^+) structures with the *n*-type depletion region width varying from 2.5 to 12.5 μm. The doping of the depletion region is generally such that the diodes are well "punched through" at breakdown; that is, the dc electric field in the depletion region just prior to breakdown is well above the saturated drift-velocity level. The device p^+ region is kept as thin as possible at 2.5 to 7.5 μm. The device diameters range from as small as 50 μm for CW operation to 750 μm at lower frequency for high-peak-power devices.

(2) **Principles of Operation.** Approximate analytical solutions for the TRAPATT mode in p^+-n-n^+ diodes have been developed by Clorfeine [121] and DeLoach [122] et. al. These analyses have shown that a high-field avalanche zone propagates through the diode and fills the depletion layer with a dense plasma of electrons and holes that become trapped in the low-field region behind the zone. A typical voltage waveform for the TRAPATT mode of an avalanche p^+-n-n^+ diode operating with as assumed square-wave current drive is shown in Fig. 6-3-5. At point A the electric field is uniform throughout the sample and its magnitude is large but less than the value required for avalanche breakdown. The current density is expressed by

$$J = \epsilon_s \frac{dE}{dt} \qquad (6\text{-}3\text{-}17)$$

where ϵ_s is the dielectric permittivity of the diode.

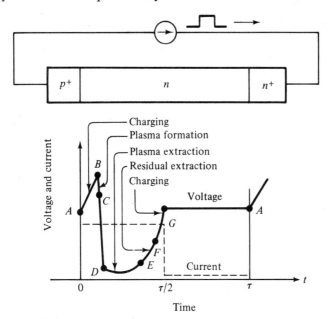

FIG. 6-3-5. Voltage and current waveforms for TRAPATT diode. *(After A. S. Clorfeine et al. [121]; reprinted by permission of RCA Laboratory.)*

At the instant of time at point A, the diode current is turned on. Since the only charge carriers present are those due to the thermal generation, the diode initially charges up like a linear capacitor, driving the magnitude of the electric field above the breakdown voltage. When a sufficient number of carriers is generated, the particle current exceeds the external current and the electric field is depressed throughout the depletion region, causing the voltage to decrease. This portion of the cycle is shown by the curve from point B to point C. During this time interval the electric field is sufficiently large for the avalanche to continue, and a dense plasma of electrons and holes is created. As some of the electrons and holes drift out of the ends of the depletion layer, the field is further depressed and "traps" the remaining plasma. The voltage decreases to point D. A long time is required to remove the plasma because the total plasma charge is large compared to the charge per unit time in the external current. At point E the plasma is removed, but a residual charge of electrons remains in one end of the depletion layer and a residual charge of holes in the other end. As the residual charge is removed, the voltage increases from point E to point F. At point F all the charge that was generated internally has been removed. This charge must be greater than or equal to that supplied by the external current; otherwise the voltage will exceed that at point A. From point F to point G the diode charges up again like a fixed capacitor. At point G the diode current goes to zero for half a period and the voltage remains constant at V_A until the current comes back on and the cycle repeats. The electric field can be expressed as

$$E(x, t) = E_m - \frac{qN_A}{\epsilon_s^2}X + \frac{Jt}{\epsilon_s} \qquad (6\text{-}3\text{-}18)$$

where N_A is the doping concentration of the n region and X is the distance.

Thus the value of t at which the electric field reaches E_m at a given distance X into the depletion region is obtained by setting $E(x, t) = E_m$, yielding

$$t = \frac{qN_A}{J\epsilon_s}X \qquad (6\text{-}3\text{-}19)$$

Differentiation of Eq. (6-3-19) with respect to time t results in

$$v_z \equiv \frac{dx}{dt} = \frac{J\epsilon_s}{qN_A} \qquad (6\text{-}3\text{-}20)$$

where v_z is the avalanche-zone velocity.

For example, for $N_A = 10^{15}$ per cubic centimeter and $J = 10^4$ A/cm², $v_z = 6 \times 10^7$ cm/s, which is much larger than the scattering-limited velocity. So the avalanche zone (or avalanche shock front) will quickly sweep across most of the diode, leaving the diode filled by a highly conducting plasma of holes and electrons whose space charge depresses the voltage to low values. Because of the dependence of the drift velocity on the field at low fields, the electrons and holes will drift at velocities determined by the low-field mobilities, and the transit time of the carriers can become much longer than

$$\tau_s = \frac{L}{v_s} \qquad (6\text{-}3\text{-}21)$$

where v_s is the saturated carrier drift velocity.

Thus the TRAPATT mode can operate at comparatively low frequencies, since the discharge time of the plasma—that is, the rate Q/I of its charge to its current—can be considerably greater than the nominal transit time τ_s of the diode at high field. Therefore the TRAPATT mode is still a transit-time mode in the real sense that the time delay of carriers in transit (i.e., the time between injection and collection) is utilized to obtained a current phase shift favorable for oscillation.

(*3*) *Power Output and Efficiency.* RF power is delivered by the diode to an external load when the diode is placed in a proper circuit with a load. The main function of this circuit is to match the diode effective negative resistance to the load at the output frequency while reactively terminating (trapping) frequencies above the oscillation frequency in order to ensure TRAPATT operation. To date, the highest pulse power of 1.2 kW has been obtained at 1.1 GHz (five diodes in series) [123], and the highest efficiency of 75% has been achieved at 0.6 GHz [124]. Table 6-3-2 shows the current state of TRAPATT diodes [125].

TABLE 6-3-2.* TRAPATT Oscillator Capabilities.

Frequency (GHz)	Peak power (W)	Average power (W)	Operating voltage (V)	Efficiency (%)
0.5	600	3	150	40
1.0	200	1	110	30
	400	2	110	35
2.0	100	1	80	25
	200	2	80	30
4.0	100	1	80	20
8.0	50	1	60	15

*After W. E. Wilson [125]; reprinted with permission from Horizon House.

The TRAPATT operation, however, is a rather complicated means of oscillation and requires good control of both device and circuit properties. In addition, the TRAPATT mode generally exhibits a considerably higher noise figure than the IMPATT mode, and the upper operating frequency appears to be practically limited to below the millimeter wave region.

6-3-4. BARITT Diodes

(*1*) *Introduction.* BARITT diodes, meaning *BARrier Injected Transit-Time diodes*, are the newest addition to the family of active microwave diodes. They have long drift regions similar to those of IMPATT diodes. The carriers traversing the drift regions of BARITT diodes, however, are generated by minority carrier injection from forward-biased junctions instead of being extracted from the plasma of an avalanche region.

Several different types of device structures have been operated as
BARITT diodes, including *p-n-p*, *p-n-v-p*, *p-n*-metal, and metal-*n*-metal. For
a *p-n-v-p* BARITT diode, the forward-biased *p-n* junction emits holes into
the *v* region. These holes drift with saturation velocity through the *v* region
and are collected at the *p*-contact. The diode will exhibit a negative resistance
for transit angles between π and 2π. The optimum transit angle is approximately 1.6π [12].

Such diodes are much less noisy than IMPATT diodes. Noise figures
are as low as 15 dB at *C*-band frequencies with silicon BARITT amplifiers.
The major disadvantages of BARITT diodes are relatively narrow bandwidth
and power outputs limited to a few milliwatts.

(2) *Principles of Operation.* A crystal *n*-type silicon wafer with 11 Ω-cm
resistivity and 4×10^{14} per cubic centimeter doping is made of a 10-μm thin
slice. Then the *n*-type silicon wafer is sandwiched between two PtSi Schottky
barrier contacts of about 1000 Å thickness. A schematic diagram of a metal-
n-metal structure is shown in Fig. 6-3-6(a).

The energy band diagram at thermal equilibrium is shown in Fig.
6-3-6(b), where ϕ_{n1} and ϕ_{n2} are the barrier heights for the metal-semiconductor contacts, respectively. For the PtSi—Si—PtSi structure mentioned
previously, $\phi_{n1} = \phi_{n2} = 0.85$ eV. The hole barrier height ϕ_{p2} for the forward-
biased contact is about 0.15 eV. Figure 6-3-6(c) shows the energy band
diagram when a voltage is applied. The mechanisms responsible for the
microwave oscillations are derived from

1. the rapid increase of the carrier injection process caused by the
   decreasing potential barrier of the forward-biased metal-semiconductor contact.
2. an apparent $3\pi/2$ transit angle of the injected carrier that traverses
   the semiconductor depletion region.

The rapid increase in terminal current with applied voltage (above 30
volts) as shown in Fig. 6-3-7 is caused by thermionic hole injection into the
semiconductor as the depletion layer of the reverse-biased contact reaches
through the entire device thickness. The critical voltage is approximately
given by

$$V_c = \frac{qNL^2}{2\epsilon_s} \qquad (6\text{-}3\text{-}22)$$

where $N =$ the doping concentration
   $L =$ the semiconductor thickness
   $\epsilon_s =$ the dielectric permittivity

The current-voltage characteristics of the silicon MSM structure (PtSi
—Si—PtSi) were measured at 77° and 300°K. The device parameters are

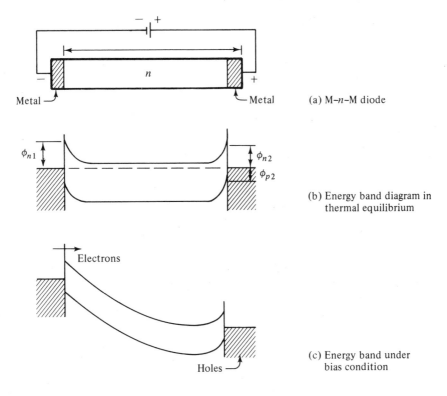

(a) M–*n*–M diode

(b) Energy band diagram in thermal equilibrium

(c) Energy band under bias condition

FIG. 6-3-6. M-n-M diode. (*After D. J. Coleman and S. M. Sze [130]; reprinted by permission of the Bell System, AT&T Co.*)

$L = 10 \ \mu\text{m}$, $N = 4 \times 10^{14} \ \text{cm}^{-3}$, $\phi_{n1} = \phi_{n2} = 0.85 \ \text{eV}$, and area $= 5 \times 10^{-4} \ \text{cm}^2$.

The current increase is not due to avalanche multiplication, as is apparent from the magnitude of the critical voltage and its negative temperature coefficient. At 77°K the rapid increase is stopped at a current of about 10^{-5} A. This saturated current is expected in accordance with the termionic emission theory of hole injection from the forward-biased contact with a hole barrier height (ϕ_{p2}) of about 0.15 eV.

(3) Microwave Performance. Continuous-wave (CW) microwave performance of the M-*n*-M-type BARITT diode was obtained over the entire *C* band of 4 to 8 GHz. The maximum power observed was 50 mW at 4.9 GHz. The maximum efficiency was about 1.8%. The FM single-sideband noise measure at 1 MHz was found to be 22.8 dB at a 7-mA bias current. This noise measure is substantially lower than that of a silicon IMPATT diode and is comparable to that of a GaAs transfer-electron oscillator. Figure 6-3-8 shows some of the measured microwave power versus current with

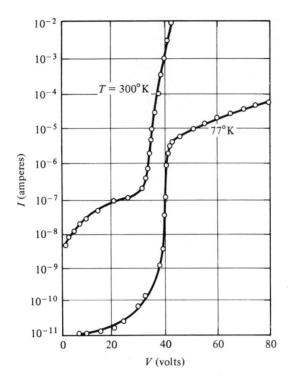

FIG. 6-3-7. Current versus voltage of a BARITT diode (Pt-Si-PtSi). (*After D. J. Coleman and S. M. Sze [130]; reprinted by permission of the Bell System, AT&T Co.*)

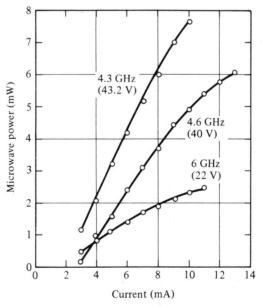

FIG. 6-3-8. Power output versus current for three Si M-n-M devices. (*After D.J. Coleman and S.M. Sze [130]; reprinted with permission of the Bell System, AT&T Co.*)

frequency of operation indicated on each curve for three typical devices tested.

The voltage marked in parenthesis for each curve indicates the average bias voltage at the diode while the diode is in oscillation. The gain-bandwith product of a 6-GHz BARITT diode was measured to be 19-dB gain at 5-mA bias current at 200 MHz. The small-signal noise measure was about 15 dB.

6-4 Quantum-Electronic Solid-State Devices

The word *maser* is an acronym for Microwave Amplification by Stimulated Emission of Radiation, and the word *laser* stands for Light Amplification by Stimulated Emission of Radiation. Unlike the operation of ordinary microwave tubes, such as the klystron and traveling-wave tubes, and semiconductor microwave devices, such as Gunn diodes and avalanche diodes, which can be analyzed in terms of classic mechanics, the electrical behavior of masers and lasers can be described only in terms of quantum electronics and statistical mechanics. In other words, the interaction of electromagnetic fields in masers and lasers results in the bound aggregates of charges rather than in free charges, as in the case of electron beams in microwave tubes. Masers and lasers are highly directional, coherent power-source devices used for the generation and amplification of radiation. Their noise figure is extremely low, and they have found wide usage in medicine, communications, space exploration, and military and metals technology. Since 1954 many types of masers and lasers have been invented and built. However, this section deals only with the ruby maser and the semiconductor lasers.

6-4-1. Introduction

(1) Evolution of Masers and Lasers. The basis of masers and lasers rests on Einstein's discovery of stimulated emission. This discovery led to a search for methods for establishing population inversion in a suitable material and hence to the construction of a radiation amplifier. In 1954 Townes and his students at Columbia University built the first microwave amplifier and oscillator based on the stimulated emission principle [135]. In 1956 Bloembergen proposed a three-level solid-state maser [136], and the first successful such maser was achieved in 1957 by Scovil, Feher, and Seidel with the paramagnetic Gd^{3+} ion at 9 GHz [137]. The first ruby laser was constructed in 1960 by Maiman [138]. It was quickly followed by the four-level lasers of Sorokin and Stevenson [139] and the He-Ne laser of Javan [140].

The common feature of masers and lasers is the use of the conversion of atomic or molecular energy to electromagnetic radiation by means of the process known as stimulated emission of radiation. When the wavelength of the emitted radiation is in the vicinity of 1 cm, the devices are called masers. When the radiated frequency approaches the visible or nearly visible range at 10^5 GHz, the devices are called optical masers or lasers. There is no sharp limit separating the regions in which the terms maser and laser are used. Both optical and microwave frequencies are shown in Table 6-4-1 and Fig. 6-4-1.

In 1962 Dumke showed that laser action was indeed possible in direct band-gap semiconductors, such as gallium arsenide (GaAs) [141]. The pulsed radiation at 8400 Å was obtained from a liquid-nitrogen-cooled, forward-biased GaAs p-n junction. Since then many new laser materials have been found. The wavelength of coherent radiation has been extended through the visible into the ultraviolet and out to the midinfrared region [142–144].

(2) *Transition Processes.* Three basic transition processes relate to the operation of masers and lasers—absorption, spontaneous emission, and stimulated emission. Figure 6-4-2 shows the three transition processes between two energy levels E_1 and E_2. The black dots indicate the state of the electrons. The initial state is at the left; and the final state, after the process has occurred, is at the right. The energy level E_1 is the ground state, and E_2 is the excited state. Any transition between these two states involves, according to Planck's law, the emission or absorption of a photon of energy with frequency f_{12} given by

$$hf_{12} = E_2 - E_1 \qquad (6\text{-}4\text{-}1)$$

where h is Planck's constant, which is 6.625×10^{-34} joule-s.

At room temperature most of the electrons are in the ground state. When a photon of energy exactly equal to hf_{12} impinges the atoms, an electron in state E_1 will absorb the photon and thereby move to the excited state E_2. This is the absorption process shown in Fig. 6-4-2(a). The excited state of the electron is unstable, and the excited electron will, after a short time and without any external stimulus, transit to the ground state by giving off a photon of energy hf_{12}. This process is called spontaneous emission and is shown in Fig. 6-4-2(b). The lifetime for spontaneous emission (i.e., the average time of the excited state) varies from 10^{-9} and 10^{-3} s, depending on various semiconductor parameters, such as bandgap and density of recombination centers. When a photon of energy hf_{12} impinges on an electron while it is still in the excited state E_2, the electron will be immediately stimulated to make its transition back to the ground state E_1 by giving off a photon of energy hf_{12}. This process is called stimulated emission and is shown in Fig. 6-4-2(c). The wave of the stimulated photon is in phase with the radiation field. If this process continues and other electrons are stimulated to emit

TABLE 6-4-1. Electromagnetic Spectrum.

I. Electromagnetic Waves

EM waves	Frequency		Wavelength	
Very long waves	3–30	kHz	100–10	km
Long waves	30–300	kHz	10–1	km
Medium waves	300–3000	kHz	1–0.1	km
Short waves	3–30	MHz	100–10	m
Ultrashort waves (meter waves)	30–300	MHz	10–1	m
Microwave (centimeter waves)	0.3–30	GHz	100–1	cm
Ultramicrowave (millimeter waves)	30–3000	GHz	10–0.1	mm
Infrared ray	0.003–0.429	MGHz	100–0.7	μm
Visible light	0.429–0.698	MGHz	0.7–0.43	μm
Ultraviolet ray	0.698–100	MGHz	0.43–0.003	μm
x ray	0.01–1000	GGHz	300–0.003	Å
Gamma ray	$1 \simeq 1$	MGGHz	3–1	μÅ

II. Radio Frequencies

Bands	Frequency		Wavelength
VLF (very low frequency)	3–30	kHz	100–10 km
LF (low frequency)	30–300	kHz	10–1 km
MF (medium frequency)	300–3000	kHz	1–0.1km
HF (high frequency)	3–30	MHz	100–10 m
VHF (very high frequency)	30–300	MHz	10–1 m
UHF (ultrahigh frequency)	300–3000	MHz	100–10 cm
SHF (superhigh frequency)	3–30	GHz	10–1 cm
EHF (extreme high frequency)	30–300	GHz	10–1 mm

III. Light Frequencies

Colors	Frequency	Wavelength	
Infrared	3×10^{12}–4.29×10^{14} Hz	100–0.7	μm
Red	4.29×10^{14}–4.92×10^{14} Hz	0.7–0.61	μm
Orange	4.92×10^{14}–5.08×10^{14} Hz	0.61–0.59	μm
Yellow	5.08×10^{14}–5.26×10^{14} Hz	0.59–0.57	μm
Green	5.26×10^{14}–6.00×10^{14} Hz	0.57–0.50	μm
Blue	6.00×10^{14}–6.67×10^{14} Hz	0.50–0.45	μm
Violet	6.67×10^{14}–6.98×10^{14} Hz	0.45–0.43	μm
Ultraviolet	6.98×10^{14}–1×10^{17} Hz	0.43–0.003	μm

IV. Band Designation

Frequency (GHz)	0.1	0.2	0.3	0.5	1	2	3	4	6	8	10	20	30	40	60	100
Old designation	VHF		UHF		L		S		C		X	Ku	K	Ka	millimeter	
New designation	A		B		C	D	E	F	G	H	I	J		K	L	M

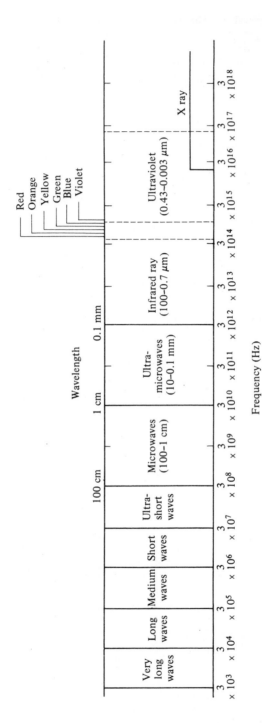

FIG. 6-4-1. Electromagnetic spectrum.

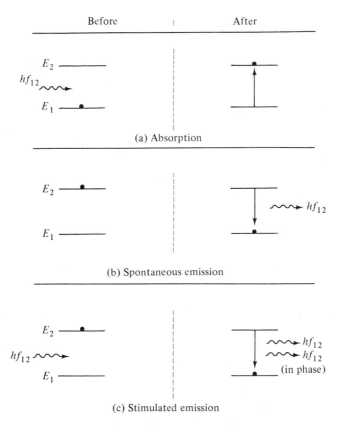

FIG. 6-4-2. Transition processes of electrons.

photons in the same fashion, a large radiation can build up. This radiation will be monochromatic, since each photon will have an energy of exactly one hf_{12}, and it will be coherent because all the photons emitted are in phase and reinforcing each other. This process of stimulated emission can be explained quantum mechanically to relate the probability of emission to the intensity of the radiation field.

For two discrete energy levels E_2 and E_1 with $E_2 > E_1$, classical gas atoms follow a Boltzmann distribution. The instantaneous population n_2 of atoms in state E_2 is related to the instantaneous population n_1 of atoms in state E_1 at thermal equilibrium by

$$\frac{n_2}{n_1} = \frac{\exp\left[-E_2/(kT)\right]}{\exp\left[-E_1/(kT)\right]} = \exp\left[-\frac{(E_2 - E_1)}{kT}\right] = \exp\left(\frac{-hf_{12}}{kT}\right) \quad (6\text{-}4\text{-}2)$$

In Eq. (6-4-2) it is assumed that the two levels have an equal number of states. The exponential term $\exp\left[-(E_2 - E_1)/(kT)\right]$ is conventionally called the *Boltzmann factor*. Since $E_2 > E_1, n_1 \gg n_2$. That is, most electrons are in the lower energy level as expected. The total transition rate from state E_2 to state E_1 in the presence of the field is given by Fig. 6-4-2(b) and (c):

$$\alpha_{21} = B_{21}n_2\rho(f) + A_{21}n_2 \qquad (6\text{-}4\text{-}3)$$

$$\underset{\substack{\text{Total} \\ \text{transition}}}{} = \underset{\substack{\text{stimulated} \\ \text{emission}}}{} + \underset{\substack{\text{spontaneous} \\ \text{emission}}}{}$$

where $\rho(f) = \dfrac{8\pi h f^3}{c^3} \dfrac{1}{e^{hf/(kT)} - 1}$ is the energy density per unit frequency

$c = 3 \times 10^8$ m/s is the velocity of light in vacuum

and B_{21} and A_{21} are constants to be determined.

At steady state the transition rate from level 2 to level 1 is equal to that from level 1 to level 2; so

$$B_{12}n_1\rho(f) = B_{21}n_2\rho(f) + A_{21}n_2 \qquad (6\text{-}4\text{-}4)$$

$$\text{Absorption} = \underset{\substack{\text{stimulated} \\ \text{emission}}}{} + \underset{\substack{\text{spontaneous} \\ \text{emission}}}{}$$

This relation was described by Einstein, and the coefficients B_{12}, A_{21}, and B_{21} are therefore called the *Einstein coefficients*. There are two special cases of interest.

Case I: At thermal equilibrium the ratio of the stimulated to spontaneous emission rate is

$$\frac{\text{Stimulated emission}}{\text{Spontaneous emission}} = \frac{B_{21}}{A_{21}}\rho(f) \qquad (6\text{-}4\text{-}5)$$

In order to have a large ratio of the stimulated emission over spontaneous emission, the photon field energy density $\rho(f)$ must be large. In the laser, this can be achieved by providing an optical resonant cavity in which the photon energy density can build up to a large value.

Case II: The ratio of the stimulated emission to absorption is given by

$$\frac{\text{Stimulated emission}}{\text{Absorption}} = \frac{B_{21}}{B_{12}} \frac{n_2}{n_1} \qquad (6\text{-}4\text{-}6)$$

This equation indicates that in order to have a large ratio of the stimulated emission rate over the absorption rate, $n_2 \gg n_1$ is required. This condition is called *population inversion* and is discussed in the next section. The list of semiconductor materials that exhibit laser action continues to grow. Table 6-4-2 shows several well-known semiconductor laser materials with their characteristics [146].

(3) *Population Inversion.* As noted, at the normal condition, most electrons are in the low energy level—that is, $n_1 \gg n_2$. However, the necessary condition for the stimulated emission rate over the absorption rate is $n_2 \gg n_1$. That is, the population is said to be inverted when there are more elec-

TABLE 6-4-2.* Semiconductor Laser Materials.

Material	Band-gap energy (eV)	Wavelength (Å)	Doping	Method of excitation
GaAs	1.43	8,680	n, p	p-n junction, electron beam
GaSb	0.7	17,700	n, p	p-n junction, electron beam
InAs	0.36	34,500	n, p	p-n junction
InP	1.26	9,850	n, p	p-n junction
InSb	0.18	69,000	n, p	p-n junction, electron beam
CdTe	1.44	8,620	n, p	electron beam
CdSe	1.67	7,430	n	electron beam
CdS	2.41	5,150	n	electron beam
ZnO	3.3	3,760	n	electron beam
ZnS	3.6	3,450	n	electron beam

*After M. R. Lorenz and M. H. Pilkuhn [146]; reprinted by permission of the IEEE, Inc.

trons in excited state E_2 than in ground state E_1. If photons of energy hf_{12} are incident on a device where the population level E_2 is inverted with respect to level E_1, stimulated emission will exceed absorption and the number of photons of energy hf_{12} leaving the device will be greater than that entering it. Such a phenomenon is called *quantum amplification*.

The inversion condition for a semiconductor laser can be analyzed by means of Fig. 6-4-3. Figure 6-4-3(a) shows the equilibrium condition at $T = 0°K$ for an intrinsic semiconductor, in which the shaded area represents

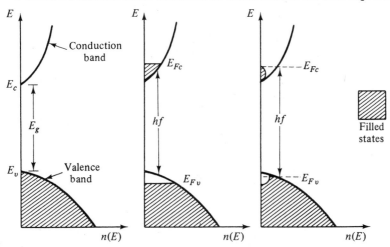

FIG. 6-4-3. Energy versus density of states in a semiconductor. (*After M. I. Nathan [147]; reprinted by permission of the IEEE, Inc.*)

the filled states. Figure 6-4-3(b) shows the situation for an inverted population at $0°K$. This inversion can be achieved by a pumping force with photon energy greater than the band gap E_g. The valence band is devoid of electrons down to an energy level E_{Fv} (Fermi level in the valence band), and the conduction band is filled up to the level E_{Fc} (Fermi level in the conduction band). Then photons with energy hf such that $E_g < hf < (E_{Fc} - E_{Fv})$ will cause downward transition and hence stimulated emission. At some finite temperatures the carrier distributions will be distorted in energy as shown in Fig. 6-4-3(c). For conduction-to-valence-band transitions in an intrinsic semiconductor, the necessary condition for stimulated emission to be dominant over absorption is

$$hf < (E_{Fc} - E_{Fv}) \qquad (6\text{-}4\text{-}7)$$

After the completion of the first ammonia maser by Townes in 1954, several new methods were proposed for obtaining population inversion in various materials. The best-known method used three levels of energy bands. A high-frequency signal incident upon the material excites the atomic system from ground state E_1 to the top excited state E_3. In this way, a population inversion can be created between the excited state E_3 and the metastable state E_2 or possibly between state E_2 and ground state E_1. This principle was first proposed by Basov and Prokhorov in 1955 [148]. Shortly thereafter Bloembergen discovered the principle independently and suggested its application to paramagnetic ions in crystals [136]. A three-level pumping system is shown in Fig. 6-4-4.

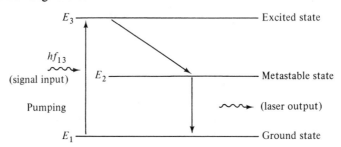

FIG. 6-4-4. Three-level pumping diagram.

In addition to the p-n junction method of excitation as described previously, other means of pumping, such as optical pumping, electron beam pumping, and avalanche breakdown pumping, are also widely used. Guillaume and Debever first utilized an electron beam with an energy level of 50 keV to bombard the semiconductor crystal [149]. A typical setup is shown in Fig. 6-4-5. The electrons in the valence band are "pumped" into the conduction band by the impinging electrons. For each electron thus excited, a hole is created so that even in a pure semiconductor the result is a degenerate electron-hole population. As the electron beam penetrates the material,

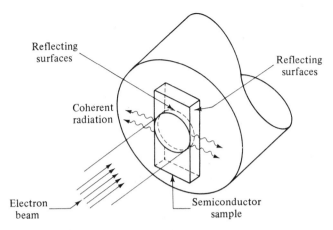

FIG. 6-4-5. Electron beam pumping scheme. (*After S. M. Sze [10]; reprinted by permission of John Wiley and Sons, Inc.*)

the electron-hole pairs form degenerate populations. Thus spontaneous emission builds up until a threshold condition is reached and lasing occurs.

(4) The Resonant Cavity. The required structure in order for laser action to occur is the resonant cavity. In the cavity, multiple reflections of light intensity are produced to achieve the laser output. Several types of resonant cavities exist, such as the Fabry–Perot cavity, cylindrical cavity, triangular cavity, and rectangular or nondirectional cavity (see in Fig. 6-4-6).

The widely used resonant cavity for producing a laser beam is the Fabry–Perot cavity shown in Fig. 6-4-6(a). A pair of parallel faces that are perpendicular to the plane of the *p-n* junction are cleaved or polished with a highly reflective material, such as aluminum (Al) or silver (Ag). The top and the bottom sides that are parallel to the junction are metalized with a conducting material. The remaining two faces are roughened or sawed to suppress the radiation. The two parallel-polished planes are used to produce multiple reflections in the cavity in order to achieve a very high intensity of laser beam. Figure 6-4-7 shows multiple reflections within a resonant cavity.

If resonance is to occur in a laser cavity, the distance between the two parallel-polished faces must be equal to an integral number of half-wavelengths in the cavity. That is,

$$L = m\frac{\lambda}{2} \qquad (6\text{-}4\text{-}8)$$

where m = any integer

$\lambda = \dfrac{c}{f}$ is the photon wavelength within the laser material

$c = 3 \times 10^8$ m/s is the velocity of light in vacuum

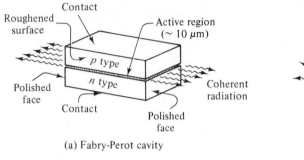

(a) Fabry-Perot cavity

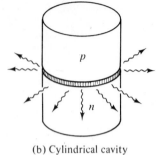

(b) Cylindrical cavity

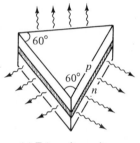

(c) Triangular cavity

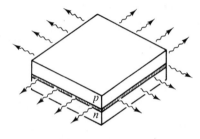

(d) Rectangular cavity

FIG. 6-4-6. Resonant cavities for laser beams. (*After S. M. Sze [10]; reprinted by permission of John Wiley and Sons, Inc.*)

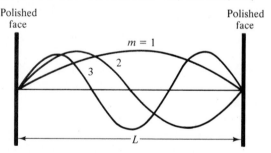

FIG. 6-4-7. Resonance within a laser cavity.

$f = $ (photon energy)$/h$ is the frequency of photon
$h = 6.63 \times 10^{-34}$ joule-s is Planck's constant

If the wavelength λ_0 in free space is used, the index of refraction n of the laser material must be considered. That is,

$$\lambda = \frac{\lambda_0}{n} \qquad (6\text{-}4\text{-}9)$$

In practice, it is not necessary to mechanically adjust the cavity length L to an integral number of half-wavelengths for resonance; Eq. (6-4-8) is automatically satisfied over some portion of the cavity because the wavelength of laser beam is very short, and many values of the integral number m will satisfy the resonant condition.

6-4-2. Solid-State Ruby Laser

During 1957 and early 1958 many masers were constructed in several laboratories utilizing Cr^{3+} in the now famous ruby crystals. The first published result was by Makhov [149] at the University of Michigan. He reported a ruby maser pumped at 24 GHz and operating at 9.3 GHz. Rubies were used in masers of greatly varying construction and performance.

(1) *Physical Description.* The first ruby laser built at the Hughes Research Laboratory in 1960 by Maiman [138] was made of a circular ruby rod having mirrored ends. A xenon flash tube surrounded the ruby rod. A bank of capacitors was discharged through the xenon tube, causing it to emit a very intense flash of light with a lifetime of several milliseconds. The original ruby developed by Maiman and its simplified cavity diagram are shown in Fig. 6-4-8.

(2) *Principles of Operation.* The ruby belongs to a family of gems consisting of corundum (Al_2O_3) with various types of impurities. For instance, a pink ruby used by Maiman contains about 0.05% chromium ion (Cr^{3+}). This ion absorbs radiation in two wide bands located in the blue and green portions of the spectrum, and the resulting excitation is followed by a very fast radiationless process. There are three basic requirements for achieving laser operation:

1. A method of excitation or pumping of atoms from ground energy level to higher energy levels
2. A sufficient large population inversion
3. A resonant cavity to produce positive feedback of the radiation

The basic operating principle for the ruby laser can be explained by three-level model theory as described earlier. The flashlight has a center frequency at a wavelength of about 5500 Å. When the pumping flashlight heats the ruby rod, chromium atoms (Cr^{3+}) in corundum are elevated from ground state E_1 to excited state E_3. The lifetime of an excited state E_3 is less than 1 μs; so the excited ions Cr^{3+} quickly lose some of their excitation energy through nonradiative transitions to the metastable state E_2. The lifetime of the metastable state E_2 is about 3 ms. This state then slowly decays by spontaneously emitting a sharp doublet, the components of which at 300°K are at 6943 and 6929 Å. Under very intense excitation the popula-

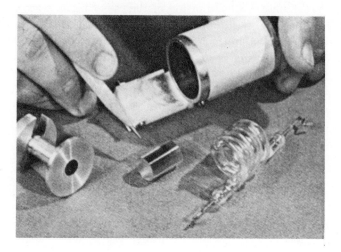

(a) Maiman's original ruby laser

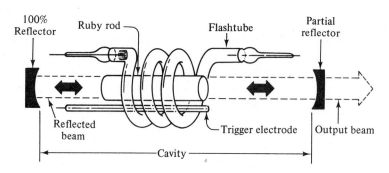

(b) Simplified diagram of Maiman's ruby laser

FIG. 6-4-8. Maiman's ruby laser and its simplified cavity diagram.

tion of this metastable state E_2 can become greater than that of the ground state E_1.

The circular ruby rod was of 1-cm dimension and coated on two faces with silver. Some photons travel longitudinally along the axis of the rod and are reflected at the ends, thus causing further stimulated emission. Population inversion is sustained by the pumping light source, and continuous photon amplification results in the generation of a very intense beam, which emerges through the partly transmissive mirror at one end of the ruby rod.

Once the stimulated emission begins, however, the population in the metastable state decreases quickly. Thus the laser output consists of an intense spike lasting from a few nanoseconds to microseconds. After the stimulated emission spike, population inversion builds up again and a second spike results. Consequently, the metastable state never reaches a highly inverted population of electrons. Such a population can be achieved by keeping the coherent photon field in the ruby rod from building up and thereby preventing stimulated emission until a larger population inversion is achieved. This process is called *Q*-switching or *Q*-spoiling, where *Q* is the quality of the resonant structure. A *Q*-switched ruby laser is shown in Fig. 6-4-9. Here one face of the ruby rod is not silver coated but instead is provided by a fast-rotating external mirror.

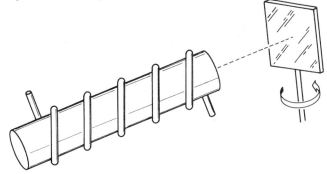

FIG. 6-4-9. Q-switched ruby laser.

When the mirror plane is aligned exactly perpendicular to the laser axis, a resonant structure exists; but as the mirror rotates away from this position, there is no laser action because of a lack of buildup of photons by multiple reflections. So a very large inverted population builds up as the mirror rotates off-axis. When the mirror again returns to the on-axis position, stimulated emission with a highly intense laser pulse occurs again. This structure is called a *Q*-switched laser. By saving the electron population for a single pulse, a large amount of energy is emitted in a very short time. For example, if the total energy in the pulse is 10 joules and the pulse width is 100 ns, the pulse power is 100 MW (megawatts).

Essentially there are two basic methods of *Q*-switching: passive and active. In passive *Q*-switching a material is used with an optical transmsision at the laser frequency that varies with intensity. Such materials include liquid cells, uranyl glass, exploding films, and semiconductor reflectors. Active *Q*-switching can be accomplished mechanically by Porro prisms, electro-optically by Kerr or Pockels cells, or by acoustic deflection. The obvious advantage of active *Q*-switching is in the precise control it offers, with the Kerr cell giving optimum control.

(3) Microwave Characteristics. All classical sources of light have the unfortunate characteristic of being polychromatic. It is possible to obtain quasimonochromaticity by using filters, but the radiation intensity is greatly

reduced. Since radiation in the oscillating modes of laser cavities is very directional, the ruby laser output beams are extremely directional as well as highly monochromatic. This situation occurs because the round-trip distance between the two silver-coated parallel faces is an integral number of wavelengths. It is this monochromaticity that gives a laser its coherence, which, in turn, contributes to the high-intensity (brightness) output. These unique characteristics exhibited by the ruby laser source—collimation (directionality), coherence, and high intensity—are found in no other radiation source. Table 6-4-3 lists the features of commercially available ruby lasers [144].

TABLE 6-4-3.*   Commercially Available Ruby Lasers.

Laser	Wavelength (μm)	Mode of operation	Power output (W)		Repetition rate (pps)	Pulse length	Processing uses
			Average	Peak			
Ruby	0.6943	pulsed	1–20		1	0.3–6 ms	welding, material removal
Ruby	0.6943	Q-switched		10^5	1	0.3–2 ms	welding, material removal
Ruby	0.6943	Q-switched		10^8	1	5–50 ns	vaporation

*From Marce Eleccion [144]; reprinted by permission of the IEEE, Inc.

6-4-3. Semiconductor Junction Lasers

Since Townes and his students invented the first maser in 1954, such lasers as solid-state lasers, neutral gas lasers, ion gas lasers, molecular gas lasers, liquid lasers, and microwave lasers have been built. During the years 1957 to 1961 Nishizawa and Watanabe [151], Basov et al. [152], and Aigrain [153] independently proposed the p-n junction semiconductor laser. In 1962 Dumke demonstrated that laser action was possible in direct bandgap semiconductors [141].

The most common type of semiconductor used for junction lasers consists of the III-V compounds listed in Table 6-4-2. The important features of semiconductor lasers are extreme monochromaticity and high directionality, features that are similiar to those of other lasers, such as the solid-state ruby laser and He-Ne gas laser. However, semiconductor junction lasers differ from other lasers in several basic respects.

1. The quantum transition of a semiconductor junction laser occurs between energy bands (conduction and valence bands) rather than between discrete energy levels, such as in the solid-state ruby laser.

2. The size of a semiconductor junction laser is very small, typically $50 \times 25 \times 50$ cubic microns

3. The characteristics of a semiconductor junction laser are strongly influenced by the properties of the junction material, such as doping and band tailing.

4. Population inversion of a semiconductor junction laser occurs in the very narrow junction region, and the pumping is supplied by a forward bias across the junction.

In this section only semiconductor junction lasers are discussed. Emphasis is on their physical description, principle of operation, and microwave characteristics.

(*1*) *Physical Description.* The semiconductor junction laser is also called an *injection laser* because its pumping method is electron-hole injection in a *p-n* junction. The semiconductor that has been extensively used for junction lasers is the GaAs. The GaAs injection laser is basically a planar *p-n* junction in a single crystal of GaAs. A typical structure of a *p-n* junction laser is shown in Fig. 6-4-10.

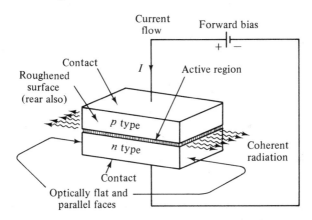

FIG. 6-4-10. Structure of a *p-n* junction laser. (*From Marce Eleccion [144]; reprinted by permission of the IEEE, Inc.*)

The two parallel, semireflective end faces form a Fabry–Perot resonant cavity that enhances the optical Q of the system. The other set of faces is roughened or sawed to suppress all but the modes propagating between the end faces.

The junction laser is made of gallium arsenide (GaAs) by the standard diffusion method. Both the *p*-type material and the *n*-type material are very highly doped up to 10^{18} cm^{-3} so that the Fermi levels are in the conduction band and the valence band for the *p*- and *n*-regions, respectively. The doping is not quite as heavy as in the tunnel diode; thus the current-voltage characteristic does not have a negative-resistance region.

The techniques for fabricating a *p-n* junction laser can be described as follows [154]. The starting material is an *n*-type GaAs wafer doped with silicon in the range of 2 to 4×10^{18} cm^{-3}. A *p*-type layer is grown on the wafer by the liquid-phase epitaxial process. The wafer is lapped to a thickness of 75 μm and the top and bottom surfaces are metalized. The wafer is then cleaved into slivers 300 to 500 μm wide. The junction depths range from 2 to 100 μm. The next step is to evaporate a reflective coating onto one of the cleaved facets of the sliver so that the laser can emit from only one facet.

(2) **Principles of Operation.** The basic principles underlying laser action can be developed in four steps.

1. Atoms (and also ions and molecules) of the material exhibit internal resonances at certain discrete characteristic frequencies. These internal atomic resonances occur at frequencies ranging from the audio up to and beyond the optical regions.
2. A signal applied to an atom at or near one of its internal resonances will cause a measurable response in the atom. Depending on the circumstances, the atom may absorb energy from the signal or it may emit energy into the signal.
3. The strength and the sign of the total response that will be obtained from a collection of many atoms of the same kind depend on the population difference—that is, the difference in population between the lower and upper quantum energy levels responsible for that particular transition. Under normal positive-temperature thermal equilibrium conditions this net response will always be absorptive.
4. If this population difference can somehow be inverted so that there are more atoms in the upper energy level than in the lower, then the total response of the collection of atoms will also invert. That is, the total response will change from a net absorptive to a net emissive condition. The atoms will then give net energy to the signal and thus amplify it.

In examining the basic principles of laser action, the question naturally arises as to whether all semiconductors can have laser action. The answer is no. Only the III-V compounds can. Semiconductors can be classified into direct and indirect types. Direct bandgap semiconductors, such as GaAs and GaSb, have the lowest conduction band minimum and the highest valence band maximum at the same wave vector in the Brillouin zone. Indirect bandgap semiconductors, such as Ge and Si, have the extrema at different wave vectors. In direct III-V compounds radiative transitions take place faster than nonradiative and impurity transitions. In indirect semiconductors radiative recombination is improbable and nonradiative recombinations occur faster than radiative recombinations. As a result, the radiation efficiency of the direct-type semiconductors is much higher than that of the indirect type. Therefore most of the *p-n* junction lasers are made of III-V compounds.

As described earlier, laser action may take place in a slab of semiconductor in the presence of population inversion and in a resonant cavity. Population inversion in the *p-n* junction is accomplished by applying a high forward-bias voltage across the junction. In order for stimulated emission to occur, the applied voltage must be at least as great as the bandgap voltage V_g, so that the electrons that have filled the conduction band on the *n*-region have energy of at least $E_v + E_g$, where E_v is the energy level of the valence band in the *n*-region and E_g is the bandgap energy, which is equal to eV_g. The energy-band diagrams for the *p-n* junction laser are shown in Fig. 6-4-11.

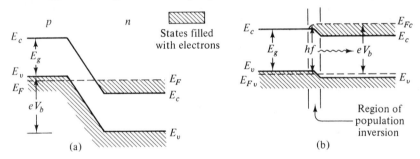

FIG. 6-4-11. Energy-band diagram of a *p-n* junction laser.

It can be seen from Fig. 6-4-11 that the height of the potential barrier energy eV_b at the *p-n* junction must be greater than the bandgap energy E_g. Otherwise if the applied voltage is equal or larger than the barrier potential V_b, the potential barrier will be reduced to zero and very excessive currents will destroy the device. In order for the bandgap energy E_g to be less than the potential barrier energy eV_b, the Fermi level must be below the valence-band energy E_v in the *p*-region and above the conduction-band energy E_c in the *n*-region, as indicated in Fig. 6-4-11(a). This is the reason why the *p-n* junction should be heavily doped in order to have high-enough currents for population inversion.

When a sufficiently high voltage is applied in the forward direction to the *p-n* junction, electrons and holes are injected into and across the transition layer in considerable concentration. As a result, the layer at the junction is far from being depleted of carriers. Consequently, this layer contains a high concentration of electrons in the conduction band and a high concentration of holes in the valence band. This change of concentrations in the two bands is called *population inversion*, and the layer at the junction over which population inversion takes place is called the *inversion region* as indicated in Fig. 6-4-11(b). Furthermore, population inversion at the junction can also be well explained by the concept of Fermi energy levels, which was shown in Fig. 6-4-3.

When electrons make transitions from the *n*-region to empty states in the valence band at the *p*-region, they emit photons with energy approximately equal to the bandgap energy E_g. In addition, holes may flow from the *p*-region to the *n*-region and emit photons by recombination with electrons. In any case, for high-enough forward-bias voltage, this active region exists in the vicinity of the *p-n* junction, and a maximum gain over a reasonable distance occurs in the plane of the junction. For this reason, the Fabry–Perot resonant cavity is extensively utilized so that a feedback path is provided directly in the plane of maximum gain.

(*3*) *Microwave Characteristics.* As described previously, the great advantages of the laser beam are its time coherence (monochromaticity) and space coherence (directionality). Time coherence makes the laser beam ideal for communication purposes, and space coherence renders it useful for space communications.

Semiconductor junction laser diodes are available with emitting junction widths of 75 μm to 1.4 mm, which produce power outputs of 2 to 70 W, respectively, at driving currents ranging from 10 to 250 A at 300°K. The power output of the laser diodes must be restriced to 1.0 to 1.5 W per 25 μm of emitting facet to prevent degradation of the diode.

At room temperature the semiconductor junction laser diode can sustain pulses on the order of 200 ns and repetition rates of 10 to 20 kHz, depending on the effectiveness of the heat sinking. Because heat generation is the factor that limits operation, a direct tradeoff can be made between pulse width and repetition rate. Thus at narrow pulse widths (less than 10 ns), repetition rates in excess of 50 kHz have been obtained with laser diodes. Average power output in the milliwatt range can be achieved by pulse operation at 300°K.

The basic concept of holography, which is more than 20 years old, is simply that the diffraction pattern of light from an object is a transform, or coded record, of the object. If such a diffraction pattern could be stored, one should be able to reconstruct an image of the object. The original problem with holography was that it is easy to record the magnitude but not the phase of the diffraction pattern. Laser beams can be used to photograph three-dimensional objects by storing both magnitude and phase information on a film (hologram). Then the hologram can be illuminated by a laser beam to reproduce the object in three dimensions [155].

A large class of laser applications depend not so much on coherence or monochromaticity as on the unprecedented brightness or energy per unit area that can be obtained by focusing a laser beam with a lens. This brightness, which is a byproduct of the beam's coherence, is a unique feature of laser light and can be many orders of magnitude greater than the brightest light produced by conventional light sources.

The energy density of the image formed by a lens in a laser beam can be used to heat, melt, or even vaporize small areas of any material. This capability will probably find wide use in the field of microelectronics. A microcapacitor can be made by using a laser beam to cut a meander path through a 0.3-μm thick gold-conducting film vapor deposited on a sapphire substrate of a microwave integrated circuit. The cut is 6 μm wide. The main advantages of laser light in the field of microelectronics are the small size of the focused image, the absence of contamination of the pure materials required in such circuits, and the precise control of the energy used. In the fabrication of microwave integrated circuits, the laser beam can also be used to weld connections between parts of the circuit. A precision resistor can be adjusted in a completely automatic machine that measures the resistance and pulses the laser to vaporize conductive material until the desired resistance is obtained.

A pulse laser beam has been used to pierce holes in diamond chips or in a alumina ceramic. The thickness of the chip is about 1.5 mm. The diameter of the holes varies between 38 to 76 μm. Twenty-four pulses each with an energy of a tenth of a joule are required to drill a hole all the way through the ceramic. The high intensity and directionality of a laser beam can also be used to align jigs for mechanical tooling by photoelectric centering of a target and to make surveys over wooded land by determining the laser beam path through the objects for laying power transmission lines or commercial telephone cables.

A small laser alarm system can be used to detect intruders or burglars when its invisible beam of light is interrupted. In operation, the laser transmitter generates a pulse beam that, in combination with a detector, can seal a path, an area, or a space against unwanted intrusion. Other applications for GaAs laser diodes include bomb fuses, secure communications, range-finding radar, data transmission, target designation, collision avoidance, and direct fire simulation. Finally, laser beams also have medical applications, such as in the restoration of detached retinas.

6-5 Parametric Devices

6-5-1. Introduction

A parametric device is one that uses a nonlinear reactance (capacitance or inductance) or a time-varying reactance. The word *parametric* is derived from the term "parametric excitation", since the capacitance or inductance, which is a reactive parameter, can be used to produce capacitive or inductive excitation. Parametric excitation can be subdivided into parametric amplification and oscillation. Many of the essential properties of nonlinear energy-storage systems were described by Faraday [156] as early as 1831 and by Lord Rayleigh [157] in 1883. The first analysis of the nonlinear capacitance was given by van der Ziel [158] in 1948. In his paper van der Ziel first suggested that such a device might be useful as a low-noise amplifier, since it was essentially a reactive device in which no thermal noise is generated. In 1949 Landon [159] analyzed and presented experimental results of such circuits used as amplifiers, converters, and oscillators. In the age of solid-state electronics microwave electronics engineers dreamed of a solid-state microwave device to replace the noisy electron beam amplifier. In 1957 Suhl [160] proposed a microwave solid-state amplifier that uses ferrite. The first realization of a microwave parametric amplifier following Suhl's proposal was made by Weiss [161] in 1957. After the work done by Suhl and Weiss, the parametric amplifier was at last discovered.

At present the solid-state junction diode is the most widely used parametric amplifier. Unlike microwave tubes, transistors, and lasers, the parametric diode is of a reactive nature and thus generates a very small amount of Johnson noise (thermal noise). One of the distinguishing features of a parametric amplifier is that it utilizes an ac rather than a dc power supply as microwave tubes do. In this respect, the parametric amplifier is analogous to the quantum amplifier laser or maser in which an ac power supply is used.

6-5-2. Nonlinear Reactance and Manley–Rowe Power Relations

A *reactance* is defined as a circuit element that stores and releases electromagnetic energy as opposed to a *resistance*, which dissipates energy. If the stored energy is predominantly in the electric field, the reactance is said to be capacitive; if the stored energy is predominantly in the magnetic field, the reactance is said to be inductive. In microwave engineering it is most convenient to speak in terms of voltages and currents rather than electric and magnetic fields. A capacitive reactance may then be a circuit element for which capacitance is the ratio of charge on the capacitor over voltage across the capacitor. Then

$$C = \frac{Q}{V} \tag{6-5-1}$$

If the ratio is not linear, the capacitive reactance is said to be nonlinear. In this case, it is convenient to define a nonlinear capacitance as the partial derivative of charge with respect to voltage. That is,

$$C(v) = \frac{\partial Q}{\partial v} \tag{6-5-2}$$

The analogous definition of a nonlinear inductance is

$$L(i) = \frac{\partial \Phi}{\partial i} \tag{6-5-3}$$

In the operation of parametric devices, the mixing effects occur when voltages at two or more different frequencies are impressed on a nonlinear reactance.

Small-Signal Method. It is assumed that the signal voltage v_s is much smaller than the pumping voltage v_p, and the total voltage across the nonlinear capacitance $C(t)$ is given by

$$v = v_s + v_p = V_s \cos \omega_s t + V_p \cos \omega_p t \tag{6-5-4}$$

where $V_s \ll V_p$. The charge on the capacitor can be expanded in a Taylor series about the point $v_s = 0$, and the first two terms are

$$Q(v) = Q(v_s + v_p) = Q(v_p) + \frac{dQ(v_p)}{dv}\bigg|_{v_s=0} v_s \tag{6-5-5}$$

For convenience, it is assumed that

$$C(v_p) = \frac{dQ(v_p)}{dv} = C(t) \tag{6-5-6}$$

where $C(v_p)$ is periodic with a fundamental frequency of ω_p. If the capacitance $C(v_p)$ is expanded in a Fourier series, the result is

$$C(v_p) = \sum_{n=0}^{\infty} C_n \cos n\omega_p t \tag{6-5-7}$$

Since v_p is a function of time, the capacitance $C(v_p)$ is also a function of time. Then

$$C(t) = \sum_{n=0}^{\infty} C_n \cos n\omega_p t \tag{6-5-8}$$

The coefficients C_n are the magnitude of each harmonic of the time-varying capacitance. In general, the coefficients C_n are not linear functions of the ac pumping voltage v_p. Since the junction capacitance $C(t)$ of a parametric diode is a nonlinear capacitance, the principle of superposition does not hold for arbitrary ac signal amplitudes.

The current through the capacitance $C(t)$ is the derivative of Eq. (6-5-5) with respect to time and it is

$$i = \frac{dQ}{dt} = \frac{dQ(v_p)}{dt} + \frac{d}{dt}[C(t)v_s] \tag{6-5-9}$$

It is evident that the nonlinear capacitance behaves like a time-varying linear capacitance for signals with amplitudes that are much smaller than the amplitude of the pumping voltage. The first term of Eq. (6-5-9) yields a current at the pump frequency f_p and is not related to the signal frequency f_s.

Large-Signal Method. If the signal voltage is not small compared with the pumping voltage, the Taylor series can be expanded about a dc bias voltage V_0 in a junction diode. In a junction diode the capacitance C is proportional to $(\phi_0 - V)^{-1/2} = V_0^{-1/2}$, where ϕ_0 is the junction barrier potential and V is a negative voltage supply. Since

$$(V_0 \mp V_p \cos \omega_p t)^{-1/2} \simeq V_0^{-1/2}\left(1 - \frac{V_p}{3V_0} \cos \omega_p t\right) \qquad \text{for } V_p \ll V_0$$

the capacitance $C(t)$ can be expressed as

$$C(t) = C_0(1 + 2\gamma \cos \omega_p t) \qquad (6\text{-}5\text{-}10)$$

The parameter γ is proportional to the pumping voltage v_p and indicates the coupling effect between the voltages at the signal frequency f_s and the output frequency f_0.

Manley–Rowe Power Relations. Manley and Rowe [162] derived a set of general energy relations regarding power flowing into and out of an ideal nonlinear reactance. These relations are useful in predicting whether power gain is possible in a parametric amplifier. Figure 6-5-1 shows an equivalent circuit for Manley–Rowe derivation.

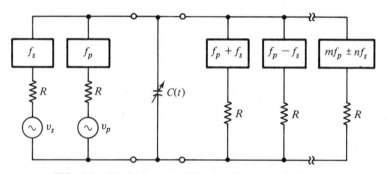

FIG. 6-5-1. Equivalent circuit for Manley-Rowe derivation.

In Fig. 6-5-1, one signal generator and one pump generator at their respective frequencies f_s and f_p, together with associated series resistances and bandpass filters, are applied to a nonlinear capacitance $C(t)$. These resonating circuits of filters are designed to reject power at all frequencies other than their respective signal frequencies. In the presence of two applied frequencies of f_s and f_p, an infinite number of resonant frequencies of $mf_p \pm nf_p$ are generated, where m and n are any integer from zero to infinity.

Each of the resonating circuits is assumed to be ideal. The power loss by the nonlinear susceptances is negligible. That is, the power entering the nonlinear capacitor at the pump frequency is equal to the power leaving the capacitor at the other frequencies through the nonlinear interaction. Manley and Rowe established the power relations between the input power at the frequencies f_s and f_p and the output power at the other frequencies $mf_p \pm nf_s$, where m and n are any integer except zero.

From Eq. (6-5-4) the voltage across the nonlinear capacitor $C(t)$ can be expressed in an exponential form as

$$v = v_p + v_s = \frac{V_p}{2}(e^{j\omega_p t} + e^{-j\omega_p t}) + \frac{V_s}{2}(e^{j\omega_s t} + e^{-j\omega_s t}) \qquad (6\text{-}5\text{-}11)$$

The general expression of the charge Q deposited on the capacitor is given by

$$Q = \sum_{m=-\infty}^{\infty} \sum_{n=-\infty}^{\infty} Q_{m,n} e^{j(m\omega_p + n\omega_s t)} \qquad (6\text{-}5\text{-}12)$$

In order for the charge Q to be real, it is necessary that

$$Q_{m,n} = Q^*_{-m,-n} \qquad (6\text{-}5\text{-}13)$$

The total voltage v can be expressed as a function of the charge Q. A similar Taylor series expansion of $v(Q)$ shows that

$$v = \sum_{m=-\infty}^{\infty} \sum_{n=-\infty}^{\infty} V_{m,n} e^{j(m\omega_p + n\omega_s t)} \qquad (6\text{-}5\text{-}14)$$

In order for the voltage v to be real, it is required that

$$V_{m,n} = V^*_{-m,-n} \qquad (6\text{-}5\text{-}14a)$$

The current flowing through $C(t)$ is the total derivative of Eq. (6-5-12) with respect to time. This is

$$i = \frac{dQ}{dt} = \sum_{m=-\infty}^{\infty} \sum_{n=-\infty}^{\infty} j(m\omega_p + n\omega_s)Q_{m,n} e^{j(m\omega_p t + n\omega_s t)}$$

$$= \sum_{m=-\infty}^{\infty} \sum_{n=-\infty}^{\infty} I_{m,n} e^{j(m\omega_p t + n\omega_s t)} \qquad (6\text{-}5\text{-}15)$$

where

$$I_{m,n} = j(m\omega_p + n\omega_s)Q_{m,n} \quad \text{and} \quad I_{m,n} = I^*_{-m,-n}$$

Since the capacitance $C(t)$ is assumed to be a pure reactance, the average power at the frequencies $mf_p + nf_s$ is

$$P_{m,n} = (V_{m,n}I^*_{m,n} + V^*_{m,n}I_{m,n})$$

$$= (V^*_{-m,-n}I_{-m,-n} + V_{-m,-n}I^*_{-m,-n}) = P_{-m,-n} \qquad (6\text{-}5\text{-}16)$$

Then conservation of power can be written

$$\sum_{m=-\infty}^{\infty} \sum_{n=-\infty}^{\infty} P_{m,n} = 0 \qquad (6\text{-}5\text{-}17)$$

Multiplication of Eq. (6-5-17) by a factor of $(m\omega_p + n\omega_s)/(m\omega_p + n\omega_s)$ and rearrangement of the resultant into two parts yield

$$\omega_p \sum_{m=-\infty}^{\infty} \sum_{n=-\infty}^{\infty} \frac{mP_{m,n}}{m\omega_p + n\omega_s} + \omega_s \sum_{m=-\infty}^{\infty} \sum_{n=-\infty}^{\infty} \frac{nP_{m,n}}{m\omega_p + n\omega_s} = 0 \qquad (6\text{-}5\text{-}18)$$

Since $I_{m,n}/(m\omega_p + n\omega_s) = jQ_{m,n}$, then $P_{m,n}/(m\omega_p + n\omega_s)$ becomes $-jV_{m,n}Q^*_{m,n} -jV_{-m,-n}Q^*_{-m,-n}$ and is independent of ω_p or ω_s. For any choice of the frequencies f_p and f_s, the resonating circuit external to that of the nonlinear capacitance $C(t)$ can be so adjusted that the currents may keep all the voltage amplitudes $V_{m,n}$ unchanged. The charges $Q_{m,n}$ are then also unchanged, since they are functions of the voltages $V_{m,n}$. Consequently, the frequencies f_p and f_s can be arbitrarily adjusted in order to require

$$\sum_{m=-\infty}^{\infty} \sum_{n=-\infty}^{\infty} \frac{mP_{m,n}}{m\omega_p + n\omega_s} = 0 \qquad (6\text{-}5\text{-}19)$$

$$\sum_{m=-\infty}^{\infty} \sum_{n=-\infty}^{\infty} \frac{nP_{m,n}}{m\omega_p + n\omega_s} = 0 \qquad (6\text{-}5\text{-}20)$$

Equation (6-5-19) can be expressed as two terms:

$$\sum_{m=0}^{\infty} \sum_{n=-\infty}^{\infty} \frac{mP_{m,n}}{m\omega_p + n\omega_s} + \sum_{m=0}^{\infty} \sum_{n=-\infty}^{\infty} \frac{-mP_{m,n}}{-m\omega_p - n\omega_s} = 0 \qquad (6\text{-}5\text{-}21)$$

Since $P_{m,n} = P_{-m,-n}$, then

$$\sum_{m=0}^{\infty} \sum_{n=-\infty}^{\infty} \frac{mP_{m,n}}{mf_p + nf_s} = 0 \qquad (6\text{-}5\text{-}22)$$

Similarly,

$$\sum_{m=-\infty}^{\infty} \sum_{n=0}^{\infty} \frac{nP_{m,n}}{mf_p + nf_s} = 0 \qquad (6\text{-}5\text{-}23)$$

where ω_p and ω_s have been replaced by f_p and f_s, respectively.

Equations (6-5-22) and (6-5-23) are the standard forms for the Manley–Rowe power relations. The term $P_{m,n}$ indicates the real power flowing into or leaving the nonlinear capacitor at a frequency of $mf_p + nf_s$. The frequency f_p represents the fundamental frequency of the pumping voltage oscillator and the frequency f_s the fundamental frequency of the signal voltage generator. The sign convention for the power term $P_{m,n}$ will follow that power flowing into the nonlinear capacitance or the power coming from the two voltage generators is positive, whereas the power leaving from the nonlinear capacitance or the power flowing into the load resistance is negative.

Consider, for instance, the case where the power output flow is allowed at a frequency of $f_p + f_s$ as shown in Fig. 6-5-1. All other harmonics are open-circuited. Then currents at the three frequencies $f_p, f_s,$ and $f_p + f_s$ are the only ones existing. Under these restrictions m and n vary from -1 through 0 to $+1$, respectively. Then Eqs. (6-5-22) and (6-5-23) reduce to

$$\frac{P_{1,0}}{f_p} + \frac{P_{1,1}}{f_p + f_s} = 0 \qquad (6\text{-}5\text{-}24)$$

$$\frac{P_{0,1}}{f_s} + \frac{P_{1,1}}{f_p + f_s} = 0 \qquad (6\text{-}5\text{-}25)$$

where $P_{1,0}$ and $P_{0,1}$ are the power supplied by the two voltage generators at the frequencies f_p and f_s, respectively, and they are considered positive. The power $P_{1,1}$ flowing from the reactance into the resistive load at a frequency of $f_p + f_s$ is considered negative.

The power gain, which is defined as a ratio of the power delivered by the capacitor at a frequency of $f_p + f_s$ to that absorbed by the capacitor at a frequency of f_s as shown in Eq. (6-5-24), is given by

$$\text{Gain} = \frac{f_p + f_s}{f_s} = \frac{f_o}{f_s} \qquad (6\text{-}5\text{-}26)$$

where $\qquad f_p + f_s = f_o \quad \text{and} \quad (f_p + f_s) > f_p > f_s.$

The maximum power gain is simply the ratio of the output frequency to the input frequency. This type of parametric device is called the *sum-frequency parametric amplifier* or *up-converter*.

If the signal frequency is the sum of the pump frequency and the output frequency, Eq. (6-5-22) predicts that the parametric device will have a gain of

$$\text{Gain} = \frac{f_s}{f_p + f_s} \qquad (6\text{-}5\text{-}27)$$

where $f_s = f_p + f_o$ and $f_o = f_s - f_p$. This type of parametric devices is called the *parametric down-converter* and its power gain is actually a loss.

If the signal frequency is at f_s, the pump frequency at f_p, and the output frequency at f_o, where $f_p = f_s + f_o$, the power $P_{1,1}$ supplied at f_p is positive. Both $P_{1,0}$ and $P_{0,1}$ are negative. In other words, the capacitor delivers power to the signal generator at f_s instead of absorbing it. The power gain may be infinite, which is an unstable condition, and the circuit may be oscillating both at f_s and f_o. This is another type of parametric device, often called a *negative-resistance parametric amplifier*.

6-5-3. Parametric Amplifiers

In a superheterodyne receiver a radio frequency signal may be mixed with a signal from the local oscillator in a nonlinear circuit (the mixer) so as to generate the sum and difference frequencies. In a parametric amplifier the local oscillator is replaced by a pumping generator and the nonlinear element by a time-varying capacitor (or inductor) as shown in Fig. 6-5-2.

In Fig. 6-5-2, the signal frequency f_s and the pump frequency f_p are mixed in the nonlinear capacitor C. Accordingly, a voltage of the fundamental frequencies f_s and f_p as well as the sum and the difference frequencies $mf_p \pm nf_s$ will appear across C. If a resistive load is connected across the terminals of the idler circuit, an output voltage can be generated across the load at the output frequency f_o. The output circuit, which does not require

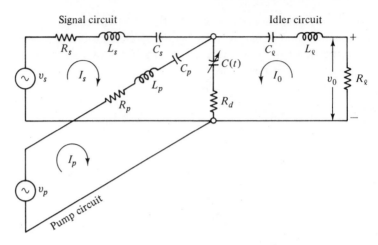

FIG. 6-5-2. Equivalent circuit for a parametric amplifier.

external excitation, is called the *idler circuit*. The output (or idler) frequency f_o in the idler circuit is expressed as the sum and the difference frequencies of the signal frequency f_s and the pump frequency f_p. That is,

$$f_o = mf_p \pm nf_s \qquad (6\text{-}5\text{-}28)$$

where m and n are positive integers from zero to infinity.

If $f_o > f_s$, the device is called a parametric *up-converter*. Conversely, if $f_o < f_s$, the device is known as a parametric *down-converter*.

Parametric Up-Converter. A parametric up-converter has the following properties:

1. The output frequency is equal to the sum of the signal frequency and the pump frequency.
2. There is no power flow in the parametric device at frequencies other than the signal, pump, and output frequencies.

POWER GAIN.   When these two conditions are satisfied, the maximum power gain of a parametric up-converter [163] is expressed as

$$\text{Gain} = \frac{f_o}{f_s} \frac{x}{(1 + \sqrt{1 + x})^2} \qquad (6\text{-}5\text{-}29)$$

where $f_o = f_p + f_s, \quad x = \frac{f_s}{f_o}(\gamma Q)^2, \quad Q = \frac{1}{2\pi f_s C R_d}$

R_d is the series resistance of a *p-n* junction diode and γQ is the figure of merit for the nonlinear capacitor. The quantity of $x/(1 + \sqrt{1 + x})^2$ may be regarded as a gain-degradation factor. As R_d approaches zero, the figure of merit γQ goes to infinity, and the gain-degradation factor becomes equal to unity. As a result, the power gain of a parametric up-converter for a lossless diode is equal to f_o/f_s, which is predicted by the Manley–Rowe relations as shown in Eq. (6-5-26). In a typical microwave diode γQ could be equal to 10. If $f_o/f_s = 15$, the maximum gain as given by Eq. (6-5-29) is 7.3 dB.

NOISE FIGURE. One advantage of the parametric amplifier over the transistor amplifier is its low-noise figure because a pure reactance does not contribute thermal noise to the circuit. The noise figure F for a parametric up-converter [163] is given by

$$F = 1 + \frac{2T_d}{T_0}\left[\frac{1}{\gamma Q} + \frac{1}{(\gamma Q)^2}\right] \qquad (6\text{-}5\text{-}30)$$

where
T_d = diode temperature in degrees Kelvin
T_o = 290°K is the ambient temperature in degrees Kelvin
γQ = figure of merit for the nonlinear capacitor

In a typical microwave diode γQ could be equal to 10. If $f_o/f_s = 10$ and $T_d = 290°K$, the minimum noise figure is 0.90 dB, as calculated by using Eq. (6-5-30).

BANDWIDTH. The bandwidth of a parametric up-converter is related to the maximum, the gain-degradation factor, and the ratio of the signal frequency and the output frequency. The bandwidth equation [163] is given by

$$BW = 2\gamma\sqrt{\frac{f_o}{f_s}} \qquad (6\text{-}5\text{-}31)$$

If $f_o/f_s = 10$ and $\gamma = 0.2$, the bandwidth (BW) is equal to 1.264.

Parametric Down-Converter. If a mode of down-conversion for a parametric amplifier is desirable, the signal frequency f_s must be equal to the sum of the pump frequency f_p and the output frequency f_o. This means that the input power must feed into the idler circuit and the output power must move out from the signal circuit as shown in Fig. 6-5-2. The down-conversion gain (actually a loss) is given by [163]

$$\text{Gain} = \frac{f_s}{f_o}\frac{x}{(1 + \sqrt{1 + x})^2} \qquad (6\text{-}5\text{-}32)$$

Negative-Resistance Parametric Amplifier. If a significant portion of power flows only at the signal frequency f_s, the pump frequency f_p, and the idler frequency f_i, a regenerative condition with the possibility of oscillation at both the signal frequency and the idler frequency will occur. The idler frequency is defined as the difference between the pump frequency and the signal frequency, $f_i = f_p - f_s$. When the mode operates below the oscillation threshold, the device behaves as a bilateral negative-resistance parametric amplifier.

POWER GAIN. The output power is taken from the resistance R_i at a frequency f_i, and the conversion gain from f_s to f_i [163] is given by

$$\text{Gain} = \frac{4f_i}{f_s} \cdot \frac{R_g R_i}{R_{Ts} R_{Ti}} \cdot \frac{a}{(1-a)^2} \qquad (6\text{-}5\text{-}33)$$

where
f_s = signal frequency
f_p = pump frequency
$f_i = f_p - f_s$ is the idler frequency
R_g = output resistance of the signal generator
R_i = output resistance of the idler generator
R_{Ts} = total series resistance at f_s
R_{Ti} = total series resistance at f_i
$a = R/R_{Ts}$
$R = \dfrac{\gamma^2}{\omega_s \omega_i C^2 R_{Ti}}$ is the equivalent negative resistance

NOISE FIGURE. The optimum noise figure of a negative-resistance parametric amplifier [163] is expressed as

$$F = 1 + 2\frac{T_d}{T_0}\left[\frac{1}{\gamma Q} + -\frac{1}{(\gamma Q)^2}\right] \qquad (6\text{-}5\text{-}34)$$

where
γQ = figure of merit for the nonlinear capacitor
$T_0 = 290°K$ is the ambient temperature in °K
T_d = diode temperature in °K

It is interesting to note that the noise figure as given by Eq. (6-5-34) is identical to that for the parametric up-converter in Eq. (6-5-30).

BANDWIDTH. The maximum gain-bandwidth of a negative-resistance parametric amplifier [163] is given by

$$BW = \frac{\gamma}{2}\sqrt{\frac{f_i}{f_s \, \text{gain}}} \qquad (6\text{-}5\text{-}35)$$

If gain $= 20$ dB, $f_i = 4f_s$, and $\gamma = 0.30$, the maximum possible bandwidth for single-tuned circuits is about 0.03.

Degenerate Parametric Amplifier. The degenerate parametric amplifier or oscillator is defined as a negative-resistance amplifier with the signal frequency equal to the idler frequency. Since the idler frequency f_i is the difference between the pump frequency f_p and the signal frequency f_s, the signal frequency is just one-half the pump frequency.

POWER GAIN AND BANDWIDTH. The power gain and bandwidth characteristics of a degenerate parametric amplifier are exactly the same as for the parametric up-converter. With $f_s = f_i$ and $f_p = 2f_s$, the power transferred from pump to signal frequency is equal to the power transferred from pump to idler frequency. At high gain the total power at the signal frequency is almost equal to the total power at the idler frequency. Hence the total power in the passband will have 3 dB more gain.

NOISE FIGURE. The noise figures for a single-sideband and a double-sideband degenerate parametric amplifier [163] are given by, respectively,

$$F_{ssb} = 2 + \frac{2\bar{T}_d R_d}{T_0 R_g} \qquad (6\text{-}5\text{-}36)$$

$$F_{dsb} = 1 + \frac{\bar{T}_d R_d}{T_0 R_g} \qquad (6\text{-}5\text{-}37)$$

where $\bar{T}_d$ = average diode temperature in °K
    $T_0 = 290°K$ is the ambient noise temperature in °K
    R_d = diode series resistance in ohms
    R_g = external output resistance of the signal generator in ohms

It can be seen that the noise figure for double-sideband operation is 3 dB less than that for single-sideband operation.

6-5-4. Applications

The choice of which type of parametric amplifier to use will depend on the microwave system requirements. The up-converter is a unilateral stable device with a wide bandwidth and low gain. The negative-resistance amplifier is inherently a bilateral and unstable device with narrow bandwidth and high gain. The degenerate parametric amplifier does not require a separate signal and idler circuit coupled by the diode and is the least complex type of parametric amplifier.

In general, the up-converter has the following advantages over the negative-resistance parametric amplifier:

1. A positive input impedance
2. Unconditionally stable and unilateral
3. Power gain independent of changes in its source impedance,
4. No circulator required
5. A typical bandwidth on the order of 5%.

At higher frequencies where the up-converter is no longer practical, the negative-resistance parametric amplifier operated with a circulator becomes the proper choice. When a low noise figure is required by a system, the degenerate parametric amplifier may be the logic choice, since its double-sideband noise figure is less than the optimum noise figure of the up-converter or the nondegenerate negative-resistance parametric amplifier. Furthermore, the degenerate amplifier is a much simpler device to build and uses a relatively low pump frequency. In radar systems the negative-resistance parametric amplifier may be the better choice, since the frequency required by the system may be higher than the X band. However, since the parametric amplifier is complicated in fabrication and expensive in production, there is a tendency in microwave engineering to replace the parametric amplifier by the field-effect transistor (FET) amplifier in airborne radar systems.

6-6 Infrared Devices and Systems

6-6-1. Introduction

Infrared (IR) radiation is an electromagnetic radiation generated by vibration and rotation of the atoms and molecules within any material at temperatures above absolute zero—that is, $0°K$ or $-273°C$. During recent years there has been an increasing emphasis on the research, design, development, and deployment of various infrared devices and systems for military applications at night or during the day when vision is diminished by fog, haze, smoke, or dust. Infrared systems are defined as those that sense the passive infrared radiation emitted by some target or source and process it to the point that a visual image of that target or source is formed.

The history of infrared discovery is an interesting one. In 1800 Sir William Herschel discovered infrared radiation when he worked for the British Royal Navy [164–166], but at that time he did not use the term "infrared." Herschel referred to the new portion of the spectrum by such names as "invisible rays," "radiant heat," "dark heat," and "the rays that occasion heat." Sir Herschel found that the heating effect increased as he moved the thermometer toward the red from the blue end of the spectrum. In 1829 Nobili made the first thermocouple, which was an improved thermometer based on the thermoelectric effect discovered by Seebeck in 1821 [167]. In 1833 Melloni invented a thermopile that was made by a number of thermocouples connected in series. More sensitive than the thermocouple, the thermopile could detect radiant heat from a person at a distance of 10 m [168]. In 1901 Langley and Abbot developed an improved bolometer that could detect radiant heat from a cow at a distance of 400 m [169]. During World War I an infrared search system could detect aircraft at a distance of 1.6 km and a person at a distance of 300 m. In 1917 Case constructed the first photoconductive sensor by using thallous sulfide [170]. Many sensitive infrared detectors, such as photon detectors and image converters, were developed during World War II. The sniperscope, consisting of an image converter and an illuminator mounted on a carbine, could enable a soldier to fire accurately at targets in night as far away as 60 m. In the late 1950s the Sidewinder and Falcon heat-seeking infrared-guided missiles were developed. Subsequently infrared devices and systems were installed in the Walleye, Redeye, and Chaparral missiles, and A-6E aircraft. Furthermore, infrared techniques became applicable to the altitude stabilization of space vehicles, measurement of planetary temperatures, earth mapping, and the early detection of cancer [176]. The pioneer and fundamental work on infrared thermal imaging systems was contributed by many dedicated scientists and engineers, such as Hudson [167, 171], Jones [174], and Johnson [177].

6-6-2. Infrared Spectrum

All bodies radiate energy throughout the infrared spectrum when their temperature is above absolute zero. If the radiating source is hot enough (above 1000°K), some of the emitted energy may be visible to the human eyes in the 0.4 to 0.7-μm range. Energy emitted at wavelengths between 0.75 and 1000 μm is defined as *infrared radiation*. It should be noted that wavelengths between 100 and 1000 μm are ultramicrowaves or millimeter waves. Figure 6-6-1 shows the infrared spectrum [167].

The infrared spectrum can be further subdivided into four divisions as shown in Table 6-6-1. The first three divisions include spectral intervals in which the earth's atmosphere is relatively transparent, the so-called atmospheric windows. It is these windows that will be utilized by any infrared sensor that must look through the earth's atmosphere. The extreme infrared, often called the ultramicrowave, is usually used only for laboratory applications where the instrument can be evacuated because the atmosphere is essentially opaque [171]. Typical targets of interest have peak emittance at the wavelength of about 10 μm; a good atmospheric window exists between 8 and 14 μm; and scattering is much lower at the wavelength of 10 μm. It is obvious, then, why the 8-to 12-μm band has been chosen consistently for thermal imaging at ranges longer than 900 m. Shorter wavelengths of 3 to 5 μm, for instance, can readily be applied to thermal imaging at ranges shorter than 900 m. Figure 6-6-2 shows the transmittance of the atmosphere for 1.8 km horizontal path at sea level in infrared range.

TABLE 6-6-1. Infrared Spectrum.

Division	Wavelength (μm)	Frequency (Hz)
Near infrared (NIR)	0.7–3	0.429×10^{14}–1×10^{14}
Middle infrared (MIR)	3–6	1×10^{14}–5×10^{13}
Far infrared (FIR)	6–15	5×10^{13}–2×10^{12}
Extreme infrared (XIR)	15–1000	2×10^{12}–3×10^{11}

6-6-3. Infrared Radiation

A fundamental law of physics states that all bodies at temperatures above absolute zero in Kelvin (i.e., $T > 0°K$ or $-273°C$) emit radiation. The amount of the infrared energy emitted depends on the absolute temperature,

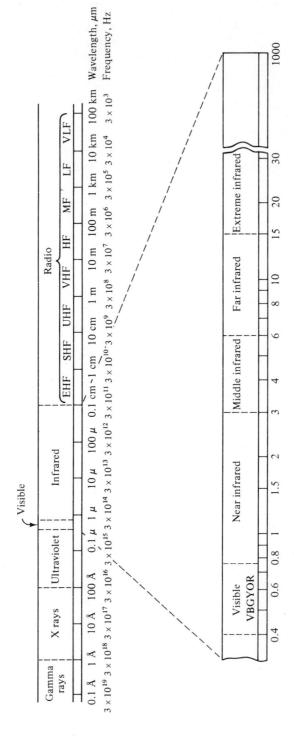

FIG. 6-6-1. Infrared spectrum. (*After R. D. Hudson, Jr. [16]; reprinted by permission of John Wiley and Sons, Inc.*)

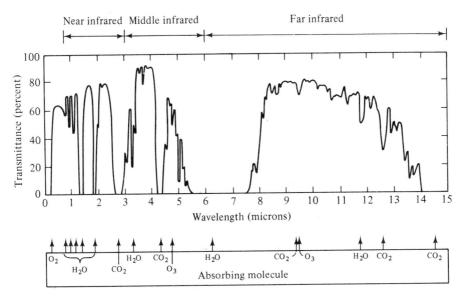

FIG. 6-6-2. Transmittance of atmosphere for 6000 ft horizontal path at sea level. (*Adapted from Gebbie et al.* [172]; *reprinted by permission of John Wiley and Sons, Inc.*)

the nature of the body, and the wavelength of the radiation. The emitting body may be a blackbody, which absorbs all incident radiation, or a graybody, which absorbs a portion of the incident radiation.

Spectral Radiant Emittance. The spectral radiant emittance of a blackbody is given by Planck's law as

$$W(\lambda) = C_1 \lambda^{-5} [e^{C_2/(\lambda T)} - 1]^{-1/2} \quad \text{watts/cm}^2/\mu\text{m} \qquad (6\text{-}6\text{-}1)$$

where $C_1 = 2\pi hc^2 = 3.7415 \times 10^4$ watts-$(\mu\text{m})^4/\text{cm}^2$

$h = 6.6256 \times 10^{-34}$ watts-sec^2 is Planck's constant

$c = 3 \times 10^{10}$ cm/s is the velocity of light in vacuum

$C_2 = ch/k = 1.4388 \times 10^4$ μm-°K

$k = 1.38 \times 10^{-23}$ watt-s/°K is Boltzmann's constant

$\lambda =$ the wavelength in micrometers

$T =$ the absolute temperature in degrees Kelvin

Radiant Emittance. The total radiant emittance into a hemisphere from a blackbody at a given absolute temperature can be obtained by integrating Eq. (6-6-1) over wavelength limits extending from zero to infinity. That is,

$$W = \int_0^\infty W(\lambda)\, d\lambda = \frac{2\pi^5 k^4}{15 c^2 h^3} T^4 = \sigma T^4 \quad \text{watts/cm}^2 \qquad (6\text{-}6\text{-}2)$$

where $\sigma = 5.6697 \times 10^{-12}$ watts/cm²/°K⁴. Equation (6-6-2) is well known as the Stefan–Boltzmann or Boltzmann law, and it indicates the radiant flux in watts per unit area of a source.

Irradiance. When the radiant flux from a source arrives at a detector, the irradiance H incident upon the surface of a detector is less than the radiant flux at the source due to the atmospheric absorption and molecular scattering in the transmission path. The irradiance H is expressed in watts per centimeter squared, and it can be measured at the detector site.

Maximum Spectral Radiant Emittance. Differentiating Eq. (6-6-1) with respect to wavelength λ and equating the derivative equal to zero result in Wien's displacement law. That is,

$$\lambda_m T = a \qquad \mu\text{m-}°\text{K} \tag{6-6-3}$$

where $a = 2898$ μm-°K

        λ_m = the wavelength in micrometers for maximum spectral radiant emittance

        T = the absolute temperature in degrees Kelvin

For example, the maximum wavelength for peak radiation from a 300°K source can be determined from Wien's displacement law as about 10 μm.

Radiant Intensity and Radiance. The radiant intensity J and radiance N of a blackbody in a hemisphere are expressed as

$$J = \frac{WA}{\pi} = NA \quad \text{watts/sr} \tag{6-6-4}$$

and

$$N = \frac{W}{\pi} = \frac{J}{A} \quad \text{watts/sr/cm}^2 \tag{6-6-5}$$

where A is the area of a radiating surface in centimeters squared.

It should be noted that a hemisphere has a solid angle of 2π. The unit of a solid angle is expressed in steradians, which is often written as sr. However, here it is assumed that the detector is a Lambertian receiver, which has an effective integrated solid angle of 1π for a hemisphere. If the source is an isotropic radiator, the radiation will emit uniformly throughout the entire solid angle of 4π. If the source is small compared with the field of view of the infrared system—that is, a point source—the irradiance will vary with the distance but not the angle about the radiator. If the source is large compared with the system field of view—namely, an extended source—the irradiance will be constant. This situation can be explained by Lambert's cosine law, which states that the radiant intensity in any direction propagating from any point of a surface is a function of the cosine of the angle θ between the said direction and the normal line to the surface at that point. In other words, the maximum radiation is in the direction to the surface and zero radiation in the tangential direction. This is why a scanning detector always receives the same amount of radiation regardless of the change of the angle θ between the detector's line of sight and the normal line to the radiating surface. As the radiating area viewed by the detector is increased, the angle θ is also increased and the value of $\cos \theta$ is decreased. Consequently, the total irradiance is constant.

Emissivity. The emissivity of a thermal radiator is a measure of its radiation efficiency. It is defined as

$$\text{Emissivity} = \frac{\text{total radiant emittance of a graybody}}{\text{total radiant emittance of a blackbody}}$$
$$\text{at the same temperature}$$

$$\epsilon = \frac{W'}{W} = \frac{\int_0^\infty \epsilon(\lambda)W(\lambda)\,d\lambda}{\int_0^\infty W(\lambda)\,d\lambda} = \frac{1}{\sigma T^4}\int_0^\infty \epsilon(\lambda)W(\lambda)\,d\lambda \qquad (6\text{-}6\text{-}6)$$

The emissivity ϵ is therefore an indication of the graybody of the thermal radiator. In other words, the lower the emissivity, the grayer the radiator; the higher the emissivity, the blacker the body. The blackbody has an emissivity of unity. Table 6-6-2 lists the emissivity of commonly used materials in total normal radiation.

TABLE 6-6-2. Emissivity of Commonly Used Materials [167].

Material	Temperature (°C)	Emissivity ϵ
Metals and their Oxides		
Aluminum		
polished sheet	100	0.05
sheet as received	100	0.09
anodized sheet, chromic acid process	100	0.55
vacuum deposited	20	0.04
Brass		
highly polished	100	0.03
rubbed with 80-grit emery	20	0.20
oxidized	100	0.61
Copper		
polished	100	0.05
heavily oxidized	20	0.78
Gold: highly polished	100	0.02
Iron		
cast, polished	40	0.21
cast, oxidized	100	0.64
sheet, heavily rusted	20	0.69

TABLE 6-6-2. (Continued)

Material	Temperature (°C)	Emissivity ϵ
Magnesium: polished	20	0.07
Nickel		
electroplated, polished	20	0.05
electroplated, no polish	20	0.11
oxidized	200	0.37
Silver: polished	100	0.03
Stainless steel		
type 18–8, buffed	20	0.16
type 18–8, oxidized at 800°C	60	0.85
Steel		
polished	100	0.07
oxidized	200	0.79
Tin: commercial tin-plated sheet iron	100	0.07
Other Materials		
Brick: red common	20	0.93
Carbon		
candle soot	20	0.95
graphite, filed surface	20	0.98
Concrete	20	0.92
Glass: polished plate	20	0.94
Lacquer		
white	100	0.92
matte black	100	0.97
Oil, lubricating (thin film on nickel base)		
nickel base alone	20	0.05
film thickness of 0.001, 0.002, 0.005 in.	20	0.27, 0.46, 0.72
thick coating	20	0.82
Paint, oil: average of 16 colors	100	0.94
Paper: white bond	20	0.93
Plaster: rough coat	20	0.91
Sand	20	0.90
Skin, human	32	0.98
Soil		
dry	20	0.92
saturated with water	20	0.95
Water		
distilled	20	0.96
ice, smooth	−10	0.96
frost crystals	−10	0.98
snow	−10	0.85
Wood: planed oak	20	0.90

Reprinted from *Hudson's Infrared System Engineering*, 1969, by permission of John Wiley & Sons, Inc.

KIRCHHOFF'S LAW.   Kirchhoff [173] discovered that at a given temperature the ratio of radiant emittance to absorptance is a constant for all materials and that it is equal to the radiant emittance of a blackbody at that temperature [167]. That is,

$$\frac{W'}{\alpha} = W \tag{6-6-7}$$

where α = absorptance of a graybody.

It is evident from Eqs. (6-6-6) and (6-6-7) that the emissivity of any graybody at a given temperature is numerically equal to its absorptance at that temperature. That is,

$$\epsilon = \alpha \tag{6-6-8}$$

Table 6-6-3 shows the values of the absorptance α, the emissivity ϵ, and the ratio α/ϵ for several materials.

TABLE 6-6-3.   Absorptance α and Emissivity ϵ of Materials [167].

Material	α	ϵ	α/ϵ
Aluminum			
polished and degreased	0.387	0.027	14.35
foil, dull side, crinkled and smoothed	0.223	0.030	7.43
foil, shiny side	0.192	0.036	5.33
sandblasted	0.42	0.21	2.00
oxide, flame sprayed, 0.001 in. thick	0.422	0.765	0.55
anodized	0.15	0.77	0.19
Fiberglass	0.85	0.75	1.13
Gold: plated on stainless steel and polished	0.301	0.028	10.77
Magnesium: polished	0.30	0.07	4.3
Paints			
Aquadag, 4 coats on copper	0.782	0.490	1.60
aluminum	0.54	0.45	1.2
Microbond, 4 coats on magnesium	0.936	0.844	1.11
TiO_2, gray	0.87	0.87	1.00
TiO_2, white	0.19	0.94	0.20
Rokide A	0.15	0.77	0.20
Stainless steel: type 18-8, sandblasted	0.78	0.44	1.77

Reprinted from *Hudson's Infrared System Engineering*, 1969, by permission of John Wiley & Sons, Inc.

6-6-4. Infrared Radiation Sources

A *blackbody* is defined as any object that completely absorbs all incident radiation. Conversely, the radiation emitted by a blackbody at a given temperature is maximum. Therefore a blackbody is an ideal radiator and absorber of radiation for all wavelengths at all temperatures, and its emissivity is equal to unity. Since the blackbody is a theoretical thermal radiator, it is generally used as a standard to calibrate all other infrared devices. Objects with emissivity less than unity are called *graybodies*, and the great majority of radiating objects belong to this group.

In general, the sources of infrared radiation can be classified into two groups, natural and artificial, as shown in Table 6-6-4. In addition to the radiation from a target or object, a certain amount of background radiation will be present. The background radiation appears in the infrared detection system as unwanted noise and must be filtered out for proper detection.

TABLE 6-6-4.   Infrared Radiation Sources.

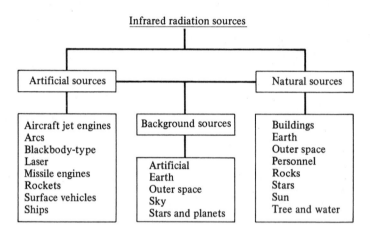

Artificial Sources of Infrared Radiation. This type of infrared radiation source includes controlled sources, such as blackbody type, and active sources, such as aircraft and rocket engines.

Most blackbody-type sources used for the calibration of infrared devices are of the cavity type with an opening of 1.27 cm or less and operate in the temperature range of 400 to 1300°K. The spectral radiant emittances of a blackbody as calculated from Planck's law in Eq. (6-6-1) in terms of temperature and wavelength are shown in Fig. 6-6-3 [167]. The dashed curve in Fig. 6-6-3 indicates the locus of these maxima as computed from Wien's displacement law in Eq. (6-6-3).

Aircraft jet engines, rockets, and missiles are powerful active sources of infrared radiation. The prime sources of infrared radiation on aircraft and missiles are the hot metal of a jet tail pipe or an engine exhaust manifold and the jet plume. Figure 6-6-4 shows the exhaust temperature contours of the turbojet engine JT4A being used on the Boeing 707 [167]. At supersonic speeds the skin of an aircraft or missile becomes an infrared radiation source because of aerodynamic heating. Figure 6-6-5 shows the equilibrium sur-

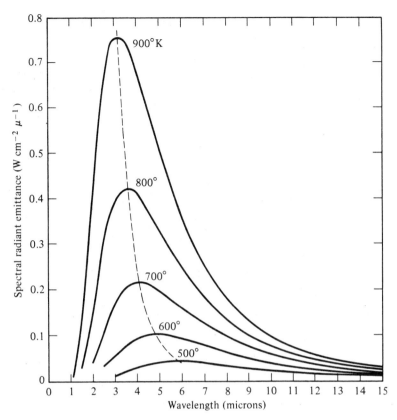

FIG. 6-6-3. Spectral radiant emittance of a blackbody. (*After R. D. Hudson, Jr.* [167]; reprinted with permission from John Wiley and Sons, Inc.)

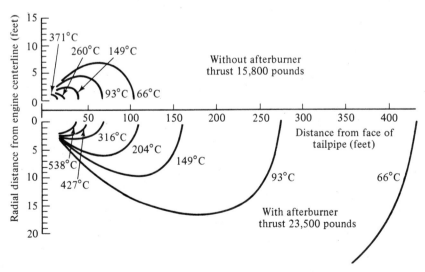

FIG. 6-6-4. Exhaust temperature contours of a Boeing 707. (*From Pratt and Whitney Aircraft Group, reprinted by permission of John Wiley and Sons, Inc.*)

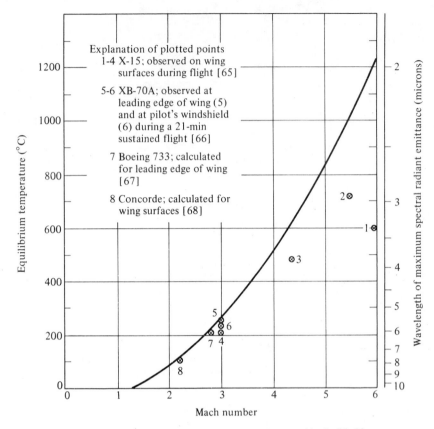

65   Various articles on the X–15 program, *Aviation Week:* **75**, 52 (November 20, 1961); **75**, 60 (November 27, 1961); **77**, 35 (August 13, 1962); **78**, 38 (June 10, 1963).

66   C.M. Plattner, XB–70A flight research – part 2. *Aviation Week* **84**, 60 (June 13, 1966).

67   C.M. Plattner, Variable-sweep wing keynotes Boeing 733 SST proposal. *Aviation Week*, **80**, 36 (May 4, 1964).

68   Size, Speed, Safety of SST are debated. *Aviation Week*, **80**, 29 (June 1, 1964).

FIG. 6-6-5. Surface temperature of SST and Concorde. (*After R. D. Hudson, Jr. [167]; reprinted by permission of John Wiley and Sons, Inc.*)

face temperature caused by aerodynamic heating in altitudes above 11.28 km and laminar flow for the Boeing 733 (i.e., Supersonic Transport—SST) and the British-French Concorde [167]. A large aircraft may emit several kilowatts of infrared energy, but the human body emits only about 2 watts.

The laser beam provides coherent sources of extremely high radiance in the portion of the spectrum extending from the ultraviolet to microwaves.

The first application of the laser in the infrared portion of the spectrum was for military operations and communication systems.

Surface vehicles may radiate sufficient infrared energy to be considered as targets. The paint used on such vehicles usually has an emissivity of 0.85 or greater. The exhaust pipes and mufflers may radiate several times as much energy as the rest of the vehicle does because of their high temperature.

Natural Sources of Infrared Radiation. This type of infrared radiation source includes terrestial sources, such as rocks, trees, earth, and water, the celestial sources, such as the sun, sky, stars, and planets, and the buildings on the ground.

The sun radiates a total radiant emittance on the order of 26.9 watts/m^2 normal to the earth's surface. In other words, the sun radiates as a 5900°K blackbody. Of this solar radiation, approximately 596 mw/m^2 is absorbed by the earth's surface from the sun at wavelengths longer than 3 μm.

During the day infrared radiation from the surface of the earth is a combination of reflected and scattered solar energy and thermal emission from the earth itself. At night, when the sun has disappeared, the spectral distribution becomes that of a graybody at the ambient temperature of the earth. Figure 6-6-6 illustrates the spectral radiance of typical terrain objects as observed during the daytime [167].

6-6-5. Infrared Optical Components

In a radar system the antenna acts as a transmitter and receiver for electromagnetic energy. However, in an infrared system the optics collect and transmit the infrared radiant flux. Therefore the optical components in an infrared system are simply analogous to the antennas in a radar system. The lenses and mirrors widely used in an infrared system are shown in Fig. 6-6-7.

Focal Points and Focal Lengths. The axis in a lens is a straight line through the geometrical center of the lens and normal to the two faces at the points of intersection as shown in Fig. 6-6-8. The primary focal point *F* lies on the axis and is defined for a positive lens as the point from which diverging rays are refracted by the lens into a parallel beam. The secondary focal point *F'* is defined by applying the principle of reversibility to the same lens. The distance between the center of a lens and either of its focal points is called its *focal length.*

Magnification and Conjugate. In any optical device the ratio between the transverse dimension of the final image and the corresponding dimension of the original object is defined as magnification of the lens. If any object is placed at the position previously occupied by its image, it will be imaged

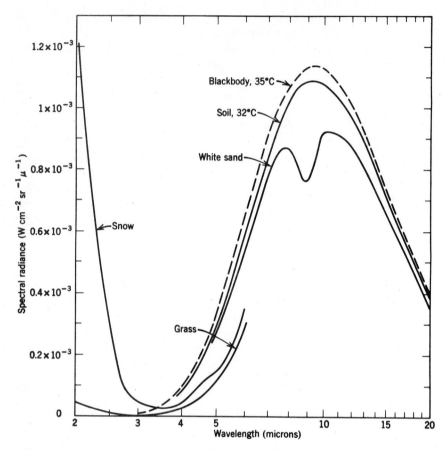

FIG. 6-6-6. Spectral radiance of terrain objects. (*After R. D. Hudson, Jr. [167]; reprinted with permission from John Wiley and Sons, Inc.*)

at the position previously occupied by the object. In this condition, the object and image are said to be *conjugate*. For a collimator, the conjugate plane is at an infinite distance.

Telescope. In an infrared system the infrared radiation from an object scene is collected by a telescope and brought out in a collimated beam with a small pupil at the scan mirror. A terrestrial-type telescope with a two-lens erecting system is used in the infrared system.

Field-Of-View and Stops. The field-of-view (FOV) in a geometrical optics determines how much of the surface of a broad object can be seen through an optical system. It is often subdivided into narrow field-of-view (NFOV), mediate field-of-view (MFOV), and wide field-of-view (WFOV)

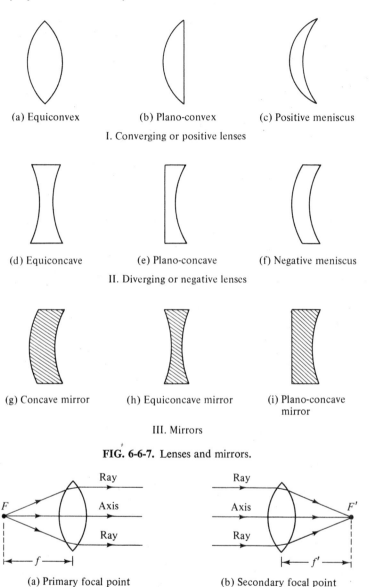

(a) Equiconvex   (b) Plano-convex   (c) Positive meniscus

I. Converging or positive lenses

(d) Equiconcave   (e) Plano-concave   (f) Negative meniscus

II. Diverging or negative lenses

(g) Concave mirror   (h) Equiconcave mirror   (i) Plano-concave mirror

III. Mirrors

FIG. 6-6-7. Lenses and mirrors.

(a) Primary focal point   (b) Secondary focal point

FIG. 6-6-8. Focal point and focal length of a lens.

according to the size of the aperture. In order to control the brightness of images, an aperture stop is placed between the lens and the focal plane in order to limit the incident bundles of rays, while the field stop in front of the focal plane determines the extent of the object, or the field, that will be produced in the image.

6-6-6. Infrared Detectors

An infrared detector is a key to the infrared imaging system for the purpose of collecting irradiance energy from a target and converting it into some other measurable form, such as electrical current or photographic image. At the wavelengths of 8 to 12 μm and at video frequencies on the order of 0.01 to 5 MHz, there are only a few detectors with sufficient sensitivity for high-performance infrared imaging systems. These detectors are Mercury Doped Germanium (Ge: Hg), Mercury Cadmium Telluride (Hg: Cd, Te), and Lead Tin Telluride (Pb: Sn, Te). The quality and characteristics of a detector are usually described by three parameters:

1. Responsivity (R)
2. Specific Detectivity (D^*)
3. Time constant (τ)

Responsivity (R). The responsivity (R) is a measure of merit for thermal detectors, and it is defined as the ratio of the detector output over input. That is,

$$R(f) = \frac{V_s}{HA_d} \qquad \text{volts/watt} \qquad (6\text{-}6\text{-}9)$$

where V_s = signal voltage in volts (RMS) is the fundamental component of the signal

H = incident radiant flux in watts per centimeters squared is the RMS value of the fundamental component of the irradiance on the detector

A_d = sensitive area of the detector in centimeters squared

Specific Detectivity (D^*). The responsivity as shown in Eq. (6-6-9) indicates only the behavior of the detector signal and does not give any information on the amount of noise in the output of the detector that will ultimately obscure the signal. In order to know the signal-to-noise ratio at the output of the detector, it is necessary to define another parameter, detectivity. Before doing so, it is desirable to define the Noise Equivalent Power (NEP) at the detector output as

$$\text{NEP} = \frac{HA_d}{V_s/V_n} = \frac{HA_dV_n}{V_s} = \frac{V_n}{R(f)} \qquad \text{watts} \qquad (6\text{-}6\text{-}10)$$

where V_n = noise voltage in RMS volts at the detector output.

The specific detectivity (D^*) is defined as directly proportional to the square root of the product of the detector area and the noise bandwidth and inversely proportional to the Noise Equivalent Power [174]. That is,

$$D^*(f) = \frac{(A_d \cdot \Delta f)^{1/2}}{\text{NEP}} \qquad \text{cm-Hz}^{1/2}/\text{watt} \qquad (6\text{-}6\text{-}11)$$

where Δf = noise bandwidth in hertz.

The detectivity (D) of a detector is related to the specific detectivity (D^*) by the following equation [174]:

$$D = D^*(A_d \cdot \Delta f)^{-1/2} = \frac{1}{\text{NEP}} \quad \text{watt}^{-1} \quad (6\text{-}6\text{-}11a)$$

This relationship was verified by Jones [174] because $DA_d^{1/2} = $ constant is well known from extensive theoretical and experimental studies. It should be noted that the quantity of the specific detectivity (D^*) refers to an electrical noise bandwidth of 1 Hz and a detector area of 1 cm². It is customary to indicate D^* by two numbers in parentheses. The first number shows the temperature of the blackbody and the second indicates the spatial frequency. For instance, D^* (400°K, 800) means a value of D^* measured with a 400°K blackbody at a spatial frequency of 800 Hz.

Time Constants. The time constant (τ) of a detector is defined as the time required for the detector output to reach 63% of its final value after a sudden change in the irradiance. The responsive time constant (τ_r) of a detector [174] is defined as

$$\tau_r = \frac{\frac{1}{4}R_m^2}{\int_0^\infty [R(f)]^2 \, df} \quad \text{seconds} \quad (6\text{-}6\text{-}12)$$

where $R_m = $ maximum value of $R(f)$ with respect to frequency.

Similarly, the detective time constant (τ_d) of a detector is expressed [174] as

$$\tau_d = \frac{\frac{1}{4}D_m^{*2}}{\int_0^\infty [D^*(f)]^2 \, df} \quad \text{seconds} \quad (6\text{-}6\text{-}13)$$

where D_m^* is the maximum value of $D^*(f)$ with respect to frequency.

MERCURY DOPED GERMANIUM (Ge: Hg). As described previously, all infrared thermal imaging systems are designed to operate at wavelengths of 8 to 12 μm. The response of the detector is then limited to that wavelength band for high performance. A variety of detectors were constructed with germanium (Ge) as a host lattice. Impurities such as gold (Au) yielded a response to about 10 μm, copper (Cu) to 30 μm, and mercury (Hg) to 14 μm. For the band range of 8 to 14 μm, Ge: Hg was found more effective than the others [175] because it had a detectivity of 4×10^{10} cm-Hz$^{1/2}$/watt and a detective time constant of 1 ns. The disadvantage of the Ge: Hg detector is that it requires cooling to 25 to 30°K. The cooling requirement and cost have prevented Ge: Hg from being continually used in the infrared imaging system.

MERCURY CADMIUM TELLURIDE (Hg: Cd, Te). The mercury cadmium telluride (Hg: Cd, Te) is an alloy consisting of a mixture of the compounds HgTe and CdTe. Its spectral response varies from 9.5 to 12 μm, and its impedance ranges from 20 to 110 Ω. The detectivity is about 1×10^{10} cm-Hz$^{1/2}$/watt, and the detective time constant is about 0.05 ns [175].

LEAD TIN TELLURIDE (Pb: Sn, Te). Lead-Tin-Telluride detectors are available only in the photovoltaic mode. This device has a detectivity of 1×10^{10} cm-Hz$^{1/2}$/watt and a detective time constant of 0.1 ns. Its operating wavelengths vary from 8 to 12 μm. Figure 6-6-9 shows the characteristic properties of the above-mentioned three detectors [175].

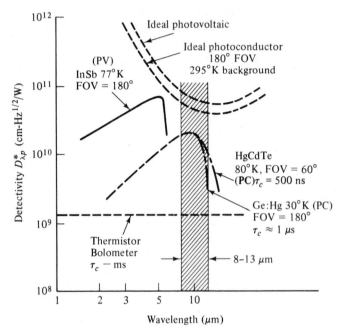

FIG. 6-6-9. Specific detectivity (D*) of leading detectors. (*After R. D. Hudson, Jr. [175]; reprinted by permission of the IEEE, Inc.*)

In addition, many other infrared detectors are available, such as

1. Thermal detectors—thermocouple, thermopile, and bolometer
2. Photon or quantum detectors—photoelectric detector, photoconductive detector, and photovoltaic or *p-n* junction detector
3. Imaging detectors—infrared film, Vidicon, and photothermionic image converter (Thermicon).

The Vidicon is a small television-type camera tube in which an electron beam scans a photoconductive target. The Thermicon is based on the thermal variation of photoemission and produces the scene image on its retina. Table 6-6-5 lists the properties of several infrared photoconductive detectors [176]. Figure 6-6-10 illustrates the infrared detectors against the infrared spectrum [171].

TABLE 6-6-5. Infrared Photoconductive Detectors.

Material	Maximum temperature for background limited operation	Long wavelength cutoff (50%) (μm)	Peak wavelength (μm)	Absorption coefficient (cm⁻¹)	Quantum efficiency	Resistance (Ω)	D* peak (cm-Hz^{1/2}/W)	Approximate response time (seconds)
InAs		3.6	3.3	$\sim 3 \times 10^3$			3×10^{11}	5×10^{-7}
InSb	110	5.6	5.3	$\sim 3 \times 10^3$	0.5–0.8	10^3–10^4	6×10^{10} -1×10^{11}	5×10^{-6}
Ge: Au	60	9	6	~ 2	0.2–0.3	4×10^5	3×10^9–10^{10}	3×10^{-8}
Ge: Au(Sb)	60	9	6			10^4	6×10^9 7×10^9	1.6×10^{-9}
Ge: Hg	35	14 14	11 10.5	~ 3 ~ 4	0.2–0.6 0.62	1.4×10^4 1.2×10^5	-4×10^{10} 4×10^{10}	$\left\{ \begin{array}{l} 3 \times 10^{-8} \\ -10^{-9} \end{array} \right.$
Ge: Hg(Sb)	35	14	11			5×10^9	1.8×10^{10}	3×10^{-10}–2×10^{-9} 3×10^{-10}–3×10^{-9}
Ge: Cu	17	27	23	~ 4	0.2–0.6	2×10^4	2.4×10^{10}	3×10^{-4}–10^{-8} 4×10^{-9}–1.3×10^{-7}
Ge: Cu(Sb)	17	27	23			2×10^5	2×10^{10}	$<2.2 \times 10^{-9}$
Hg: Cd, Te x = 0.2		14	12	$\sim 10^3$	0.05–0.3	60-400 20-200	10^{10} 6×10^{10}	$<10^{-8}$ $<4 \times 10^{-4}$
Pb: Sn, Te x = 0.17–0.2		11 15	10 14	$\sim 10^4$		42 52	3×10^8 1.7×10^{10}	1.5×10^{-8} 1.2×10^{-4}

*(After Levinstein and Mudar [176]; reprinted with permission from the IEEE, Inc.).

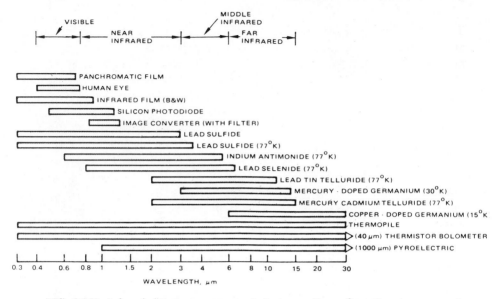

FIG. 6-6-10. Infrared detectors versus spectral ranges. *Note*: *Operating temperature is 300°K if not indicated. (After R. D. Hudson, Jr. [171]; reprinted with permission from the IEEE, Inc.)*

6-6-7. Infrared Systems

The infrared system is one consisting of the optical unit, the detector array, the signal processors, and the display indicators (see Fig. 6-6-11).

Since a simple infrared system is passive, it can be used for the detection, recognition, and identification of a target by sensing the radiation emitted by the target, but it provides no information on the distance to the target. If an infrared system has some illuminating devices, such as the laser transmitter and receiver, built in, the entire infrared system may become active with a capability of determining the range to the target. The target is the object of interest for which the infrared system is designed and built.

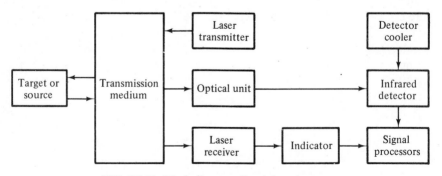

FIG. 6-6-11. Block diagram of an infrared system.

The radiated energy varies with the temperature of the target, its emissivity, and the viewing angle. The atmosphere is not a very favorable transmission medium. However, if the operating wavelength is chosen in the range of the atmospheric window as shown in Fig. 6-6-2, the atmospheric absorptance may reduce to a minimum. The optical unit collects the radiation emitted by the target and delivers it to the detector just like an antenna in the radar system. Before reaching the detector, the information radiation from the target may be recovered by an optical demodulator from the unwanted emission in the background. The detector then converts the information radiation into an electrical signal. Since a wide field-of-view (WFOV) and high resolution require a large number of detector elements, the *Forward-Looking Infra-Red (FLIR)* unit uses the newest detection techniques based on discrete detector arrays to scan object space and generate a video signal. Finally, the signal processor amplifies the electrical signal and sends the coded target information to the indicator for display. In addition, the infrared detector must be cooled down by a cryogenic cooler to a specified operating level of temperature, 30°K for Ge: Hg detector, for high performance. This factor is necessary because the internal radiation emitted by the detector itself must be reduced to a minimum. Otherwise it is difficult, if not impossible, to separate the two radiations with equal magnitude from the target and the detector itself.

The performance of an infrared imaging system is described by the following three parameters:

1. the Noise Equivalent Differential Temperature (NEΔT)
2. the Minimum Resolvable Temperature (MRT)
3. the Modulation Transfer Function (MTF).

Noise Equivalent Differential Temperature (NEΔT). The signals from the detectors in an infrared imaging system respond to the variation in the irradiance at the entrance pupil of the optical system as the detectors are scanned across the target scene. Small differences in irradiance or temperature can be detected as an ac video signal while the high ambient irradiance or temperature is presented as a dc level or white noise. Then the noise signal is subtracted by a single RC lowpass filter. The Noise Equivalent Differential Temperature (NEΔT) is defined as the temperature difference required at the input of the detector to produce a peak signal-to-noise (RMS) ratio of unity at the detector preamplifier output [174]. That is,

$$\text{NE}\Delta T = \frac{V_n}{V_s} \Delta T \qquad \text{degrees in Centigrade} \qquad (6\text{-}6\text{-}14)$$

where V_s = ac signal voltage in volts (peak to peak)
V_n = noise voltage in volts (RMS)
ΔT = target temperature difference in degrees Centigrade

In Eq. (6-6-14) it is assumed that the reflectance of a collimator is 100%. The peak signal voltage from a target may be measured by an oscilloscope. The noise voltage may be recorded by turning off the scanner, covering the optical aperture with a flat black cover or an opaque one, and reading the noise on an RMS voltmeter.

***Minimum Resolvable Temperature* (*MRT*).** The Minimum Resolvable Temperature (MRT) is a measure that is sensitive to the thermal imaging system-human observer combination. Its output is a function describing the minimum temperature difference necessary to resolve various spatial-frequency patterns projected into the input port of the infrared system under test. A square 4-bar pattern, 7 to 1 aspect ratio, is currently the standard pattern used for such measurements.

JOHNSON'S IMAGING MODEL.   One of the earliest experimental attempts to relate threshold resolution to the visual discrimination of images of real scenes is attributed to Johnson [177]. The basic experimental scheme was to move a real scene object, such as a car, out of range until it could just barely be discerned on a detector's display at a given discrimination level, such as detection, recognition, or identification. Then the real scene object was replaced by a bar pattern of contrast similar to that of the scene object. The number of bars per minimum object dimension in the pattern was then increased until the bars could barely be individually resolved [178]. Figure 6-6-12 shows the Johnson's imaging model. In Fig. 6-6-12 the car is replaced by a bar pattern. According to Johnson's hypothesis, if the car is to be barely detected, the bar pattern width ρ should be equal to $\frac{1}{2}$ the car's minimum dimension. For simple recognition, the bar width should be $\frac{1}{8}$, whereas for identification the bar width should be $\frac{1}{13}$ the car's minimum dimension.

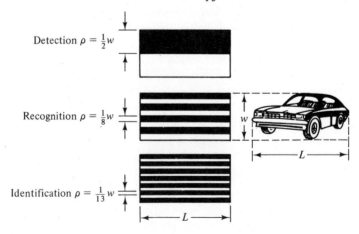

FIG. 6-6-12. Johnson's imaging model.

The length of the bars was unspecified, but it was assumed that the length is to be equal to the car's longest dimension.

In the laboratory measurement, the bars are normally oriented vertically, but 45° and horizontal orientations are sometimes also used. The observer is allowed to adjust the display brightness and contrast controls for the most resolvable pictures at some minimum temperature ΔT. This process is repeated for other spatial frequencies. It is important that the ambient temperature during the measurement must not be changed by more than $\pm 0.20°C$ from its initial values. The smaller the resolvable temperature, the better the infrared system.

The characteristic spatial frequency f of a 4-bar pattern, 7 to 1 aspect ratio, is defined as

$$f = \frac{1}{2\theta} = \frac{L}{2\rho} \qquad \text{cycles/mrd} \qquad (6\text{-}6\text{-}15)$$

where L = focal length of a collimator in inches

ρ = bar width in mills is the smallest resolution dimension of the target that is to be viewed

θ = arctan (ρ/L) is the angle subtended by target (or bar) width in milliradians

Figure 6-6-13 shows schematically the relationship between the bar width and focal length of a collimator. For instance, if the bar width of a 4-bar pattern is 10 mills, and the focal length of the collimator is 100 inches, the angle subtended by the bar width is 0.1 mrd (milliradian) because the tangent of an angle is equal to the angle if the angle is much smaller than 1°. In this case, the characteristic spatial frequency f is 5 cycles/mrd. In practice, the Minimum Resolvable Temperature (MRT) is usually plotted against some normalized spatial frequency f_0 instead of the characteristic spatial frequency f.

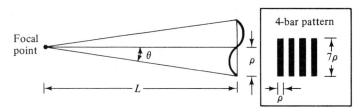

FIG. 6-6-13. Relationship between bar width and focal length.

Modulation Transfer Function (MTF). From electrical system theory the total system transfer function is the product of the individual system element transfer functions. An infrared thermal imaging system collects the thermal energy emitted by a target and converts it to an electrical current or voltage for display. Because an infrared imaging system is composed of optical, physical, chemical, mechanical, electrical, and electronic elements, it is more complicated than any ordinary electrical or electronic system. Therefore it is extremely difficult, if not impossible, to calculate analytically the infrared system transfer function.

From Fourier theory any analytical function can be transferred to a Fourier series. The Modulation Transfer Function (MTF) of an infrared system can be defined as the modulus of the Fourier transform of the one-dimensional spatial impulse response of the system. For a 4-bar pattern, 7 to 1 aspect ratio, the object consists of alternate light and dark bands that vary sinusoidally. The distribution of brightness, which is a spatial function, may be resolved into a Fourier transform as

$$B(x) = B_0 + B_1 \cos 2\pi f x \qquad (6\text{-}6\text{-}16)$$

where $f =$ the spatial frequency of the brightness variation in cycles per milliradian

$B_0 =$ the dc or average level of brightness

$B_1 =$ the magnitude of the variable brightness

$X =$ the spatial coordinate in milliradians

Figure 6-6-14 shows the energy response of a 4-bar pattern. For simplicity, the third and higher harmonic terms are neglected. The modulation of the object pattern at the input port of the system is given by Smith [179] as

$$M_0 = \frac{(B_0 + B_1) - (B_0 - B_1)}{(B_0 + B_1) + (B_0 - B_1)} = \frac{B_1}{B_0} \qquad (6\text{-}6\text{-}17)$$

where $(B_0 + B_1)$ is the maximum brightness and $(B_0 - B_1)$ is the minimum brightness.

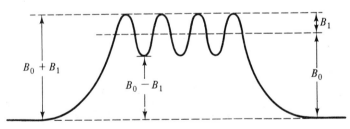

FIG. 6-6-14. Object energy response of 4-bar pattern.

When the object brightness pattern passes through the infrared system, the image pattern will be affected by the system transfer function as shown in Fig. 6-6-15. The modulation of the image pattern at the output port of the system is given by Smith [179] as

$$M_i = \frac{[B_0 + B_1(\text{MTF})] - [B_0 - B_1(\text{MTF})]}{[B_0 + B_1(\text{MTF})] + [B_0 - B_1(\text{MTF})]} = \frac{B_1}{B_0}(\text{MTF}) = M_0(\text{MTF})$$

$$(6\text{-}6\text{-}18)$$

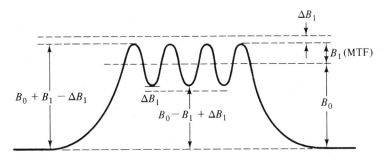

FIG. 6-6-15. Image energy response of 4-bar pattern.

Therefore the modulation transfer function (MTF) of an infrared thermal imaging system is written

$$\text{MTF} = \frac{M_i}{M_0} \tag{6-6-19}$$

In practice, the quantities of the image modulation M_i and the object modulation M_0 are unknown. In the infrared system laboratory, the maximum and minimum image pattern can be measured from an oscilloscope as shown in Fig. 6-6-15. Then Eq. (6-6-19) can be expressed as

$$\text{MTF} = \frac{M_i}{M_0} = \frac{B_0}{B_i} M_i = \frac{B_0}{B_i} \cdot \frac{(B_0 + B_1 - \Delta B_1) - (B_0 - B_1 + \Delta B_1)}{(B_0 + B_1 - \Delta B_1) + (B_0 - B_1 + \Delta B_1)}$$

$$= \frac{B_1 - \Delta B_1}{B_1} \tag{6-6-20}$$

where $\Delta B_1 =$ the differential portion decreased by the effect of the system transfer function

$B_1 =$ the magnitude of the variable object brightness

$(B_0 + B_1 - \Delta B_1) =$ the maximum image pattern

$(B_0 - B_1 + \Delta B_1) =$ the minimum image pattern

In conclusion, the higher the modulation transfer function, the better the infrared imaging system.

System Measurement Techniques. The infrared thermal imaging system is currently based on discrete detectors and optical scanning mechanisms for image display. The basic principles of operation and measurement are shown in Fig. 6-6-16.

FOR SYSTEM OPERATION. The infrared thermal energy emitted by a target is collected by the telescope of the infrared system and brought out in a collimated beam with a small pupil at the scan mirror. The scan mirror provides the scan motion to generate one angular dimension of the sensor field of view. The scanned energy at the smaller aperture is collected by the detector lens and focused on the infrared detectors. The signals from the

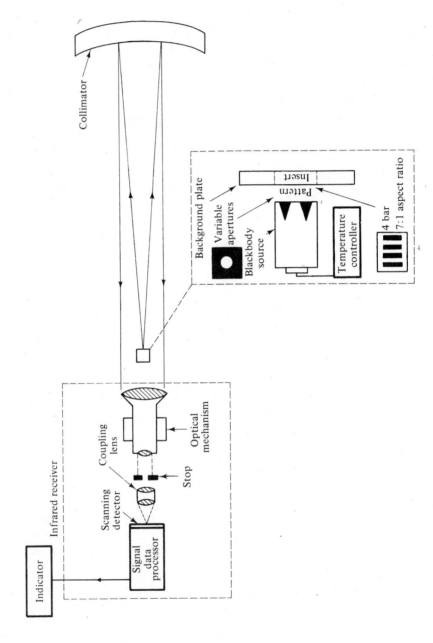

FIG. 6-6-16. Schematic diagram of infrared system.

Collimator

Background plate

Variable apertures

Pattern Insert

Blackbody source

Temperature controller

4 bar
7:1 aspect ratio

Infrared receiver

Coupling lens

Optical mechanism

Stop

Scanning detector

Signal data processor

Indicator

detectors and preamplifier are processed through the data processors for image display.

FOR SYSTEM MEASUREMENT. A square 4-bar pattern, 7 to 1 aspect ratio, is inserted into the plate slot. The temperature of the source is reduced until a match is obtained between the background and target. When a differential temperature ΔT is increased the 4-bar image is just distinguished. The differential temperature ΔT value is the target temperature difference for the tested pattern. The same procedure is repeated for various spatial frequencies. When the best image of each spatial frequency appears on the oscilloscope, the maximum, minimum, and average levels of brightness are recorded. The noise voltage or power can be measured by covering the aperture of the infrared receiver with an opaque [180, 181].

6-6-8. Infrared Applications

The infrared thermal imaging system has unlimited application potential in many areas. Some current applications are listed below.

Military: military fire control at night or during the day when vision is diminished due to fog, smoke, or haze. Detection and tracking of ships, aircraft, missiles, surface vehicles, and personnel; submarine detection and range finding.

Medical: Early detection and identification of cancer, obstacle detection for the blind, location of blockage in a vein, and early diagnosis of incipient stroke.

Scientific: Satellite and space communications, environmental survey and control, detection of life and vegetation on other planets, and measurement of lunar and planetary temperature.

Industrial: Aircraft landing aid and traffic counting, forest fire detection, and natural source detection.

6-7 Summary

The topic of microwave solid-state devices has been discussed. In particular, Gunn diodes and the Read diode were described in detail because their principles of operation are fundamental to the newest solid-state devices. The common feature of these devices is their negative resistance. The power outputs and efficiencies that have been demonstrated in the laboratories are shown in Figs. 6-7-1 and 6-7-2. The power outputs of some commercially available solid-state diodes and oscillators are illustrated in Fig. 6-7-3.

Microwave Transistors, Tunnel Diodes, and Microwave Field-Effect Transistors: As analyzed in Section 6-1-1, microwave transistors have power-

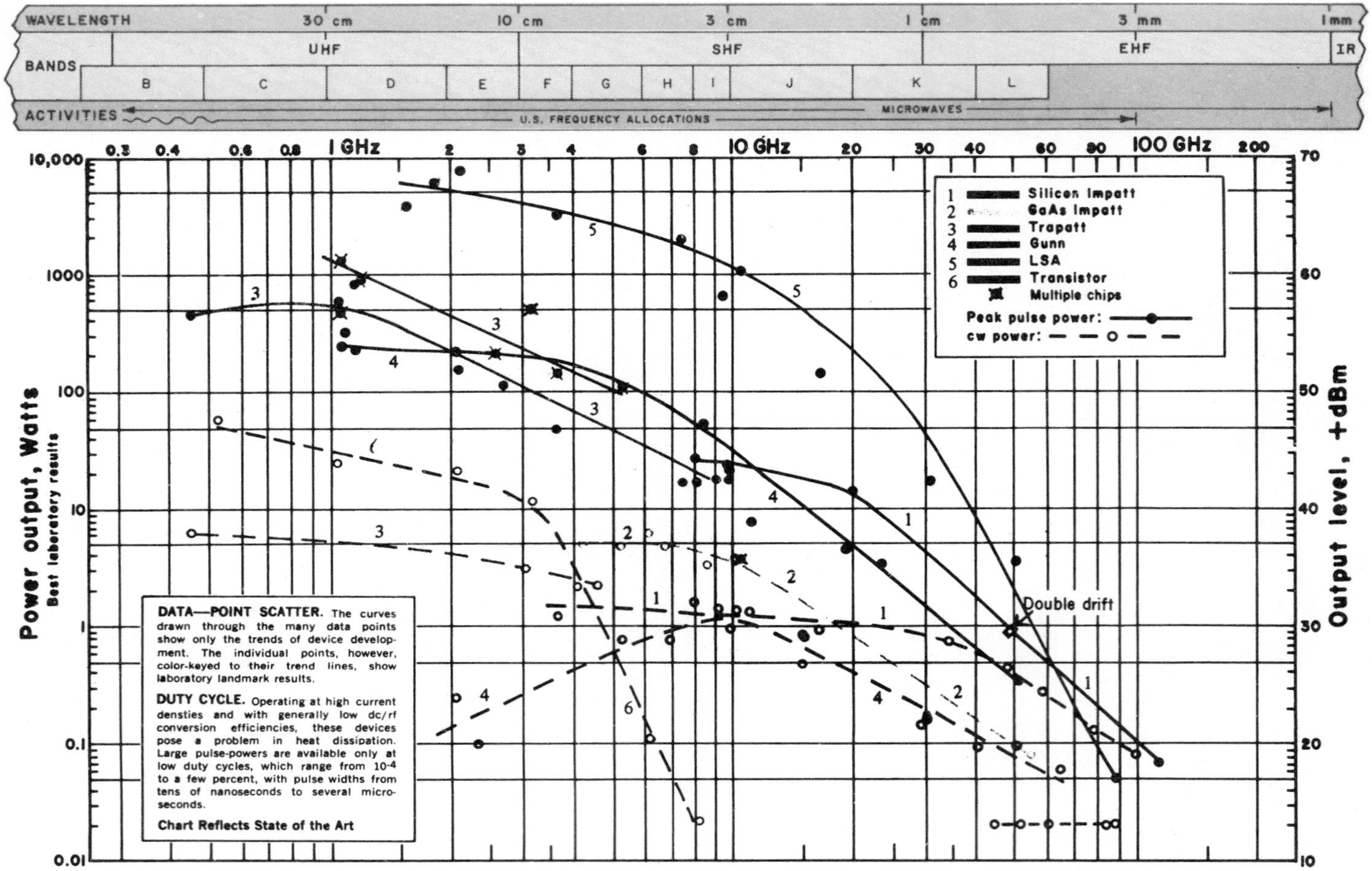

FIG. 6-7-1. Power outputs in watts of microwave solid-state devices.

402

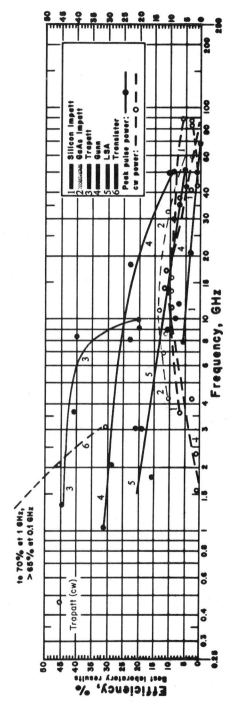

FIG. 6-7-2. Efficiency of microwave solid-state devices.

COMMERCIALLY AVAILABLE PACKAGED OSCILLATORS
Representative characteristics of significant sources

COMMERCIALLY AVAILABLE DIODES
Representative characteristics of significant diodes

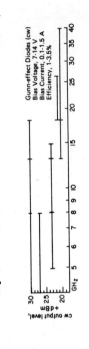

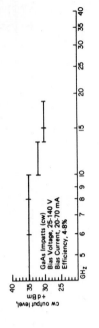

Notes:

F=fixed-tuned at a frequency within the range shown.

T=tunable over a small range either side of a fixed frequency within the range shown.

T_m=mech.; T_e=elec. (w/varactor or yig). Tuning bandwidths are typically 0.1 to 0.2 GHz, 0.5 GHz, or 0.8 to 1 GHz.

FIG. 6-7-3. Power output of commercially available microwave solid-state diodes and oscillators.

frequency limitations in the microwave frequency range just as conventional vacuum tubes do. It can be clearly seen from Fig. 5-4-1 that the power output of microwave transistors decreases rapidly while their operating frequencies are above 3 GHz. Microwave transistors are available for frequencies up to about 12 GHz with power levels from about 1 mW to 10 W. The primary use of these devices is in stripline-type applications for local oscillator operation. Efficiencies are generally high, ranging from about 15% and up. Microwave tunnel diodes are in the low milliwatt range, with frequency of operation up to about 8 GHz. Tunnel diodes find particular use in many signal circuits because of their high speed, high efficiency, frequency stability, and low noise. Since the dynamic negative resistance of a tunnel diode is exhibited at a very low voltage range of 0.1 to 0.3 volt, its power output is limited to the region of milliwatts. Microwave Schottky barrier-gate field-effect transistors have the capability of amplifying small signals up to the frequency range of X band with low noise figure and have been used recently in replacing the parametric amplifier in airborne radar systems because the latter is complicated in fabrication and expensive in production.

Gunn-Effect Oscillators: These oscillators cover the frequency range from about 4 to 60 GHz. Their power outputs vary from 10 mW to 700 mW. Gunn-effect diodes are also referred to as Transferred Electron Devices (TEDs). They have the advantages of broadband, low-noise, low-voltage and simple power supply, and are linear as an amplifier. These crystals will oscillate by themselves in a microwave cavity.

IMPATT Oscillators: The IMPATT devices, also referred to as Avalanche Transit-Time Devices, generally operate from about 1 to 60 GHz, with power outputs ranging from 20 mW to about 5 watts. Their special features are high power output and high efficiency, but their disadvantages are high noise, narrow band, nonlinear amplification, and they require a high applied voltage. Some experimental work has demonstrated the IMPATT devices to 150 GHz.

LSA Devices: These devices offer promise to 90 GHz. At present LSA devices appear best for pulsed-power sources with the capability of much higher peak power than IMPATT or Gunn diodes.

Quantum-Electronic Solid-State Devices: A basic study of the operating principles of masers and lasers was given in this chapter. Also discussed in detail were the solid-state ruby laser and semiconductor junction lasers. In order for laser action to occur, three factors are required:

1. A pumping (or excitation) mechanism to stimulate the atoms from a lower energy level to higher energy levels
2. A large population inversion
3. A resonant cavity to produce back and forth the reflection of radiation.

Parametric Amplifiers: A parametric amplifier is a device using a time-varying capacitance or inductance for signal amplification with a low noise figure. However, since the parametric amplifier is complicated to fabricate and expensive to produce, there is a tendency in microwave engineering to replace it by the microwave field-effect transistor (FET) amplifier in airborne radar systems.

Infrared Devices and Systems: Infrared devices are those that sense the passive infrared radiation emitted by some target or source and process it to the point where a visual image of that target or source is formed. In recent years there has been an increase in the research, design, development, and production of various infrared devices and systems for military applications at night—or during the day when vision is diminished by fog, haze, smoke, or dust. It has been demonstrated that a missile can be fired accurately at a target at night by using an infrared device as far away as 1.8 km.

REFERENCES

[1] PERLMAN, B. S. et al., Microwave properties and applications of negative conductance transferred electron devices. *Proc. IEEE*, **59**, No. 8 1222–1237, August 1971.

[2] STERZER, F., Transferred electron (Gunn) amplifiers and oscillators for microwave applications. *Proc. IEEE*, **59**, No. 8, 1155–1163, August 1971.

[3] COOKE, HARRY F., Microwave transistors: Theory and design. *Proc. IEEE*, **59**, No. 8, 1163–1181, August 1971.

[4] JOHNSON, E. O., Physical limitations on frequency and power parameters of transistors. *RCA Review*, **26**, No. 6, 163–177, June 1965.

[5] STERZER, F., Tunnel diode devices in Leo Young (Ed.), *Advances in Microwaves*, by vol. 2, pp. 1–41. Academic Press, New York, 1967.

[6] SZE, S. M., and R. M. RYDER, Microwave avalanche diodes. *Proc. IEEE*, **59**, No. 8, 1140–1154, August 1971.

[7] EVANS, W. J., and D. L. SCHARFETTER, Characterization of avalanche diode TRAPATT oscillators. *IEEE Trans. on Electron Devices*, **ED-17**, No. 5, 397–404, May 1970.

[8] TOWNES, C. H., et al., The Maser—A new type of microwave amplifier, frequency standard, and spectrometer. *Physical Review*, **99**, No. 4, 1264–1274, August 15, 1955.

[9] BURNS, G., and M. I. NATHAN, *p-n* Junction lasers. *Proc. IEEE*, **52**, 770–794, July 1964.

[10] SZE, S. M., *Physics of Semiconductor Devices*, pp. 20–21, 284. John Wiley & Sons, New York, 1969.

[11] KELLER, R. E., LSA Transmitters. *Microwave Journal*, **14**, No. 6, 50–54, June 1971.

[12] Sobol, H., and F. STERZER, Solid-state microwave power sources. *IEEE Spectrum*, **9**, 32, April 1972.

[13] SHOCKLEY, W., The theory of *p-n* junction in semiconductors and *p-n* junction transistors. *Bell System Tech. J.*, **28**, 435 (1949).

[14] JOHNSON, E. O., et al., Simple general analysis of amplifier devices with emitter, control, and collector functions. *Proc. IRE*, **47**, 407–418, March 1959.

[15] EDWARDS, R., Fabrication control is key to microwave performance. *Electronics*, pp. 109–113, February 1968.

[16] SZE, S. M., and H. K. GUMMEL, Appraisal of semiconductor-metal-semiconductor transistors. *Solid State Electronics*, **9**, 751 (1966).

[17] EARLY, J. M., Maximum rapidly switchable power density in junction triodes. *IRE Trans. on Electron Devices*, **ED-6**, 322–325 (1959).

[18] DELOACH, B. C., JR., Recent advances in solid state microwave generators. *Advances in Microwaves*, vol. 2. Academic Press, New York, 1967.

[19] ESAKI, L., New phenomenon in narrow germanium *p-n* junction. *Physical Review*, **109**, 603–604 (1959).

[20] BURRUS, C. A., Millimeter wave Esaki diode oscillator. *Proc. IRE*, **48**, 2024–2030 (1960).

[21] SCHNEIDER, M. V., A 5-gigacycle tunnel diode oscillator with 9 milliwatt output from a single diode. *Bell System Tech. J.*, **42**, 2972–2974 (1963).

[22] RCA, A 10-milliwatt X-band tunnel diode oscillator. *Third Quarter Progr. Report*, Contract No. DA 36-039 AMC-03195 (E) for U.S. Army Electronics Material Agency, 1964.

[23] RCA, A 10-milliwatt X-band tunnel diode oscillator. *Final Report*, Contract No. DA 36-039 AMC-03195 (E) for U.S. Army Electronics Material Agency, 1964.

[24] YOUNG, D. T., C. A. BURRUS, and R. C. SHAW, High efficiency millimeter wave tunnel diode oscillators. *Proc. IEEE*, **52**, 1260–1265 (1964).

[25] SHOCKLEY, W., A unipolar "field-effect" transistor. *Proc. IRE*, **40**, No. 11, 1365–1376, November 1952.

[26] MEAD, C. A., Schottky barrier gate field-effect transistor. *Proc. IEEE*, **54**, No. 2, 307–308, February 1966.

[27] HOOPER, W. W., et al., An epitaxial GaAs field-effect transistor. *Proc. IEEE*, **55**, No. 7, 1237–1238, July 1967.

[28] WOLF, P., Microwave properties of Schottky-barrier field-effect transistors. *IBM. J. Res. Develop.*, **14**, 125–141, March 1970.

[29] ZULEEG, R., and K. LEHOVEC, High frequency and temperature characteristics of GaAs junction field-effect transistors in the hot electron range. *Proc. Symp. GaAs, Institute of Physics Conf. Series No. 9*, pp. 240–245, 1970.

[30] HOWER, P. L., and N. G. BECHTEL, Current saturation and small-signal characteristics of GaAs field-effect transistors. *IEEE Trans. on Electron Devices*, **ED-20**, No. 3, 213–220, March 1973.

[31] LEHOVEC, K., and R. ZULEEG, Voltage-current characteristics of GaAs J-FET's in the hot electron range. *Solid-State Electronics*, vol. 13, pp. 1415–1426. Pergamon Press, Great Britain, 1970.

[32] DRANGEID, K. E., and R. SOMMERHALDER, Dynamic performance of Schottky-Barrier field-effect transistors. *IBM. J. Res. Develop.*, **14**, 82–94, March 1970.

[33] HOWER, P. L., et al., The Schottky barrier Gallium Arsenide field-effect transistor. *Proc. Symp. GaAs, Institute of Physics and Physical Society Conf. Series No. 7*, pp. 187–194, 1968.

[34] DAS, M. B., and P. SCHMIDT, High-frequency limitations of abrupt-junction FET's. *IEEE Trans. on Electron Devices*, **ED-20**, No. 9, 779–792, September 1973.

[35] BAECHTOLD, W., Noise behavior of Schottky barrier gate field-effect transistors at microwave frequencies. *IEEE Trans. on Electron Devices*, **ED-18**, No. 2, 97–104, February 1971.

[36] BAECHTOLD, W., Noise behavior of GaAs field-effect transistors with short gate lengths. *IEEE Trans. on Electron Devices*, **ED-19**, No. 5, 674–680, May 1972.

[37] VAN DER ZIEL, A., Gate noise in field-effect transistors at moderately high frequencies. *Proc. IEEE*, **51**, No. 3, 461–467, March 1963.

[38] VAN DER ZIEL, A., and ERO, J. W., Small-signal, high-frequency theory of field-effect transistors. *IEEE Trans. on Electron Devices*, **ED-11**, No. 2, 128–135, February 1964.

[39] VAN DER ZIEL, A., Thermal noise in field-effect transistors. *Proc. IRE*, **50**, 108–112, January 1962.

[40] SHOCKLEY, W., Negative resistance arising from transit time in semiconductor diodes. *Bell System Tech. J.*, **33**, 799–826, July 1954.

[41] RIDLEY, B. K., and T. B. WATKINS, The possibility of negative resistance effects in semiconductors. *Proc. Phys. Soc.*, **78**, 293–304, August 1961.

[42] HILSUM, C., Transferred electron amplifiers and oscillators. *Proc. IEEE*, **50**, 185–189, February 1962.

[43] GUNN, J. B., Microwave oscillations of current in III-V semiconductors. *Solid-state Communications*, **1**, 89–91, September 1963.

[44] RIDLEY, B. K., Specific negative resistance in solids. *Proc. Phys. Soc. (London)*, **82**, 954–966, December 1963.

[45] KROEMER, HERBERT, Theory of the Gunn effect. *Proc. IEEE*, **52**, 1736 (1964).

[46] GUNN, J. B., Microwave oscillations of current in III-V semiconductors. *Solid State Communications*, **1**, 88–91 (1963).

[47] GUNN, J. B., and B. J. ELLIOTT, *Phys. Letters*, **22** (1966).

[48] GUNN, J. B., Instabilities of current in III-V semiconductors. *IBM J. Res. Develop.*, **8**, 141–159, April 1964.

[49] GUNN, J. B., Instabilities of current and of potential distribution in GaAs and InP. 7th Int. Conf. on Physics of Semiconductor "Plasma Effects in Solids," pp. 199–207, Tokyo, 1964.

[50] KROEMER, HERBERT, Proposed negative-mass microwave amplifier. *Physical Review*, **109**, No. 5, 1856, March 1, 1958.

[51] KROEMER, HERBERT, The physical principles of a negative-mass amplifier. *Proc. IRE*, **47**, 397–406, March 1959.

[52] COPELAND, JOHN A., Bulk negative-resistance semiconductor devices. *IEEE Spectrum*, May 1967.

[53] BUTCHER, P. N., and W. FAWCETT, Calculation of the velocity-field characteristics of gallium arsenide. *Phys. Letters*, **21**, 498 (1966).

[54] CONWELL, E. M., and M. O. VASSELL, High-field distribution function in GaAs. *IEEE Trans. on Electron Devices*, **ED-13**, 22 (1966).

[55] RUCH, J. G., and G. S. KINO, Measurement of the velocity-field characteristics of gallium arsenide. *Appl. Phys. Letters*, **10**, 50 (1967).

[56] KROEMER, HERBERT, Negative conductance in semiconductors. *IEEE Spectrum*, **5**, No. 1, 47, January 1968.

[57] SCHOTTKY, W., Naturwiss., **26**, 843, 1938.

[60] COPELAND, JOHN A., Characterization of bulk negative-resistance diode behavior. *IEEE Trans. on Electron Devices*, **ED-14**, No. 9, September 1967.

[61] ELLIOTT, B. J., J. G. GUNN, and J. C. McGRODDY, Bulk negative differential conductivity and traveling domains in *n*-type germanium. *Appl. Phys. Letters*, **11**, 253 (1967).

[62] COPELAND, J. A., Stable space-charge layers in two-valley semiconductors. *J. Appl. Phys.*, **37**, No. 9, 3602, August 1966.

[63] GUNN, J. B., Effect of domain and circuit properties on oscillations in GaAs. *IBM J. Res. Develop.*, pp. 310–320, July 1966.

[64] HOBSON, G. S., Some properties of Gunn-effect oscillations in a biconical cavity. *IEEE Trans. on Electron Devices*, **ED-14**, No. 9, 526–531, September 1967.

[65] THIM, H. W., Computer study of bulk GaAs devices with random one-dimensional doping fluctuations. *J. Appl. Phys.*, **39**, 3897 (1968).

[66] THIM, H. W., and M. R. BARBER, Observation of multiple high-field domains in *n*-GaAs. *Proc. IEEE*, **56**, 110 (1968).

[67] UENOHARA, M., Bulk gallium arsenide devices in H. A. Watson (Ed.), *Microwave Semiconductor Devices and Their Circuit Application*, Chapter 16, McGraw-Hill, New York, 1969.

[70] DOW, D. G., et al., Pulsed GaAs oscillators. Intern. Electron Devices Meeting, Washington, D.C. 1965.

[71] HAKKI, B. W., CW microwave oscillations in GaAs. Intern. Solid State Circuits Conf., Philadelphia, Pa., 1965.

[72] REYNOLDS, J. F., et al., High-efficiency transferred—electron oscillators. *Proc. IEEE*, **57**, 1692 (1969).

[73] QUIST, T. M., and A. F. FOYT, S-band GaAs Gunn effect oscillators. MIT Lincoln Lab. *Solid State Res. Repts.*, **1**, 21–23 (1965).

[74] HAKKI, B. W., Private communication.

[75] BUTCHER, P. N., and C. HILSUM, Bulk negative-resistance effects in semi-conductors. Intern. Electron Devices Meeting, Washington, D.C., 1965.

[76] CALIFANO, F. P., High-efficiency X-band oscillators. *Proc. IEEE*, **57**, 251 (1969).

[77] NARAYAN, S. Y., et al., High-power CW transferred electron oscillators. *Electron Letters*, **6**, 17 (1970).

[78] EDRIDGE, A. L., et al., Gunn-effect oscillators. Paper presented at 1969 European Microwave Conf., London.

[79] FRANK, F. B., and G. F. DAY, High CW power K-band Gunn oscillators. *Proc. IEEE*, **57**, 339 (1969).

[80] HILSUM, C., New developments in transferred electron effects. Proc. 3rd Conference on High Frequency Generation and Amplification. Devices and Applications. August 17–19, 1971, Cornell University.

[81] HAKKI, B. W., GaAs post-threshold microwave amplifier, mixer, and oscillator. *Proc. IEEE (Letters)*, **54**, 299–300, February 1966.

[82] THIM, H. W., Linear microwave amplification with Gunn oscillators. *IEEE Trans. on Electron Devices*, **ED-14**, No. 9, September 1967.

[83] PERLMAN, B. S., et al., Microwave properties and applications of negative conductance transferred electron devices. *Proc. IEEE*, **59**, No. 8, August 1971.

[84] COPELAND, JOHN A., CW operation of LSA oscillator diodes—44 to 88 GHz. *Bell System Tech. J.*, **46**, 284–287, January 1967.

[85] WILSON, W. E., Pulsed LSA and TRAPATT sources for microwave systems. *Microwave Journal*, **14**, No. 8, August 1971.

[90] G. B. L., Three-level oscillator in indium phosphide, *Physics Today*, **23**, 19–20, December 1970.

[91] COLLIVER, D., and BRIAN PREW, Indium phosphide: Is it practical for solid state microwave sources? *Electronics*, pp. 110–113, April 10, 1972.

[92] TAYLOR, BRIAN C., and DAVID J. COLLIVER, Indium phosphide microwave oscillators. *IEEE Trans. on Electron Devices*, **ED-18**, No. 10, 835–840, October 1971.

[93] FOYT, A. G., and A. L. MCWHORTER, The Gunn effect in polar semiconductors. *IEEE Trans. on Electron Devices*, **ED-13**, 79–87, January 1966.

[94] LUDWIG, G. W., R. E. HALSTED, and M. AVEN, Current saturation and instability in CdTe and ZnSe. *IEEE Trans. on Electron Devices*, **ED-13**, 671, August–September 1966.

[95] LUDWIG, G. W., Gunn effect in CdTe. *IEEE Trans. on Electron Devices*, **ED-14**, No. 9, 547–551, September 1967.

[96] BUTCHER, P. N., and W. FAWCETT, *Proc. Phys. Soc. (London)*, **86**, 1205 (1965).

[97] OLIVER, M. R., and A. G. FOYT, The Gunn effect in *n*-CdTe. *IEEE Trans. on Electron Devices*, **ED-14**, No. 9, 617–618, September 1967.

[100] READ, W. T., A proposed high-frequency negative-resistance diode. *Bell System Tech. J.*, **37**, 401–446 (1958).

[101] COLEMAN, D. J., JR., and S. M. SZE, A low-noise metal-semiconductor-metal (MSM) microwave oscillator. *Bell System Tech. J.*, **50**, 1695–1699, May–June 1971.

[102] LEE, C. A., et al., The Read diode, an avalanche, transit-time, negative-resistance oscillator. *Appl. Phys. Letters*, **6**, 89 (1965).

[103] JOHNSTON, R. L., B. C. DELOACH, and G. B. COHEN, A silicon diode microwave oscillator. *Bell System Tech. J.*, **44**, 369–372, February 1965.

[104] GILDEN, M., and M. E. HINES, Electronic tuning effects in the Read microwave avalanche diode. *IEEE Trans. on Electron Devices*, **ED-13**, January 1966.

[105] CHANG, K. K. N., Avalanche diodes as UHF and *L*-band sources. *RCA Review*, **30**, No. 1, 3 (1969).

[106] IGLESIAS, D. E., and W. J. EVANS, High-efficiency CW IMPATT operation, *Proc. IEEE*, **56**, 1610 (1968).

[107] JOHNSON, R. L., et al., High-efficiency oscillations in germanium avalanche diodes below transit-time frequency. *Proc. IEEE*, **56**, 1611 (1968).

[108] IGLESIAS, D. E., Circuit for testing high-efficiency IMPATT. *Proc. IEEE*, **55**, 2065 (1967).

[109] MELICK, D. R., High-frequency pulsed GaAs avalanche diodes. *Proc. IEEE*, **55**, 435 (1967).

[110] GILDEN, M., and W. MORONEY, High power pulsed avalanche diode oscillators for microwave frequencies. *Proc. IEEE*, **55**, 1617 (1967).

[111] SWAN, C. B., et al., Composite avalanche diode structure for increased power capacity. *IEEE Trans. on Electron Devices*, **ED-14**, 584 (1967).

[112] LIU, S. G., GaAs avalanche microwave oscillator with 1-W power output. *Proc. IEEE*, **55**, 689 (1967).

[113] IRWIN, J. C., GaAs avalanche microwave oscillators. *IEEE Trans. on Electron Devices*, **ED-13**, 208 (1966).

[114] EDWARDS, R., et al., Millimeter-wave silicon IMPATT diodes. Proc. IEEE Conference on Electron Devices, Washington, D.C., 1969.

[120] PRAGER, H. J., et al., High-power, high-efficiency silicon avalanche diodes at ultra high frequencies. *Proc. IEEE (Letters)*, **55**, 586–587, April 1967.

[121] CLORFEINE, A. S., et al., A theory for the high-efficiency mode of oscillation in avalanche diodes. *RCA Review*, **30**, 397–421, September 1969.

[122] DELOACH, B. C., JR., and D. L. SCHARFETTER, Device physics of TRAPATT oscillators. *IEEE Trans. on Electron Devices*, **ED-17**, 9–21, January 1970.

[123] LIU, S. G., and J. J. RISKA, Fabrication and Performance of kilowatt L-band avalanche diodes. *RCA Review*, **31**, 3, March 1970.

[124] KOSTICHACK, D. F., UHF avalanche diode oscillator providing 400 watts peak power and 75 percent efficiency. *Proc. IEEE (Letters)*, **58**, 1282–1283, August 1970.

[125] WILSON, W. E., Pulsed LSA and TRAPATT sources for microwave systems. *Microwave Journal*, **14**, No. 8, 33–41, August 1971.

[130] COLEMAN, D. J., JR., and S. M. SZE, A low-noise metal-semiconductor-metal (MSM) microwave oscillator. *Bell System Tech. J.*, pp. 1675–1695, May–June 1971.

[135] GORDON, J. P., H. J. ZEIGER, and C. H. TOWNES, Molecular microwave oscillator and new hyperfine structure in the microwave spectrum of NH_3. *Phys. Rev.*, **95**, 282–284, 1954.

[136] BLOEMBERGEN, N., Proposal for a new type solid-state maser. *Phys. Rev.*, **104**, No. 2, 324–327, October 15, 1956.

[137] SCOVIL, H. E. D., G. FEHER, and H. SEIDEL, Operation of a solid-state maser. *Phy. Rev.*, **105**, 762–763, 1957.

[138] MAIMAN, T. H., Stimulated optical radiation in ruby masers. *Nature*, **187**, 493–494, August 6, 1960.

[139] SONOKIN, P. P., and M. J. STEVENSON, Stimulated infrared emission from trivalent uranium. *Phys. Rev. Letters*, **5**, 577 (1960).

[140] JAVAN, A., et al., Population inversion and continuous optical maser oscillation in a gas discharge, containing a He-Ne mixture. *Phys. Rev. Letters*, **6**, 106 (1961).

[141] DUMKE, W. P., Interband transitions and maser action. *Phys. Rev.*, **127**, 1559 (1962).

[142]　BURNS, G., and M. I. NATHAN, *p-n* junction lasers. *Proc. IEEE*, **52**, 770–794, July 1964.

[143]　BERGH, A. A., and P. J. DEAN, Light-emitting diodes. *Proc. IEEE*, **60**, No. 2, 156–224, February 1972.

[144]　ELECCION, MARCE, The family of lasers: A survey. *IEEE Spectrum*, **9**, No. 4, 23–40, March 1972.

[145]　LEVINE, A. K., Lasers. *American Scientists*, **51**, 14 (1963).

[146]　LORENZ, M. R., and M. H. PILKUHN, Semiconductor-diode light sources. *IEEE Spectrum*, **4**, No. 4, 87–96, April 1967.

[147]　NATHAN, M. I., Semiconductor lasers. *Proc. IEEE*, **54**, 1276 (1966).

[148]　BASOV, N. G., and A. M. PROKHOROV, Zh. Eksperim. i Tear. Fiz., **28**, 249 (1955) (English transl.: *Soviet Phys.—JETP* 1, 184 (1955)).

[149]　GUILLAUME, G. B., and J. M. DEBEVER, Solid-state communications. **2**, 145 (1965), also in Symposium on radiative recombination, 255 (1964), Dunod, Paris, 1964.

[150]　MAKHOV, G., et al., Maser action in ruby. *Phys. Rev.*, **109**, 1399–1400 (1958).

[151]　NISHIZAWA, J. I., and Y. WATANABE, *Electronics*. p. 117, December 11, 1967.

[152]　BASOV, N. G., et al., *Soviet Physics Uspekhi*, **3**, 7 (1961).

[153]　AIGRAIN, P., Unpublished lecture at the "International Conference on Solid-State Physics in Electronics and Telecommunications." Brussels, 1958.

[154]　GLICKSMAN, R., Technology and Design of GaAs Laser and Non-Coherent IR-Emitting Diodes.
Part I.　Solid State Technology, vol. 13, No. 9, pp. 29–35, September 1970.
Part II.　Solid State Technology, vol. 13, No. 10, pp. 39–44, October 1970.

[155]　HERRIOTT, D. R., Applications of Laser Light. *Scientific American*, **219**, No. 3, 141–156, September 1968.

[156]　FARADAY, M., On a peculiar class of acoustical figures; and certain forms assumed by a group of particles upon vibrating elastic surface. *Phil. Trans. Roy. Soc. (London)*, **121**, 299–318, May 1831.

[157]　LORD RAYLEIGH, and J. W. STRUTT, On the crispations of fluid resting upon a vibrating support. *Phil. Mag.*, **16**, 50–53, July 1883.

[158]　VAN DER ZIEL, A., On the mixing properties of nonlinear capacitances. *J. Applied Phys.*, **19**, p. 999–1006, November 1948.

[159]　LANDON, V. D., The use of ferrite cored coils as converters, amplifiers, and oscillators. *RCA Review*, **10**, p. 387–396, September 1949.

[160]　SUHL, H., Proposal for a ferromagnetic amplifier in the microwave range. *Phys. Rev.*, **106**, 384–385, April 15, 1957.

[161]　WEISS, M. T., A solid-state microwave amplifier and oscillator using ferrites. *Phys. Rev.*, **107**, 317, July 1957.

[162] MANLEY, J. M., and H. E. ROWE, Some general properties of nonlinear elements—Part I, General Energy Relations. *Proc. IRE*, **44**, 904–913, July 1956.

[163] BLACKWELL, L. A., and K. L. KOTZEBUE, *Semiconductor-Diode Parametric Amplifiers,* pp. 41, 42, 45, 53, 57, 62, 70. Prentice-Hall, Englewood Cliffs, N.J., 19´1.

[164] HERSCHEL, W., Investigation of the powers of the prismatic colours to heat and illuminate objects: With remarks that prove the different refrangibility of radiant heat. *Phil. Trans. Roy. Soc. (London)*, Pt. II, **90**, 255, 1800.

[165] HERSCHEL, W., Experiments on the refrangibility of the invisible rays of the sun. *Phil. Trans. Royal Soc. (London)*, Pt. II, **90**, 284, 1800.

[166] HERSCHEL, W., Experiments on the solar, and on the terrestrial rays that occasion heat. *Phil. Trans. Royal Soc. (London)*, Pt. II, **90**, 293, 437, 1800.

[167] HUDSON, R. D., JR., *Infrared System Engineering*, pp. 6, 21, 36, 40, 43, 45, 92, 102, 107, and 115. John Wiley & Sons, New York, 1968.

[168] BARR, E. S., The infrared pioneers—II. Macedonia Melloni. *Infrared Phys.*, p. 67, 1962.

[169] BARR, E. S., The infrared pioneers—III. Pierpont Langley. *Infrared Phys.* 3, p. 195, 1963.

[170] CASE, T. W., Notes on the change of resistance of certain substances in light. *Phys. Rev.*, **9**, 305–310, 1917.

[171] HUDSON, R. D., JR., and J. W. HUDSON, The military application of remote sensing by infrared. *Proc. IEEE*, **63**, No. 1, 104–128, January 1975.

[172] GEBBIE, H. A., et al., Atmospheric transmission in the 1 to 14 μm region. *Proc. Roy. Soc.*, **A206**, 87 (1951).

[173] PLANCK, M., *Theory of Heat Radiation*. Dover, New York. (A reprint of the 1910 edition.)

[174] JONES, R. C., Phenomenological description of the response and detecting ability of radiation detectors. *Proc. IRE*, **47**, No. 9, September 1959.

[175] RICHTER, J. J., Infrared thermal imaging. *Proc. IEEE Southeast Region 3 Conference*. Orlando, Fla. April 29, 1974.

[176] LEVINSTEIN, H., and J. MUDAR, Infrared detectors in remote sensing. *Proc. IEEE*, **63**, No. 1, 6–14, January 1975.

[177] JOHNSON, J., Analysis of image forming systems. *Proc. of Image Intensifier Symposium*, Ft. Belvoir, Va., AD220160, October 1958.

[178] ROSELL, F. A., Levels of visual discrimination for real scene objects vs. bar pattern resolution for aperture and noise limited imagery. *Proc. the National Aerospace and Electronics Conference*, pp. 327–334. Dayton, Ohio. January 10–12, 1975.

[179] SMITH F. D., Optical image evaluation and the transfer function. *Applied Optics*, **2**, No. 4, 335–350, April 1963.

[180] LIAO, SAMUEL Y., System performance of the detecting and ranging set (DRS) for the Navy A-6E TRAM aircraft. Report for Hughes Aircraft Company, El Segundo, Ca., August 1978.

[181] WOOD, J. T., Test and evaluation of thermal imaging systems. Northeast Electronics Research and Engineering Meeting. Record part 3, November 1973.

SUGGESTED READINGS

1. BLACKWELL, L. A., and K. L. KOTZEBUE, *Semiconductor-Diode Parametric Amplifiers*. Prentice-Hall, Englewood Cliffs, N.J., 1961.

2. CHANG, K. K. N., *Parametric and Tunnel Diodes*. Prentice-Hall, Englewood Cliffs, N.J., 1964.

3. EASTMAN, L. F., *Gallium Arsenide Microwave Bulk and Transit-Time Devices*. Artech House, Dedham, Mass., 1973.

4. HADDAD, GEORGE, *Avalanche Transit-Time Devices*. Artech House, Dedham, Mass., 1973.

5. HEWLETT PACKARD, *S-Parameter Design*, Application Note 154, April 1972.

6. HEWLETT PACKARD, *S-Parameters . . . Circuit Analysis and Design*, Application Note 95, September 1968.

7. HUDSON, R. D., JR., *Infrared System Engineering*. John Wiley & Sons, New York, 1969.

8. HUDSON, R. D., JR., and J. W. HUDSON, (Eds.), *Infrared Detectors*. Dowden, Hutchinson, and Ross, Stroudsburg, Pa., 1975.

9. IEEE Proceedings, Special Issue on Infrared Technology for Remote Sensing. vol. 63, No. 1, January 1975.

10. IRE Proceedings, Special Issue on Infrared Physics and Technology. vol. 47, No. 9, September 1959.

11. PENFIELD, P., and R. P. RAFUSE, *Varactor Applications*. MIT Press, Cambridge, Mass., 1962.

12. SMITH, W. V., *Laser Applications*. Artech House, Dedham, Mass., 1972.

13. SOOHOO, RONALD F., *Microwave Electronics*, Chapters 11, 12, and 13. Addison-Wesley Publishing Company, Reading, Mass., 1971.

14. STREETMAN, BEN G., *Solid State Electronics Devices*, Chapters 7 and 12. Prentice-Hall, Englewood Cliffs, N.J., 1972.

15. SZE, S. M., *Physics of Semiconductor Devices*, Chapters 5, 8, 13, and 14. Wiley-Interscience, New York, 1969.

16. WATSON, H. A., *Microwave Semiconductor Devices and Their Circuit Applications*, Chapters 14, 15, 16, 17, and 18. McGraw-Hill Book Company, New York, 1969.

17. YARIV, A., and J. E. PEARSON, Parametric Processes. Vol. 1, Part 1 of *Progress in Quantum Electronics*. Pergamon Press, New York, 1969.

Microstrip Transmission Lines

7-0   Introduction

In Chapters 5 and 6 microwave tubes and microwave solid-state devices were described and discussed in detail. All microwave tubes are high-
output power devices, and their circuits consist of
the conventional transmission links, such as coaxial
lines and waveguides. The transmission-line theory
pertaining thereto may be found in standard refer-
ences [1]. The microwave solid-state device is usually fabricated as a semiconducting chip with a
volume on the order to 500 to 5000 cubic milli-
inches. The method of applying signals to and
extracting output power from the chips is entirely
different from that used for tubes. Microwave
integrated circuits with microstrip lines are commonly incorporated with the chips. Here we are

chapter

7

417

concerned with microstrip lines. Prior to 1965 nearly all microwave equipment utilized coaxial, waveguide, or stripline circuits. In recent years, with the advent of microwave integrated circuits, microstrip lines have been extensively used, for they provide one free and accessible surface on which solid-state devices can be placed. Therefore a microstrip line is also called an *open-strip* line. The next chapter covers the application of microstrip lines to the technology and design of hybrid integrated microcircuits.

Modes on microstrip lines are only quasi-Tranverse Electric and Magnetic (TEM). Thus the theory of TEM coupled lines applies only approximately. Radiation loss in microstrip lines is a problem, particularly at such discontinuities as short-circuit posts, corners, and so on. However, the use of thin high-dielectric materials considerably reduces the radiation loss of the open strip. A microstrip line has an advantage over the balanced-strip line because the open strip has better interconnection features and easy fabrication. The circuit analysis of a microstrip line mounted on an infinite dielectric substrate over an infinite ground plane has been made by several researchers [2] to [5]. The numerical analysis of microstrip lines, however, requires large digital computers, whereas microstrip line problems can generally be solved by conformal transformations without requiring complete numerical calculation.

7-1 Characteristic Impedance of Microstrip Lines

Microstrip lines are extensively used to interconnect high-speed logic circuits in digital computers because they can be fabricated by automated techniques and they provide the required uniform signal paths. Figure 7-1-1 shows the cross-sectional geometry of several types of transmission lines for comparison.

From part (d) it can be seen that the characteristic impedance of a microstrip line is a function of the strip-line width, the strip-line thickness, the distance between the line and the ground plane, and the relative dielectric constant of the board material. There are different methods for determining the characteristic impedance of a microstrip line. The field equation method was used by several authors [3] to [5] for calculating an accurate value of the characteristic impedance. However, it requires the use of a large digital computer and is complicated. Another method is to derive the characteristic impedance equation of a microstrip line from a certain well-known equation with some changes [2]. This method may be called a comparative or an indirect method, and it is described as follows. The well-known equation of the characteristic impedance of a wire-over-ground transmission line as shown in Fig. 7-1-1(b) is given by

$$Z_0 = \frac{60}{\sqrt{\epsilon_r}} \, \ell n \, \frac{4h}{d} \qquad \text{for } h \gg d \qquad (7\text{-}1\text{-}1)$$

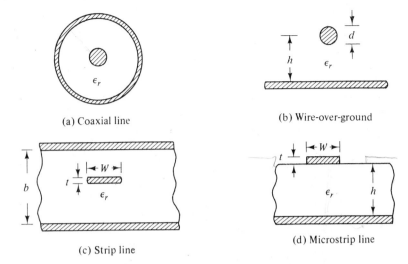

(a) Coaxial line   (b) Wire-over-ground

(c) Strip line   (d) Microstrip line

FIG. 7-1-1. Cross sections of several types of transmission lines.

where ϵ_r = relative dielectric constant of the ambient medium
h = distance between the center of the wire and the ground plane,
d = diameter of the wire

If effective or equivalent values of the relative dielectric constant ϵ_r of the ambient medium and the diameter d of the wire can be determined for the microstrip line, the characteristic impedance of the microstrip line can be calculated.

7-1-1. Effective Dielectric Constant ϵ_{re}

For a homogeneous dielectric medium, the propagation-delay time per unit length is

$$T_d = \sqrt{\mu\epsilon} \qquad (7\text{-}1\text{-}2)$$

where μ = permeability of the medium
ϵ = permittivity of the medium

In free space, the propagation-delay time is given by

$$T_{df} = \sqrt{\mu_0\epsilon_0} = 3.333 \text{ ns/m} = 1.016 \text{ ns/ft} \qquad (7\text{-}1\text{-}3)$$

where $\mu_0 = 4\pi \times 10^{-7}$ henry/m $= 3.83 \times 10^{-7}$ henry/ft
$\epsilon_0 = 8.85 \times 10^{-12}$ farad/m $= 2.69 \times 10^{-12}$ farad/ft

In transmission lines used for interconnection purposes, the relative permeability is unity. Consequently, the propagation-delay time for a line in a nonmagnetic medium is

$$T_d = 1.016\sqrt{\epsilon_r} \qquad \text{ns/ft} \qquad (7\text{-}1\text{-}4)$$

The effective relative dielectric constant for a microstrip line can be related to the relative dielectric constant of the board material. DiGiacomo and coworkers [6] discovered an empirical equation for the effective relative dielectric constant of a microstrip line by measuring the propagation-delay time and the relative dielectric constant of several board materials, such as fiber-glass-epoxy and nylon phenolic. His empirical equation, as shown in Fig. 7-1-2, is given by

$$\epsilon_{re} = 0.475\epsilon_r + 0.67 \tag{7-1-5}$$

where ϵ_r = relative dielectric constant of the board material
           ϵ_{re} = effective relative dielectric constant for a microstrip line

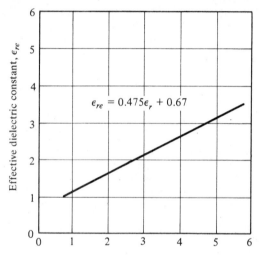

FIG. 7-1-2. Effective dielectric constant as a function of relative dielectric constant for a microstrip line. (*After H. R. Kaupp [2]; reprinted by permission of the IEEE, Inc.*)

7-1-2. Transformation of Rectangular Conductor into an Equivalent Circular Conductor.

The cross section of a microstrip line is rectangular. It is desirable to transform the rectangular conductor into an equivalent circular conductor. Springfield [13] discovered an empirical equation for the transformation. His equation is

$$d = 0.67 \, w\left(0.8 + \frac{t}{w}\right) \tag{7-1-6}$$

where d = diameter of the wire over ground
           W = width of the microstrip line
           t = thickness of the microstrip line

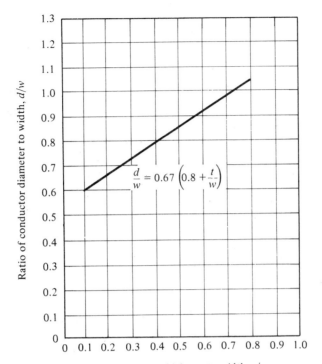

FIG. 7-1-3. Relationship between a round and rectangular conductor that is far from its ground plane. (*After H. R. Kaupp [2]; reprinted by permission of the IEEE, Inc.*)

The limitation of the ratios of thickness to width is between 0.1 and 0.8 as indicated in Fig. 7-1-3.

7-1-3. Characteristic Impedance Equation

Substitution of Eq. (7-1-3) for the dielectric constant and Eq. (7-1-6) for the equivalent diameter in Eq. (7-1-1) yields

$$Z_0 = \frac{87}{\sqrt{\epsilon_r + 1.41}} \ln\left(\frac{5.98h}{0.8w + t}\right) \qquad \text{for } h < 0.8w \qquad (7\text{-}1\text{-}7)$$

where ϵ_r = relative dielectric constant of the board material
h = distance between the microstrip line and ground
W = width of the microstrip line
t = thickness of the microstrip line

Equation (7-1-7) is the characteristic impedance equation for a narrow microstrip line. The velocity of propagation is

$$v = \frac{c}{\sqrt{\epsilon_{re}}} = \frac{3 \times 10^8}{\sqrt{\epsilon_r}} \qquad \text{m/s} \qquad (7\text{-}1\text{-}8)$$

The characteristic impedance for a wide microstrip line was derived by Assadourian and others [7], and it is expressed by

$$Z_0 = \frac{h}{w}\sqrt{\frac{\mu}{\epsilon}} = \frac{377}{\sqrt{\epsilon_r}}\frac{h}{w} \qquad \text{for } w \gg h \qquad (7\text{-}1\text{-}9)$$

7-1-4. Limitations of Equation (7-1-7)

Most microstrip lines are made from boards of copper with a thickness of 1.4 or 2.8 mils (1 or 2 ounces of copper per square foot). The narrowest widths of lines in production are about 0.005 to 0.010 inch. Line widths are usually less than 0.020 inch; consequently, the ratios of thickness to width of less than 0.1 are uncommon. The straight-line approximation from Eq. (7-1-6) limits the ratio of thickness to width to be between 0.1 and 0.8 for a very accurate value of characteristic impedance.

Since the dielectric constant of materials of interest does not vary excessively with frequency, the dielectric constant of a microstrip line can be considered independent of frequency. The validity of Eq. (7-1-7) is doubtful for values of dielectric thickness h less than 80% of the line width w. The typical values for the characteristic impedance of a microstrip line vary from 50 to 150 Ω if the values of parameters vary from $\epsilon_r = 5.23$, $t = 2.8$ mils, $w = 10$ mils, and $h = 8$ mils to $\epsilon_r = 2.9$, $t = 2.8$ mils, $w = 10$ mils, and $h = 67$ mils [2].

7-2 Losses in Microstrip Lines

Microstrip transmission lines consisting of a conductive ribbon attached to a dielectric sheet with conductive backing (see Fig. 7-2-1), are widely used in both microwave and computer technology. Because such lines are easily fabricated by printed-circuit manufacturing techniques, they have undoubted economic and technical merits.

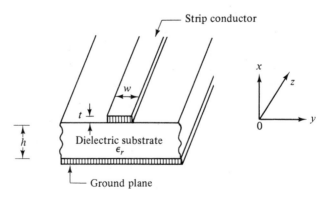

FIG. 7-2-1. Diagram for a microstrip line.

The characteristic impedance and wave-propagation velocity of a microstrip line have been analyzed in the previous section. The other characteristic of the microstrip line is its attenuation. The attenuation constant of the dominant microstrip mode depends on geometrical factors, on the electronic properties of the substrate and conductors, and on the frequency. For a nonmagnetic dielectric substrate, there are two types of losses in the dominant microstrip mode, namely dielectric loss in the substrate and ohmic skin loss in the strip conductor and the ground plane. The sum of these two losses may be expressed as losses per unit length in terms of an attenuation factor α. From ordinary transmission-line theory, the power carried by a wave traveling in the positive z direction is given by

$$P = \frac{1}{2} VI^* = \frac{1}{2}(V_+ e^{-\alpha z} I_+ e^{-\alpha z}) = \frac{1}{2} \frac{|V_+|^2}{Z_0} e^{-2\alpha z} = P_0 e^{-2\alpha z} \quad (7\text{-}2\text{-}1)$$

where $P_0 = |V_+|^2/(2Z_0)$ is the power at $z = 0$.

The attenuation constant α can be expressed as

$$\alpha = -\frac{dP/dz}{2P(z)} = \alpha_d + \alpha_c \quad (7\text{-}2\text{-}2)$$

where α_d is the dielectric attenuation constant and α_c is the ohmic attenuation constant.

The gradient of power in the z direction in Eq. (7-2-2) can be further expressed in terms of the power loss per unit length dissipated by the resistance and the power loss per unit length in the dielectric. That is,

$$-\frac{dP(z)}{dz} = -\frac{d}{dz}\left(\frac{1}{2} VI^*\right) = \frac{1}{2}\left(-\frac{dV}{dz}\right)I^* + \frac{1}{2}\left(-\frac{dI^*}{dz}\right)V$$

$$\doteq \frac{1}{2}(RI)I^* + \frac{1}{2}\sigma V^* V$$

$$= \frac{1}{2}|I|^2 R + \frac{1}{2}|V|^2 \sigma = P_c + P_d \quad (7\text{-}2\text{-}3)$$

where σ is the conductivity of the dielectric.

Substitution of Eq. (7-2-3) in (7-2-2) results in

$$\alpha_d \simeq \frac{P_d}{2P(z)} \quad \text{nepers/cm} \quad (7\text{-}2\text{-}4)$$

and

$$\alpha_c \simeq \frac{P_c}{2P(z)} \quad \text{nepers/cm} \quad (7\text{-}2\text{-}5)$$

7-2-1. Dielectric Losses

As described in Section 2-5-3, when the conductivity of a dielectric cannot be neglected, the electric and magnetic fields in the dielectric are no longer in time phase. The dielectric attenuation constant as shown in Eq.

(2-5-20) is given by

$$\alpha_d = \frac{\sigma}{2}\sqrt{\frac{\mu}{\epsilon}} \qquad \text{nepers/cm} \qquad (7\text{-}2\text{-}6)$$

This dielectric constant can be expressed in terms of dielectric loss tangent as shown in Eq. (2-5-17):

$$\tan \theta = \frac{\sigma}{\omega \epsilon} \qquad (7\text{-}2\text{-}7)$$

Then the dielectric attenuation constant is expressed by

$$\alpha_d = \frac{\omega}{2}\sqrt{\mu\epsilon} \tan \theta \qquad \text{nepers/cm} \qquad (7\text{-}2\text{-}8)$$

Since the microstrip line is a nonmagnetic mixed dielectric system, the upper dielectric above the microstrip ribbon is air with no loss. Welch and Pratt [8] derived an expression for the attenuation constant of a dielectric substrate. Later on Pucel and coworkers [9] modified Welch's equation. The result is

$$\alpha_d = 4.34 \frac{q\sigma}{\sqrt{\epsilon_{re}}}\sqrt{\frac{\mu_0}{\epsilon_0}} = 1.634 \times 10^3 \frac{q\sigma}{\sqrt{\epsilon_{re}}} \qquad \text{dB/cm} \qquad (7\text{-}2\text{-}9)$$

In Eq. (7-2-9) the conversion factor of one neper equal to 8.686 dB has been replaced, ϵ_{re} is the effective dielectric constant of the substrate as given by Eq. (7-1-5), and q denotes the dielectric filling factor, defined by Wheeler [3] as

$$q = \frac{\epsilon_{re} - 1}{\epsilon_r - 1} \qquad (7\text{-}2\text{-}10)$$

Usually the attenuation constant is expressed per wavelength as

$$\alpha_d = 27.3 \left(\frac{q\epsilon_r}{\epsilon_{re}}\right)\frac{\tan \theta}{\lambda_g} \qquad \text{dB}/\lambda_g \qquad (7\text{-}2\text{-}11)$$

where $\lambda_g = \dfrac{\lambda_0}{\sqrt{\epsilon_{re}}}$ and λ_0 is the wavelength in free space or

$\lambda_g = \dfrac{c}{f\sqrt{\epsilon_{re}}}$ and c is the velocity of light in vacuum

It should be noted that if the loss tangent $\tan \theta$ is frequency independent, the dielectric attenuation per wavelength is also frequency independent. On the other hand, if the substrate conductivity is frequency independent, as for a semiconductor, the dielectric attenuation per unit is also frequency independent. Since q is a function of ϵ_r and w/h, the filling factors for the loss tangent $q\epsilon_r/\epsilon_{re}$ and for the conductivity $q/\sqrt{\epsilon_{re}}$ are also functions of these quantities. Figure 7-2-2 shows the loss tangent filling factor against w/h for a range of dielectric constants suitable for microwave integrated circuits. For most practical purposes, this factor can be approximated by unity. Figure 7-2-3 illustrates the product $\alpha_d \rho$ against w/h for two semiconducting substrates that are being used for integrated microwave circuits, such as

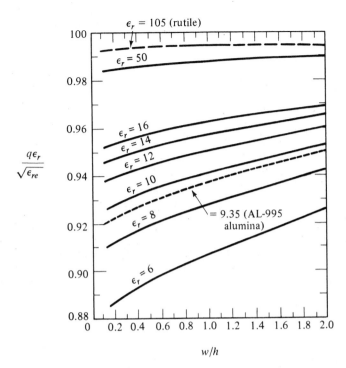

FIG. 7-2-2. Filling factor for loss tangent of microstrip substrate as a function of w/h. (*After R. A. Pucel et al. [9]; reprinted by permission of the IEEE, Inc.*)

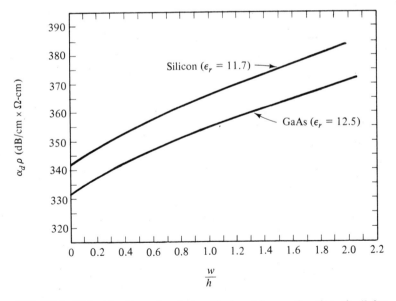

FIG. 7-2-3. Dielectric attenuation factor of microstrip as a function of w/h for silicon and gallium arsenide substrates. (*After R. A. Pucel et al. [9]; reprinted by permission of the IEEE, Inc.*)

silicon and gallium arsenide. For design purposes, the conductivity filling factor exhibits only a mild dependence on w/h, which can probably be ignored in practice.

7-2-2. Ohmic Losses

In a microstrip line over a low-loss dielectric substrate, the predominant sources of losses at microwave frequencies are the nonperfect conductors. The current density in the conductors of a microstrip line is concentrated in a sheet approximately a skin depth deep inside the conductor surface exposed to the electric field. Both the strip conductor thickness t and the ground plane thickness are assumed to be at least three or four skin depths thick. The current density in the strip conductor and the ground conductor is not uniform in the transverse plane. The microstrip conductor contributes the major part of the ohmic loss. A diagram of the current density J for a microstrip line is shown in Fig. 7-2-4.

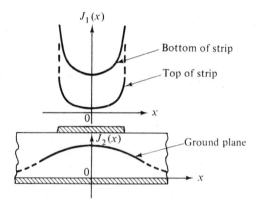

FIG. 7-2-4. Current distribution on microstrip conductors. (*After R. A. Pucel et al. [9]; reprinted with permission from the IEEE, Inc.*)

Because of mathematical complexity, exact expressions for the current density of a microstrip line with nonzero thickness have never been derived [9]. Several researchers [7] assumed, for simplicity, that the current distribution is uniform and equal to I/w in both conductors and confined to the region $|x| < w/2$. With this assumption, the conducting attenuation constant of a wide microstrip line is given by

$$\alpha_c \simeq \frac{8.68 R_s}{Z_0 w} \quad \text{dB/cm} \quad \text{for } \frac{w}{h} > 1 \qquad (7\text{-}2\text{-}12)$$

where $R_s = \sqrt{\dfrac{\pi f \mu}{\sigma}}$ is the surface skin resistivity or

$R_s = \dfrac{1}{\delta \sigma}$ is in ohms per square centimeter

$\delta = \sqrt{\dfrac{1}{\pi f \mu \sigma}}$ is the skin depth in centimeters

However, for a narrow microstrip line with $w/h < 1$, Eq. (7-2-12) is not applicable. This is because the current distribution in the conductor is unknown—that is, not uniform as assumed.

Pucel and coworkers [9, 10] derived the following three formulas from the results of Wheeler's work [3]:

$$\frac{\alpha_c Z_0 h}{R_s} = \frac{8.68}{2\pi}\left[1 - \left(\frac{w'}{4h}\right)^2\right]\left[1 + \frac{h}{w'} + \frac{h}{\pi w'}\left(\ell n\,\frac{4\pi w}{t} + \frac{t}{w}\right)\right] \qquad \text{for } \frac{w}{h} \leq \frac{1}{2\pi}$$

(7-2-13)

$$\frac{\alpha_c Z_0 h}{R_s} = \frac{8.68}{2\pi}\left[1 - \left(\frac{w'}{4h}\right)^2\right]\left[1 + \frac{h}{w'} + \frac{h}{w'}\left(\ell n\,\frac{2h}{t} - \frac{t}{h}\right)\right] \qquad \text{for } \frac{1}{2\pi} < \frac{w}{h} \leq 2$$

(7-2-14)

$$\frac{\alpha_c Z_0 h}{R_s} = \frac{8.68}{\{w'/h + 2/\pi\,\ell n\,[2\pi e(w'/(2h) + 0.94)]\}^2}\left[\frac{w'}{h} + \frac{w'/(\pi h)}{w'/(2h) + 0.94}\right]$$

$$\cdot\left[1 + \frac{h}{w'} + \frac{h}{\pi w'}\left(\ell n\,\frac{2h}{t} - \frac{t}{h}\right)\right] \qquad \text{for } 2 \leq \frac{w}{h}$$

(7-2-15)

where α_c is in decibels per centimeter and

$$e = 2.718$$

$$w' = w + \Delta w \tag{7-2-16}$$

$$\Delta w = \frac{t}{\pi}\left(\ell n\,\frac{4\pi w}{t} + 1\right) \qquad \text{for } \frac{2t}{h} < \frac{w}{h} \leq \frac{1}{2\pi} \tag{7-2-17}$$

$$\Delta w = \frac{t}{\pi}\left(\ell n\,\frac{2h}{t} + 1\right) \qquad \text{for } \frac{w}{h} \geq \frac{1}{2\pi} \tag{7-2-17a}$$

The values of α_c computed from Eqs. (7-2-13) through (7-2-15) are plotted in Fig. 7-2-5. For purposes of comparison, α_c based on Assadourian and Rimai's Eq. (7-2-12) is also shown in the same diagram.

7-2-3. Radiation Losses

In addition to the conductor and dielectric losses, a microstrip line also has radiation losses. The radiation loss depends on the substrate's thickness and dielectric constant, as well as the geometry. Lewin [11] has calculated the radiation for several discontinuities with the following approximations:

1. TEM transmission
2. Uniform dielectric in the neighborhood of the strip, equal in magnitude to an effective value
3. Neglect of radiation from the transverse electric (TE) field component parallel to the strip
4. Substrate thickness much less than the free-space wavelength

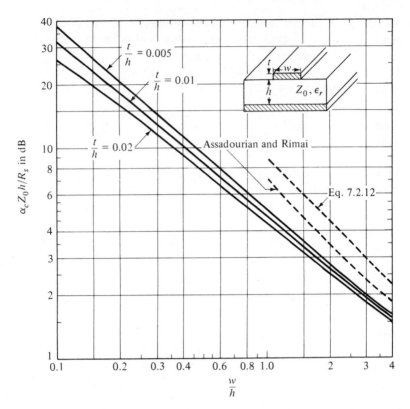

FIG. 7-2-5. Theoretical conductor attenuation factor of microstrip as a function of w/h. (*After R. A. Pucel et al. [9]; reprinted with permission from the IEEE, Inc.*)

From the results given by Lewin, the ratio of radiated power to total dissipated power for an open-circuited microstrip line is

$$\frac{P_{\text{rad}}}{P_t} = 240\pi^2 \left(\frac{h}{\lambda_0}\right)^2 \frac{F(\epsilon_{re})}{Z_0} \tag{7-2-18}$$

where $F(\epsilon_{re})$ is a radiation factor given by

$$F(\epsilon_{re}) = \frac{\epsilon_{re} + 1}{\epsilon_{re}} - \frac{(\epsilon_{re} - 1)}{2\epsilon_{re}\sqrt{\epsilon_{re}}} \ln \frac{\sqrt{\epsilon_{re}} + 1}{\sqrt{\epsilon_{re}} - 1} \tag{7-2-19}$$

in which ϵ_{re} equals the effective relative dielectric constant and $\lambda_0 = c/f$ is the free-space wavelength.

It can be seen that the radiation factor decreases with increasing substrate dielectric constant. Alternatively, Eq. (7-2-18) can be written

$$\frac{P_{\text{rad}}}{P_t} = \frac{R_r}{Z_0} \tag{7-2-20}$$

where R_r is the radiation resistance of an open-circuited microstrip and is given by

$$R_r = 240\pi^2 \left(\frac{h}{\lambda_0}\right)^2 F(\epsilon_{re}) \qquad (7\text{-}2\text{-}21)$$

The ratio of the radiation resistance R_r to the real part of the characteristic impedance Z_0 of the microstrip line is equal to a small fraction of the power radiated from a single open-circuit discontinuity. In view of Eq. (7-2-18), the radiation loss decreases when the characteristic impedance increases. For lower dielectric-constant substrates, radiation is significant at higher impedance levels. For higher dielectric-constant substrates, radiation becomes significant until very low impedance levels are reached.

7-3 The Quality Factor Q of Microstrip Lines

Many microwave integrated circuits require very high Q resonant circuits. The quality factor Q of a microstrip line is very high, but it is limited by the radiation losses of the substrates and with low dielectric constant. For a uniform current distribution in the microstrip line, the ohmic attenuation constant of a wide microstrip line is given by Eq. (7-2-12) as

$$\alpha_c = \frac{8.68R_s}{Z_0 w} \quad \text{dB/cm} \qquad (7\text{-}3\text{-}1)$$

The characteristic impedance of a wide microstrip line as shown in Eq. (7-1-9) is

$$Z_0 = \frac{h}{w}\sqrt{\frac{\mu}{\epsilon}} = \frac{377}{\sqrt{\epsilon_r}}\frac{h}{w} \quad \text{ohms} \qquad (7\text{-}3\text{-}2)$$

The wavelength in the microstrip line is given by

$$\lambda_g = \frac{30}{f\sqrt{\epsilon_r}} \quad \text{cm} \qquad (7\text{-}3\text{-}3)$$

where f is the frequency in gigahertz (GHz).

Since Q_c is related to the conductor attenuation constant by

$$Q_c = \frac{27.3}{\alpha_c} \qquad (7\text{-}3\text{-}4)$$

where α_c is in dB/λ_g, the Q_c of the wide microstrip line is expressed as

$$Q_c = 39.5\left(\frac{h}{R_s}\right) f_{\text{GHz}} \qquad (7\text{-}3\text{-}5)$$

where h is in centimeters and R_s is given by

$$R_s = \sqrt{\frac{\pi f \mu}{\sigma}} = 2\pi\sqrt{\frac{f_{\text{GHz}}}{\sigma}} \quad \text{(MKS units)} \qquad (7\text{-}3\text{-}6)$$

Finally, the quality factor Q_c of a wide microstrip line is

$$Q_c = 0.63h\sqrt{\sigma f_{\mathrm{GHz}}} \qquad (7\text{-}3\text{-}7)$$

where σ is the conductivity in per ohm-meter.

For a copper strip, the conductivity σ is 5.8×10^7 mhos per meter whence Q_c becomes

$$Q_{\mathrm{cop}} = 4780h\sqrt{f_{\mathrm{GHz}}} \qquad (7\text{-}3\text{-}8)$$

For a 25-mil alumina at 10 GHz, the maximum Q_c achievable from wide microstrip lines is 954 [12].

Similarly, there is a quality factor Q_d that is related to the dielectric attenuation constant by

$$Q_d = \frac{27.3}{\alpha_d} \qquad (7\text{-}3\text{-}9)$$

where α_d is in dB/λ_g.

Substitution of Eq. (7-2-11) in (7-3-9) yields

$$Q_d = \frac{\lambda_0}{\sqrt{\epsilon_{re}}\,\tan\theta} \simeq \frac{1}{\tan\theta} \qquad (7\text{-}3\text{-}10)$$

where λ_0 is the free-space wavelength in centimeters. It should be noted that the Q_d that is due to the dielectric attenuation constant of a microstrip line is approximately the reciprocal of the dielectric loss $\tan\theta$ and relatively constant with frequency.

7-4 Summary

Microstrip lines aie widely used in both microwave and computer technology. Because they are easily fabricated by printed-circuit munufacturing techniques, they have unquestionable economic and technical merits.

Characteristic Impedance of Microstrip Lines: The characteristic impedance for a wide microstrip line is given by

$$Z_0 = \frac{h}{w}\sqrt{\frac{\mu}{\epsilon}} = \frac{377}{\sqrt{\epsilon_r}}\frac{h}{w} \qquad \text{for } w \gg h$$

where ϵ_r = relative dielectric constant of the board material
h = distance between the microstrip line and ground
w = width of the microstrip line

There are three types of losses in microstrip lines: dielectric, ohmic, and radiation losses.

Dielectric Losses: The dielectric loss in the microstrip line is expressed by

$$\alpha_d = 4.34\frac{q}{\sqrt{\epsilon_{re}}}\sqrt{\frac{\mu_0}{\epsilon_0}}\sigma = 1.634 \times 10^3\frac{q\sigma}{\sqrt{\epsilon_{re}}} \qquad \text{dB/cm}$$

where 　σ = conductivity of the microstrip conductor

　　　ϵ_{re} = effective relative dielectric constant of the substrate

　　　$q = \dfrac{\epsilon_{re} - 1}{\epsilon_r - 1}$ is the dielectric filling factor

Ohmic Losses: The ohmic loss (or copper loss) of a wide microstrip line is given by

$$\alpha_c \simeq \frac{8.68 R_s}{Z_0 w} \qquad \text{dB/cm} \qquad \text{for } \frac{w}{h} > 1$$

where 　$R_s = \sqrt{\dfrac{\pi f \mu}{\sigma}}$ is the surface skin resistivity or

　　　$R_s = \dfrac{1}{\delta \sigma}$ is in ohms per square

　　　$\delta = \sqrt{\dfrac{1}{\pi f \mu \sigma}}$ is the skin depth in centimeters

Radiation Losses: The radiation loss is a major problem for microstrip lines. The loss can be expressed in terms of the radiation resistance of an open-circuit microstrip line—namely,

$$R_r = 240\pi^2 \left(\frac{h}{\lambda_0}\right)^2 F(\epsilon_{re})$$

where $F(\epsilon_{re})$ is a radiation factor given by

$$F(\epsilon_{re}) = \frac{\epsilon_{re} + 1}{\epsilon_{re}} - \frac{(\epsilon_{re} - 1)}{2\epsilon_{re}\sqrt{\epsilon_{re}}} \ell n \frac{\sqrt{\epsilon_{re}} + 1}{\sqrt{\epsilon_{re}} - 1}$$

REFERENCES

[1] JOHNSON, W. C., Transmission lines and networks. McGraw-Hill Book Company, Inc., 1950.

[2] KAUPP, H. R., Characteristics of microstrip transmission lines. *IEEE Trans. Electronic Computers*, **EC-16**, No. 2, 185–193, April 1967.

[3] WHEELER, H. A., Transmission-line properties of parallel strips separated by a dielectric sheet. *IEEE Trans. Microwave Theory and Techniques*, **MTT-3**, No. 3, 172–185, March 1965.

[4] BRYANT, T. G., and J. A. WEISS, Parameters of microstrip transmission lines and of coupled pairs of microstrip lines. *IEEE Trans. Microwave Theory and Techniques*, **MTT-6**, No. 12, 1021–1027, December 1968.

[5] STINEHELFER, H. E., An accurate calculation of uniform microstrip transmission lines. *IEEE Trans. Microwave Theory and Techniques*, **MTT-16**, No. 7, 439–443, July 1968.

[6]   DiGiacomo, J. J., et al., Design and fabrication of nanosecond digital equipment. RCA publication, March 1965.

[7]   Assadourian, F., and E. Rimai, Simplified theory of microwave transmission systems. *Proc. IRE*, **40**, 1651–1657, December 1952.

[8]   Welch, J. D., and H. J. Pratt, Losses in microstrip transmission systems for integrated microwave circuits. *NEREM Rec.*, **8**, 100–101, 1966.

[9]   Pucel, Robert A., Daniel J. Masse, and Curtis P. Hartwig, Losses in microstrip. *IEEE Trans. Microwave Theory and Techniques*, **MTT-16**, No. 6, 342–350, June 1968.

[10] Pucel, Robert A., Daniel J. Masse, and Curtis P. Hartwig, Correction to "Losses in Microstrip." *IEEE Trans. Microwave Theory and Techniques*, **MTT-16**, No. 12, 1064, December 1968.

[11] Lewin, L., Radiation from discontinuities in strip-line. *IEEE Monograph No. 358E*, February 1960.

[12] Vendeline, George D., Limitations on stripline Q. *Microwave Journal*, pp. 63–69, May 1970.

[13] Springfield, W. K., Multilayer printed circuitry in computer applications, presented at Fall Meeting of Printed Circuits, Chicago, Ill., October 6, 1964.

Microwave Integrated Circuits

8-0   Introduction

As described in Section 7-0, microwave integrated circuits with microstrip lines are commonly incorporated with microwave solid-state device chips. In Chapter 7 we discussed microstrip lines. Here microwave integrated circuits are analyzed.

Microwave integrated circuits are a combination of active and passive elements that are manufactured by successive diffusion processes on a semiconductor substrate. The active elements are generally silicon planar chips. The passive elements are either thin or thick film components. In thin films, a thin film of conducting (resistor) or nonconducting (capacitor) material is deposited on a passive insulated substrate, such as ceramic, glass, or silicon dioxide, by vacuum deposition. Thick

chapter

8

film applies to films more than several thousand angstroms thick. Such films are used almost exclusively to form resistors, and the pattern is usually defined by silk screening.

Electronic circuits can be classified into three categories according to circuit technology as shown in Fig. 8-0-1.

FIG. 8-0-1. Discrete circuit, integrated circuit, and microwave integrated circuit. (*From M. Caulton et al. [8]; reprinted by permission of the IEEE, Inc.*)

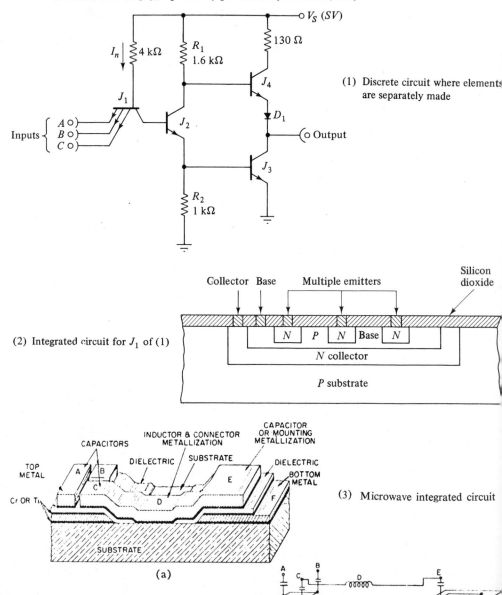

(1) Discrete circuit where elements are separately made

(2) Integrated circuit for J_1 of (1)

(3) Microwave integrated circuit

1. Discrete circuit (DC): The conventional electrical or electronic circuit, in which the elements are separately manufactured and then connected together by conducting wires, is now referred to as a *discrete circuit*. The word discrete literally means separately distinct.

2. Integrated circuit (IC): The integrated circuit consists of a single-crystal chip of semiconductor, typically 50×50 mils in cross section, containing both active and passive elements and their interconnections. This type of integrated circuit is also called a *monolithic integrated circuit*. The word monolithic is derived from the Greek *monos*, meaning single, and *lithos*, meaning stone. Thus a monolithic integrated circuit is built on a single crystal. Such circuits are produced by the processes of epitaxial growth, masked impurity diffusion, oxide growth, and oxide etching.

3. Microwave integrated circuit (MIC): Prior to 1965 nearly all microwave equipment utilized waveguide, coaxial, or stripline circuits. In recent years the conventional technology of integrated circuits has been introduced into microwave frequencies. Microwave integrated circuits, like conventional integrated circuits, can be made in monolithic or hybrid form. However, MICs are quite different from the conventional ICs. The conventional ICs contain very high packing densities, whereas the packing density of a typical MIC is quite low. An MIC whose elements are formed on an insulating substrate, such as glass or ceramic, is called a *film integrated circuit*. A microwave integrated circuit, which consists of a combination of two or more integrated circuit types, such as monolithic or film, or one IC type together with discrete elements, is referred to as a *hybrid microwave integrated circuit*.

Microwave integrated circuits offer the following advantages over discrete circuits:

1. Low cost (due to the large quantitites processed)
2. Small size
3. Light weight
4. High reliability (all components are fabricated simultaneously, and there are no soldered joints)
5. Improved reproducibility
6. Improved performance

MICs are suitable for space and military applications because they meet the requirements for shock, temperature conditions, and severe vibration. A major factor in the success of MICs has been the advances in the development of microwave solid-state devices as described in Chapter 6. In this chapter the basic materials and processes necessary for fabrication of MICs are described. Three general types of circuits can be utilized for hybrid MICs: distributed microstrip lines, lumped-element (inductors and capacitors) circuits, and thin-film circuits. These three types are discussed in the following sections.

8-1 Materials

The basic materials for microwave integrated circuits, in general, are subdivided into four categories:

1. Substrate materials—alumina, beryllia, ferrite/garnet, GaAs, glass, rutile, and sapphire
2. Conductor materials—aluminum, copper, gold, and silver
3. Dielectric films—Al_2O_3, SiO, SiO_2, Si_3N_4, and Ta_2O_5
4. Resistive films—Cr, Cr-SiO, NiCr, Ta, and Ti

8-1-1. Substrate Materials

A substrate of microwave integrated circuits is a piece of substance on which microwave devices are built. The ideal substrate materials should have the following characteristics [1]:

1. High dielectric constant (9 or higher)
2. Low dissipation factor or loss tangent
3. Dielectric constant should remain constant over the frequency range of interest and over the temperature range of interest
4. High purity and constant thickness
5. High surface smoothness
6. High resistivity and dielectric strength
7. High thermal conductivity

Table 8-1-1 lists the properties of some popular substrates that have been used for MICs [2, 3]. The selection of a substrate material also depends on the expected circuit dissipation, the circuit function, and the type of circuit to be used.

8-1-2. Conductor Materials

The ideal conductor materials for microwave integrated circuits should have the following properties [1]:

1. High conductivity
2. Low temperature coefficient of resistance
3. Good adhesion to the substrate
4. Good etchability and solderability
5. Easily deposited or electroplated

TABLE 8-1-1.* Properties of Substrates

Material	$\tan\theta \times 10^4$ at 10 GHz	Relative dielectric constant (ϵ_r)	Thermal conductivity $K\,(W/cm\,^\circ C)$	Applications
Alumina	2	10	0.3	microstrip, suspended substrate
Beryllia	1	6	2.5	compound substrate
Ferrite/garnet	2	13–16	0.03	microstrip, coplanar, compound substrate
GaAs	16	13	0.03	high frequency, microstrip, monolithic MIC
Glass	4	5	0.01	lumped element
Rutile	4	100	0.02	microstrip
Sapphire	1	9.3–11.7	0.4	microstrip, lumped element

*From H. Sobol [3]; reprinted by permission of the IEEE, Inc.

Table 8-1-2 shows the properties of some widely used conductor materials for microcircuits [3]. These materials not only have excellent conductivity, but they can also be deposited by a number of methods and are capable of being photoetched. They are used to form both the conductor pattern and the bottom ground plane. The conductor thickness should be equal to at least four skin depths, to include 98% of the current density. It can be seen from Table 8-1-2 that good electrical conductors have poor substrate adhesion, whereas poor electrical conductors have good substrate adhesion. Aluminum has relatively good conductivity and good adhesion. It is possible to obtain good adhesion with high-conductivity materials by using a very thin film of one of the poorer conductors between the substrate and the good conductor. Some typical combinations are Cr-Au, Cr-Cu, and Ta-Au. A typical adhesion layer may have a surface resistivity ranging from 500 to 1000 Ω/square without loss. The choice of conductors is usually determined by compatibility with other materials required in the circuit and the processes required. For small losses, the conductors should be of the order of 3 to 5 skin depths in the thickness. That is, thick films of the good conductor (about 10 μm thick) are required. Films of this thickness can be achieved by evaporation or plating or by any of the standard thick-film processes.

8-1-3. Dielectric Films

Dielectric films are used in microwave integrated circuits for blockers, capacitors, and some couple-line structures. The properties of dielectric films should be

TABLE 8-1-2.* Properties of Conductors

Material	Skin depth δ at GHz (μm)	Surface resistivity ($\Omega/sq \times 10^{-7}\sqrt{f}$)	Coefficient of thermal expansion ($\alpha_t/°C \times 10^6$)	Adherence to dielectric film or substrate	Method of deposition
Ag	1.4	2.5	21	poor	evaporation, screening
Cu	1.5	2.6	18	very poor	evaporation, plating
Au	1.7	3.0	15	very poor	evaporation, plating
Al	1.9	3.3	26	very good	evaporation
W	2.6	4.7	4.6	good	sputtering, vapor phase, electron-beam evaporation
Mo	2.7	4.7	6.0	good	electron-beam evaporation, sputtering
Cr	2.7	4.7	9.0	good	evaporation
Ta	4.0	7.2	6.6	very good	electron-beam sputtering

*After H. Sobol [3]; reproduced with permission from the IEEE, Inc.

1. Reproducibility
2. Capability of withstanding high voltages
3. Ability of undergoing processes without developing pin holes
4. Low RF dielectric loss

Some of the dielectrics used in microcircuits are shown in Table 8-1-3, in which SiO, SiO_2, and Ta_2O_5 are the most commonly used. Thin-film SiO_2 with high-dielectric Q can be obtained by growing the pyrolitic deposition of SiO_2 from silane and then densifying it by heat treatment. SiO_2 can also be deposited by sputtering. With proper processing, SiO_2 capacitors with Qs in excess of 100 have been fabricated with good success. Capacitors fabricated with SiO_2 films have capacitances in the range of 0.02 to 0.05 pF/square mil. Thin-film SiO is not very stable and can be used only in noncritical applications, such as bypass capacitors. In power microwave integrated circuits, capacitors may require breakdown voltages in excess of 200 volts. Such capacitors can be achieved with films on the order of 0.5 to 1.0 μm thick with low probability of pin holes or shorts.

TABLE 8-1-3.* Properties of Dielectric Films

Material	Method of deposition	Relative dielectric constant (ϵ_r)	Dielectric strength (V/cm)	Microwave Q
SiO	evaporation	6–8	4×10^5	30
SiO$_2$	deposition	4	10^7	100–1000
Si$_3$N$_4$	vapor phase, sputtering	7.6 6.5	10^7 10^7	
Al$_2$O$_3$	anodization, evaporation	7–10	4×10^6	
Ta$_2$O$_5$	anodization, sputtering	22–25	6×10^6	100

*After H. Sobol [3]; reprinted by permission of the IEEE, Inc.

8-1-4. Resistive Films

Resistive films are used in microwave integrated circuits for bias networks, terminations, and attenuators. The properties required for a good microwave resistor are similar to those required for low-frequency resistors and should be [4]

1. Good stability
2. Low temperature coefficient of resistance (TCR)
3. Adequate dissipation capability
4. Sheet resistivities in the range of 10 to 1000 Ω per square

Table 8-1-4 lists some of the thin-film resistive materials used in microwave integrated circuits. Evaporated nichrome and tantalum nitride are the most widely used materials. The exact temperature coefficient of

TABLE 8-1-4.* Properties of Resistive Films

Material	Method of deposition	Resistivity $(\Omega/square)$	TCR $(\%/°C)$	Stability
Cr	evaporation	10–1000	$-0.100-+0.10$	poor
NiCr	evaporation	40–400	$+0.002-+0.10$	good
Ta	sputtering	5–100	$-0.010-+0.01$	excellent
Cr–SiO	evaporation or cermet	–600	$-0.005--0.02$	fair
Ti	evaporation	5–2000	$-0.100-+0.10$	fair

*From H. Sobol [3]; reproduced by permission of the IEEE, Inc.

resistance achieved depends on film formation conditions. Thick-film resistors may be utilized in circuits incorporating chip components. The thickness of the thick-film is in the range of 1 to 500 μm. The term "thick-film" refers to the process used, not to the film thickness. Thick-film techniques involves silk screening through a mask, such as the printing and screening of silver or gold in a glass frit, which is applied on the ceramic and fired at 850°C. Microwave "thick-film" metals are sometimes several micrometers thick, which are thicker than those of low-frequency integrated circuits.

8-2   Fabrication

Like lower-frequency integrated circuits, microwave integrated circuits can be made in monolithic or hybrid form. In a monolithic circuit active devices are grown on or in a semiconducting substrate, and passive elements are either deposited on the substrate or grown in it. In the hybrid circuit active devices are attached to a glass, ceramic, or substrate, which contains the passive circuitry. Monolithic ICs have been successful in digital and linear applications in which all required circuit components can be simultaneously fabricated. In most cases, the same device, such as bipolar or MOS transistors, can be used for amplifiers, diodes, resistors, and capacitors with no loss in performance. Many digital circuits used in computers require large arrays of identical devices. Thus the conventional ICs contain very high packing densities. On the other hand, very few applications of microwave integrated circuits require densely packed arrays of identical devices, and there is little opportunity to utilize active devices for passive components.

Monolithic technology is not well suited to microwave integrated circuits because the processing difficulties, low yields, and poor performance have seriously limited their applications. To date, the hybrid form of technology is used almost exclusively for microwave integrated circuits in the frequency range of 1 to 15 GHz. Hybrid MICs are fabricated on a high-quality ceramic, glass, or ferrite substrate. The passive circuit elements are deposited on the substrate, and active devices are mounted on the substrate and connected to the passive circuit. The active devices may be utilized in chip form, in chip carriers, or in small plastic packages. The resistivity of microwave integrated circuits should be much greater than 1000 Ω-cm for good circuit performance.

The fabrication procedures for MICs are depicted in Fig. 8-2-1. The diagram shows a three-layer metal-dielectric-metal structure that is suitable for both distributed and lumped circuits. If only chip capacitors are used with microstrip, there will be only one layer. Figure 8-2-2 illustrates the fabrication procedures for MICs in four steps.

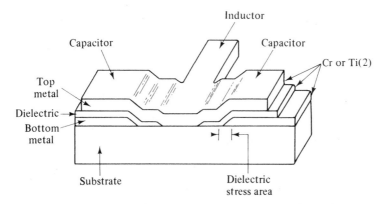

FIG. 8-2-1. MIC: Three-layer metal-dielectric-metal. (*After M. Caulton and H. Sobol [2]; reprinted with permission from the IEEE, Inc.*)

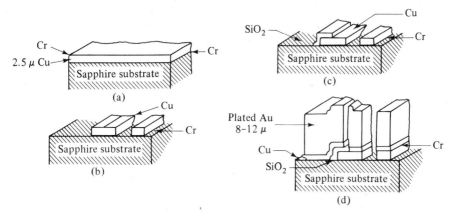

FIG. 8-2-2. Fabrication procedures for MIC. (*After M. Caulton and H. Sobol [2]; reprinted with permission from the IEEE, Inc.*)

Step 1. Diagram (a) shows three thin-film layers of Cr-Cu-Cr that are deposited in one vacuum evaporation on a sapphire substrate.

Step 2. Diagram (b) demonstrates that a photoresist etch defines bottom layers of capacitors, leaving chrome on top, and grounding connections without chrome.

Step 3. Diagram (c) shows that SiO_2 is deposited and defined for capacitors using a thick photoresist to protect the metal sides from being etched.

Step 4. Diagram (d) illustrates the last step. After densifying the oxide, a thin top layer of metal is deposited, a thick photoresist form is prepared, and the gold is selectively plated. The thin layer of unplated copper is then etched to form the circuits.

Several basic fabrication techniques used for microwave integrated circuits
need further description.

8-2-1. Epitaxy

The word epitaxy is derived from the Greek $\epsilon\pi\iota$ (upon) and $\tau\epsilon\acute{\iota}\nu\epsilon\nu$
(arranged). An epitaxial growth consists of extending the single-crystal
substrate by exposing it to a gaseous concentration at a temperature lower
than that for diffusion. In this formation the single-crystal structure is continued in the epitaxially formed layer on the top of the substrate. The
thickness of the formed layer is determined by the nature of the gaseous
vapor. Figure 8-2-3 shows the main process steps in epitaxial selective
growth.

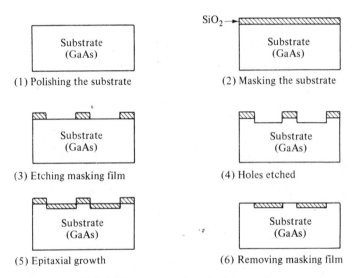

FIG. 8-2-3. Processes in epitaxial growth.

Step 1. The substrate is chemically polished.
Step 2. A masking layer is then deposited. SiO_2 or Si_3N_4 with thickness
of 300 to 500 Å is used as mask.
Step 3. The desired deposit pattern is etched into this masking mate-
rial, using photolithographical methods and etches.
Step 4. Holes are etched into the substrate by using the masking film
to protect the remaining surface of the slice.
Step 5. The masked slice is then processed through an epitaxial
deposition.
Step 6. The masking film SiO_2 is removed by hydrofluoric acid.

8-2-2. Diffusion

The process of diffusion consists of diffusing impurities into a pure metal in order to alter the basic electrical characteristics of the pure material. Although several methods are available for diffusion, the solid-state diffusion process is the standard one. Solid-state diffusion is practical at elevated temperature. Usually impurities from a gaseous atmosphere diffuse into the surface of a pure metal. Because of the high temperature, atoms of the pure metal will be highly excited, and many will leave their lattice sites and move into new positions in the crystal. Consequently, atoms of the impurity diffuse into the pure material and occupy vacancies in the crystal lattice left by the excited atoms in the pure material. Gold is sometimes used as a dopant in MICs. The reason for its use is not to create new conductivity regions but rather to enhance the recombination of carriers. This effect results in faster switching speeds in digital circuits.

8-2-3. Deposition

Three methods, vacuum evaporation, electron-beam evaporation, and dc sputtering, are commonly used for making microwave integrated circuits.

1. Vacuum evaporation: Here the impurity material to be evaporated is placed in a metallic boat through which a high current is passed. The substrate with a mask on it and the heated boat are located in a glass tube in which a high vacuum at a pressure of 10^{-6} to 10^{-8} torr is maintained. The substrate is heated slightly while the heat is evaporating the impurities, and the impurity vapor deposits itself on the substrate, forming a polycrystalline layer on it.

2. Electron-beam evaporation: In another method of evaporating the impurity a narrow beam of electrons is generated to scan the substrate in the boat so as to vaporize the impurity.

3. dc sputtering: The third method of vacuum deposition is known as dc sputtering or cathode sputtering. In a vacuum, the crucible containing the impurity is used as the cathode and the substrate as the anode of a diode. A slight trace of argon gas is introduced into the vacuum. When the applied voltage between cathode and anode is high enough, a glow discharge of argon gas is formed. The positive argon ions are accelerated toward the cathode, where they dislodge atoms of the impurity. The impurity atoms have enough energy to reach the substrate and adhere to it.

8-2-4. Etching

In the processes of making MICs, a selective removal of SiO_2 is required to form openings through which impurities can be diffused. The photoetching method used for this removal is shown in Fig. 8-2-4.

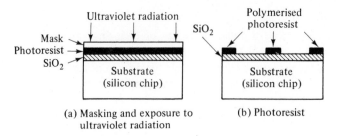

(a) Masking and exposure to   (b) Photoresist
ultraviolet radiation

FIG. 8-2-4. Photoetching processes.

During the photolithographic process the substrate is coated with a uniform film of Kodak photoresist (KPR), which is a photosensitive emulsion. A mask for the desired openings is placed over the photoresist, and ultraviolet light exposes the photoresist through the mask as shown in Fig. 8-2-4(a). A polymerized photoresist is developed, and the unpolymerized portions are dissolved by using trichloroethylene after the mask is removed [see Fig. 8-2-4(b)]. SiO_2, which is not covered by the photoresist, can be removed by hydrofluoric acid. The thick-film process usually involves the printing and silk screening of silver or gold through a metal mask in a glass frit, which is applied on the ceramic and fired at 850°C. After firing, the initial layer may be covered with gold.

8-3   Hybrid Microwave Integrated Circuits

Two general classes of circuits can be used for hybrid MICs:

   1. Distributed transmission-line circuits
   2. Lumped-element (inductor and capacitors) circuits

In some instances, one circuit may incorporate both distributed and lumped components [5]. Hybrid MICs are fabricated on a high-quality ceramic, glass, or ferrite substrate. The passive circuit elements are deposited on the substrate, and active devices are mounted on the substrate and connected to the passive circuit. The active devices may be in the form of chip, chip carriers, or small plastic packages.

8-3-1. Distributed-Line MICs

Four types of distributed transmission lines, such as microstrip line, suspended-substrate line, slot line, and coplanar waveguide, are used in microwave integrated circuits (see Fig. 8-3-1).

The microstrip line, which is illustrated in Fig. 8-3-1(a) and described in Chapter 7, is the most commonly used structure of circuit for MICs. The line

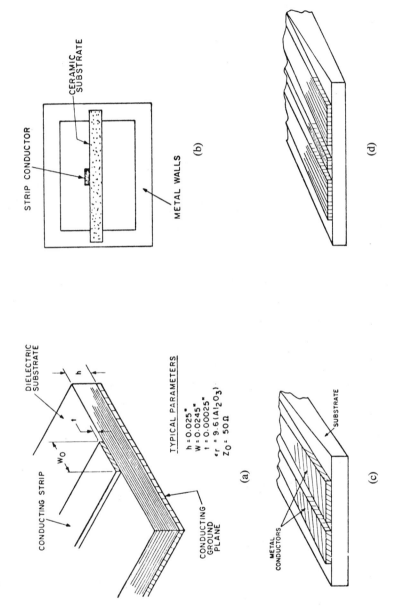

FIG. 8-3-1. Four types of distributed transmission lines. *(After H. Sobol [3]; reprinted by permission of the IEEE, Inc.)*

445

consists of a strip conductor that is separated from the ground plane by a dielectric layer. The function of the circuit is performed in the plane of the strip conductor. The impedance and length of the line determine the circuit properties. The major part of the propagating field is confined to the region of the dielectric below the strip conductor.

The suspended-substrate transmission line [see Fig. 8-3-1(b)] is another distributed MIC that is also used frequently. A strip conductor is placed on the upper surface of a ceramic substrate, and the substrate is then suspended in a metal enclosure. The major portion of the propagating field for this structure is in the air between the ceramic and ground. The transmission characteristics of this line are determined by the substrate thickness, dielectric constant, air-space height, and width of the strip conductor.

Two other structures of distributed circuits that have lately been used for MICs are the slot line and the coplanar waveguide [see Fig. 8-3-1(c) and (d)]. Because both lines have conductors on only one side of the substrate, they generally permit shunt mounting of devices without requiring holes through the substrate, as in the case of microstrip lines. These two lines also have longitudinal as well as transverse RF magnetic fields and polarization properties that are useful for nonreciprocal ferrite devices. Such lines, however, are not yet as widely used as microstrip or suspended-substrate lines.

Since it is impossible to discuss all these transmission lines in detail, the microstrip line will be analyzed in some detail in order to illustrate the principles previously described in Chapter 7. The fundamental propagation mode of a microstrip line is TEM (Transverse Electric and Magnetic). At low microwave frequencies in the range of 1 to 3 GHz, most of the energy propagates in the dielectric below the strip conductor. At frequencies above 3 GHz the propagation in microstrip line is a mixed mode of TE (Transverse Electric) and TM (Transverse Magnetic) distribution. The characteristic impedance and wavelength are shown in Fig. 8-3-2.

8-3-2. Lumped-Element MICs

The term *lumped element* means true lumped—that is, no variation of L and C with frequency, nor any variation of phase over the element. The lumped element can be much smaller than equivalent distributed circuits at microwave frequencies. The lumped-element form of microwave integrated circuits consists of several components that are a fraction of a wavelength in size and perform as capacitors, inductors, or resistors. The values of the components are independent of frequency. This type of circuit was not feasible at microwave frequencies in the past because conventional fabrication techniques could not provide coils and capacitors in such small size that they could behave as true lumped elements. Today, however, it is possible

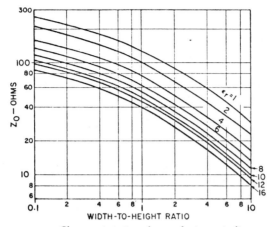

Characteristic impedance of microstrip line.

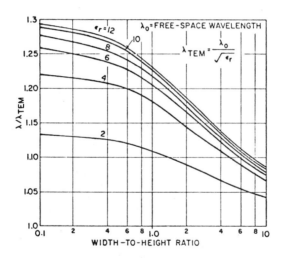

FIG. 8-3-2. Characteristic impedance and wavelength of microstrip line. (*After H. Sobol [3]; reproduced with permission from the IEEE, Inc.*)

to fabricate such elements through microwave integrated circuit technology. In view of fabrication technology, two types of lumped-element circuits are used in MICs.

   Type I. The chip devices are mounted on a substrate. The chip thin-film inductors, capacitors, and resistors are interconnected to the chip devices by wire bonding.

Type II. The interconnections among the passive elements are deposited simultaneously with the circuit elements themselves. The only wire bonds in this type of MIC are the chip devices. This type results in greater reproducibility than type I.

The lumped-element fabrication always results in smaller circuits than the distributed approach. The key advantages of small size are low cost and the adaptability of the lumped-circuit component in a hybrid subsystem. Because of the small size of lumped-element circuits, many devices can be fabricated simultaneously on a ceramic wafer. Because optimum use is made of batch fabrication, the cost of lumped-element circuits is less than that of microstrip lines, which are usually made one at a time.

Any microwave circuit can be characterized by an equivalent transmission line as shown in Fig. 8-3-3. All parameter values are specified per

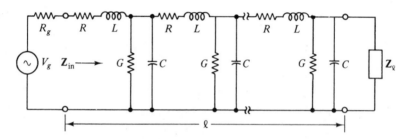

FIG. 8-3-3. Transmission-line model for lumped elements.

unit length of line. A very short length of line can be considered as a lumped element. From transmission-line theory, a TEM mode is assumed propagating along the line. The driving-point impedance of the line is given by

$$Z_{\text{in}} = Z_0 \frac{Z_\ell \cosh \gamma\ell + Z_0 \sinh \gamma\ell}{Z_0 \cosh \gamma\ell + Z_\ell \sinh \gamma\ell} \qquad (8\text{-}3\text{-}1)$$

where $\quad Z_0 = \sqrt{\dfrac{R + j\omega L}{G + j\omega C}}$ is the characteristic impedance

$\gamma = \sqrt{(R + j\omega L)(G + j\omega C)}$ is the propagation constant

Equation (8-3-1) can be used to derive equivalent circuits for various types of lumped elements.

(1) Lumped Inductor. The driving-point impedance of a very small length ($\gamma\ell \ll 1$) of a short-circuited ($Z_\ell = 0$) line is expressed as

$$Z_{\text{in}} \approx R\ell + j\omega L\ell \qquad (8\text{-}3\text{-}2)$$

Equation (8-3-2) is the equation of the equivalent circuit for a lumped inductor, which is simply the series connection of the total resistance and inductance in the series arms of a transmission line. The resistance R per unit length of a strip inductor (in ohms per mil) can be determined by assuming that the RF current flow is confined within a skin depth at the upper and lower surfaces of the strip as follows[4]:

$$R = \frac{\pi K}{w}\sqrt{\rho f} \tag{8-3-3}$$

where $f =$ the frequency in gigahertz
$\rho =$ the resistivity in ohm-cm
$w =$ the width of the strip in centimeters
$K =$ a factor between 1 and 2, which accounts for current crowding at the corners of the strip [6]

The inductance per unit length of a straight strip in nanohenries per mil is given by [7].

$$L = 5.08 \times 10^{-3}\left(\ell n\,\frac{\ell}{w} + 1.193 + 0.224\,\frac{w}{\ell}\right) \tag{8-3-4}$$

where ℓ is the length of the strip inductor in mils and w is the width of the strip inductor in mils.

Since the resistance R per unit length of the strip varies as the square root of frequency, the quality Q of the strip inductor is then proportional to the square root of frequency and is written

$$Q = \frac{\omega L}{R} \simeq \sqrt{f} \tag{8-3-5}$$

A strip inductor used for MICs usually requires inductances of 0.5 to 4 nH. If the width of the strip is much less than the diameter of the turn, a single-turn coil can be used. Strip conductors have been used for lumped inductors at 2 GHz with Q values of 100 [6].

(2) *Lumped Capacitor.* The driving-point impedance of a very short length ($\gamma\ell \ll 1$) of an open-circuited ($Z_\ell = \infty$) line is given by

$$Z_{\text{in}} \simeq \frac{2R_0}{3} + \frac{1}{Q_d\omega C_0} - \frac{1}{j\omega C_0} \tag{8-3-6}$$

where $R_0 = R\ell = \dfrac{2R_s}{w}\,\ell$

$C_0 = C\ell = \dfrac{\epsilon w}{h}\,\ell$

$Q_d = \dfrac{\omega C}{g}$ is the dielectric Q

$R_s = 2.61 \times 10^{-7} \times f^{1/2}$ is the surface resistance for copper

Equation (8-3-6) is the equation of equivalent circuit for a lumped capacitor, which simply is two resistors in series with a capacitor. One resistor represents conductor loss, and the other dielectric loss. The total Q of the capacitor is given by

$$Q_{\text{total}} = \frac{Q_d Q_s}{Q_d + Q_s} \qquad (8\text{-}3\text{-}7)$$

where

$$Q_s = \frac{3}{2\omega C_0 R_0}$$

In general, Q_d of most material is independent of frequency in the microwave range, whereas Q_s varies inversely as the product of frequency to the three-halves power and capacitance. When the capacitor is used in a circuit, the parasitic inductance L caused by the connection to the capacitor plates must be accounted for. The effective capacitance C_{eff} is then given by

$$C_{\text{eff}} \simeq C_0 \left(1 + \frac{\omega^2}{\omega_0^2} \right) \qquad (8\text{-}3\text{-}8)$$

where

$$\omega_0^2 = \frac{1}{LC_0}$$

Microwave capacitors constructed with SiO_2 dielectric layers that are 0.5 to 1.0 μm thick are typical of the type used in lumped-element MICs. Capacitances ranging from 0.25 to 100 pF have been achieved. Capacitors in the range of 1 to 10 pF have exhibited total Qs of several hundreds.

(3) *Lumped Resistor.* The driving-point impedance for a resistor connected to ground by neglecting L and G is given by

$$Z_{\text{in}} \simeq \frac{R_0}{1 + j\omega C_0 R_0 / 3} \qquad (8\text{-}3\text{-}9)$$

where

$$R_0 = R\ell \quad \text{and} \quad C_0 = C\ell$$

Equation (8-3-9) is the equation of equivalent circuit for a lumped microwave resistor, which is a resistor of $R\ell$ shunted by a capacitor of $\frac{1}{3}C\ell$.

8-4   Summary

In this chapter microwave integrated circuits have been described. The materials and fabrication processes in making microwave integrated circuits are summarized below.

Materials: The basic materials for microwave integrated circuits are subdivided into four types.

1. Substrate materials: alumina, beryllia, ferrite/garnet, GaAs, glass, rutile, and sapphire
2. Conductor materials: aluminum, copper, gold, and silver
3. Dielectric films: Al_2O_3, SiO, SiO_2, Si_3N_4, and Ta_2O_5
4. Resistive films: Cr, Cr-SiO, NiCr, Ta, and Ti

Fabrication Processes: Several basic fabrication techniques used for microwave integrated circuits are

1. Epitaxy
2. Diffusion
3. Deposition
4. Etching

REFERENCES

[1] KEISTER, FRANK Z., An evaluation of materials and processes for integrated microwave circuits. *IEEE Trans. on Microwave Theory and Techniques,* **MTT-16**, No. 7, 469–475, July 1968.

[2] CAULTON, MARTIN, and HAROLD SOBOL, Microwave integrated-circuit technology—A survey. *IEEE Journal of Solid-State Circuits,* **SC-5**, No. 6, 292–303, December 1970.

[3] SOBOL, HAROLD, Applications of integrated circuit technology to microwave frequencies. *Proc. IEEE,* **59**, No. 8, 1200–1211, August 1971.

[4] SOBOL, H., Technology and design of hybrid microwave integrated circuits. *Solid State Technology,* **13**, No. 2, 49–57, February 1970.

[5] CAULTON, M., J. J. HUGHES, and H. SOBOL, Measurements on the properties of microstrip transmission lines for microwave integrated circuits. *RCA Review,* **27**, 377–391, September 1966.

[6] CAULTON, M., S. P. KNIGHT, and D. A. DALY, Hybrid integrated lumped-element microwave amplifier. *IEEE Trans. on Electron Devices.* **ED-15**, No. 7, pp. 399–406, July 1968.

[7] TERMAN, F. E., *Radio Engineer's Handbook*, p. 51. McGraw-Hill Book Company, New York, 1943.

[8] CAULTON, MARTIN et al.: Status of lumped elements in microwave integrated circuits—present and future. *IEEE Trans. on Microwave Theory and Techniques,* **MTT-19**, No. 7, 1971.

SUGGESTED READINGS

1. FITCHER, F. C., *Electronic Integrated Circuits and Systems*, Chapter 2. D. Van
   Nostrand Reinhold Company, New York, 1970.

2. LYNN, D. K., et al. (Ed.), *Analysis and Design of Integrated Circuits*, Chapters 1
   and 2. McGraw-Hill Book Company, New York, 1967.

Microwave Enclosures

9-0 Introduction

So far we have described those microwave devices and circuits in wide use, and by now, the reader should have a good knowledge of their theory and applications. If asked to take some measurements for microwave devices and circuits in the laboratory, however, he or she may not be quite sure that the measurements will be accurate. Here the microwave enclosures that are specially designed for the purpose of microwave measurements are discussed. A microwave enclosure is a physical structure designed to prevent the electromagnetic energy inside the enclosure from escaping and, conversely, to prevent the electromagnetic energy outside the enclosure from entering.

The accuracy of microwave measurements is

chapter

9

a major concern of electronics engineers. Theoretically, a transmitter should only transmit electric energy, and a receiver should only receive energy. This situation hardly ever happens in the ordinary laboratory because a piece of electronics equipment can either transmit or receive electric energy. Furthermore, the walls of and the objects in the ordinary laboratory can reflect a fraction of the electric energy that they have received. Therefore electronic equipment in the ordinary laboratory is constantly subjected to unwanted sources of energy and is constantly producing energy that adjacent equipment does not want. An ideally designed piece of equipment should not radiate any unwanted energy, nor should it be susceptible to any unwanted energy. To achieve this goal, a medium would have to enclose the equipment so that unwanted energy either leaving or attempting to enter the equipment would be effectively attenuated. Shielded rooms and anechoic chambers are the ideal enclosures for microwave measurements. In this chapter electromagnetic compatibility (EMC) is described, and plane wave propagation in microwave enclosures, such as shielded rooms and anechoic chambers, is analyzed.

9-1   Electromagnetic Compatibility

Electromagnetic compatibility (EMC) is the science of containing the damaging effects of electromagnetic interference (EMI). When electromagnetic interference comes from within a system or equipment, it is called *intrasystem interference*. When electromagnetic interference develops between two systems, it is called *intersystem interference*. Both intra- and intersystem interferences may exist simultaneously. The central theme of electromagnetic compatibility is to contain, control, and/or eliminate electromagnetic interference. Electronic equipment or systems must operate in conjunction with other equipment without causing malfunction or degradation of operation of any of the equipment or systems. The area of electromagnetic compatibility consists of various subareas, such as the EMV (Electromagnetic Vulnerability) test and EMS (Electromagnetic Susceptibility) test. The definitions of various EMC measurements are described below.

> EMC—Electromagnetic Compatibility is defined as the ability of communications-electronics (C-E) equipment, subsystems, and systems to operate in their intended operational environment without suffering from or causing unacceptable degradation to any other system because of unintentional electromagnetic radiation or response [1]. In other words, the equipment or system should not adversely affect the operation of, or be adversely affected by, any other equipment or system.

EMV—Electromagnetic Vulnerability is defined as the operation of equipment or systems uncompromised by electromagnetic environment.

EMS—Electromagnetic Susceptibility is defined as the tolerance of a system and/or subsystem to all sources of extraneous signal or unwanted electromagnetic energy.

EMI—Electromagnetic Interference is defined as the impairment of the reception of a wanted electromagnetic signal caused by an electromagnetic disturbance.

EME—Electromagnetic Environment is defined as the electromagnetic field or fields existing in a transmission medium.

9-2 Plane Wave Propagation in Shielded Rooms

As described in Section 2-4-1, in free space the plane waves are uniform in the far field. However, this is not true when microwave measurements are taken in the ordinary laboratory because too much electromagnetic interference is caused by unwanted energies that may be either generated by neighboring electronic equipment or reflected by the walls and objects in the laboratory. Electromagnetic interference is a serious problem in extremely sensitive electronic equipments, such as missile guidance and radar tracking systems. One solution is to use shielded rooms, which provide ideal enclosures for microwave measurements. Shielding theory was originally developed by Schelkunoff [2, 3]. Jarva and Vasaka also did some pioneer work on shielding theory [4, 5, 6]. Schulz and coworkers later conducted very extensive and excellent research on shielding theory that has been applied to practical problems [7, 8, 9]. A shielded room made of high-mu welded copper or steel simulates the environment of free space by suppressing unwanted electromagnetic energy. If the shield is perfect, the isolation between the interior and exterior is perfect. In practice, however, the degree of isolation for most practical shields is less than perfect and often lower than 100 dB.

The degree of isolation between the interior and exterior of a shield is called *shielding effectiveness* or *insertion loss*. Shielding effectiveness is defined as the total attenuation of radiated RF electromagnetic energy obtained when the energy attempts to pass through the shield. Figure 9-2-1 shows that a uniform plane wave in air is normally incident upon boundary 1, passes through the shield of thickness t, and emerges from boundary 2 into the air. At boundary 1 the boundary conditions require that

$$E_{x1}^+ + E_{x1}^- = E_{x2}^+ \qquad (9-2-1)$$

and

$$H_{y1}^+ + H_{y1}^- = H_{y2}^+ \qquad (9-2-2)$$

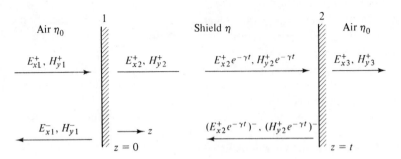

FIG. 9-2-1. Wave propagation in a shield.

As described in Section 2-3-3, the reflection and transmission coefficients of the electric and magnetic components are defined in terms of the intrinsic impedances as

$$\Gamma_{e1} = \frac{E_{x1}^-}{E_{x1}^+} = \frac{\eta - \eta_0}{\eta + \eta_0} \qquad (9\text{-}2\text{-}3)$$

$$\Gamma_{h1} = \frac{H_{y1}^-}{H_{y1}^+} = \frac{\eta_0 - \eta}{\eta + \eta_0} = -\Gamma_{e1} \qquad (9\text{-}2\text{-}4)$$

$$T_{e1} = \frac{E_{x2}^+}{E_{x1}^+} = \frac{2\eta}{\eta + \eta_0} = 1 + \Gamma_{e1} \qquad (9\text{-}2\text{-}5)$$

$$T_{h1} = \frac{H_{y2}^+}{H_{y1}^+} = \frac{2\eta_0}{\eta + \eta_0} = 1 - \Gamma_{e1} \qquad (9\text{-}2\text{-}6)$$

From transmission-line theory, if the shield is a lossy material, the plane wave traveling from boundary 1 to boundary 2 is attenuated by a factor of $e^{-\gamma t}$. The reflection loss at boundary 2 is given by

$$\Gamma = \Gamma_{e1}\Gamma_{e2}e^{-2\gamma t} = \rho e^{-2\gamma t} \qquad (9\text{-}2\text{-}7)$$

where

$$\rho = \Gamma_{e1}\,\Gamma_{e2}$$

At boundary 2 the boundary conditions require that

$$E_{x2}^+e^{-\gamma t} + (E_{x2}^+e^{-\gamma t})^- = E_{x3}^+ \qquad (9\text{-}2\text{-}8)$$

and

$$H_{y2}^+e^{-\gamma t} + (H_{y2}^+e^{-\gamma t})^- = H_{y3}^+ \qquad (9\text{-}2\text{-}9)$$

Therefore

$$(E_{x2}^+e^{-\gamma t})\frac{2\eta_0}{\eta + \eta_0} = E_{x3}^+(1 - \rho e^{-2\gamma t}) \qquad (9\text{-}2\text{-}10)$$

and

$$(H_{y2}^+e^{-\gamma t})\frac{2\eta_0}{\eta + \eta_0} = H_{y3}^+(1 - \rho e^{-2\gamma t}) \qquad (9\text{-}2\text{-}11)$$

The overall electric and magnetic transmission coefficients through the shield are defined as

$$T_e = \frac{E_{x3}}{E_{x1}} = \frac{E_{x2}}{E_{x1}} \cdot \frac{E_{x3}}{E_{x2}} = \tau_e(1 - \rho e^{-2\gamma t})^{-1} e^{-\gamma t} \qquad (9\text{-}2\text{-}12)$$

and

$$T_h = \frac{H_{y3}}{H_{y1}} = \frac{H_{y2}}{H_{y1}} \cdot \frac{H_{y3}}{H_{y2}} = \tau_h(1 - \rho e^{-2\gamma t})^{-1} e^{-\gamma t} \qquad (9\text{-}2\text{-}13)$$

where

$$\tau_e = \tau_h = \frac{4\eta\eta_0}{(\eta + \eta_0)^2},$$

and

$$\rho = \Gamma_{e1}\Gamma_{e2} = \left(\frac{\eta - \eta_0}{\eta + \eta_0}\right)\left(\frac{\eta_0 - \eta}{\eta + \eta_0}\right)$$

Finally,

$$T = T_e = T_h = \tau(1 - \rho e^{-2\gamma t})^{-1} e^{-\gamma t} \qquad (9\text{-}2\text{-}14)$$

The transmission loss or the shielding effectiveness is defined as

$$\text{SE} = -20 \log_{10}|T| = 20 \log_{10}|e^{\gamma t}| - 20 \log_{10}|\tau| + 20 \log_{10}|1 - \rho e^{-2\gamma t}|$$
$$= A + R + C \quad \text{dB} \qquad (9\text{-}2\text{-}15)$$

where
$A = 20 \log_{10}|e^{\gamma t}| = 20(\alpha t) \cdot \log_{10} e$ is the absorption or attenuation loss in decibels inside the shield

$R = -20 \log_{10}|\tau|$ is the reflection loss in decibels from the multiple boundaries of the shield

$C = $ correction term in decibels required to account for multiple internal reflections when absorption loss A is much less than 10 dB for electrically thin film

9-2-1. Shielding Effectiveness

The shielding effectiveness of a shield consists of three components: attenuation loss, reflection loss, and the correction term.

(1) Attenuation Loss A. The propagation constant of a uniform plane wave in a good conductor as shown in Eq. (2-5-7) is given by

$$\gamma = \alpha + j\beta = (1 + j)\sqrt{\pi f \mu \sigma} \qquad \text{nepers/unit length} \qquad (9\text{-}2\text{-}16)$$

Thus the attenuation loss in a shield of thickness t is

$$A = 20 \log_{10} e^{\alpha t} = 20(\alpha t) \log_{10} e = 20(0.4343) \cdot \alpha t$$
$$= 8.686\alpha t = 8.686 t \sqrt{\pi f \mu \sigma} \quad \text{dB} \qquad (9\text{-}2\text{-}17)$$

where t represents the thickness of the shield in meters and 1 neper has been replaced by 8.686 dB.

It can be seen that the attenuation loss is directly related to the frequency of a plane wave and to the conductivity and permeability of a shield. The materials used for shielding, such as copper, which has high conductivity, are excellent shields for high-impedance fields over the entire frequency spectrum because the attenuation increases as the frequency increases.

(2) *Reflection Loss R.* From transmission-line theory, a change of the line impedance introduces a mismatch, which, in turn, causes a reflection loss. The intrinsic impedance of a good conductor as shown in Eq. (2-5-11) is

$$\eta = \sqrt{\frac{j\omega\mu}{\sigma}} = (1+j)\sqrt{\frac{\pi f\mu}{\sigma}} = (1+j)\frac{1}{\sigma\delta} \quad \text{ohms} \quad (9\text{-}2\text{-}18)$$

where σ equals the conductivity in mhos per meter and $\delta = 1/\sqrt{\pi f\mu\sigma}$ is the skin depth in meters.

Let K be the ratio of the intrinsic impedance of the incident plane wave in free space to the intrinsic impedance of the wave in the shield. Then

$$K = \frac{\eta_0}{\eta} \quad (9\text{-}2\text{-}19)$$

The reflection loss, which is defined as the product of the current and voltage transmission coefficients existing across the boundaries of the shield, becomes

$$R = 20 \log_{10} \frac{|1+K|^2}{4|K|} \quad \text{dB} \quad (9\text{-}2\text{-}20)$$

The reflection loss of Eq. (9-2-20) can be further simplified to

$$R = 20 \log_{10}\left(\frac{1}{4|K|}\right) \quad \text{for } |K| \ll 1 \quad (9\text{-}2\text{-}21)$$

and

$$R = 20 \log_{10}\left(\frac{|K|}{4}\right) \quad \text{for } |K| \gg 1 \quad (9\text{-}2\text{-}22)$$

(3) *Correction Term C.* The correction term C is then expressed by

$$C = 20 \log_{10}|1 - pe^{-2\gamma t}| \quad (9\text{-}2\text{-}23)$$

The factor $e^{-2\gamma t}$ can be written

$$e^{-2\gamma t} = e^{-2\alpha t}e^{-j2\alpha t} = e^{-2\alpha t}(\cos 2\alpha t - j\sin 2\alpha t)$$

From Eq. (9-2-17)

$$2\alpha t = \frac{2A}{8.686} = 0.23A \quad \text{and} \quad -A = 20 \log_{10} e^{-\alpha t}$$

Then

$$\frac{-A}{10} = \log_{10} e^{-2\alpha t} \quad \text{or} \quad 10^{-A/10} = e^{-2\alpha t}$$

Hence

$$e^{-2\gamma t} = 10^{-A/10}[\cos(0.23A) - j\sin(0.23A)]$$

Finally, the correction term becomes

$$C = 20 \log_{10}|1 - p10^{-A/10}(\cos 0.23A - j\sin 0.23A)| \quad (9\text{-}2\text{-}24)$$

9-2-2. High-Impedance Field

When undesired energy is transmitted in space, it is the properties of the electromagnetic fields that determine the degree of impedance mismatch at the shield. These fields are either high impedance or low impedance in nature. A high-impedance field is one that has an impedance greater than the intrinsic impedance of the dielectric in which it exists. Most of its energy is stored in the electric field. A low-impedance field, however, is one that has an impedance less than the intrinsic impedance of the dielectric in which it exists. Most of its energy is stored in the magnetic field. Often high-impedance fields are called electric fields, and low-impedance fields are referred to as magnetic fields. The field equations of an high-impedance electric-dipole antenna are given by [10]

$$E_\theta = \frac{j\eta\beta^2}{4\pi} I\ell \sin\theta e^{-j\beta r}\left(\frac{1}{\beta r} - \frac{j}{\beta^2 r^2} - \frac{1}{\beta^3 r^3}\right) \tag{9-2-25}$$

$$E_\phi = 0$$

$$E_r = \frac{\eta\beta^2}{4\pi} I\ell \cos\theta e^{-j\beta r}\left(\frac{1}{\beta^2 r^2} - \frac{j}{\beta^3 r^3}\right) \tag{9-2-26}$$

$$H_\theta = 0$$

$$H_\phi = \frac{j\beta^2}{4\pi} I\ell \sin\theta e^{-j\beta r}\left(\frac{1}{\beta r} - \frac{j}{\beta^2 r^2}\right) \tag{9-2-27}$$

$$H_r = 0$$

where $\beta = \dfrac{\omega}{v} = \omega\sqrt{\mu\epsilon} = \dfrac{2\pi}{\lambda}$

$\eta = \sqrt{\dfrac{\mu}{\epsilon}} = \mu v$

$v = \dfrac{1}{\sqrt{\mu\epsilon}}$

$\lambda = \dfrac{v}{f}$

$I =$ current in dipole antenna
$\ell =$ dipole length
$r =$ the distance in meters from the source to the shield

In the induction-field region $r \ll 1/\beta$, and the dominant field terms are

$$E_\theta = \frac{I\ell}{j4\pi\omega\epsilon r^3}\sin\theta \tag{9-2-28}$$

$$E_r = \frac{I\ell}{2\pi\omega\epsilon r^3}\cos\theta \tag{9-2-29}$$

$$H_\phi = \frac{I\ell}{4\pi r^2}\sin\theta \tag{9-2-30}$$

For an electric-dipole source, the wave impedance of the induction fields in the radial direction in free space is defined as

$$\eta_r = \frac{E_\theta}{H_\phi} = \frac{1}{j\omega\epsilon_0 r} \quad \text{ohms} \quad \text{for } r \ll \frac{1}{\beta} \tag{9-2-31}$$

In the radiation-field region $r \gg 1/\beta$, and the dominant field terms become

$$E_\theta = \frac{j\eta_0\beta I\ell}{4\pi r} \sin\theta \tag{9-2-32}$$

$$H_\phi = \frac{j\beta I\ell}{4\pi r} \sin\theta \tag{9-2-33}$$

where

$$\eta_0 = \sqrt{\frac{\mu_0}{\epsilon_0}} = 120\pi = 377 \quad \text{ohms}$$

For an electric-dipole source, the wave impedance of the radiation fields in the radial direction in free space is given by

$$\eta_0 = \frac{E_\theta}{H_\phi} = \frac{1}{c\epsilon_0} = 377 \quad \text{ohms} \quad \text{for } r \gg \frac{1}{\beta} \tag{9-2-34}$$

where $c = 3 \times 10^8$ m/s is the velocity of light in vacuum. This impedance is often referred to as the intrinsic impedance of electromagnetic plane waves in free space for the far field. The electric and magnetic wave are always uniform, in phase, and normal to each other.

9-2-3. Low-Impedance Field

The field equations of a low-impedance magnetic-dipole or loop antenna are given by [11]

$$E_\theta = 0$$

$$E_\phi = \frac{\eta\beta^3 IA}{4\pi} \sin\theta e^{-j\beta r}\left(\frac{1}{\beta r} - \frac{j}{\beta^2 r^2}\right) \tag{9-2-35}$$

$$E_r = 0$$

$$H_\theta = \frac{-\beta^3 IA}{4\pi} \cos\theta e^{-j\beta r}\left(\frac{1}{\beta r} - \frac{j}{\beta^2 r^2} - \frac{1}{\beta^3 r^3}\right) \tag{9-2-36}$$

$$H_\phi = 0$$

$$H_r = \frac{j\beta^3 IA}{2\pi} \cos\theta e^{-j\beta r}\left(\frac{1}{\beta^2 r^2} - \frac{j}{\beta^3 r^3}\right) \tag{9-2-37}$$

where $A = $ area of the loop antenna.

In the induction-field region $r \ll 1/\beta$, and the field equations become

$$E_\phi = \frac{-j\eta\beta IA}{4\pi r^2} \sin\theta \tag{9-2-38}$$

$$H_\theta = \frac{IA}{4\pi r^3} \sin \theta \qquad (9\text{-}2\text{-}39)$$

$$H_r = \frac{IA}{2\pi r^3} \cos \theta \qquad (9\text{-}2\text{-}40)$$

The wave impedance of the induction fields in the radial direction in free space for a magnetic-dipole or loop source is expressed as

$$\eta = -\frac{E_\phi}{H_\theta} = j\omega\mu_0 r \qquad \text{ohms} \qquad \text{for } r \ll \frac{1}{\beta} \qquad (9\text{-}2\text{-}41)$$

In the radiation-field region $r \gg 1/\beta$, and the dominant field terms are

$$E_\phi = \frac{\eta\beta^2 IA}{4\pi r} \sin \theta \qquad (9\text{-}2\text{-}42)$$

$$H_\theta = \frac{-\beta^2 IA}{4\pi r} \sin \theta \qquad (9\text{-}2\text{-}43)$$

The wave impedance of the radiation fields in the radial direction in free space for a magnetic-dipole or loop source is written

$$\eta_0 = -\frac{E_\phi}{H_\theta} = c\mu_0 = 377 \qquad \text{ohms} \qquad \text{for } r \gg \frac{1}{\beta} \qquad (9\text{-}2\text{-}44)$$

A second important factor in determining the nature of the field impedance is the distance from the field to the source. The energy that constitutes a net flow away from the source is contained in the radiation field, whereas the energy in the induction field is stored for one-quarter of a cycle and returned to the source in the next quarter cycle. At a distance close to the source, most of the energy is contained in the induction field. This region is known as the *near field*, or *Fresnel zone*. If the parameter r_n is defined to be the distance from the source at which 99 % of the total energy is contained in the induction field, then, for the practical purpose used by many electronics engineers, the distance is given by

$$r_n \le \frac{0.01v}{\omega} = 0.0016\lambda \qquad \text{meters} \qquad (9\text{-}2\text{-}45)$$

where $v =$ phase velocity.

At a distance far from the source, most of the energy is contained in the radiation field. This region is known as the *far field*, or *Fraunhofer zone*. The wave is a plane wave, and the energy contained in this field is divided equally between the electric and magnetic components. If the parameter r_f is defined to be the distance from the source to where 99 % of the total energy is contained in the radiation field, then, for all practical purposes, the distance is given by

$$r_f > \frac{100v}{\omega} = 16\lambda \qquad \text{meters} \qquad (9\text{-}2\text{-}46)$$

TABLE 9-2-1. Fields in Terms of Frequency and Distance

| | Distance from source (meters) | |
Frequency	Radiation field	Induction field
1 kHz	4,780,000	478
1 MHz	4,780	0.478
1 GHz	4.78	0.000478
1000 GHz	0.00478	0.000000478

Table 9-2-1 shows the distance of the radiation and induction fields from a source at different frequencies.

The reflection loss R for a high-impedance induction field in the near zone is found as follows. Substitution of Eqs. (9-2-18) and (9-2-31) in (9-2-19) yields

$$K = \frac{\eta}{\eta_r} = (j - 1)(2\pi^{3/2}\epsilon_0)\sqrt{\frac{f^3 r^2 \mu}{\sigma}} \qquad (9\text{-}2\text{-}47)$$

The magnitude of Eq. (9-2-47) becomes

$$|K| = 1.4 \times 10^{-10}\left(\frac{f^3 r^2 \mu}{\sigma}\right)^{1/2} \qquad \text{for } \eta \ll \eta_r \qquad (9\text{-}2\text{-}48)$$

Substitution of Eq. (9-2-48) in (9-2-21) gives the reflection loss in the near field for a high impedance; thus

$$R = 186.4 + 10\log_{10}\left(\frac{\sigma}{f^3 r^2 \mu}\right) \qquad \text{dB} \qquad \text{for near field} \quad (9\text{-}2\text{-}49)$$

It can be seen from Eq. (9-2-49) that since the reflection loss R is inversely proportional to the cube of the frequency, R will increase to its highest value as the frequency decreases to its lowest value. Low-impedance or magnetic fields are not easily shielded at low frequencies. The difficulty is that excellent shield materials for high-impedance or electric fields offer no attenuation to magnetic fields at low frequencies.

Shielding in the far field is similar to shielding in the near field. The reflection loss is derived as follows. Substitution of Eqs. (9-2-18) and (9-2-44) in (9-2-19) yields

$$K = \frac{\eta}{\eta_0} = \frac{(1 + j)}{377}\sqrt{\frac{\pi f \mu}{\sigma}} \qquad (9\text{-}2\text{-}50)$$

The magnitude of Eq. (9-2-50) becomes

$$|K| = 0.666 \times 10^{-2}\sqrt{\frac{f \mu}{\sigma}} \qquad \text{for } \eta \ll \eta_0 \qquad (9\text{-}2\text{-}51)$$

Substitution of Eq. (9-2-51) in (9-2-21) gives the reflection loss as

$$R = 31.5 + 10\log_{10}\left(\frac{\sigma}{f \mu}\right) \qquad \text{dB} \qquad \text{for far field} \qquad (9\text{-}2\text{-}52)$$

9-3 Plane Wave Propagation in Anechoic Chambers

9-3-1. Anechoic Materials

An anechoic chamber is an enclosure lined by anechoic material that has a high absorption coefficient. The more electromagnetic energy is absorbed by the anechoic material, the less the energy is reflected, and the more the environment looks like free space to the wave. The commonly used absorbers are made of curled animal hair uniformly impregnated with a mixture of carbon black neoprene. The front surface of the absorber is cut into a series of pyramids. The most important property that an absorber must have is its ablility to absorb electromagnetic energy over a wide range of frequencies. The absorption property of the material is expressed in decibels of reflectivity at a given frequency of operation. The reflectivity is defined as

$$\text{Reflectivity} = 10 \log_{10} \frac{W_{\text{ref}}}{W_{\text{inc}}} \quad \text{dB} \qquad (9\text{-}3\text{-}1)$$

where W_{ref} is the reflected electromagnetic energy and W_{inc} is the incident electromagnetic energy.

An absorption of 20 to 60 dB represents 1 to 0.0001 % reflection from the material. An absorber must be able to dissipate the energy incident upon it. The materials usually have power ratings in the range from 0.1 to 3 W/in². In MKS units, these same power ratings range from 15.5 to 465 mW/cm².

9-3-2. Quiet Zone

The quiet zone of an anechoic chamber is a defined volume within the chamber where electromagnetic waves reflected from the walls, floor, and ceiling are stated to be below a certain specified minimum level. A quiet zone performance rating of 40 dB, for example, means that the reflected energy arriving at any point within a specific volume will be 40 dB below the direct wave arriving at the same point. The quiet zone may be a sphere centered at the test stand, or it may be specified as a cylinder with its axis coincident with the central axis of the chamber. In either case, the performance of the quiet zone is specified as that value where the reflected wave from any surface bounded by the chamber is less than the direct wave by a specified amount, expressed in decibels [12]. Figure 9-3-2-1 shows the quiet zone in a rectangular anechoic chamber. Since the quiet zone within a chamber has been simulated like free space, any devices or electronic systems to be tested should be placed within the quiet zone. If the electromagnetic energy level is known to be acceptable to the devices or systems under test, the transmitting antenna should be placed at a distance of R meters away from the quiet zone in order to ensure that the electric intensity will be uniform and acceptable at the test site. This distance R is given by

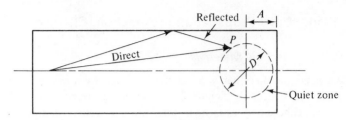

(a) Spherical quiet zone

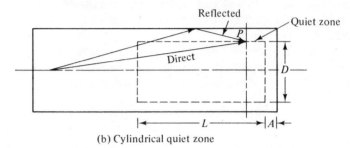

(b) Cylindrical quiet zone

(c) Rectangular anechoic chamber with cylindrical quiet zone

FIG. 9-3-2-1. Quiet zones in rectangular anechoic chambers. (*Courtesy Hayden Publishing Co., Inc.*)

$$R \geq \frac{2D^2}{\lambda} = \frac{A}{2\lambda} \qquad \text{meters} \qquad (9\text{-}3\text{-}2)$$

where D = the largest dimension of the antenna in meters
         λ = the wavelength of the signal in meters
         A = the antenna aperture in square meters

EXAMPLE 9-3-1:   Consider the electromagnetic energy transmission for the following test in an anechoic chamber as showm in Fig. 9-3-2-2. Determine the electric field intensity.

Solution.   The power density at the test site in the quiet zone is given by

$$p_d = \frac{p_t g_a g_c}{4\pi R^2} \qquad \text{watts/m}^2 \qquad (9\text{-}3\text{-}3)$$

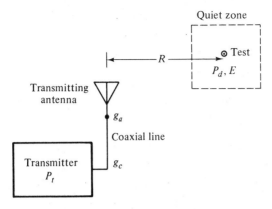

FIG. 9-3-2-2. Test in anechoic chamber.

where p_t = the transmitter power in watts
g_c = the coaxial-line loss (numerical)
g_a = the antenna gain (numerical)
R = the distance between antenna and test site in meters

The electric field intensity is computed from Eq. (9-3-3) to be

$$E = 19.4\sqrt{p_d} \qquad \text{volts/m} \tag{9-3-4}$$

Both the power density and electric intensity can be written

$$P_d = -11 \text{ dB} - 20 \log (R) + P_t + G_c + G_a \qquad \text{dBW/m}^2 \tag{9-3-5}$$

and $E = 14.8 \text{ dB} - 20 \log (R) + P_t + G_c + G_a \qquad \text{dBV/m} \tag{9-3-6}$

Subtraction of Eq. (9-3-5) from (9-3-6) results in the conversion of power density in dBW/m² to field intensity in dBV/m and vice versa. That is,

$$E = 25.8 \text{ dB} + P_d(\text{dBW/m}^2) \qquad \text{dBV/m} \tag{9-3-7}$$

where P_t = the transmitter power in dBW
G_a = the antenna gain in dB
G_c = the coaxial-line loss in dB
R = the range between the antenna and test site in meters

9-4 Summary

We have described the theory and design of shielded rooms and anechoic chambers. Some basic equations for shielded rooms are restated below.
    Shielding Effectiveness Equation:

$$\text{SE} = A + R + C \qquad \text{dB}$$

where A, the attenuation loss, is expressed by

$$A = 8.686t\sqrt{\pi f \mu \sigma} \qquad \text{dB}$$

where t = thickness of the shield in meters
μ = permeability of the shielding material
σ = conductivity of the shielding material

R is the reflection loss and is expressed by

$$R = 20 \log_{10} \frac{|1 + K|^2}{4|K|} \qquad \text{dB}$$

in which K is the ratio of the intrinsic impedance to the radial impedance for waves normally incident upon the shield—that is,

$$K = \frac{\eta}{\eta_r} = \frac{(1 + j)}{j\omega\mu_0 r} \sqrt{\frac{\pi f \mu}{\sigma}}$$

where r = radial distance in meters from the source to the shield
$\mu_0 = 4\pi \times 10^{-7}$ henry/m is the permeability of free space

C is the correction term and it is expressed by

$$C = 20 \log_{10} |1 - 10^{-A/10} [\cos (0.23A) - j \sin (0.23A)]|$$

where

$$\rho = \left(\frac{Z_\ell - Z_0}{Z_\ell + Z_0}\right)^2$$

REFERENCES

[1] Department of Defense Directive Number 3222.3, July 5, 1967, Department of Defense Electromagnetic Compatibility Program.

[2] SCHELKUNOFF, S. A., The electromagnetic theory of coaxial transmission lines and cylindrical shields. *Bell System Tech. J.*, **13**, No. 4, 532–79, October 1934.

[3] SCHELKUNOFF, S. A., The impedance concept and its application to problems of reflection, refraction, shielding and power absorption. *Bell System Tech. J.*, **17**, No. 1, 17–48, January 1938.

[4] JARVA, W., Shielding efficiency calculation methods for screening, waveguide ventilation panels and other perforated shields. Proceedings of the 7th Conference on Radio Interference Reduction and Electronic Compatibility, Illinois Institute of Technology, Chicago, Ill., November 1961.

[5] VASAKA, C. S., Technical note—Problems in shielding electric and electronic equipments. *Report NADC-ED-N5507*, U.S. Naval Air Development Center, Johnsville, Pa., 1954.

[6] VASAKA, C. S., Theory, design, and engineering evaluation of radio frequency shielded rooms. *Report NADC-EL-54129*, U.S. Naval Air Development Center, Johnsville, Pa., August 13, 1956.

[7]   SCHULZ, R. B. et al., Shielding theory and practice. Proceedings of the 9th Tri-Service Conference on Electromagnetic Compatability, IIT Research Institute, pp. 597–636, October 1963.

[8]   SCHULZ, R. B., ELF and VLF shielding effectiveness of high permeability materials. IEEE International Convention, March 1964.

[9]   SCHULZ, R. B., et al., Low-frequency shielding resonance. IEEE International Convention, New York, March 1965.

[10] SCHELKUNOFF, S. A., *Electromagnetic Waves.* p. 304. D. Van Nostrand Company, Princeton, N. J., 1944.

[11] SCHELKUNOFF, S. A., *Electromagnetic Waves.* D. Van Nostrand Company, Princeton, N.J., 1957.

[12] GALAGAN, STEVE., Understanding microwave absorbing materials and anechoic chambers.
      Part 1. *Microwaves*, December 1969.
      Part 2. *Microwaves*, January 1970.
      Part 3. *Microwaves*, April 1970.
      Part 4. *Microwaves*, May 1970.

SUGGESTED READINGS

1.   APPEL-HANSEN, J., Reflectivity level of radio anechoic chambers. *IEEE Trans. Antennas and Propagation*, **AP-21**, No. 4, 490–498, July 1973.

2.   EMERSON, WILLIAM H., Electromagnetic wave absorbers and anechoic chambers through the years. *IEEE Trans Antennas and Propagation*, **AP-21**, No. 4, 484–490, July 1973.

Measurements and Computations
of Microwave Electric Field
and Power Density*

chapter

10

10-0   Introduction

The problem of electric field intensity measurements and computations is becoming increasingly
important in the microwave area. Many electronic systems, such as radar or missiles, require
a field intensity in microvolts per centimeter, $\mu V/cm$.
However, many of the field intensity meters at
the receiver position read voltages in μV or in
$dB\mu V$. Then it is necessary to convert the intensity
meter reading in μV or $dB\mu V$ to $\mu V/cm$. Although
the receiver reads power in watts or milliwatts, it is
still necessary to convert the receiving power to

power density in milliwatts per square centimeter or to field intensity in volts per centimeter. Such problems confront the electronics engineer every day. This chapter briefly presents the basic field equations and then describes the field intensity measurements and computations in detail.

10-1 Basic Field Equations

According to field theory, the electric and magnetic field intensities at the far field or Fraunhofer region in free space are always in phase and mutually perpendicular to each other. The power density carried by these two waves at the observation point is expressed by the Poynting vector [1]

$$\mathbf{P} = \text{Re}\,(\mathbf{E} \times \mathbf{H}^*) \qquad \text{watts/unit area} \qquad (10\text{-}1\text{-}1)$$

where $\mathbf{E}$ = the electric field intensity in volts (RMS) per unit length (complex quantity)

      $\mathbf{H}$ = the magnetic field intensity in amperes (RMS) per unit length (complex quantity)

      Re = the real part and

      $*$ = the complex conjugate

The total power at the observation point can be obtained by integrating the power density function over an imaginary spherical surface through the observation point with a radius of R from the source at the center of the sphere [2]. That is,

$$P = \iint \mathbf{P} \cdot d\mathbf{a} = \text{Re} \iint (\mathbf{E} \times \mathbf{H}^*) \cdot (R^2 \sin\theta\, d\theta\, d\phi)\mathbf{u}$$

$$= \frac{E^2}{\eta_0} \int_{\phi=0}^{2\pi} \int_{\theta=0}^{\pi} R^2 \sin\theta\, d\theta\, d\phi = \frac{E^2}{\eta_0}(4\pi R^2) \qquad \text{watts} \qquad (10\text{-}1\text{-}2)$$

where R = the distance between the observation point and source

      $\eta_0 = 377\,\Omega$ is the intrinsic impedance of free space

The energy stored in the electric field is given by [3]

$$W_e = \frac{\epsilon_0}{2} \iiint_v |E|^2\, dv \qquad \text{joules} \qquad (10\text{-}1\text{-}3)$$

where $\epsilon_0 = 8.854 \times 10^{-12} = 1/(36\pi) \times 10^{-9}$ farad/m is the permittivity or capacitivity of vacuum or air

      $|E|$ = the magnitude of the electric field intensity

The energy stored in the magnetic field is expressed by [4]

$$W_m = \frac{\mu_0}{2} \iiint_v |H|^2\, dv \qquad \text{joules} \qquad (10\text{-}1\text{-}4)$$

where $\mu_0 = 4\pi \times 10^{-7}$ henry/m is the permeability or inductivity of a vacuum or air

        $|H| =$ the magnitude of the magnetic field intensity

At the far field, the electric and magnetic fields are related by the relationship [5]

$$\eta_0 = \frac{E}{H} \quad \text{ohms} \tag{10-1-5}$$

10-2   Units of Measurement

In field intensity measurements the units of measurement and the conversion of one unit to another are the essential parts of the process. A few widely used units are described below.

1. dB—The decibel (dB) is a dimensionless number that expresses the ratio of two power levels. It is defined as

$$dB \equiv 10 \log_{10}\left(\frac{P_2}{P_1}\right) \tag{10-2-1}$$

The two power levels are relative to each other. If power level P_2 is higher than power level P_1, dB is positive and vice versa. Since $P = V^2/R$, when their voltages are measured across the same or equal resistors, the number of dB is given by

$$dB \equiv 20 \log_{10}\left(\frac{V_2}{V_1}\right) \tag{10-2-2}$$

The voltage definition of dB has no meaning at all unless the two voltages under consideration appear across equal impedances. Above 10 GHz the impedance of waveguides varies with frequency, and the dB calibration is limited to power levels only. Table 10-2-1 shows the conversion of voltage and power ratios to dB.

2. dBW—The decibel above 1 watt (dBW) is another useful measure for expressing power level P_2 with respect to a reference power level P_1 of 1 watt. Similarly, if the power level P_2 is lower than 1 watt, the dBW is negative.

3. dBm—The decibel above 1 milliwatt (dBm) is also a useful measure of expressing power level P_2 with respect to a reference power level P_1 of 1 milliwatt (mW). Since the power level in the microwave region is quite low, the dBm unit is very useful in that frequency range. It is customary to designate milli by a lower case letter m and mega by an upper case letter M.

4. dBV—The decibel above 1 volt (dBV) is a dimensionless voltage ratio in dB referred to a reference voltage of 1 volt.

5. dBμV—The decibel above 1 microvolt (dBμV) is another dimensionless voltage ratio in dB referred to a reference voltage of 1 microvolt

TABLE 10-2-1. Conversion of Voltage and Power Ratios to dB

Voltage ratio	Power ratio	−dB+	Voltage ratio	Power ratio	Voltage ratio	Power ratio	−dB+	Voltage ratio	Power ratio
1.000	1.000	0	1.000	1.000	.596	.355	4.5	1.679	2.818
.989	.977	.1	1.012	1.023	.589	.347	4.6	1.698	2.884
.977	.955	.2	1.023	1.047	.582	.339	4.7	1.718	2.951
.966	.933	.3	1.035	1.072	.575	.331	4.8	1.738	3.020
.955	.912	.4	1.047	1.096	.569	.324	4.9	1.758	3.090
.944	.891	.5	1.059	1.122	.562	.316	5.0	1.778	3.162
.933	.871	.6	1.072	1.148	.556	.309	5.1	1.799	3.236
.923	.851	.7	1.084	1.175	.550	.302	5.2	1.820	3.311
.912	.832	.8	1.095	1.202	.543	.295	5.3	1.841	3.388
.902	.813	.9	1.109	1.230	.537	.288	5.4	1.862	3.467
.891	.794	1.0	1.122	1.259	.531	.282	5.5	1.884	3.548
.881	.776	1.1	1.135	1.288	.525	.275	5.6	1.905	3.631
.871	.759	1.2	1.148	1.318	.519	.269	5.7	1.928	3.715
.861	.741	1.3	1.161	1.349	.513	.263	5.8	1.950	3.802
.851	.724	1.4	1.175	1.380	.507	.257	5.9	1.972	3.890
.841	.708	1.5	1.189	1.413	.501	.251	6.0	1.995	3.981
.832	.692	1.6	1.202	1.445	.496	.246	6.1	2.018	4.074
.822	.676	1.7	1.216	1.479	.490	.240	6.2	2.042	4.159
.813	.661	1.8	1.230	1.514	.484	.234	6.3	2.065	4.265
.804	.646	1.9	1.245	1.549	.479	.229	6.4	2.089	4.365
.794	.631	2.0	1.259	1.585	.473	.224	6.5	2.113	4.467
.785	.617	2.1	1.274	1.622	.468	.219	6.6	2.138	4.571
.776	.603	2.2	1.288	1.660	.462	.214	6.7	2.163	4.677
.767	.589	2.3	1.303	1.698	.457	.209	6.8	2.188	4.786
.759	.575	2.4	1.318	1.738	.452	.204	6.9	2.215	4.898
.750	.562	2.5	1.334	1.778	.447	.200	7.0	2.239	5.012
.741	.550	2.6	1.349	1.820	.442	.195	7.1	2.265	5.129
.733	.537	2.7	1.365	1.862	.437	.191	7.2	2.291	5.248
.724	.525	2.8	1.380	1.905	.432	.186	7.3	2.317	5.370
.716	.513	2.9	1.390	1.950	.427	.182	7.4	2.344	5.495
.708	.501	3.0	1.413	1.995	.422	.178	7.5	2.371	5.623
.700	.490	3.1	1.429	2.042	.417	.174	7.6	2.399	5.754
.692	.479	3.2	1.445	2.089	.412	.170	7.7	2.427	5.888
.684	.468	3.3	1.462	2.138	.407	.166	7.8	2.455	6.026
.676	.457	3.4	1.479	2.188	.403	.162	7.9	2.483	6.166
.668	.447	3.5	1.496	2.239	.398	.159	8.0	2.512	6.310
.661	.437	3.6	1.514	2.291	.394	.155	8.1	2.541	6.457
.653	.427	3.7	1.531	2.344	.389	.151	8.2	2.570	6.607
.646	.417	3.8	1.549	2.399	.385	.148	8.3	2.600	6.761
.638	.407	3.9	1.567	2.455	.380	.145	8.4	2.630	6.918
.631	.398	4.0	1.585	2.512	.376	.141	8.5	2.661	7.079
.624	.389	4.1	1.603	2.570	.372	.138	8.6	2.692	7.244
.617	.380	4.2	1.622	2.630	.367	.135	8.7	2.723	7.413
.610	.372	4.3	1.641	2.692	.363	.132	8.8	2.754	7.586
.603	.363	4.4	1.660	2.754	.359	.129	8.9	2.786	7.762

TABLE 10-2-1. (Continued)

Voltage ratio	Power ratio	−dB+	Voltage ratio	Power ratio	Voltage ratio	Power ratio	−dB+	Voltage ratio	Power ratio
.355	.126	9.0	2.818	7.943	.211	.0447	13.5	4.732	22.39
.351	.123	9.1	2.851	8.128	.209	.0437	13.6	4.786	22.91
.347	.120	9.2	2.884	8.318	.207	.0427	13.7	4.842	23.44
.343	.118	9.3	2.917	8.511	.204	.0417	13.8	4.898	23.99
.339	.115	9.4	2.951	8.710	.202	.0407	13.9	4.955	24.55
.335	.112	9.5	2.985	8.913	.200	.0398	14.0	5.012	25.12
.331	.110	9.6	3.020	9.120	.197	.0389	14.1	5.070	25.70
.327	.107	9.7	3.055	9.333	.195	.0380	14.2	5.129	26.30
.324	.105	9.8	3.090	9.550	.193	.0372	14.3	5.188	26.92
.320	.102	9.9	3.126	9.772	.191	.0363	14.4	5.248	27.54
.316	.100	10.0	3.162	10.000	.188	.0355	14.5	5.309	28.18
.313	.0977	10.1	3.199	10.23	.186	.0347	14.6	5.370	28.84
.309	.0955	10.2	3.236	10.47	.184	.0339	14.7	5.433	29.51
.306	.0933	10.3	3.273	10.72	.182	.0331	14.8	5.495	30.20
.302	.0912	10.4	3.311	10.96	.180	.0324	14.9	5.559	30.90
.299	.0891	10.5	3.350	11.22	.178	.0316	15.0	5.623	31.62
.295	.0871	10.6	3.388	11.48	.176	.0309	15.1	5.689	32.36
.292	.0851	10.7	3.428	11.75	.174	.0302	15.2	5.754	33.11
.288	.0832	10.8	3.467	12.02	.172	.0295	15.3	5.821	33.88
.283	.0813	10.9	3.508	12.30	.170	.0288	15.4	5.888	34.67
.282	.0794	11.0	3.548	12.59	.168	.0282	15.5	5.957	35.48
.279	.0776	11.1	3.589	12.88	.166	.0275	15.6	6.026	36.31
.275	.0759	11.2	3.631	13.18	.164	.0269	15.7	6.095	37.15
.272	.0741	11.3	3.673	13.49	.162	.0263	15.8	6.166	38.02
.269	.0724	11.4	3.715	13.80	.160	.0257	15.9	6.237	38.90
.266	.0708	11.5	3.758	14.13	.159	.0251	16.0	6.310	39.81
.263	.0692	11.6	3.802	14.45	.157	.0246	16.1	6.383	40.74
.260	.0676	11.7	3.846	14.79	.155	.0240	16.2	6.457	41.69
.257	.0661	11.8	3.890	15.14	.153	.0234	16.3	6.531	42.66
.254	.0646	11.9	3.936	15.49	.151	.0229	16.4	6.607	43.65
.251	.0631	12.0	3.981	15.85	.150	.0224	16.5	6.683	44.67
.248	.0617	12.1	4.027	16.22	.148	.0219	16.6	6.761	45.71
.246	.0603	12.2	4.074	16.60	.146	.0214	16.7	6.839	46.77
.243	.0589	12.3	4.121	16.98	.145	.0209	16.8	6.918	47.86
.240	.0575	12.4	4.169	17.38	.143	.0204	16.9	6.998	48.98
.237	.0562	12.5	4.217	17.78	.141	.0200	17.0	7.079	50.12
.234	.0550	12.6	4.266	18.20	.140	.0195	17.1	7.161	51.29
.232	.0537	12.7	4.315	18.62	.138	.0191	17.2	7.244	52.48
.229	.0525	12.8	4.365	19.05	.137	.0186	17.3	7.328	53.70
.227	.0513	12.9	4.416	19.50	.135	.0182	17.4	7.413	54.95
.224	.0501	13.0	4.467	19.95	.133	.0178	17.5	7.499	56.23
.221	.0490	13.1	4.519	20.42	.132	.0174	17.6	7.586	57.54
.219	.0479	13.2	4.571	20.89	.130	.0170	17.7	7.674	58.88
.216	.0468	13.3	4.624	21.38	.129	.0166	17.8	7.762	60.26
.214	.0457	13.4	4.677	21.88	.127	.0162	17.9	7.852	61.66

TABLE 10-2-1. (Continued)

Voltage ratio	Power ratio	−dB+	Voltage ratio	Power ratio	Voltage ratio	Power ratio	−dB+	Voltage ratio	Power ratio
.126	.0159	18.0	7.943	63.10	.106	.0112	19.5	9.441	89.13
.125	.0155	18.1	8.035	64.57	.103	.0110	19.6	9.550	91.20
.123	.0151	18.2	8.128	66.07	.104	.0107	19.7	9.661	93.33
.122	.0148	18.3	8.222	67.61	.102	.0105	19.8	9.772	95.50
.120	.0145	18.4	8.318	69.18	.101	.0102	19.9	9.886	97.72
.119	.0141	18.5	8.414	70.79	.100	.0100	20.0	10.000	100.00
.118	.0138	18.6	8.511	72.44		10^{-3}	30		10^3
.116	.0135	18.7	8.610	74.13	10^{-2}	10^{-4}	40	10^2	10^4
.115	.0132	18.8	8.710	75.86		10^{-5}	50		10^5
.114	.0129	18.9	8.811	77.62	10^{-3}	10^{-6}	60	10^3	10^6
.112	.0126	19.0	8.913	79.43		10^{-7}	70		10^7
.111	.0123	19.1	9.016	81.28	10^{-4}	10^{-8}	80	10^4	10^8
.110	.0120	19.2	9.120	83.18		10^{-9}	90		10^9
.108	.0118	19.3	9.226	85.11	10^{-5}	10^{-10}	100	10^5	10^{10}
.107	.0115	19.4	9.333	87.10		10^{-11}	110		10^{11}
					10^{-6}	10^{-12}	120	10^6	10^{12}

(μV). The field intensity meters used for the measurements in the microwave region often have a scale in dBμV, since the power levels to be measured are usually extremely low.

    6. μV/m—Microvolts per meter (μV/m) are units of 10^{-6} volt per meter, expressing the electric field intensity.

    7. dBμV/m—The decibel above 1 microvolt per meter (dBμV/m) is a dimensionless electric field intensity ratio in dB relative to 1 μV/m. This unit is also often used for field intensity measurements in the microwave region.

    8. μV/m/MHz—The microvolts per meter per megahertz (μV/m/MHz) are units of 10^{-6} volt-second per broadband electric field intensity distribution. This is a two-dimensional distribution, in space and in frequency.

    9. dBμV/m/MHz—The decibel above 1 microvolt per meter per megahertz (dBμV/m/MHz) is a dimensionless broadband electric field intensity distribution ratio with respect to 1 μV/m/MHz.

    10. μV/MHz—Microvolts per megahertz per second of bandwidth (μV/MHz) are units of 10^{-6} volt-second of broadband voltage distribution in the frequency domain. The use of this unit is based on the assumption that the voltage is evenly distributed over the bandwidth of interest.

Figure 10-2-1 shows the conversion of dB scales in power and voltage.

$$dBW = -30 + dBm \qquad dBV = -60 + dBmW$$
$$= -60 + dB\mu W \qquad\quad = -120 + dB\mu V \qquad (10\text{-}2\text{-}3)$$

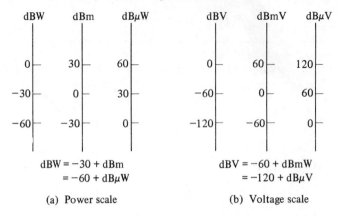

dBW = −30 + dBm
= −60 + dBμW

dBV = −60 + dBmW
= −120 + dBμV

(a) Power scale

(b) Voltage scale

FIG. 10-2-1. Conversion of dB scale in power and in voltage.

10-3   Free-Space Attenuation

Free-space attenuation is different from the dissipative attenuation of a medium such as air that absorbs energy from the wave. The power density in a spherical wave must decrease with distance as the energy in the wave spreads out over an ever-increasing surface area as the wave progresses. Figure 10-3-1 illustrates the situation.

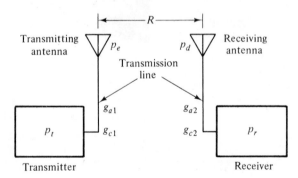

FIG. 10-3-1. Free-space attenuation.

The receiver power is equal to the power density at the antenna head times the aperture of the receiving antenna, the antenna gain, and the cable loss. The power density at the receiving antenna is equal to the transmitter power times the antenna gain and the cable loss and is then divided by the spherical surface area. Therefore the receiving power can be written

$$p_r = \left(\frac{p_e}{4\pi R^2}\right)\left(\frac{\lambda^2}{4\pi}\right)g_{a2}g_{c2} = p_t g_{a1}g_{c1}g_{a2}g_{c2}\left(\frac{\lambda}{4\pi R}\right)^2 \quad \text{watts} \quad (10\text{-}3\text{-}1)$$

where $p_r =$ the receiving power in watts

 $p_d =$ the receiving power density in watts/m^2

 $p_t =$ the transmitter power in watts

 $p_e =$ the effective radiated power in watts

 $g_{a1} =$ the transmitting antenna gain (numeric)

 $g_{c1} =$ the transmitting line loss (numeric)

 $g_{a2} =$ the receiving antenna gain (numeric)

 $g_{c2} =$ the receiving line loss (numeric)

 $R =$ the range between two antennas in meters

 $\lambda =$ the wavelength in meters

The factor $\lambda^2/(4\pi)$ is the antenna aperture [6]. It does not imply that a higher-frequency wave decreases in magnitude more rapidly than a lower-frequency wave. It is simply a result of the fact that, for a given gain, the aperture of a higher-frequency antenna is smaller than that of a lower-frequency antenna and so it intercepts a smaller amount of the power from the wave.

If the transmitter power is in dBW, the antenna gains and the cable losses in dB, the receiving power given by Eq. (10-3-1) can be expressed in dBW:

$$P_r = P_t + G_{a1} + G_{a2} - G_{c1} - G_{c2} - 20 \log_{10} \left(\frac{4\pi R}{\lambda}\right) \qquad \text{dBW} \quad (10\text{-}3\text{-}2)$$

The last term is called the free-space attenuation. That is,

$$\text{Free-space attenuation} = -20 \log_{10} \left(\frac{4\pi R}{\lambda}\right) \qquad \text{dB} \qquad (10\text{-}3\text{-}3)$$

The free-space attenuation can be found from the standard nomogram, which is shown in Fig. 10-3-2. For instance, if the wavelength is 0.30 m and the range is 3 m, the free-space attenuation is about 42 dB. Table 10-3-1 shows the free-space attenuations in dB for different distances in feet. However, if the distances are in meters, all values in dB should be increased by 10. It should be noted that the uppercase letters in Eq. (10-3-2) are designated for the values in dB.

10-4 Conversion of Transmitting Power to Electric Field Intensity

The electric field intensity at a distance R in meters from the transmitting antenna may be computed from the transmitter power. From Fig. 10-3-1 the power density at the point of R from the transmitting antenna is given by

$$p_d = \frac{p_t g_{c1} g_{a1}}{4\pi R^2} \qquad \text{watts/m}^2 \qquad (10\text{-}4\text{-}1)$$

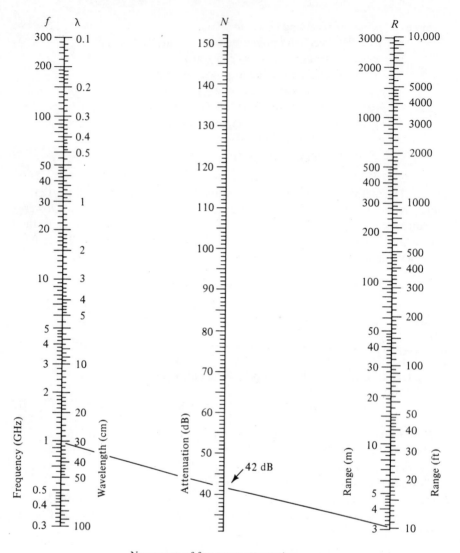

Nomogram of free-space attenuation

FIG. 10-3-2. Nomogram of free-space attenuation.

From field theory, the average power density carried by an electromagnetic plane wave at the far field in free space is given by Eq. (10-1-2) as

$$p = \frac{E^2}{\eta_0} = \frac{E^2}{120\pi} \qquad \text{watts/m}^2 \qquad (10\text{-}4\text{-}2)$$

In Eq. (10-4-2) the electric field intensity is assumed to be in volts per meter.

TABLE 10-3-1.   Free-Space Attenuation in dB

Frequency (MHz)	Distance in feet							
	10	50	100	500	1000	5000	10,000	50,000
100	22	36	42	56	62	76	82	96
200	28	42	48	62	68	82	88	102
300	32	46	52	66	72	86	92	106
400	34	48	54	68	74	88	94	108
500	36	50	56	70	76	90	96	110
600	38	52	58	72	78	92	98	112
700	39	53	59	73	79	93	99	113
800	40	54	60	74	80	94	100	114
900	41	55	61	75	81	95	101	115
1000	42	56	62	76	82	96	102	116
1500	46	60	66	80	86	100	106	120
2000	48	62	68	82	88	102	108	122
2500	50	64	70	84	90	104	110	124
3000	52	66	72	86	92	106	112	126
3500	53	67	73	87	93	107	113	127
4000	54	68	74	88	94	108	114	128
4500	55	69	75	89	95	109	115	129
5000	56	70	76	90	96	110	116	130
5500	57	71	77	91	97	111	117	131
6000	58	72	78	92	98	112	118	132
7000	59	73	79	93	99	113	119	133
8000	60	74	80	94	100	114	120	134
9000	61	75	81	95	101	115	121	135
10,000	62	76	82	96	102	116	122	136
15,000	66	80	86	100	106	120	126	140

Note: If the distance is in meters, all numbers in decibels should be increased by 10.

Therefore the field intensity E can be computed from Eqs. (10-4-2) and (10-4-1), and it is given by

$$E = \sqrt{120\pi p_d} = 19.4\sqrt{p_d} = \frac{5.48}{R}(p_t g_{c1} g_{a1})^{1/2} \qquad \text{volts/m} \qquad (10\text{-}4\text{-}3)$$

It is often desirable to express the power density and the field intensity in terms of dB. That is,

$$P_d = -11 \text{ dB} - 20 \log_{10}(R) + P_t \text{ (dBW)}$$
$$+ G_{a1} \text{ (dB)} - G_{c1} \text{ (dB)} \qquad \text{dBW/m}^2 \qquad (10\text{-}4\text{-}4)$$
$$E = 14.8 \text{ dB} - 20 \log_{10}(R) + P_t \text{ (dBW)}$$
$$+ G_{a1} \text{ (dB)} - G_{c1} \text{ (dB)} \qquad \text{dBV/m} \qquad (10\text{-}4\text{-}5)$$

Subtraction of Eq. (10-4-4) from (10-4-5) results in the conversion of power density in dBW/m² to field intensity in dBV/m and vice versa. That is,

$$E = 25.8 \text{ dB} + P_d \, (\text{dBW/m}^2) \qquad \text{dBV/m} \qquad (10\text{-}4\text{-}6)$$

It should be noted that if values per square centimeter and per centimeter are desired, a term of −40 dB must be added to each of the preceding three equations. Table 10-4-1 tabulates the values of electric field intensity and power density conversions.

TABLE 10-4-1.   Conversion of Field Intensity and Power Density (refer to free-space resistance 377 ohms)

Field intensity			*Power density*		
volts/m	*dBV/m*	*dBµV/m*	*watts/m²*	*dBW/m²*	*dBm/cm²*
1×10^{-6}	−120	0	27×10^{-16}	−146	−156
2×10^{-6}	−114	6	11×10^{-15}	−140	−150
3×10^{-6}	−110	10	24×10^{-15}	−136	−146
4×10^{-6}	−108	12	42×10^{-15}	−134	−144
5×10^{-6}	−106	14	66×10^{-15}	−132	−142
7×10^{-6}	−103	17	13×10^{-14}	−129	−139
8×10^{-6}	−102	18	17×10^{-14}	−128	−138
10×10^{-6}	−100	20	27×10^{-14}	−126	−136
20×10^{-6}	−94	26	11×10^{-13}	−120	−130
30×10^{-6}	−90	30	24×10^{-13}	−116	−126
40×10^{-6}	−88	32	42×10^{-13}	−114	−124
50×10^{-6}	−86	34	66×10^{-13}	−112	−122
70×10^{-6}	−83	37	13×10^{-12}	−109	−119
80×10^{-6}	−82	38	17×10^{-12}	−108	−118
100×10^{-6}	−80	40	27×10^{-12}	−106	−116
200×10^{-6}	−74	46	11×10^{-11}	−100	−110
300×10^{-6}	−70	50	24×10^{-11}	−96	−106
400×10^{-6}	−68	52	42×10^{-11}	−94	−104
500×10^{-6}	−66	54	66×10^{-11}	−92	−102
700×10^{-6}	−63	57	13×10^{-10}	−89	−99
800×10^{-6}	−62	58	17×10^{-10}	−88	−98
1×10^{-3}	−60	60	27×10^{-10}	−86	−96
10×10^{-3}	−40	80	27×10^{-8}	−66	−76
100×10^{-3}	−20	100	27×10^{-6}	−46	−56
1	0	120	27×10^{-4}	−26	−36
10	20	140	27×10^{-2}	−6	−16
100	40	160	27	14	4
1×10^{3}	60	180	27×10^{2}	34	24
10×10^{3}	80	200	27×10^{4}	54	44

10-5 Conversion of Receiving Power to Electric Field Intensity

From Fig. 10-3-1 the power density at the receiving antenna is equal to the receiver power divided by the antenna gain, the cable loss, and the antenna aperture. That is,

$$p_d = \frac{4\pi p_r}{\lambda^2 g_{a2} g_{c2}} \qquad \text{watts/m}^2 \qquad (10\text{-}5\text{-}1)$$

Substitution of Eq. (10-5-1) in (10-4-3) yields the field intensity at that point as

$$E = \frac{68.77}{\lambda} \left(\frac{p_r}{g_{a2} g_{c2}}\right)^{1/2} \qquad \text{volts/m} \qquad (10\text{-}5\text{-}2)$$

Similarly, the power density and the field intensity can be expressed in dB:

$$P_d = 11 \text{ dB} - 20 \log_{10}(\lambda) + P_r \text{ (dBW)} - G_{a2} \text{ (dB)}$$
$$- G_{c2} \text{ (dB)} \qquad \text{dBW/m} \qquad (10\text{-}5\text{-}3)$$

$$E = 36.8 \text{ dB} - 20 \log_{10}(\lambda) + P_r \text{ (dBW)} - G_{a2} \text{ (dB)}$$
$$- G_{c2} \text{ (dB)} \qquad \text{dBV/m} \qquad (10\text{-}5\text{-}4)$$

where λ is the wavelength in meters.

10-6 Conversion of Receiving Voltage to Electric Field Intensity

Generally the field intensity meter at the receiving position reads voltages in either μV or dBμV. A conversion of the receiving voltage to field intensity is necessary. Figure 10-6-1 shows a diagram for computing the field intensity

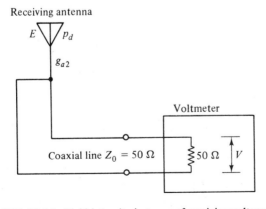

FIG. 10-6-1. Field intensity in terms of receiving voltage.

from the receiving voltages. The input impedance of the field intensity receiving meter is normally specified by the manufacturer to be 50 Ω, since the coaxial line connected to the meter usually has a characteristic impedance of 50 Ω. This means that the input impedance of the meter perfectly matches the coaxial line. The power input to the intensity meter is given by

$$p_r = \frac{V^2}{50} \quad \text{watts} \tag{10-6-1}$$

where V is the intensity meter reading in volts (RMS).

If the input impedance is specified by another value, it is necessary to choose a coaxial line with the same value of characteristic impedance as the input impedance for impedance matching. The power density at the antenna aperture is computed from Eq. (10-5-1) as

$$p_d = \frac{0.251 V^2}{\lambda^2 g_{a2} g_{c2}} \quad \text{watts/m}^2 \tag{10-6-2}$$

where λ is the wavelength in meters.

The field intensity can be computed from Eq. (10-4-3) as

$$E = \frac{9.7V}{\lambda} \left(\frac{1}{g_{a2} g_{c2}} \right)^{1/2} \quad \text{volts/m} \tag{10-6-3}$$

Similarly, the power density and the field intensity can be expressed in dB as

$$P_d = -6 \text{ dB} - 20 \log_{10}(\lambda) + V\,(\text{dBV}) - G_{a2} \quad (\text{dB})$$
$$- G_{c2}\,(\text{dB}) \quad \text{dBW/m}^2 \tag{10-6-4}$$
$$E = 19.8 \text{ dB} - 20 \log_{10}(\lambda) + V\,(\text{dBV}) - G_{a2} \quad (\text{dB})$$
$$- G_{c2}\,(\text{dB}) \quad \text{dBV/m} \tag{10-6-5}$$

To simplify the computation, the manufacturer often specifies the antenna factor for a specific antenna at a given wavelength. From Eq. (10-6-5) the antenna factor is defined as

$$\text{AF (Antenna Factor)} = 19.8 \text{ dB} - 20 \log_{10}(\lambda) - G_{a2}\,(\text{dB}) \quad \text{dB} \tag{10-6-6}$$

Thus the antenna factor is a quantity that is used to convert the receiver reading in dBV or dBμV to the field intensity value in dBV/m or dBμV/m for one-meter spacing. It should be noted that for a given frequency the antenna factor for wavelengths in centimeters is less than the one for wavelengths in meters by 40 dB.

> **EXAMPLE 10-1:** The receiving antenna has a gain of 10 dB, and the coaxial line connecting the antenna to the receiving intensity meter is assumed lossless. The characteristic impedance of the coaxial line and the input impedance of the meter are both 50 Ω. The signal frequency is 3 GHz, and the meter reading is 100 μV. Determine the power density and the field intensity at the receiving antenna aperture.

Solution.

1. The intensity meter reading is 40 dBμV or -80 dBV.
2. The power input to the receiver with an input impedance of 50 Ω is computed from Eq. (10-6-1) as

$$p_r = \frac{V^2}{R} = \frac{(100 \times 10^{-6})^2}{50} = 200 \times 10^{-12} \quad \text{watt}$$

3. The power density at the antenna aperture is determined from Eq. (10-5-1) or (10-6-2) to be

$$p_d = \frac{0.251 V^2}{\lambda^2 g_{a2} g_{c2}} = \frac{(0.251)(100 \times 10^{-6})^2}{(0.1)^2(10)} = 251 \times 10^{-10} \quad \text{watt/m}^2$$

4. The field intensity is computed from Eq. (10-4-3) or (10-6-3) as

$$E = \sqrt{120\pi p_d} = \sqrt{(120\pi)(251 \times 10^{-10})} = 3076.7 \times 10^{-6} \quad \text{volt/m}$$

5. The power density can be expressed in dB by use of Eq. (10-6-4).

$$p_d = -6 \text{ dB} - 20 \log_{10}(0.1) - 80 \text{ (dBV)} - 10 \text{ (dB)}$$

$$= -76 \quad \text{dBW/m}^2$$

$$= -6 \text{ dB} - 20 \log_{10}(10) - 20 \text{ (dB}\mu\text{V)} - 10 \text{ (dB)}$$

$$= -56 \quad \text{dB}\mu\text{W/cm}^2$$

$$= -6 \text{ dB} - 20 \log_{10}(0.1) - 20 \text{ (dB}\mu\text{V)} - 10 \text{ (dB)}$$

$$= -16 \quad \text{dB}\mu\text{W/m}^2$$

6. The field intensity can be expressed in dB by means of Eq. (10-6-5).

$$E = 19.8 \text{ dB} - 20 \log_{10}(0.1) - 80 \text{ (dBV)} - 10 \text{ (dB)}$$

$$= -50.2 \quad \text{dBV/m}$$

$$= 19.8 \text{ dB} - 20 \log_{10}(10) + 40 \text{ (dB}\mu\text{V)} - 10 \text{ (dB)}$$

$$= 29.8 \quad \text{dB}\mu\text{V/cm}$$

$$= 19.8 \text{ dB} - 20 \log_{10}(0.1) + 40 \text{ (dB}\mu\text{V)} - 10 \text{ (dB)}$$

$$= 69.8 \quad \text{dB}\mu\text{V/m}$$

7. The antenna factor is computed from Eq. (10-6-6) as

$$\text{AF} = 19.8 \text{ dB} - 20 \log_{10}(10) - 10 \text{ (dB)} = -10.2 \text{ dB for } \lambda \text{ in centimeter}$$

$$= 19.8 \text{ dB} - 20 \log_{10}(0.1) - 10 \text{ (dB)} = 29.8 \text{ dB for } \lambda \text{ in meters}$$

10-7 Summary

Conversion equations for converting the transmitting power to electric field intensity and power density and for changing the receiving power or voltage to electric field intensity and power density, both in power scale or in dB units, have been derived and discussed. If antenna and cable parameters are known, the field intensity or power density can be easily determined by means of a specific conversion equation for a given signal frequency at a certain range. It may be concluded that the developed conversion equations may aid electronics engineers to compute the electric field intensity and power density rapidly [7].

REFERENCES

[1] to [5]   HARRINGTON, ROBERT F., *Time-Harmonic Electromagnetic Fields*. McGraw-Hill Book Company, New York, 1961.

[6]   KRAUS, JOHN D. *Antennas*. McGraw-Hill Book Company, New York, 1950.

[7]   LIAO, SAMUEL Y., Measurements and Computations of Microwave Electric Field Intensity and Power Density. *IEEE Trans. on Instrumentation and Measurements*, Vol. **IM-25**, No. 1, March 1977.

Constants of Materials

1. Conductivity σ in mhos per meter

Conductor	σ	*Insulator*	σ
Silver	6.17×10^7	Quartz	10^{-17}
Copper	5.80×10^7	Polystyrene	10^{-16}
Gold	4.10×10^7	Rubber (hard)	10^{-15}
Aluminum	3.82×10^7	Mica	10^{-14}
Tungsten	1.82×10^7	Porcelain	10^{-13}
Zinc	1.67×10^7	Diamond	10^{-13}
Brass	1.50×10^7	Glass	10^{-12}
Nickel	1.45×10^7	Bakelite	10^{-9}
Iron	1.03×10^7	Marble	10^{-8}
Bronze	1.00×10^7	Soil (sandy)	10^{-5}
Solder	0.70×10^7	Sands (dry)	2×10^{-4}
Steel (stainless)	0.11×10^7	Clay	10^{-4}
Nichrome	0.10×10^7	Ground (dry)	10^{-4}–10^{-5}
Graphite	7.00×10^4	Ground (wet)	10^{-2}–10^{-3}
Silicon	1.20×10^3	Water (distilled)	2×10^{-4}
Water (sea)	3–5	Water (fresh)	10^{-3}
		Ferrite (typical)	10^{-2}

2. Dielectric Constant—Relative Permittivity ϵ_r

Material	ϵ_r	*Material*	ϵ_r
Air	1	Sands (dry)	4
Alcohol (ethyl)	25	Silica (fused)	3.8
Bakelite	4.8	Snow	3.3
Glass	4–7	Sodium chloride	5.9
Ice	4.2	Soil (dry)	2.8
Mica (ruby)	5.4	Styrofoam	1.03
Nylon	4	Teflon	2.1
Paper	2–4	Water (distilled)	80
Plexiglass	2.6–3.5	Water (sea)	20
Polyethylene	2.25	Water (dehydrated)	1
Polystyrene	2.55	Wood (dry)	1.5–4
Porcelain (dry process)	6	Ground (wet)	5–30
Quartz (fused)	3.80	Ground (dry)	2–5
Rubber	2.5–4	Water (fresh)	80

3. Relative Permeability μ_r

Diamagnetic material	μ_r	Ferromagnetic material	μ_r
Bismuth	0.99999860	Nickel	50
Paraffin	0.99999942	Cast iron	60
Wood	0.99999950	Cobalt	60
Silver	0.99999981	Machine steel	300
		Ferrite (typical)	1,000
Paramagnetic material	μ_r	Transformer iron	3,000
		Silicon iron	4,000
Aluminum	1.00000065	Iron (pure)	4,000
Beryllinum	1.00000079	Mumetal	20,000
Nickel chloride	1.00004	Supermalloy	100,000
Manganese sulphate	1.0001		

4. Properties of Free Space

Velocity of light in vacuum c	2.997925 meters per second
Permittivity ϵ_0	8.854×10^{-12} farad per meter
Permeability μ_0	$4\pi \times 10^{-7}$ henry per meter
Intrinsic impedance η_0	377 or 120π ohms

5. Physical Constants

Charge of electron	e	1.60×10^{-19} coulomb
Mass of electron	m	9.1×10^{-31} kilogram
Charge-to-mass ratio of electron	$\dfrac{e}{m}$	1.76×10^{11} coulombs per kilogram

Bessel Functions

appendix

B

First-Order Bessel Function Values

x	$J_1(x)$	x	$J_1(x)$	x	$J_1(x)$	x	$J_1(x)$	x	$J_1(x)$
.00	0.000	.92	0.413	1.86	0.582	2.86	0.389	3.84	−0.003
.02	+.010	.94	.420	1.88	.5815	2.88	.3825	3.86	.011
.04	.020	.96	.427	1.90	.581	2.90	.375	3.88	.019
.06	.030	.98	.4335	1.92	.5805	2.92	.368	3.90	.027
.08	.040	1.00	.440	1.94	.580	2.94	.361	3.92	.035
.10	.050	1.02	.4465	1.96	.579	2.96	.354	3.94	.043
.12	.060	1.04	.453	1.98	.578	2.98	.3465	3.96	.051
.14	.070	1.06	.459	2.00	.577	3.00	.339	3.98	.058
.16	.080	1.08	.465	2.02	.575	3.02	.3315	4.00	.066
.18	.090	1.10	.471	2.04	.574	3.04	.324	4.10	.103
.20	.0995	1.12	.477	2.06	.572	3.06	.316	4.20	.139
.22	.109	1.14	.482	2.08	.570	3.08	.309	4.30	.172
.24	.119	1.16	.488	2.10	.568	3.10	.301	4.40	.203
.26	.129	1.18	.493	2.12	.566	3.12	.293	4.50	.231
.28	.139	1.20	.498	2.14	.564	3.14	.285	4.60	.2565
.30	.148	1.22	.503	2.16	.561	3.16	.277	4.70	.279
.32	.158	1.24	.508	2.18	.559	3.18	.269	4.80	.2985
.34	.1675	1.26	.513	2.20	.556	3.20	.261	4.90	.315
.36	.177	1.28	.5175	2.22	.553	3.22	.253	5.00	.3275
.38	.187	1.30	.522	2.24	.550	3.24	.245	5.05	.334
.40	.196	1.32	.526	2.26	.547	3.26	.237	5.10	.337
.42	.205	1.34	.5305	2.28	.543	3.28	.229	5.16	.341
.44	.215	1.36	.534	2.30	.540	3.30	.221	5.20	.343
.46	.224	1.38	.538	2.32	.536	3.32	.212	5.26	.345
.48	.233	1.40	.542	2.34	.532	3.34	.204	5.30	.346
.50	.242	1.42	.5455	2.36	.5285	3.36	.196	5.32	.346
.52	.251	1.44	.549	2.38	.524	3.38	.1865	5.34	.346
.54	.260	1.46	.552	2.40	.520	3.40	.179	5.36	.346
.56	.269	1.48	.555	2.42	.516	3.42	.171	5.38	.346
.58	.278	1.50	.558	2.44	.511	3.44	.1625	5.40	.345
.60	.287	1.52	.561	2.46	.507	3.46	.154	5.47	.343
.62	.295	1.54	.563	2.48	.502	3.48	.146	5.50	.341
.64	.304	1.56	.566	2.50	.497	3.50	.137	5.56	.3375
.66	.312	1.58	.568	2.52	.492	3.52	.129	5.60	.334
.68	.321	1.60	.570	2.54	.487	3.54	.121	5.66	.3285
.70	.329	1.62	.572	2.56	.482	3.56	.112	5.70	.324
.72	.337	1.64	.5735	2.58	.476	3.58	.104	5.80	.311
.74	.345	1.66	.575	2.60	.471	3.60	.0955	5.90	.295
.76	.353	1.68	.5765	2.62	.465	3.62	.087	6.00	.277
.78	.361	1.70	.578	2.64	.4595	3.64	.079	6.10	.256
.80	.369	1.72	.579	2.66	.454	3.66	.070	6.20	.233
.82	.3765	1.74	.580	2.68	.448	3.68	.062	6.30	.208
.84	.384	1.76	.5805	2.70	.442	3.70	.054	6.40	.182
.86	.3915	1.78	.581	2.72	.435	3.72	.0455	6.60	.125
.88	.399	1.80	.5815	2.74	.429	3.74	.037	6.70	.095
.90	.406	1.82	.582	2.76	.423	3.76	.029	6.80	.065
		1.84	.582	2.78	.416	3.78	.021	6.90	.035
				2.80	.410	3.80	.013	7.00	.005
				2.82	.403	3.82	.005	7.01	.000
				2.84	.396	3.83	.000		

Microwave Devices
and Circuits Laboratory

Experiment 1   Power-Frequency
Limitations of Conventional
Vacuum Tubes

Purpose

The purpose of this experiment is to verify
the power-frequency limitations of conventional
vacuum tubes in the microwave frequency region.

Theory

At microwave frequencies, the performance
of conventional vacuum tubes is impaired by the
effects of lead inductance, electron transit time, and
gain-bandwidth product limitation. The power
output of a conventional vacuum tube is decreased
when the signal frequency is increased above the
microwave region.

Procedures

1. Set up the circuit as shown in Fig. C-1.

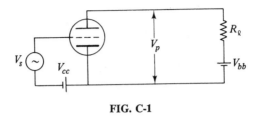

FIG. C-1

2. Set $V_{bb} = 150$ to 200 V with an increment of 10 V, $V_{cc} = -3$ V, and $R_\ell = 10$ kohms. Keep all leads as short as possible. Set the signal voltage $V_s = 0.2 \sin \omega t$ at 1.8 GHz.

3. Check the output waveform of V_p with an oscilloscope and measure the output power with a power meter.

4. Repeat parts 2 and 3 for frequencies of 3, 4, 5, and 6 GHz.

Experiment 2 Mode Characteristics of Reflex Klystron

Purpose

The purpose of this experiment is to demonstrate the mode characteristics of a reflex klystron.

Theory

Reflex klystron is a single-cavity klystron oscillator with a negative repeller voltage. If the resonant cavity is tuned to any one frequency, a number of repeller voltages will provide oscillation at the same frequency. The number of repeller modes are generated by different repeller voltages.

Procedures

1. Assemble the apparatus as shown in Fig. C-2.

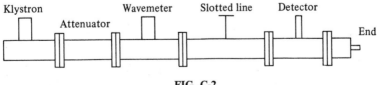

FIG. C-2

2. Set the repeller voltage V_r to zero and connect a sinusoidal voltage V_s in series to the repeller voltage.

3. Apply the detector output to the vertical input of an oscilloscope and the sinusoidal voltage V_s to the horizontal input of the same scope.
4. By adjustment of the repeller voltage determine the number of repeller modes on the scope. Then adjust the repeller voltage V_r and sinusoidal voltage V_s so that all modes of oscillation are displayed.
5. Tune the wavemeter until a maximum waveform is obtained on the oscilloscope and record the frequency.
6. Adjust the values of V_r and V_s so that only one mode curve is displayed. Determine the instantaneous repeller voltage and wavelengths corresponding to the peak and the end points of the mode curve.
7. Repeat part 6 for the other modes.
8. Plot mode curves of repeller voltage vs. wavelength for maximum, minimum, and optimum repeller voltage for the various modes.

Experiment 3   Measurement of Frequency and Wavelength

Purpose

The purpose of this experiment is to determine the frequency and wavelength in a waveguide.

Theory

When a transmission line or waveguide is mismatched by a load, a standing wave is created in the line or the waveguide. The distance between two adjacent maxima or minima is one half of a wavelength. The frequency can be determined from the measured wavelength. Also the frequency and wavelength can be found by using a wavemeter.

Procedures

1. Assemble the apparatus as shown in Fig. C-3.

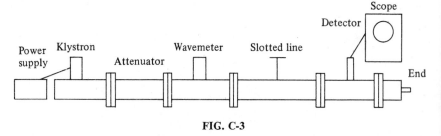

FIG. C-3

2. Set the attenuator at zero dB and switch the modulation dial to Ref.
3. Adjust the repeller voltage V_r and modulation voltage V_m until a better maximum square wave output is on the scope.

4. Tune the wavemeter until a maximum waveform is observed on the scope and read the frequency from the wavemeter.

5. Carefully move the detector of the carriage back and forth and read the distance between two adjacent maxima and minima.

6. Compare the measured values of frequency and wavelength from parts 4 and 5.

Experiment 4   Measurement of Voltage Standing-Wave Ratio (VSWR)

Purpose

The purpose of this experiment is to examine some of common methods of measuring voltage standing-wave ratios (VSWR).

Theory

The instrument most commonly used to measure VSWR is a slotted section of transmission line. Usually the sampling circuit employed with a slotted line is a detector with a probe which can be moved along the slot in the line. Most microwave detectors have a characteristic of "square-law." This means that their output current is proportional to the square of their input voltage.

$$i = kV^2$$

where i = reading of the current meter,

V = voltage on the line at the probe, and

k = constant introduced by the detector and the probe coupling.

Then

$$\text{VSWR} = \frac{V_{\max}}{V_{\min}} = \frac{(I_{\max})^{1/2}}{(I_{\min})^{1/2}}$$

Procedures

1. Assemble the apparatus as shown in Fig. C-4.

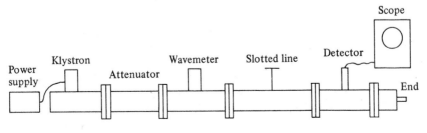

FIG. C-4

2. Set the attenuator at zero dB and switch the modulation dial to Ref.
3. Adjust the repeller voltage V_r and the modulation voltage V_m until a better maximum square wave is on the scope.
4. Carefully move the detector of the carriage back and forth and read the values of maxima and minima.
5. Repeat part 4 using a power meter, a microampere meter, or a VSWR meter.
6. Compare the measured values from parts 4 and 5.

Experiment 5   Negative Resistance of Tunnel Diode

Purpose

The purpose of this experiment is to measure the negative resistance of a tunnel diode.

Theory

The tunnel diode is an extremely heavily doped PN junction diode whose current-voltage characteristic exhibits negative resistance behavior. Quantum mechanical tunneling of majority carriers through the voltage barrier at the juntion is used to explain the operation of the diode. The negative resistance is capable of operating as an amplifier or as an oscillator.

Procedures

1. Connect the circuit as shown in Fig. C-5.

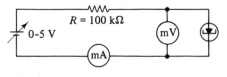

FIG. C-5

2. Vary the supply voltage from zero to 40 mV with an increment of 10 mV each step and from 40 to 200 mV with an increment of 20 mV each step. Record the voltage and current each time.
3. Plot the measured current-voltage characteristic curve, indicate the negative resistance region, mark the valley voltage, valley current, peak voltage, and pcak current on the plotted curve.
4. Discuss the three regions: monostable state, astable state, and bistable state.

Experiment 6   Measurement of Microwave Power

Purpose

The purpose of this experiment is to measure microwave power by using a thermistor with a power meter.

Theory

When microwave power is absorbed by a material, the temperature of the material is increased. If the material has a nonzero temperature coefficient, its resistance will change. Such a material is called a *bolometer*. There are two types of bolometers in popular use. One is the *barretter*, which is composed of a thin metallic wire (platinum). The other common type is the *thermistor*.

Procedures

1. Assemble the apparatus as shown in Fig. C-6.

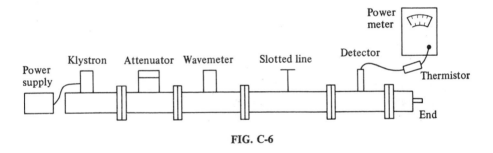

FIG. C-6

2. Set the attenuator at zero dB and switch the modulation dial to Ref.

3. Adjust the repeller voltage V_r and the modulation voltage V_m until a better maximum square wave is on the scope.

4. Set the range selector to *Coarse zero*, turn off the RF power, and adjust the meter to zero.

5. Select the ranges from 0.01 to 10 mw (−20 to +10 dBm). *Coarse zero* setting is used to zero meter with no power applied to thermistor mount.

6. The power input to the thermistor mount in dBm is the sum of the range reading and the reading on the dBm scale of the meter.

7. Reset the attenuator at 5, 10 and 15 dB, and repeat parts 3 to 6.

8. Check the measured values with the attenuator readings.

Experiment 7   Directional Couplers

Purpose

The purpose of this experiment is to measure the coupling factor and directivity for a two-hole directional coupler (Hewlett Packard model 779D).

Theory

A directional coupler is a junction device between two pairs of terminals as shown in Fig. C-7-1. When all ports are terminated in their characteristic

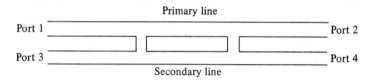

Primary line

Port 1 ————————————————————— Port 2

Port 3 ————————————————————— Port 4

Secondary line

FIG. C-7-1

impedances, there is free transmission of power, without reflection, between ports 1 and 2 and no transfer of power between ports 1 and 3, or between ports 2 and 4. The degree of coupling between ports 1 and 4 is defined as

$$\text{Coupling factor} = 10 \log \frac{P_1}{P_4} \quad \text{dB}$$

The coupling factor is a measure of the ratio of power levels in the primary and secondary lines. The directivity is defined as

$$\text{Directivity} = 10 \log \frac{P_4}{P_3} \quad \text{dB}$$

The directivity is a measure of how well the forward traveling wave in the primary line couples only to a specific port of the secondary line.

Procedures

1. Assemble the apparatus as shown in Fig. C-7-2.

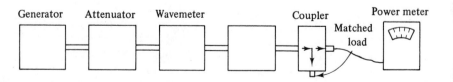

Generator   Attenuator   Wavemeter   Coupler   Power meter
                                                             Matched
                                                             load

FIG. C-7-2

2. Match port 2 with a 50-ohm load or a matched load.
3. Measure the input power at port 1 and the output power at port 4.
4. Determine the coupling factor in dB.
5. Connect the directional coupler in the reverse direction. That means port 2 is connected to the input power.
6. Measure the output power at port 3.
7. Determine the directivity in dB.

Experiment 8   Plane Wave Propagation in Medium

Purpose

The purpose of this experiment is to measure the reflectivity and transmittance of microwave in medium.

Theory

When an electromagnetic plane wave propagates from free space through a medium to free space, one part of the energy will be reflected and/or absorbed by the medium and the rest of the energy will be transmitted through the medium.

	Intrinsic impedance η (Ω)	*Relative dielectric constant* ϵ_r	*Conductivity* σ ($\mho/m$)	*Attenuation and phase constant*
Free space	377	1		
Polystyrene	603	2.56		
Plywood	477	1.60		
Glass		5	10^{-12}	
Copper	$\eta = \sqrt{\dfrac{j\omega\mu}{\sigma}}$		5.8×10^{-7}	$\alpha = \beta = \sqrt{\pi f \mu \sigma}$

Procedures

1. Assemble the apparatus as shown in Fig. C-8.

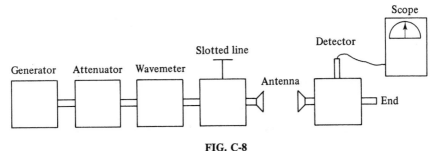

FIG. C-8

2. Adjust the power supply and the set up until a maximum power output shown on the oscilloscope or power meter in dB.
3. Insert the polystyrene slab between the two antennas and measure the transmitted power in dB.
4. Determine the reflectivity and transmittance of the polystyrene.
5. Repeat parts 3 and 4 for plywood, glass, and copper sheet.

Experiment 9   Measurement of Antenna Radiation Pattern

Purpose

The purpose of this experiment is to measure the radiation pattern of a pyramidal horn antenna.

Theory

The antenna radiation pattern is different for each type of antenna. The transmitting antenna is fixed in position, and the antenna under test is rotated on a vertical axis by the antenna support shaft. The distance between the two antennas must be such that the antenna under test is in the far field of the transmitting antenna.

Procedures

1. Assemble the apparatus as shown in Fig. C-9.

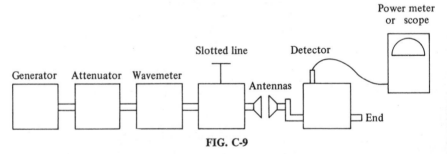

FIG. C-9

2. Line up the two antennas (face to face) and adjust the power supply until a maximum power output is shown on the oscilloscope or power meter.
3. Record the meter reading when the antenna under test is in the position of part 1 which is assumed to be zero degree.
4. Rotate the antenna under test 20 degrees from the horizontal position and record the power-meter reading.
5. Repeat part 4 for 40, 60, 80, and 90 degrees.

Problems

Chapter 1

1-1 At time $t = t_0$ an electron is emitted from a planar diode with zero initial velocity and the anode voltage is $+15$ volts. At time $t = t_1$ the electron is midway between the plates and the anode voltage changes discontinuously to -30 volts.

a. Determine which electrode the electron will strike.

b. Compute the kinetic energy of the electron in electron volts (eV) when it strikes the electrode.

1-2 A cylindrical cavity is constructed of a center conductor and an outer conductor. The inner center conductor has a radius of 1 cm and is grounded. The outer conductor has a radius of 10 cm and is connected to a power supply of $+10$ kV. The electrons are emitted from the cathode at the center conductor and move toward the anode at the outer conductor.

a. Determine the magnetic flux density **B** in webers/m² in the axial direction so that the electrons just graze the anode and return to the cathode again.

b. If the magnetic flux density **B** is fixed at 4 milliwebers/m², find the required supply voltage so that the electrons just graze the anode and return to the cathode again.

Chapter 2

2-1 In a certain homogeneous medium the group velocity as measured by the propagation time of a pulse is found to be proportional to the square root of the frequency ($v_g = \sqrt{A\omega}$ where A is a constant) over a particular frequency range. It is assumed the medium is a nonmagnetic insulator.

a. Determine the relationship between the phase and group velocities.
b. Derive an expression for the relative dielectric constant of this medium.

2-2 Show that $\dfrac{\partial}{\partial z} = \beta_g$ and $\dfrac{\partial}{\partial t} = j\omega$ for a sinusoidal wave propagating in the z direction.

2-3 The electric field of a plane wave propagating in free space is given in complex notation by

$$\mathbf{E} = 10^{-4}e^{j(\omega t + 20\pi z)}u_x + 10^{-4}e^{j(\omega t + \pi/2 + 20\pi z)}u_y$$

where u_x and u_y are unit vectors in the x and y directions of a right-hand coordinate system.

a. In which direction is the wave propagating?
b. Find the frequency of the propagating signal.
c. Determine the type of wave polarization (linear, circular, or elliptical).
d. Express the magnetic field intensity **H** of the propagating wave.
e. Calculate the average power flow per square meter normal to the direction of the propagation.

2-4 Determine the permittivity of a slab of dielectric material that will reflect 20% of the energy in a plane wave striking the slab at normal incidence. Neglect the reflections from the back face of the slab.

2-5 The reflection and refraction of microwave propagating in the ionosphere are determined by the electron density in the ionosphere. If the electron density is assumed to be 10^{14} electrons per cubic meter, determine the critical frequency for vertical incidence so that the signal wave will be reflected back to the earth.

2-6 The conductivity σ of copper is 5.8×10^7 mhos/m and its relative permeability is unity. Calculate the magnitudes of reflectivity of copper for vertical and horizontal polarizations against the grazing angle ψ of 0 to 90° at a frequency range of 1 to 40 GHz. The increment of the angle ψ is 10° each step and the increment of the frequency is 10 GHz each step.

2-7 At the Brewster angle there is no reflected wave when the incident wave is vertically polarized. If the incident wave is not entirely vertically polarized there will be some reflection but the reflected wave will be entirely of horizontal polarization. Verify Eq. (2-5-24b) for the Brewster angle in terms of the relative dielectrics.

2-8 Determine the pseudo-Brewster angle ψ in terms of ν_1, ν_2, η_1 and η_2 for a good conductor [Hint: Start from Eq. (2-5-13b).]

2-9 Calculate the pseudo-Brewster angles for seawater, dry sands, concrete cement, and dry ground.

2-10 Determine the pseudo-Brewster angle ψ in terms of ϵ_r and X for a lossy dielectric [Hint: Start from Eq. (2-5-26).]

2-11 Bulk gold has a conductivity of 4.1×10^7 mhos/m, a resistivity of 2.44×10^{-8} ohm-m and an electron mean-free path of 570 Å. Calculate the surface conductivity, surface resistivity, and the surface resistance of fold film for thicknesses of 10 to 100 Å with an increment of 10 Å for each step.

2-12 Silver has a conductivity of 0.617×10^8 mhos/m, a resistivity of 1.620×10^{-8} ohm-m and an electron mean-free path of 570 Å. Repeat problem 2-11 for silver film.

2-13 Seawater has a conductivity of 4 mhos/m and a relative dielectric constant of 20 at a frequency of 4 GHz. Compute

    a. the intrinsic impedance,
    b. the propagation constant, and
    c. the phase velocity.

2-14 Repeat Prob. 2-13 for dry sands ($\sigma = 2 \times 10^{-4}$ mho/m and $\epsilon_r = 4$) and copper ($\sigma = 5.8 \times 10^7$ mhos/m).

2-15 A uniform plane wave is normally incident upon the surface of seawater. The electric intensity of the incident wave is 100×10^{-3} Volt/m at a frequency of 5 GHz in the vertical polarization. Calculate

    a. the electric intensity of the reflected wave, and
    b. the electric intensity of the transmitted wave.

2-16 Repeat Prob. 2-15 for an angle of incidence of 30°.

2-17 Dry ground has a conductivity of 5×10^{-4} mhos/m and a relative dielectric constant of 10 at a frequency of 500 MHz. Compute

    a. the intrinsic impedance,
    b. the propagation constant, and
    c. the phase velocity.

2-18 Copper has a conductivity of 5.8×10^7 mhos/m and is considered a ideal material for shielding. A shield is made of copper with a thickness of 1 mm. If a uniform plane wave is normally incident upon the copper shield, compute the absorption loss in dB by the copper at $f = 10$ MHz.

2-19 A radar transmitter has an output power of 100 kW average. Calculate the power density in dBW/m² at a range of 3000 meters and the free-space attenuation in dB at $f = 10$ GHz.

2-20 Write a complete FORTRAN program to compute the magnitudes of reflectivity in vertical polarization against a grazing angle of seawater. The frequency varies from 0.1 to 40 GHz with an increment of 0.1 GHz between 0.1 to 1 GHz, 1 GHz between 1 to 10 GHz, and 5 GHz between 10 to 40 GHz. Use F10.5 format for numerical outputs and Hollerith format for character outputs. Print the outputs in three columns such as frequency (GHz), grazing angle (degrees), and gamma (vertical).

2-21 Write a complete FORTRAN program to compute the magnitudes of reflectivity in horizontal polarization against a grazing angle for seawater. (Refer to problem 2-20 for specifications).

2-22 Write a complete FORTRAN program to compute the light transmittance and light reflection of a gold-film coating on a nonabsorbing plastic glass for thicknesses of 10 to 100 Å with an increment of 10 Å each step. The wavelength varies from 2000 to 10000 Å with an increment of 500 Å each step. The values of the refractive index n and the extinction index k of a gold-film coating on a nonabsorbing plastic glass deposited in a vacuum are listed in Table 2-6-2. The refractive index n of the nonabsorbing plastic glass is 1.5. Use F10.5 format for numerical outputs, Hollerith format for character outputs, and Data statements to read in the input values.

2-23 Write a complete FORTRAN program to compute the light transmittance and light reflection for an aluminum-film coating on a nonabsorbing plastic substrate for thicknesses of 10 to 100 Å and print out the results in percent. Use F10.5 format for numerical outputs, Hollerith format for character outputs, and Data statements to read in the input values. Print the outputs in column form with proper head-letters and units. The refractive index n of the nonabsorbing plastic glass is 1.50. The refractive index n and extinction index k for aluminum film are tabulated in Table P2-23. (Refer to Problem 2-22 for specifications).

TABLE P2-23

Wavelength (Å)	Refractive index n	Extinction index k	Wavelength (Å)	Refractive index n	Extinction index k
2000	0.11	2.20	4500	0.51	5.00
2200	0.13	2.40	5000	0.64	5.50
2400	0.16	2.54	5500	0.82	5.99
2600	0.19	2.85	6000	1.00	6.50
2800	0.22	3.13	6500	1.30	7.11
3000	0.25	3.33	7000	1.55	7.00
3200	0.28	3.56	7500	1.80	7.12
3400	0.31	3.80	8000	1.99	7.05
3600	0.34	4.01	8500	2.08	7.15
3800	0.37	4.25	9000	1.96	7.70
4000	0.40	4.45	9500	1.75	8.50

2-24 Repeat Problem 2-22 for a silver-film coating on a nonabsorbing plastic glass except that the wavelength varies from 2000 to 3700 Å with an increment of 100 Å each step. (Refer to Table 2-6-2 for the values of n and k.)

2-25 Repeat Problem 2-22 for a copper-film coating on a nonabsorbing plastic glass except that the wavelength varies from 4500 to 10000 Å with an increment of 500 Å each step.

2-26 Repeat Problems 2-20 and 2-21 for dry sands.

2-27 Repeat Problems 2-20 and 2-21 for concrete cement.

2-28 Repeat Problems 2-20 and 2-21 for dry ground.

Chapter 3

3-1 A transmission line has a characteristic impedance of 300 ohms and is terminated in a load of $300 - j300$ ohms. The propagation constant of the line is $0.054 + j3.53$ per meter. Determine

   a. the reflection coefficient at the load,
   b. the transmission coefficient at the load, and
   c. the reflection coefficient at a point 2 meters away from the load.

3-2 A lossless transmission line has a characteristic impedance of 50 ohms and is terminated in a load of 100 ohms. The magnitude of a voltage wave incident to the line is 20 volts RMS. Determine

   a. the VSWR on the line,
   b. the maximum voltage V_{max} and minimum voltage V_{min} on the line,
   c. the maximum current I_{max} and minimum current I_{min} on the line, and
   d. the power transmitted by the line.

3-3 A lossless line has a characteristic impedance of 75 ohms and is terminated in a load of 300 ohms. The line is energized by a generator which has an open-circuit output voltage of 20 volts RMS and output impedance of 75 ohms. The line is assumed to be $2\frac{1}{4}$ wavelengths long.

   a. Find the sending-end impedance.
   b. Determine the magnitude of the receiving-end voltage.
   c. Calculate the receiving-end power at the load.

3-4 A lossless transmission line has a characteristic impedance of 100 ohms and is terminated in a load of 75 ohms. The line is 0.75 wavelength long. Determine

   a. the sending-end impedance, and
   b. the reactance which, if connected across the sending end of the line, will make the input impedance a pure resistance.

3-5 A coaxial line with a solid polyethylene dielectric is to be used at a frequency of 3 GHz. Its characteristic impedance Z_0 is 50 ohms and its attenuation constant α is 0.0156 neper/meter. The velocity constant which is defined as the ratio of phase velocity over velocity of light in free space is 0.660. The line is 100 meters long and is terminated in its characteristic impedance. A generator, which has an open-circuit voltage of 50 volts (RMS) and an internal impedance of 50 ohms, is connected to the sending end of the line. The frequency is tuned at 3 GHz. Compute

a. the magnitude of the sending-end voltage and of the receiving-end voltage,
b. the sending-end power and the receiving-end power, and
c. the wavelengths of the line.

3-6   An open-wire transmission line has $R = 5$ ohms/meter, $L = 5.2 \times 10^{-8}$ henry/meter, $G = 6.2 \times 10^{-3}$ mho/meter, and $C = 2.13 \times 10^{-10}$ farad/meter. The signal frequency is 4 GHz. Calculate

a. the characteristic impedance of the line in both rectangular form and polar form,
b. the propagation constant of the line,
c. the normalized impedance of a load $100 + j100$,
d. the reflection coefficient at the load, and
e. the sending-end impedance if the line is assumed a quarter-wavelength long.

3-7   A quarter-wave lossless line has a characteristic impedance of 50 ohms and is terminated in a load of 100 ohms. The line is energized by a generator of 20 volts RMS with an internal resistance of 50 ohms. Calculate

a. the sending-end impedance,
b. the magnitude of the receiving-end voltage, and
c. the power delivered to the load.

3-8   A lossless transmission line is terminated in an open circuit. The sending end is energized by a generator which has an open-circuit output voltage of V_g RMS and an interval impedance equal to the characteristic impedance of the line. Show that the sending-end voltage is equal to the output voltage of the generator.

3-9   A lossless transmission line has a characteristic impedance of 300 ohms and it is operated at a frequency of 10 GHz. The observed standing-wave ratio on the line is 5.0. It is proposed to use a short-circuited stub to reduce the standing wave.

a. Determine the distance in cm from a voltage minimum at which the stub should be located. (Two possible solutions)
b. Find the length of the stub in cm. (Two possible solutions)

3-10   A lossless line has a characteristic impedance of 50 ohms and is loaded by $60 - j60$ ohms. One stub is at the load, and the second one is $3\lambda/8$ distance away from the first one.

a. Determine the lengths in cm of the two short-circuited stubs when a match is achieved.
b. Locate and crosshatch the *forbidden region* of the normalized admittance for possible match.

3-11   A lossless transmission line has a characteristic impedance of 300 ohms and is terminated by an impedance Z_ℓ. The observed standing-wave ratio on the line is 6, and the distance of the first voltage minimum from the load is 0.166λ.

a. Determine the load Z_ℓ.

b. Find the lengths in λ of two shorted stubs, one at the load and one at $\lambda/4$ from the load, which are required to match the load to the line.

3-12 A single-stub tuner is to match a lossless line of 400 ohms to a load of 800 $-$ j300 ohms. The frequency is 3 GHz.

    a. Find the distance in meters from the load to the tuning stub.

    b. Determine the length in meters of the short-circuited stub.

3-13 A halfwave-dipole antenna has a driving point impedance of 73 + j42.5 ohms. A lossless transmission line connected to a TV set has a character-istic impedance of 300 ohms. The problem is to design a shorted stub with same characteristic impedance to match the antenna to the line. The stub may be placed at a location closest to the antenna. The reception is assumed to be Channel 83 at a frequency of 0.88525 GHz.

    a. Determine the susceptance contributed by the stub.

    b. Calculate the length in cm of the stub.

    c. Find the distance in cm between the antenna and the point where the stub is placed. (Note: There are two sets of solutions.)

3-14 A lossless transmission line has a characteristic impedance of 100 ohms and is loaded by 100 + j100 ohms. A single shorted stub with the same charac-teristic impedance is inserted at $\lambda/4$ from the load to match the line. The load current is measured to be 2 amperes. The length of the stub is $\lambda/8$.

    a. Determine the magnitude and the phase of the voltage across the stub location.

    b. Find the magnitude and the phase of the current flowing through the end of the stub.

3-15 A double-stub matching line is shown in Fig. P3-15. The characteristic resis-tances of the lossless line and stubs are 100 ohms, respectively. The spacing between the two stubs is $\lambda/8$. The load is 100 + j100. One stub is located at the load. Determine

    a. the reactances contributed by the stubs, and

    b. the lengths of the two shorted double-stub tuners. (Note: There are two sets of solutions.)

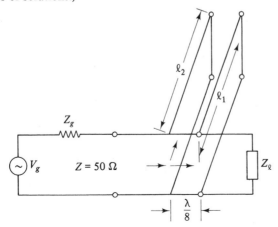

FIG. P3-15

3-16 A lossless transmission line has a characteristic impedance Z_0 of 100 ohms and is loaded by an unknown impedance. Its voltage standing-wave ratio is 4 and the first voltage maximum is $\lambda/8$ from the load.

   a. Find the load impedance.

   b. To match the load to the line, a quarter section of a different line with a characteristic impedance $Z_{01} < Z_0$ is to be inserted somewhere between (in cascade with) the load and the original line. Determine the minimum distance between the load and matching section, and the characteristic impedance Z_{01} in terms of Z_0.

3-17 A lossless transmission line has a characteristic impedance of 100 ohms and is loaded by an unknown impedance. The standing-wave ratio along the line is 2. The first two voltage minima are located at $z = -10$ and -35 cm from the load where $z = 0$. Determine the load impedance.

3-18 A matched transmission line is shown in Fig. P3-18.

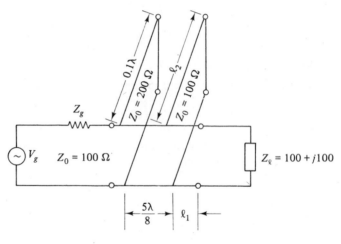

FIG. P3-18

   a. Find ℓ_1 and d which provide a proper match.

   b. With the line and load properly matched determine the VSWR on the section of line between the stubs.

3-19 An air-filled rectangular waveguide has dimensions of $a = 4$ cm and $b = 6$ cm. The signal frequency is 3 GHz. Compute the following for the TE_{01}, TE_{10}, TE_{11}, and TM_{11} modes.

   a. cutoff frequency

   b. wavelength in the waveguide

   c. phase constant and phase velocity in the waveguide

   d. group velocity and wave impedance in the waveguide

3-20 Show that the TM_{01} and TM_{10} modes in a rectangular waveguide do not exist.

3-21 The dominant mode TE_{01} is propagated in a rectangular waveguide of dimensions $a = 4$ cm and $b = 6$ cm. The distance between a maximum and a minimum is 4.47 cm. Determine the signal frequency of the dominant mode.

3-22 A TE_{11} mode of 10 GHz is propagated in an air-filled rectangular waveguide. The magnetic field in the z direction is given by

$$H_z = H_0 \cos \frac{\pi x}{\sqrt{6}} \cos \frac{\pi y}{\sqrt{6}} \qquad \text{ampere/meter}$$

The phase constant is $\beta = 1.0367$ radians per cm, and the quantities x and y are expressed in cm. Determine the cutoff frequency f_c, phase velocity v_g, guided wavelength λ_g and the magnetic field intensity in the y direction.

3-23 A rectangular waveguide is designed to propagate the dominant mode TE_{01} at a frequency of 5 GHz. The cutoff frequency is 0.8 of the signal frequency. The ratio of the guide height to width is 0.5. The time-average power flowing through the guide is 1 kW. Determine the magnitudes of electric and magnetic intensities in the guide and indicate where these occur in the guide.

3-24 An air-filled rectangular waveguide has dimensions of $a = 4$ cm and $b = 6$ cm. The guide transports energy in the dominant mode TE_{01} at a rate of 1 horse power (746 joules). If the frequency is 20 GHz, what is the peak value of electric field occurring in the guide?

3-25 An impedance of $(0.5 - j0.4)Z_0$ is connected to a rectangular waveguide. A capacitive window with a susceptance $jB = j0.4Y_0$ is located at a distance of 0.2λ from the load.

   a. Determine the VSWR on the line in the absence of the window.
   b. Find the VSWR on the line in the presence of the window.

3-26 An air-filled rectangular waveguide with dimensions of 1 cm $\times$ 3 cm operates in the TE_{01} mode at 10 GHz. The waveguide is perfectly matched and the maximum E field existing everywhere in the guide is 10^3 volts/meter. Determine the voltage, current, and wave impedance in the waveguide.

3-27 The dominant mode TE_{01} is propagated in a rectangular waveguide of dimensions $a = 1$ cm and $b = 2.25$ cm. Assume an air dielectric with a breakdown gradient of 30 kV/cm and a frequency of 10 GHz. There are no standing waves in the guide. Determine the maximum average power that can be carried by the guide.

3-28 A rectangular waveguide is terminated in an unknown impedance at $z = 25$ cm. A dominant mode TE_{01} is propagated in the guide and its VSWR is measured as 2.8 at a frequency of 8 GHz. The adjacent voltage minima are located at $z = 9.46$ cm and $z = 12.73$ cm.

   a. Determine the value of the load impedance in terms of Z_0.
   b. Find the position closest to the load where an inductive window is placed in order to obtain a VSWR of unity.

    c. Determine the value of the window admittance.

3-29   A rectangular waveguide is filled by dielectric material of $\epsilon_r = 9$ and has inside dimensions of 3.5×7 cm. It operates in the dominant TE_{01} mode.

    a. Determine the cutoff frequency
    b. Find the phase velocity in the guide at a frequency of 2 GHz
    c. Find the guided wavelength λ_g at the same frequency

3-30   The electric field intensity of the dominant TE_{01} mode in a lossless rectangular waveguide is

$$E_x = E_0 \sin \frac{\pi y}{b} e^{-j\beta_g z} \qquad \text{for } f > f_c$$

    a. Find the magnetic field intensity **H**.
    b. Compute the cutoff frequency and the time-average transmitted power.

3-31   An air-filled circular waveguide is to be operated at a frequency of 6 GHz and is to have dimensions such that $f_c = 0.8f$ for the dominant mode. Determine

    a. the diameter of the guide and
    b. the wavelength λ_g and the phase velocity v_g in the guide.

3-32   An air-filled cylindrical waveguide of 2 cm inside radius is operated in the TE_{01} mode.

    a. Compute the cutoff frequency.
    b. If the guide is to be filled with a dielectric material of $\epsilon_r = 2.25$, to what value must its radius be changed in order to maintain the cutoff frequency at its original value?

3-33   An air-filled cylindrical waveguide has a radius of 1.5 cm and is to carry energy at a frequency of 10 GHz. Find all TE and TM modes for which transmission is possible.

3-34   A TE_{11} wave is propagating through a cylindrical waveguide. The diameter of the guide is 10 cm, and the guide is air-filled.

    a. Find the cutoff frequency.
    b. Find the wavelength λ_g in the guide for a frequency of 3 GHz.
    c. Determine the wave impedance in the guide.

3-35   An air-filled cylindrical waveguide has a diameter of 4 cm and is to carry energy at a frequency of 10 GHz. Determine all TE_{nm} modes for which transmission is possible.

3-36   A circular waveguide has a cutoff frequency of 9 GHz in dominant mode.

    a. Find the inside diameter of the guide if it is air-filled.
    b. Determine the inside diameter of the guide if the guide is dielectric-filled. The dielectric constant is $\epsilon_r = 4$.

Chapter 4

4-1 A coaxial resonator is constructed of a section of coaxial line and is open-circuited at both ends. The length of the resonator is 5 cm long and is filled with dielectric of $\epsilon_r = 9$. The inner conductor has a radius of 1 cm and the outer conductor has a radius of 2.5 cm.

   a. Find the resonant frequency of the resonator.
   b. Determine the resonant frequency of the same resonator with one end open and one end shorted.

4-2 An air-filled cylindrical waveguide has a radius of 3 cm and is used as a resonator for TE_{01} mode at 10 GHz by placing two perfectly conducting plates at its two ends. Determine the minimum distance between the two end plates.

4-3 A four-port circulator is constructed of two magic Tees and one phase shifter as shown in Fig. 3-3-4. The phase shifter produces a phase shift of 180°. Explain how this circulator works.

4-4 A coaxial resonator is constructed of a section of coaxial line 6 cm long and is short-circuited at both ends. The cylindrical cavity has an inner radius of 1.5 cm and an outer radius of 3.5 cm. The line is dielectric-filled with $\epsilon_r = 2.25$.

   a. Determine the resonant frequency of the cavity for TEM_{001}.
   b. Calculate the quality Q of the cavity.

4-5 A rectangular-cavity resonator has dimensions of $a = 2\,cm$, $b = 5\,cm$ and $d = 15\,cm$. Compute

   a. the resonant frequency of the dominant mode for an air-filled cavity and
   b. the resonant frequency of the dominant mode for a dielectric-filled cavity of $\epsilon_r = 2.56$.

4-6 An undercoupled resonant cavity is connected to a lossless transmission line as shown in Fig. P4-6. The directional coupler is assumed to be ideal and matched on all arms. The unloaded Q of the cavity is 1,000 the VSWR at resonance is 2.5.

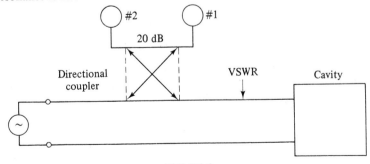

FIG. P4-6

a. Calculate the loaded Q_ℓ of the cavity.
b. Find the reading of bolometer No. 2 if bolometer No. 1 reads 4 mW.
c. Compute the power dissipated in the cavity.

4-7 A microwave transmission system consists of a generator, an overcoupled cavity, two ideal but not identical dual directional couplers with matched bolometers, and a load Z_ℓ. The lossless transmission line has a characteristic impedance Z_0. The readings of the 4 bolometers (No. 1, No. 2, No. 3, and No. 4) are 2 mW, 4 mW, 0, and 1 mW, respectively. The system is shown in Fig. P4-7.

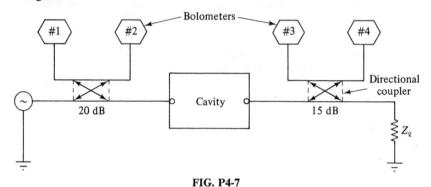

FIG. P4-7

a. Find the load impedance Z_ℓ in terms of Z_0.
b. Calculate the power dissipated by Z_ℓ.
c. Compute the power dissipated in the cavity.
d. Determine the VSWR on the input transmission line.
e. Find the ratio of Q_ℓ/Q_0 for the cavity.

4-8 A symmetrical directional coupler has an infinite directivity and a forward attenuation of 20 dB. The coupler is used to monitor the power delivered to a load Z_ℓ as shown in Fig. P4-8.

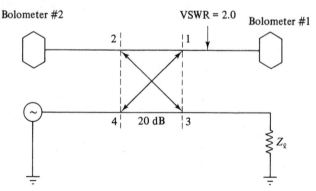

FIG. P4-8

Bolometer No. 1 introduces a VSWR of 2.0 on arm 1, bolometer No. 2 is matched to arm 2. If bolometer No. 1 reads 8 mW and bolometer No. 2 reads 2 mW,

a. find the amount of power dissipated in the load Z_ℓ, and
b. determine the VSWR on arm 3.

4-9   A semicircular-cavity resonator has a length of 5 cm and a radius of 2.5 cm.

a. Calculate the resonant frequency for the dominant mode if the cavity is air-filled.
b. Repeat part a if the cavity is loaded by a dielectric with a relative constant of 9.

4-10   The impedance matrix of a certain lumped-element network is given by

$$[z_{ij}] = \begin{bmatrix} 4 & 2 \\ 2 & 4 \end{bmatrix}$$

Determine the scattering matrix by using S-parameter theory and indicate the values of the components.

$$[S_{ki}] = \begin{bmatrix} S_{11} & S_{12} \\ S_{21} & S_{22} \end{bmatrix}$$

4-11   A hybrid waveguide is constructed of two identical waveguides across each other and works as a four-port device. Write a general scattering matrix and then simplify it as much as possible by inspection of geometrical symmetry and using the known phases of the electric waves.

4-12   A helix slow-wave structure has a pitch P of 2 mm and a diameter of 4 cm. Calculate the wave velocity in the axial direction of the helix.

Chapter 5

Vacuum Tubes

5.1-1   A vacuum pentode tube has five grids: a cathode, a control grid, a screen grid, a suppressor grid, and an anode plate as shown in Fig. P5-1-1.

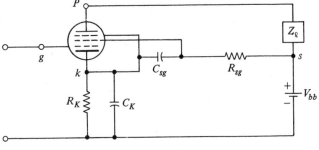

FIG. P5-1-1.

    a. Sketch the equivalent circuit.
    b. Derive an expression for the input impedance Z_{in} in terms of the angular frequency ω and the circuit parameters.
    c. Determine the transit-angle effect.

Klystrons

5.2-1   The parameters of a two-cavity amplifier klystron are:

$$V_0 = 1200 \text{ volts}$$
$$I_0 = 28 \text{ mA}$$
$$f = 8 \text{ GHz}$$

    Gap spacing in either cavity: $d = 1$ mm
    Spacing between the two cavities: $L = 4$ cm
    Effective shunt resistance: $R_{sh} = 40 \text{ K}\Omega$

    a. Find the input microwave voltage V_1 to generate a maximum output voltage V_2.
    b. Determine the voltage gain.
    c. Calculate the efficiency of the amplifier.
    d. Compute the beam loading conductance and show that it is justified to neglect it in the above calculations.

5.2-2   A two-cavity amplifier klystron has the following characteristics:

    Voltage gain is 15 dB
    Input power is 5 mW (milliwatts)
    Total shunt impedance of the input cavity R_{sh} is 30 KΩ
    Total shunt impedance of the output cavity R_{sh} is 40 KΩ
    The load impedance at the output cavity R_ℓ is 40 KΩ

    Determine:

    a. the input voltage (RMS),
    b. the output voltage (RMS), and
    c. the power delivered to the load in watts.

5.2-3   A two-cavity amplifier klystron has the following parameters:

Beam voltage	$V_0 = 900$ V
Beam current	$I_0 = 30$ mA
Frequency	$f = 8$ GHz
Gap spacing in either cavity	$d = 1$ mm
Spacing between centers of cavities	$L = 4$ cm
Effective shunt impedance	$R_{sh} = 40$ KΩ

    Determine:

    a. the electron velocity,
    b. the dc transit time of electron,
    c. the input voltage for maximum output voltage, and
    d. the voltage gain in dB.

5.2-4   Derive Eq. (5-2-35).

Reflex Klystrons

5.2-5 A reflex klystron operates at the peak mode of $n = 2$ with

Beam voltage	$V_0 = 300$ V
Beam current	$I_0 = 20$ mA
Signal voltage	$V_1 = 40$ V (RMS)

Determine

a. the input power in watts,
b. the output power in watts, and
c. the efficiency.

5.2-6 A reflex klystron operates under the following conditions:

$$V_0 = 500 \text{ volts}$$
$$R_{\text{sh}} = 20 \text{ k}\Omega$$
$$f_r = 8 \text{ GHz}$$
$$d = 1 \text{ mm is the spacing between repeller and cavity.}$$

The tube is oscillating at f_r at the peak of the $n = 2$ mode or $1\frac{3}{4}$ mode. Assume that the transit time through the gap and beam loading can be neglected.

a. Find the value of repeller voltage V_r.
b. Find the direct current necessary to give microwave gap voltage of 200 volts.
c. Calculate the electronic efficiency.

5.2-7 A reflex klystron operates at the peak of the $n = 2$ mode. The dc power input is 40 mW and $V_1/V_0 = 0.278$. If 20% of the power delivered by the beam is dissipated in the cavity walls, find the power delivered to the load.

5.2-8 A reflex klystron operates at the peak of the $n = 1$ or 3/4 mode. The dc power input is 40 mW and the ratio of V_1 over V_0 is 0.278.

a. Determine the efficiency of the reflex klystron.
b. Find the total output power in milliwatts, and
c. If 20 per cent of the power delivered by the electron beam is dissipated in the cavity walls, find the power delivered to the load.

Traveling-Wave Tubes (TWTs)

5.2-9 A traveling-wave tube (TWT) has the following characteristics:

Beam voltage	$V_0 = 2$ kV
Beam current	$I_0 = 4$ mA
Frequency	$f = 8$ GHz
Turns of the helix	$N = 50$ turns
Characteristic impedance	$Z_0 = 20$ ohms

Determine:

a. the gain parameter C, and
b. the power gain in dB.

5.2-10   A TWT operates under the following parameters:

Beam current	$I_0 = 50$ mA
Beam voltage	$V_0 = 2.5$ kV
Characteristic impedance of helix	$Z_0 = 6.75$ ohms
Turns of the helix	$N = 100$ turns
Frequency	$f = 8$ GHz

Determine:

a. the gain parameter C,
b. the output power gain A_p in dB,
c. all four propagation constants, and
d. the wave equations for all four modes in exponential form.

5.2-11   An O-type traveling-wave tube operates at 2 GHz. The slow-wave structure has a pitch angle of 5.7 degrees. Determine the propagation constant of the traveling wave in the tube. It is assumed that the tube is lossless.

5.2-12   An O-type helix traveling-wave tube operates at 8 GHz. The slow-wave structure has a pitch angle of 4.4 degrees and an attenuation constant of 2 nepers per meter. Determine the propagation constant γ of the traveling wave in the tube.

5.2-13   In an O-type traveling-wave tube, the acceleration voltage (beam voltage) is 3000 volts. The characteristic impedance is 10 ohms. The operating frequency is 10 GHz and the beam current is 20 mA. Determine the propagation constants of the four modes of the traveling waves.

5.2-14   Describe the structure of an O-type traveling-wave tube (TWT) and its characteristics; then explain how it works.

5.2-15   In an O-type traveling-wave tube, the accleration voltage is 4000 volts and the magnitude of the axial electric field is 4 V/m. The phase velocity on the slow-wave structure is 1.10 times the average electron beam velocity. The operating frequency is 2 GHz. Determine the magnitude of velocity fluctuation.

Gridded Traveling-Wave Tubes (GTWTs)

5.2-16   The current I_r caused by the returning electrons at an overdepression voltage of -11.5 KV in a GTWT is about 0.973 amp as shown by the spent beam curve in Fig. 5-2-4-9b.

a. Calculate the number of electrons returned per second.
b. Determine the energy in eV associated with these returning electrons in 1 ms for part a.
c. Find the power in watts for the returning electrons.

5.2-17   The output iron pole piece of a GTWT has the following characteristics:

1. The specific heat H at 20°C is 0.108 calories/gram °C.
2. A factor for converting joules to calories is 0.238.

3. The mass of the heated iron pole piece is assumed to be 203.05 milligrams.
4. The duration time t of the collector depression transient voltage of -11.5 KV is 15 ms.
5. The melting point of iron is $1535°C$.

   a. Calculate the heat in calories associated with the returning electrons at an overdepression voltage of -11.5 KV.
   b. Compute the temperature T in $°C$ for the output iron pole piece [Hint: $T = 0.238VIt/(\text{mass} \times \text{specific heat})$].
   c. Determine whether the output iron pole piece is melted.

5.2-18 The efficiency of a gridded traveling-wave tube (GTWT) is expressed

$$\eta = \frac{\text{RF } P_{\text{ac}}}{P_{\text{dc}}} = \frac{\text{RF } P_{\text{ac}}}{V_0 I_0}$$

If the cathode voltage is -18 KV and the collector voltage is depressed to -7.5 KV determine the efficiency of the GTWT.

Magnetrons

5.3-1 Describe the principle of operation for a normal cylindrical magnetron and its characteristics.

5.3-2 A normal cylindrical magnetron has the following parameters:

Inner radius	$R_a = 0.15$ meter
Outer radius	$R_b = 0.45$ meter
Magnetic flux density	$B_0 = 1.2$ milliwebers/m^2

   a. Determine the Hull cutoff voltage.
   b. Determine the cutoff magnetic flux density if the beam voltage V_0 is 6000 volts.

5.3-3 It is assumed that in a normal cylindrical magnetron the inner cylinder of radius a carries a current of I_0 in the z direction (i.e. $\mathbf{I} = I_0 \mathbf{u}_z$) and the anode voltage is V_0. The outer radius is b. Determine the differential equation in terms of the anode voltage V_0 and the current I_0.

5.3-4 Compare the cutoff conditions for an inverted cylindrical magnetron (i.e. the inner cathode voltage is V_0 and the outer anode is grounded) with a normal cylindrical magnetron. It is assumed that the magnetic field does no work on the electrons.

5.3-5 It is assumed that electrons in an inverted cylindrical magnetron leave the interior of the coaxial cathode with initial velocity due to thermal voltage V_t in volts. Find the initial velocity required for the electrons to just hit the anode at the center conductor.

5.3-6 It is assumed that electrons in an inverted cylindrical magnetron leave the interior of the coaxial cathode with zero initial velocity. Find the minimum velocity for an electron to just graze the anode at the center conductor.

5.3-7   In a linear magnetron the electric and magnetic field intensities as shown in
Fig. P5.3-7 are given by

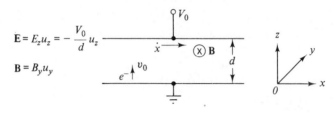

FIG. P5-3-7.

Determine the trajectory of an electron with an initial velocity v_0 in the
z direction.

Chapter 6

Microwave Transistors, Tunnel Diodes, and Microwave FETs

6.1-1   The S-parameters of a certain microwave transistor measured at 1 GHz
with a 50-Ω resistance matching the input and output are $S_{11} = 0.606\underline{/20°}$,
$S_{22} = 0.40\underline{/-15°}$ and $S_{21} = 6\underline{/180°}$. The transistor is to be used as an
amplifier device in an amplifier circuit as shown in Fig. 6-1-7(d).

  a. Calculate and plot the input and output constant power gain circles on a
  Smith chart.
  b. Overlap two Smith charts to facilitate the design procedures. Use the
  original chart to read the impedance and the overlaid chart to read the
  admittance.
  c. Determine the capacitance and inductance of the output matching
  network.
  d. Find the capacitance and inductance of the input matching network.

6.1-2   A certain silicon microwave transistor has a maximum electric field intensity
E_m of 2×10^5 volts/cm and a maximum possible saturated drift velocity
v_s of 6×10^6 cm/s. The emitter-collector length L is 6 μm.

  a. Calculate the maximum allowable applied voltage.
  b. Compute the average time for a charge to transverse the emitter-collector
  distance L.
  c. Determine the maximum possible cutoff frequency in GHz.

6.1-3   A negative-resistance device is connected through a 1-kΩ resistor in series
and a 0.01-μF capacitor in parallel to a combination of a supply voltage V
of 10 V and a signal source V_s as shown in Fig. P6.1-3(a). Its V-I characteristic curve is shown in Fig. P6.1-3(b).

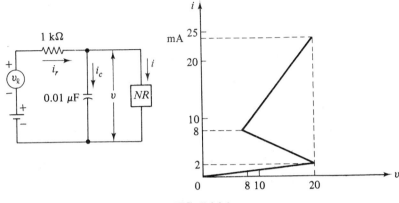

FIG. P6-1-3.

   a. Find the negative resistance and forward resistances of the device.
   b. Draw the load line on the V-I curve.
   c. Determine the quiescent operating point of the circuit by the values of
   voltage and current.
   d. Determine the new operating point by the values of voltage and current
   when a signal voltage of 14 volts is applied to the circuit.
   e. Draw the new load line on the V-I curve.
   f. Find the time constant of the circuit.
   g. Compute v, i, i_c and i_r as a function of time after the triggering signal is
   applied and before the transition takes place.
   h. Find the transition time T in microseconds.
   i. Calculate, v, i, i_c and i_r immediately after transition.

6.1-4   A microwave tunnel diode has a negative resistance R_n and the resonant
circuit has a circuit resistance R_ℓ. Derive an equation for the gain of a microwave tunnel-diode amplifier.

6.1-5   A certain microwave tunnel diode has a negative resistance of $69 + j9.7$
ohms. Determine the resonant-circuit impedance so that the microwave
tunnel-diode amplifier will produce a power gain of 15 dB.

6.1-6   A certain microwave GaAs FET has a gate length 6.6 μm Calculate the
maximum frequency of oscillation f_{max} for the GaAs FET.

6.1-7   Describe the advantages and disadvantages of a microwave FET.

6.1-8   The S-parameters of a certain silicon Schottky barrier-gate FET are S_{11}
$= 0.50\underline{/20°}$, $S_{22} = 0.4\underline{/-15°}$, $S_{21} = 5.5\underline{/180°}$, and $S_{12} = 0.5\underline{/-180°}$.

a. Calculate the maximum available power gain $G_{\max}$.

b. Compute the unilateral power gain G_u for $\Gamma_s = 0.1$ and $\Gamma_\ell = 0.1$.

c. Determine the maximum unilateral power gain $G_{u\,\max}$.

d. Find the intrinsic noise figure in dB if $\gamma = 2.3$ and $E = 1500$ volts/m.

Transferred Electron Devices (TEDs)

6.2-1 **a.** Spell out the following abbreviated terms: LSA, InP, and CdTe.

      b. Describe in detail the principles of the following terms: Gunn effect, high-field domain theory, two-valley theory, and three-valley theory.

      c. Discuss the differences between the transferred electron devices and avalanche transit-time devices.

6.2-2 Describe the operating principles of tunnel diode, Gunn diode and LSA diodes.

6.2-3 Derive Eq. (6-2-11).

6.2-4 Describe the modes of operation for Gunn diodes.

Avalanche Transit-Time Devices

6.3-1 Spell out the following abbreviated terms: IMPATT, TRAPATT, and BARITT.

6.3-2 Derive Eq. (6-3-15).

6.3-3 Describe the operating principles of Read diode, IMPATT diode, TRAPATT diode, and BARITT diode.

Laser Devices

6.4-1 **a.** Spell out the following abbreviated terms: laser, maser and Cr.

      b. Describe in detail the principles of the population inversion and laser operation.

6.4-2 **a.** Describe the operating principles of ruby laser and semiconductor junction laser.

      b. Describe the microwave characteristics of a laser.

Parametric Devices

6.5-1 **a.** Describe the advantages and disadvantages of the parametric devices.

      b. Describe the applications of the parametric amplifiers.

6.5-2 The figure of merit for a diode nonlinear capacitor in an up-converter parametric amplifier is 8 and the ratio of the output (or idler) frequency f_0 over the signal frequency f_s is 8. The diode temperature is 290°K.

      a. Calculate the maximum power gain in dB.

      b. Compute the noise figure F in dB.

      ç. Determine the bandwidth (BW) for $\gamma = 0.2$.

6.5-3 A negative-resistance parametric amplifier has the following parameters:

$$f_s = 2 \text{ GHz} \qquad R_l = 1 \text{ K}\Omega \qquad \gamma = 0.35$$

$$f_p = 12 \text{ GHz} \qquad R_g = 1 \text{ K}\Omega \qquad \gamma Q = 10$$

$$f_l = 10 \text{ GHz} \qquad R_{Ts} = 1 \text{ K}\Omega \qquad T_d = 290°\text{K}$$

$$f_i = 5f_s \qquad R_{Ti} = 1 \text{ K}\Omega \qquad C = 0.01 \, pF$$

a. Calculate the power gain in dB.
b. Compute the noise figure F in dB.
c. Determine the bandwidth (BW).

Infrared Devices

6.6-1 Describe the three most important parameters of an infrared imaging system: the modulation transfer function (MTF), the minimum resolvable temperature (MRT) and the noise equivalent differential temperature ($NE \, \Delta T$).

6.6-2 A square 4-bar pattern, 7 to 1 aspect ratio, is currently the standard pattern used for measuring the minimum resolvable temperature (MRT). (Refer to Fig. 6-6-13.) The bar width of the 4-bar pattern is 5 mils and the focal length of the collimator is 50 inches.

a. Determine the angle subtended by the bar width in milliradians.
b. Calculate the characteristic spatial frequency f_0 in cycles per milliradian.

6.6-3 One of the earliest experimental attempts to relate threshold resolution to the visual discrimination of images of real scenes is attributed to Johnson. Explain Johnson's imaging model for thermal imaging measurements.

6.6-4 A blackbody emits radiation into a hemisphere at 427°C with an emitting area of 100 cm^2.

a. Calculate the radiant emittance of the blackbody.
b. Compute the irradiance of the blackbody.
c. Determine the radiant intensity.
d. Find the radiance.
e. Determine the maximum spectral radiant emittance.

6.6-5 The quality and characteristics of a detector are usually described by three parameters: responsivity (R), specific detectivity (D^*) and time constant (τ). Describe the quality and characteristics of the detectors: Ge: Hg; Hg: Cd, Te; and Ph: Sn, Te.

6.6-6 Describe how to measure the noise equivalent differential temperature ($NE \, \Delta T$), the minimum resolvable temperature (MRT), and the modulation transfer function (MTF) of an infrared imaging system.

6.6-7 The modulation transfer function (MTF) of an infrared thermal imaging system is the product of the individual system element transfer functions. In an infrared imaging system laboratory, the differential brightness ΔB_1 decreased by the effect of the system transfer function, the magnitude of the variable object brightness B_1 and the dc or average level of brightness B_0 responding to the characteristic spatial frequency f_0 are 0.1, 0.4, and 1, respectively. (Refer to Fig. 6-6-15.) Determine the value of the modulation transfer function of the system at the characteristic spatial frequency f_0.

6.6-8   In an infrared imaging system laboratory, the temperature of the source is reduced until a match is obtained between the background and target. The differential temperature ΔT, the noise voltage, and the ac signal voltage were measured to be 2°C, 2 volts (RMS), and 10 volts (peak to peak), respectively. Calculate the noise equivalent differential temperature ($NE\,\Delta T$) of the system.

Chapter 7

7-1   A certain microstrip line has the following parameters:

$\epsilon_r = 5.23$ is the relative dielectric constant of the fiber-glass-epoxy board material

$h = 8$ mils

$t = 2.8$ mils

$w = 10$ mils

Write a FORTRAN program to compute the characteristic impedance Z_0 of the line. Use a READ statement to read in the input values, the F10.5 format for numerical outputs, and the Hollerith format for character outputs.

7-2   Since modes on microstrip lines are only quasi-transverse electric and magnetic (TEM), theory of TEM coupled lines applies only approximately. Derive from the basic theory of a lossless line that the inductance L and capacitance C of a microstrip line are given by

$$L = v/Z_0 = c/(Z_0\sqrt{\epsilon_r})$$
$$C = 1/(Z_0 v) = \sqrt{\epsilon_r}/(Z_0 c)$$

where Z_0 = characteristic impedance of the microstrip line
   v = wave velocity in the microstrip line
   $c = 3 \times 10^8$ m/s is the velocity of light in vacuum
   ϵ_r = relative dielectric constant of the board material

7-3   A microstrip line is constructed of a perfect conductor and a lossless dielectric board. The relative dielectric constant of the fiber-glass-epoxy board is 5.23 and the line characteristic impedance is 50 ohms. Calculate the line inductance and the line capacitance.

7-4 A certain microstrip line is constructed of a copper conductor and nylon phenolic board. The relative dielectric constant of the board material is 4.19 measured at 25 GHz and its thickness is 0.4836 mm (19 mils). The line width is 0.635 mm (25 mils) and the line thickness is 0.071 mm (2.8 mils).

  a. Compute the characteristic impedance Z_0 of the microstrip line.
  b. Calculate the dielectric filling factor q.
  c. Compute the dielectric attenuation constant α_d.
  d. Find the surface skin resistivity R_s of the copper conductor at 25 GHz.
  e. Determine the conductor attenuation constant α_c.

7-5 A certain microstrip line is made of a copper conductor 0.254 mm (10 mils) wide on a G-10 fiber-glass-epoxy board 0.20 mm (8 mils) in height. The relative dielectric constant ϵ_r of the board material is 4.8 measured at 25 GHz. The microstrip line of 0.035 mm (1.4 mils) thick is to be used for 10 GHz.

  a. Calculate the characteristic impedance Z_0 of the microstrip line.
  b. Compute the surface resistivity R_s of the copper conductor.
  c. Calculate the conductor attenuation constant α_c.
  d. Determine the dielectric attenuation constant α_d.
  e. Find the quality factor Q_c and Q_d.

Chapter 8

8-1 Describe the advantages of microwave integrated circuits over discrete circuits.

8-2 **a.** List the basic materials for microwave integrated circuits.
  b. List the basic characteristics required for an ideal substrate material.

8-3 List the basic properties provided by the ideal conductor material, the dielectric film, and the resistive film used in microwave integrated circuits.

8-4 Describe the basic fabrication techniques used for microwave integrated circuits: epitaxy, diffusion, deposition, and etching.

Chapter 9

9-1 A power density of 15.50 mW/cm² (0.1 W/in²) is the maximum safe power level that can be applied to the absorber of an anechoic chamber without generating enough heat for combustion. A short-wire antenna with a gain of 2 dB has an effective aperture $3\lambda^2/(8\pi)$ at 1 GHz. The performance of the absorber is 30 dB at 1 GHz. The shortest distance between the antenna and the absorber is 2 meters. Determine the maximum power level from the transmitter that will not damage the absorber.

9-2 Derive Eq. (9-2-24).

9-3   Write a FORTRAN program to compute and plot the electric or magnetic field pattern of a short-dipole antenna as shown in Eqs. (9-2-32) and (9-2-33). The angle varies from 0 to 360° with an increment of 10° each step. Use F5.1 format for the angle and F7.3 for the magnitude. Print the angle and magnitude in parallel with the plot of the field pattern. The normalized value of the field intensity is assumed to be unity.

9-4   A shield is made of a copper sheet 1 mm thick for a frequency of 2 MHz. Calculate the shielding effectiveness of the shield.

9-5   A transmitter has an output power of 10 watts at a frequency of 3 GHz. The coaxial line connected to the transmitter has a loss of 3 dB and the antenna connected to the other end of the line has a gain of 5 dB. The test location in the quiet zone of an anechoic chamber is 3 meters away from the transmitting antenna.

a. Determine the electric field intensity in volts/m and power density in watts/m² at the test site.

b. Convert the electric field intensity and power density into dBV/m and dBW/m².

Chapter 10

10-1   A transmitting antenna has a gain of 10 dB and the coaxial line connecting the antenna to the transmitter is assumed lossless. The characteristic impedance of the coaxial line and the input impedance of the transmitter are both 50 ohms. The signal frequency is 2 GHz and the transmitting power is 2 kW. Determine the power density in dBm/m² and the electric field intensity in dBμV/m at the receiving antenna aperture for a distance of 1 meter.

10-2   A standing-wave detector is equiped with a crystal having a law of $n = 1.8$. The meter reads 50 μA at a maximum and 10 μA at a minimum, as the probe is moved along the waveguide. Determine the VSWR in the waveguide.

10-3   In order to determine the crystal response law n over a limited range of induced voltages it is sufficient to measure the guided wavelength λ_g and the distance L between half-current points as shown in Fig. P10-3.

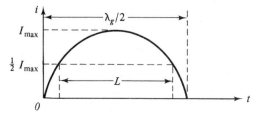

FIG. P10-3.

a. Derive the crystal response law

$$n = \frac{\log (1/2)}{\log [\cos (\pi L/\lambda_g)]}.$$

b. Plot n vs. $2L/\lambda_g$.

10-4 The receiving antenna has a gain of 15 dB and the coaxial line connecting the antenna to the receiving field intensity meter has a loss of 3 dB. The characteristic impedance of the coaxial line and the input impedance of the meter are both 50 ohms. The signal frequency is 5 GHz and the meter reading is 500 μV.

a. Convert the intensity meter reading 500 μV to dBμV and dBV.

b. Calculate the power input to the receiver in watts.

c. Determine the power density at the antenna aperture in watts/m².

d. Convert the power density from watts/m² to dBW/m², dBμW/m², and dBμW/cm².

e. Compute the electric field intensity at the antenna aperture in volts/m.

f. Convert the field intensity from volts/m to dBV/m and dBμV/cm.

g. Find the antenna factor.

Index